Rheology Applied in Polymer Processing

Rheology Applied in Polymer Processing

Dr. B. R. Gupta
Prof. (Retired)
Rubber Technology Centre
IIT Kharagpur
West Bengal, India.

CRC Press
Taylor & Francis Group
Boca Raton London New York

CRC Press is an imprint of the
Taylor & Francis Group, an **informa** business

Manakin
PRESS

First published 2023
by CRC Press
4 Park Square, Milton Park, Abingdon, Oxon, OX14 4RN

and by CRC Press
6000 Broken Sound Parkway NW, Suite 300, Boca Raton, FL 33487-2742

© 2023 B.R. Gupta and Manakin Press

CRC Press is an imprint of Informa UK Limited

British Library Cataloguing-in-Publication Data
A catalogue record for this book is available from the British Library

ISBN: 9781032384191 (hbk)
ISBN: 9781032384207 (pbk)
ISBN: 9781003344971 (ebk)

DOI: 10.1201/9781003344971

Typeset in Arial, Calibri, Cambria Math, Century Schoolbook, MT-Extra, Cabria Math, Symbol MT, Tahoma, Verdana, Palatino, Monotype Corsiva, Euclid Extra, KozGoPr6N, Minion Pro, Symbol, Times New Roman, Wingdings
by Manakin Press, Delhi

Manakin
PRESS

Preface

Polymers and rubbers find numerous industrial applications for the manufacture of different components. It requires polymer melts to be processed in different processing equipment. These melts are subjected to varying degree of shear stress and rate of shear and extensional deformations. The melt has to pass through the different flow channels under high pressure and temperature. The volumetric flow rate-pressure drop relationship, which depends on the rheological behaviour of the polymers is of considerable importance from the point of view of energy consumption in processing. The study of rheological behaviour therefore forms an essential part of course work in polymer science and engineering, plastic technology, plastic engineering, rubber technology, paint technology etc. The science of rheology has developed from two different approaches, (i) Theoretical Rheology and (ii) Applied Rheology. The theoretical approach requires the knowledge of stress, strain and rate of strain tensors, which is normally beyond the scope of practicing polymer and rubber technologists, as it does not form an essential subject of polymer science in different universities and institutions. The applied rheology on the other hand though based on observations and interpretations of experimental results and empirical correlations also includes the explanations from the theoretical rheology and therefore makes the understanding of the processing problems relatively easy. The present book is designed to impart an in depth understanding of science of rheology applied to polymer processing. It is expected to be suitable for both students and practicing technologist and engineers.

The book is divided in to eight chapters. *Chapter 1* gives introduction outlining briefly different processing steps involved in polymer processing with emphasis on the importance of the science of rheology in processing of polymers. *Chapter 2* covers the basic principles, parameters and applied mathematical models used in rheological studies. The melt flow analysis of non-Newtonian fluids, velocity distribution, laminar flow, pressure drop-volumetric flow rate relationship, transition between laminar and turbulent flow and modified Reynolds number for power law fluids etc. are included in *Chapter 3*. *Chapter 4* outlines the effect of different parameters like temperature, pressure, MW, molecular weight distribution, filler loading etc. on purely viscous rheological response of polymer melts and solutions. *Chapter 5* discusses the principles of rheometery and different equipment like capillary viscometers, cone and plate, cylindrical and sliding-plate viscometers used in rheological measurements. The static and dynamic viscoelastic response of polymer melts and solutions, linear viscoelastic models like Maxwell, Voigt-Kelvin and Zener models and Boltzmann superposition principles are given in *Chaptr 6*. It also includes the discussion on the molecular structure and viscoelastic response of the polymeric systems and some theoretical background of linear and non-linear viscoelasticity. *Chapter 7* includes the effect of various process, material, and molecular parameters on the dynamic viscoelastic functions like storage and loss moduli and compliances and loss tangent etc. The role of rheology in understanding problems associated with processing of polymers in different equipment like calenders, mixers, extruders, injection moulding, film blowing extruder dies etc. including a discussion on the die swell and melt fracture phenomenon are included in *Chapter 8*.

The book extensively covers the aspects of rheology applied to polymer processing. A good number of solved examples and exercise problems are included in different chapters to help the students to understand the polymer rheology better. Annexures I(a), I(b), II and III and IV contain respectively the modified Power-law model constants, two-range power law constants, rheology software program in Java, comparison of different rheological models using rheology software and answers to problems given at the end of different chapters. It is the considered view of the author that the book will be of immense help to both students and practicing rheologists.

Author
B.R. Gupta

Brief Contents

Annexure

Detailed Contents

Introduction

Polymers play a very important role in everyday life so much so that their presence is felt everywhere as an omnipresent entity. The term polymer is derived from ancient Greek word Polus meaning many and meros meaning parts and refer to structure composed of multiple repeat units from which originate characteristics of high relative molecular mass and different properties [http://goldbook.iupac.org/M03667.html; Oct. 2012]. There are both naturally occuring biopolymers like DNA and proteins and natural rubber, and synthetic polymers like polyethylenes and other plastics. Polymers cover a wide range of properties like toughness, elasticity, viscous nature, capability to form semicrystalline structures etc. Recently a new type of polymeric material has been developed at Eindhoven University of Technology and Kent State University. The material developed is capable of undulating and thus propelling itself under the influence of light [AZOM, June 29, 2017]. It is a small device of the size of a paper clip, considered to be the first machine in the world that will help convert light directly in to walking, by using just one fixed light source. The movement of the new material is explained on the basis of expansion of one side and contraction of the other side due to the effect of light.

The polymers must be processed in suitable processing equipment before they are used to produce various components. The polymer processing involves a number of steps depending on the type and application of the component. The polymer components may be manufactured either from (a) low viscosity liquid polymers or their solutions and emulsions like monomer solutions in the reaction injection moulding or rubber latexes for making latex goods or (b) from polymer or rubber melts to manufacture various extruded or moulded components.

The processing stages and equipment used in the manufacture of latex and solution based components are not highly energy intensive due to the low viscosity of the starting fluids. On the other hand solid polymers (both rubbers and plastics) are used for a large number of component manufacture and the energy required for their processing is very high because of their high melt viscosity. The elastic nature of the rubbers presents additional processing problems. The important equipments used in the polymer processing are mixers, calenders, extruders and injection and other moulding systems. During mixing the rubber is mixed with fillers and other ingredients in a two-roll mill or an internal mixer. The calenders are used for topping, frictioning or coating the fabric with the rubber based compounds or other polymer solutions. Extruders and moulds are used for forming and curing the compounds respectively.

During processing through these stages the rubber and polymer melts are subjected to different stresses like shear, compressive, tensile stresses and are deformed accordingly under high temperature, pressure and shear rate conditions. The compound flows through varying cross sections while passing through different equipment thereby requiring high-energy consumption. The viscosity and therefore the rheological behaviour of these compounds plays a very important role in optimizing the energy consumption in these highly energy intensive processes. The subsequent sections present briefly the flow situations in different processing equipment and help in crystallizing the need for the study of science of rheology of such materials.

Fig. 1.1   Different type of impellers used in agitators[1]
(with permission from Jiangsu Ruixu Mixing Equipment Co. Ltd.,
Jiangsu, China. URL-www.liquid-mixing.com)

1.1 LIQUID POLYMER PROCESSING

Liquid polymer systems like latexes, monomers and solutions are used to produce various items like adhesives, paints and varnishes, gloves, prophylactics, balloons, catheters, seamless football bladders etc. and rubber and plastic foams. Different chemicals are required to be mixed with these liquid polymers and latexes using agitated vessels of different designs so as to achieve complete and homogeneous mixing. Some of the examples of such equipment are mixing tanks fitted with different types of agitators, which are fitted with variety of impellers (Fig. 1.1). These agitators produce various flow patterns as shown in Fig. 1.2(a), and (b). In comparison to simple agitators the helical blade impellers produce a highly complex flow patterns (Fig. 1.2(c)). The flow patterns obtained during the flow of melts through a two-roll mill, an orifice die, a calender system and an extruder are shown in figures 1.2d, 1.2e, 1.2f, and 1.2g respectively. The rotation of agitatotors and use of other mixing equipment for preparation

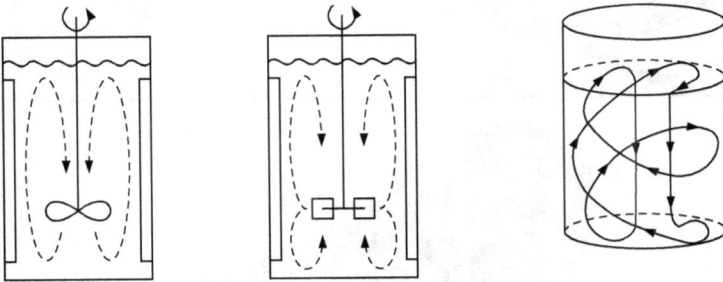

(a) Axial flow agitator   (b) Radial flow agitator   (c) Flow field by a helical impeller[1]

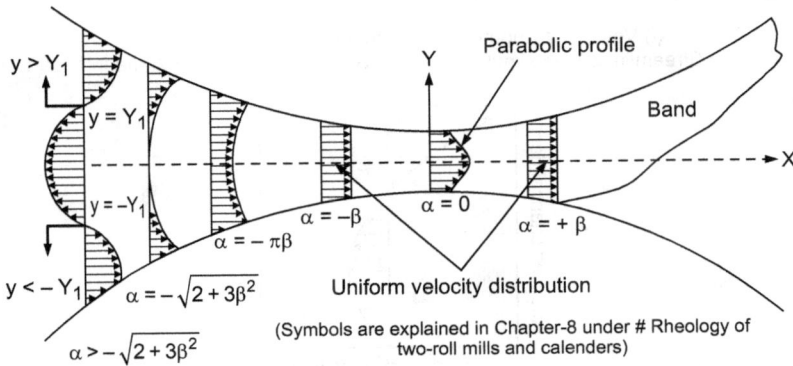

(d) Velocity distribution on a two-roll mill

(Symbols are explained in Chapter-8 under # Rheology of two-roll mills and calenders)

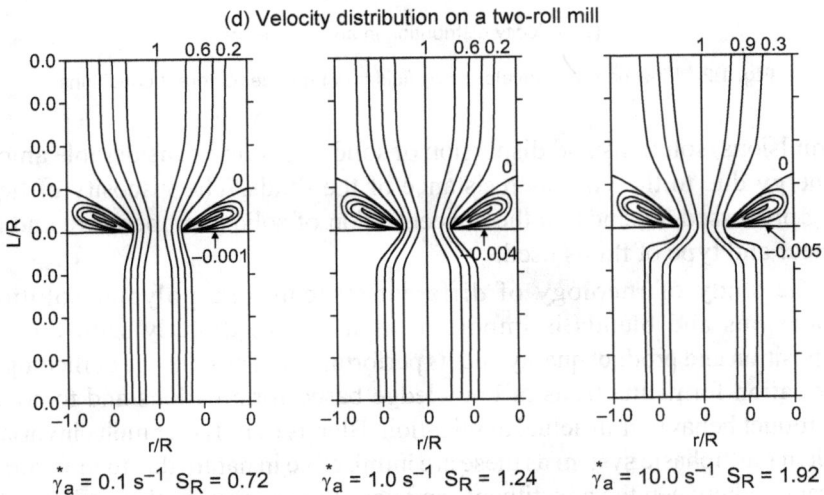

$\dot{\gamma}_a = 0.1\ \mathrm{s}^{-1}\ S_R = 0.72$ $\dot{\gamma}_a = 1.0\ \mathrm{s}^{-1}\ S_R = 1.24$ $\dot{\gamma}_a = 10.0\ \mathrm{s}^{-1}\ S_R = 1.92$

(e) Flow pattern of LDPE melt through an orifice die[11] and [12]

$S_R = $ (First normal stress difference)/(2 × shear stress at wall) $= N_1/2\tau_w$

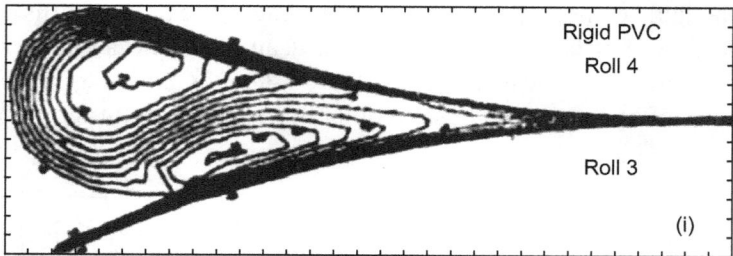

(f) Velocity distribution in calendering of PVC with slip[12]
(i) Streamlines from simulations[13] (ii) Streamlines from experiment[14]

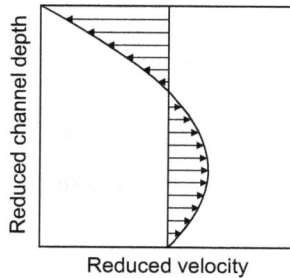

(g) Velocity distribution in an extruder[15]

Fig. 1.2   Flow patterns developed by fluid flowing under different conditions.

of emulsions, solutions and dispersion of solids consume considerable amount of energy due to the viscous resistance of the fluids. The viscosity of liquid being processed depends on the concentration of solids, temperature and rate of shear and type of fluids used.

The study of rheology of different systems like polymer solutions, suspensions and blends is important as it relates directly with the feed composition and product quality and its performance. Thus the use of rheological information forms the basis of knowledge based formulations and to control the product behaviour in actual application. Blends of polymer melts invariably result in multiphasic system as these are immiscible in nature due to unfavorable interaction between the constituents and very low entropy of their mixing. It is threrfore essential to have the information about the rheological behaviour of polymers to precisely control their processing. Different rheological principles involved in the study of flow and deformation of polymer melts are discussed in Chapter 2.

Low molecular weight liquid monomers and other chemical constituents are mixed and filled into the moulds at high pressure in different processes like monomer casting, thermoset injection moulding (TIM) or reaction injection moulding (RIM). The diagrams of simple moulding systems and an injection moulding machines are shown in Figs. 1.3 and 1.4.

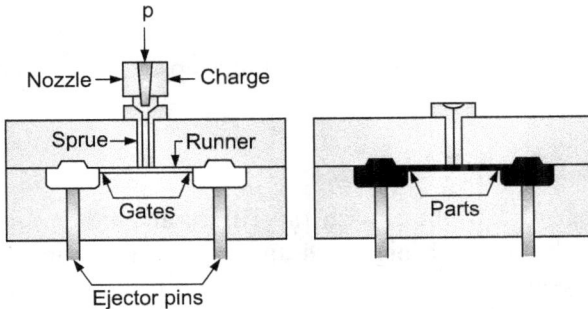

Fig. 1.3 A simple moulding system [en.wikipedia.org][2]

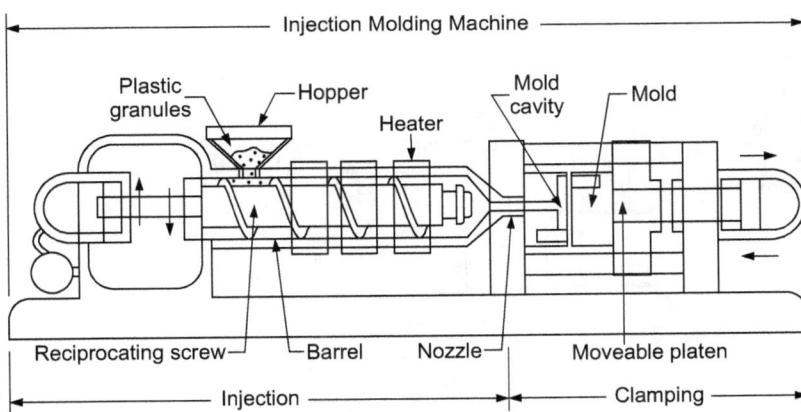

Fig. 1.4 An Injection Moulding System [en.wikipedia.org][2]

The liquid polymeric system may constitute a solution of two or more polymers in suitable solvents. The solutions, which are used for the manufacture of various polymeric products also have to flow through different processing equipment and consume energy.

The rheological behaviour of such systems may vary depending on type of liquids mixed, their concentration and mutual interaction between them. The liquid reactants after filling the moulds polymerize inside the moulds to stabilize the shape of the components. In these processes the chemical reactants are low viscosity liquids, which are premixed in specially designed mix-heads[7] (for example parallel-stream reaction injection mix-heads). The reaction vessels are low-pressure re-circulation systems fitted with stirrer whereas the mix-heads are supplied with high-pressure piston pumps, which pump the metered quantity of reactants at pre-determined rate (Fig. 1.5). The rheological behavior of the liquid constituents plays an important role in their extent of mixing, rate of polymerization, flow characteristics and the overall energy requirement in such operations.

1.2. SOLID POLYMERS/RUBBERS PROCESSING

The processing of solid polymers and rubbers to manufacture different components involves a number of processing steps. These are briefly outlined in the following sections.

1.2.1 Mixing

The mixing of solid rubbers with the fillers and other ingredients and that of thermoplastics with pigments and some times with the processing aids are highly energy intensive processes. Solid rubbers are handled in a two-roll mill (Fig. 1.6) or an internal mixers like Banbury mixer (Fig. 1.7) or in continuous mixers like intermix where as the thermoplastics are processed in plastic extruders.

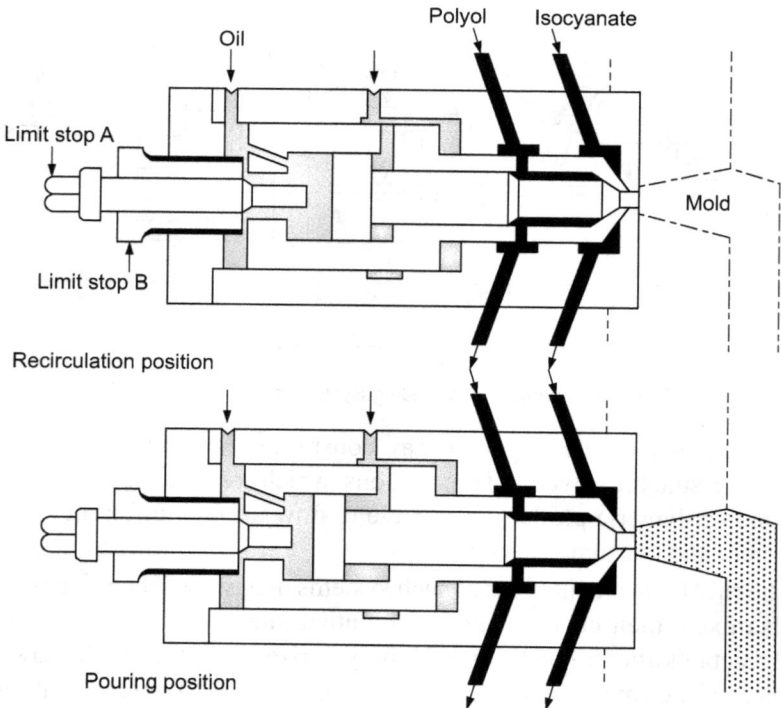

Fig. 1.5   Parrelel stream reaction injection mix-head[7, 8]

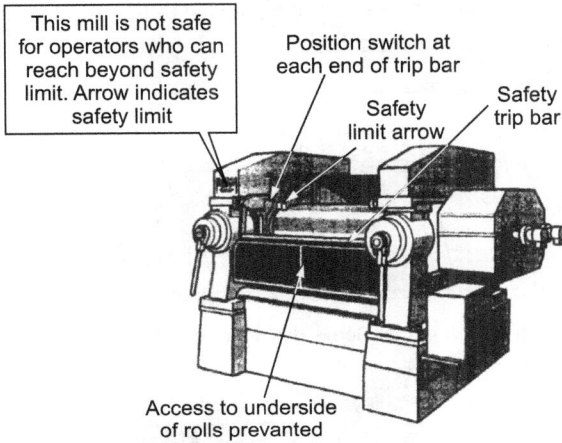

Fig. 1.6   A Two Roll Mixing Mill [hse.gov.uk]

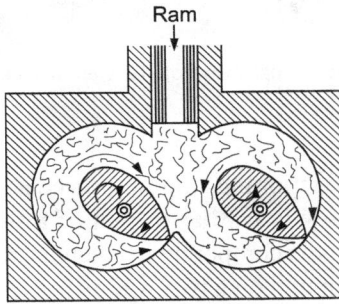

Fig. 1.7   A Banbury Internal Mixer[6,8]

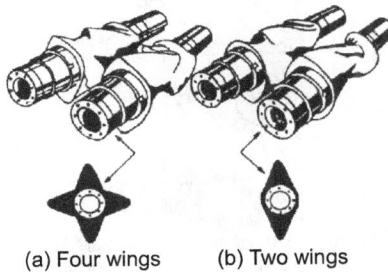

(a) Four wings   (b) Two wings

Fig. 1.8   Farrel Bridge Banbury Rotors (a) Four Wing rotor and (b) Two Wing Rotor[10]

In the operations of all these equipment the melt flow involves the high degree of shearing deformation of polymer matrix achieved by making it pass through small cross-section area like (1) nip region in a two-roll mill (Fig. 1.2(d)) [9] associated with the manual cutting and folding operations, (2) rotor tip and the chamber wall area (in an internal mixer) associated with cross flows due to design features of rotors i.e. two wing or four wing rotors[10] (Fig. 1.8) and (3) screw tip and the barrel wall in the intermix[5] along with the cross flows obtained due to special screw[4] and barrel cross-section designs[5] as in a barrier screw (Fig. 1.9) or a transfermix design (Fig. 1.10).

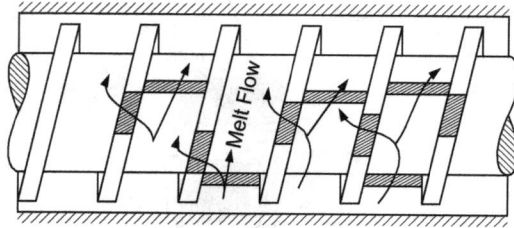

Fig. 1.9   A barrier Screw design for continuous mixer[7]

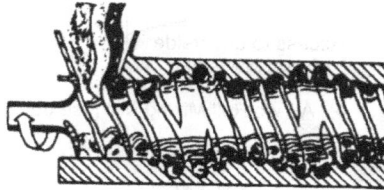

Fig. 1.10   Schematic view of Transfermix (shear mix)[11]

1.2.2 Calendering

Calendering is used for frictioning, topping or coating of the fabrics[9]. The pattern of the material flow is similar to that of a two-roll mill (Fig. 1.2(d)) except that the viscosity of the compound in calendering operations (Fig. 1.2(f)) is lower than rubber compound on a two-roll mill. Fig. 1.11 shows the material flow through the 3-Roll and 4-Roll calender configurations used for different operations like frictioning, topping, sheeting and double topping of fabrics[10].

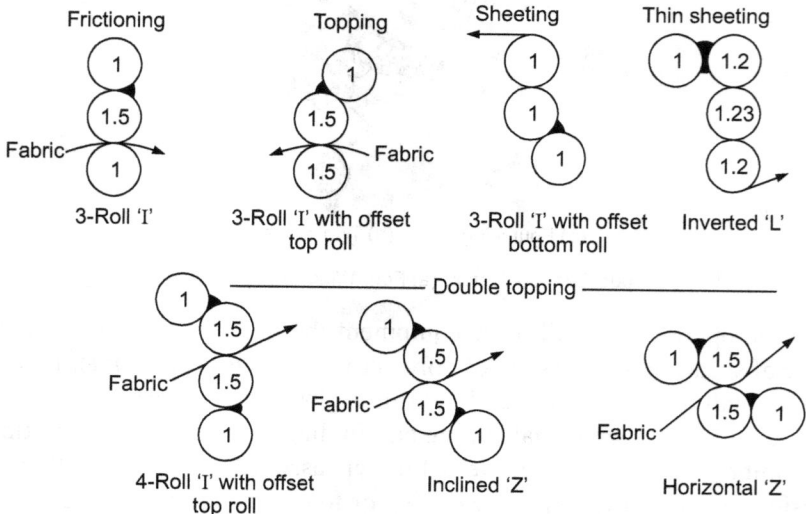

Fig. 1.11   Application based calender configurations with friction ratio indicated within the rolls[10]

1.2.3 Extrusion

The extrusion plays a very important role in the processing of plastic and rubber melts. It is used to produce pre-designed shapes like tire tread, tubes, hoses etc. in rubber industry. In addition, the extruders are extensively used in plastic film blowing, blow molding, wire coating, injection molding etc. The extruders may be classified as (i) Ram extruders (discontinuous or reciprocating), (ii) single screw extruders (Fig. 1.12) and (iii) twin- screw extruders (Fig. 1.13&1.14),

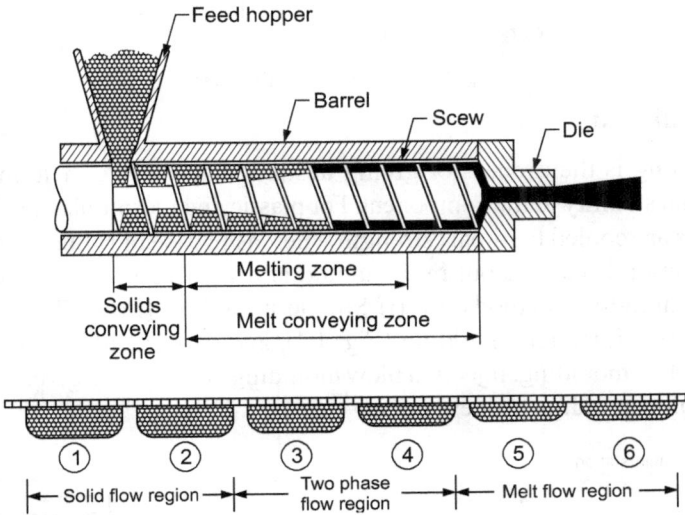

Fig. 1.12   Schematics of Single Screw Plasticating Extruder [14,16]

(discontinuous or reciprocating or rotating or intermeshing or non-intermeshing type) and (iv) Buss-Kneader etc. The polymer melt flows through the gap between the screw flight top and barrel surface in a screw extruder whereas it is pushed by a plunger in a ram extruder. The flow of melt through screw extruder is complex in nature due to the screw design. The melt is subjected to high pressure and shear stress in a continuously varying cross-section area. Because of high pressure there may result a back flow (Fig.1.2(g)) of the melt and also a considerable heat generation. At the end of the extruder the melt enters the die region and as the die cross-section is much smaller as compared to that of the barrel the melt is subjected to the flow contraction and circulation (Fig. 1.2(e)) and therefore the knowledge of rheological behavior of the melt is very essential for the analysis of such flow problems.

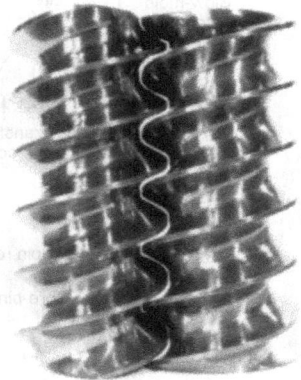

Fig. 1.13   Screws of a Twin Screw extruder [clextal.com]

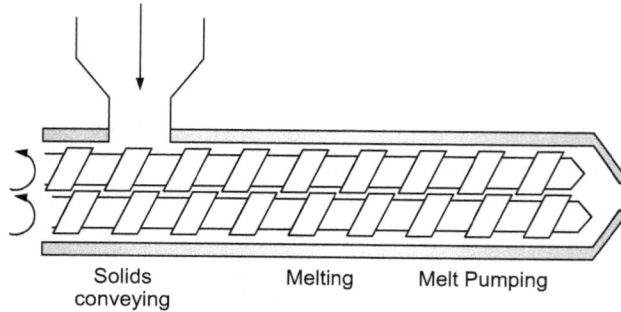

Fig. 1.14   Twin Screw Extruder

1.2.4 Moulding

The moulding is the process of giving desired shape to the melt and to give dimensional stability to the component. For plastic and thermoplastic elastomers the moulds are cooled to impart strength whereas in case of rubber components curing reaction is carried out in the mould at high pressure and temperature. Different moulding methods are (i) Simple moulding (Fig. 1.3), (ii) transfer moulding, (iii) injection moulding (Fig. 1.4), and blow moulding for example extrusion blow moulding, injection blow moulding etc. Fig. 1.15 gives a complete injection blow molding process layout[17].

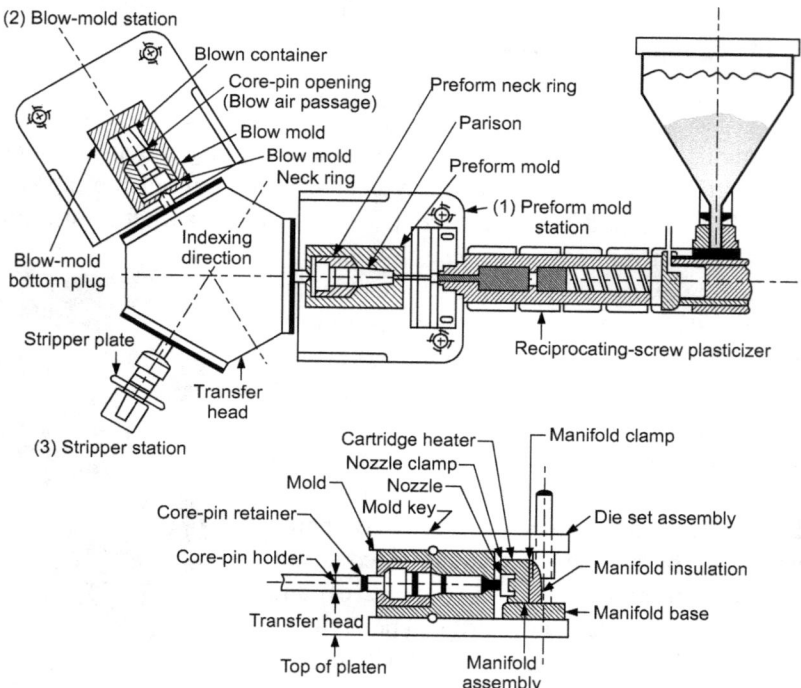

Fig. 1.15   Complete Injection Blow Molding Process[14, 17, 17a]

Irrespective of the method adopted, the precondition for the satisfactory moulding operation is that the polymer melt must completely fill the mould cavity. For the plastic melts as the viscosity of melt is comparatively low, it does not present any major problem if the mold temperature is properly maintained at required level. Rubber molds however require a high pressure and temperature conditions to ensure uniform flow of the melt.

1.3 WHY RHEOLOGY

The Rheology is defined as the science of deformation and flow. Both common man and scientist and engineers experience rheology in:

- Non-flow of mayonnaise even under long time application of stress and instantaneous flow of honey and other liquids under the applied stress.
- Silly putty- It bounces as elastic material and also flows as viscous fluid.
- Different response of flour-water solutions and doughs under the applied stress.
- Strange behaviour of corn starch in cold water when poked slowly it deforms easily around the finger, but when punched rapidly it make the fist bounce off the surface.

These behaviours in day to day life and flow and deformations of fluids in different processing equipment are the consequence of different rheological behaviours of the fluids. The scientists, engineers and technical persons will be concerned to understand the reasons behind these effects, as well as associated effects like yield stress, viscoelastic behaviour, memory effects, shear thinning and shear thickening characteristics. All these behaviours are as a result of applied stress and strains.

The polymer solutions and melts must flow under varying conditions of pressure, temperature, shear rate and shear stress through different cross sections subjecting the material to the elongational or the shear deformations. Depending on the fluid properties, volumetric flow rate and the flow cross section, the flow may be laminar (for high viscosity systems) or turbulent (for low viscosity systems) in nature. The rubber and polymer melts and solutions are known to display different rheological behaviours from a simple Newtonian to a highly complex time dependent non-Newtonian nature associated with yield stress (see chapter 2) due to the complex nature of the polymer molecules. The polymer molecules have very high molecular weight and very long chain length, which results in highly entangled chain configuration. Such a molecular state when subjected with stresses, provides higher resistance to the deformation resulting in the increase in the viscosity and consequent decrease in processability. Therefore the study of rheology of polymers is very important to

1. Understand the flow characteristics of these materials,
2. Establish the flow rate-energy consumption relationships,

3. Establish the raw material, and end product quality control and
4. Predict the performance of materials.

In addition the rheology plays an important role in

- Polymer processing design, costs and production rates.
- Product quality like surface distortions, anisotropy, strength, structure development etc.
- End uses like food texture, consistency of blood and injections fluids, lubrication, usefulness of different utility and engineering products, and ease of flow of petroleum products like petrol, diesel, aviation fuels and engine oils etc.

The science of rheology deals with the study of deformation and flow under applied stresses for all types of materials like elastic, viscous as well as viscoelastic in nature. For purely elastic materials it describes the relationship between stress and strain, for purely viscous systems between stress and rate of strain whereas for viscoelastic fluids the stress is a complex function of stain, rate of strain and time. The equations of motion and continuity form the basis of derivations of theoretical rheological relations. The theoretical analyses invariably result in very complex and impractical equations and only the simplified cases are useful, therefore a large number of rheologists rely on developing empirical relationships between various parameters for different flow situations.

The subsequent chapters describe the basic definition of different rheological terms; theoretical and empirical rheological equations for elastic, viscous and viscoelastic polymers; analysis of non-Newtonian fluids under laminar and turbulent flow conditions; effect of various physical and molecular parameters of polymeric materials on polymer rheology, rheometery, dynamic rheological response of the polymers and principles of rheology applied to different processing steps.

REFERENCES

1. Jiangsu Ruixu Mixing Equipment Co., Ltd., Jiangsu China. www.liquid-mixing.com
2. Platic Extrusion, Wikipedia Encyclopedia.
3. Barakos, G. and Mitsoulis, E. "Numerical Simulation of Extrusionthrough Orifice Dies and prediction of Bagley correction for an IUPAC-LDPE Melt" J. Rheology, 39, 193-209, 1995.
4. Mitsoulis E., Vlachopoulos, J. and Mirza F. A., "Calandering Analysis without Lubrication Approximation, Polym. Eng. Sci. 25, 6-18, 1985.
5. Agast J. F.,and Epsy M., "Theoretical and experimental study of molten polymer flow in Calander Bank, Polym. Eng. Sci. 25, 113-121, 1985.

6. J.M. Coulson and J. F. Richardson, Chemical Engineering Vol.3, The English Language Book Society and Pergamon Press Ltd., Oxford, U K, 1971, Pp. 465-469.

7. B. R. Gupta, Applied rheology in Polymer Processing, Asian Books Pvt. Ltd., 2005.

8. C. W. Macosko, RIM Fundamentals of Reaction Injection Molding, Hanser Publishers, Munich, 1989, pp. 87-10.

9. B. R. Gupta, Rubber Processing on a Two-Roll Mill, Allied Publishers Ltd., New Delhi, 1998, p.101.

10. P.K. Freakley, Rubber Processing and Production Organization, Plenum Press, New York, 1985, p. 52.

11. Z. Tadmor and C. G. Gogos, Principles of Polymer Processing, John-Wiley & Sons Inc., New York, 1979, p. 410.

12. J. L. White, Twin Screw Extrusion Technology and Principles, Hanser Publishers, Munich, 1991.

13. Annual Transactions of Nordic Rheological Society,10, 3-10 2002.

14. B. R. Gupta Ed. Polymer Processing Technology, Asian Books Pvt. Ltd., 2008. (From Vlachopoulos, J. and D. Strutt " An Overview of Polymer Processing, www.polydynamics.com/Overview_of_Polymer-Processing.PDF).

15. Rauwendaal, C., Polymer Extrusion, Hanser Publishers, Munich (a) 1986 and 4th Ed., 2001.

16. Vlachopoulos, J.,and D. Strutt, An Overview of polymer Processing, www. polydynamics.com/Overview_of_Polymer_Processing.PDF

17. Berins M. L. (ed.), "SPI Plastic Engineeering Handbook" Society of Plastics Industry, Kluwer Academic Publishers, 2000. [https://external-content.duckduckgo com/iu/?u=https%3A%2F%2Ftse4mm.bing.net%2Fth%3Fid%3DOIP.dXoZ5fEGQcx9FJp7NQAzDwHaEB%26pid%3DApi&f=1]

❑ ❑ ❑

Rheological Principles

2.1 HISTORICAL BACKGROUND

The science of rheology deals with the study of deformation of materials under applied stresses. It describes the interrelation between the extent and rate of deformation with magnitude of stress. Hooke in 1676 developed the first scientific relationship between the extent of deformation and applied force based on the experimental observations of change in length of an elastic metallic rod subjected to the elongational force in the direction of length (Fig. 2.1). The relationship is known as Hooke's law and states that in the linear range the deformation is directly proportional to the applied force. The materials, which follow this relationship, are known as the Hookean solids. Newton in 1686 studied the deformation of simple liquids and established that the rate of deformation of liquids is directly proportional to the applied shear stress (Fig. 2.2). This relationship is well known as Newton's law of viscosity and all fluids, which follow this relationship, are known as purely viscous Newtonian fluids. The purely elastic materials are capable of storing all the energy used in the deformation and releasing it as useful work as soon as the applied force is removed. These materials undergo reversible deformation. The purely viscous fluids on the other hand dissipate all the energy as heat due to viscous resistance. These materials start flowing as soon as subjected to the shear stress and do not go back to their original state i.e. they flow irreversibly. Purely elastic and purely viscous natures form two extreme cases of deformational behaviours and there are a large number of examples available for both types. In 1867 Maxwell suggested that all substances must possess both these characteristics to a varying degree. In fact a large number of materials particularly the macromolecular systems like high molecular weight polymer melts, their solutions, suspensions, blend and filled systems show both characteristics when deformed implying thereby that a part of the deformational energy is stored as elastic energy and remaining is dissipated as heat due to the viscous drag. Such systems are known as viscoelastic materials. The stored elastic energy manifests itself in different forms depending on the application for example, in melt extrusion it shows up as die swell (Fig. 2.3 (a)) where the extrudate diameter is larger than die diameter or the rod climbing effect where stirred polymer solution climbs the stirrer rod

(Fig. 2.3(b)) as against the Newtonian fluids, which show a downward parabolic profile (Fig. 2.3(c)).

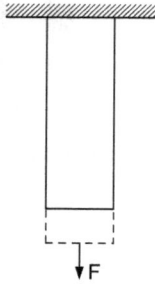

Fig. 2.1   Elastic deformation   Fig. 2.2   Viscous deformation

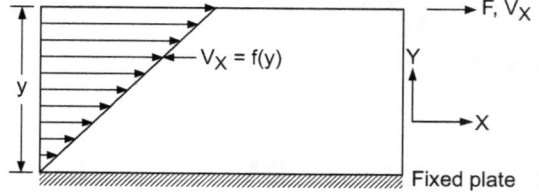

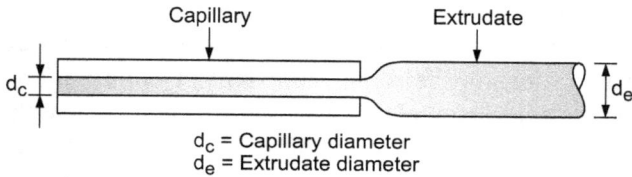

d_c = Capillary diameter
d_e = Extrudate diameter

(a) Die swell during extrusion of polymer melts.

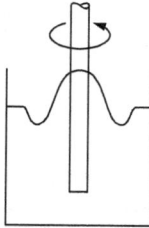

(b) Rod climbing effect in viscous solutions[6].   (c) Parabolic profile in Newtonian Fluids[6].

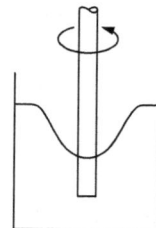

Fig. 2.3

The polymers contain long chain molecules in coiled spatial structure, which during flow or rotation tend to uncoil and orient in the direction of application of the force resulting in stressing of the chains and storing elastic energy. This stored elastic energy is the cause of developing normal stresses giving rise to die swell or the rod climbing effects. This is known as the stress relaxation process in which the long chain molecules tend to acquire their original spatial configurations indicating that these fluids possess some sort of memory, which results in such a behaviour. Therefore these fluids are also known as the memory fluids.

Thus the three basic types of material deformations are purely elastic, purely viscous and viscoelastic deformation. All solids do not follow the Hooke's law when deformed, such materials are known as non-Hookean solids. Similarly all fluids do not obey Newton's law and these are known as non-Newtonian fluids (Table 2.1). Amongst the non-Newtonian purely viscous liquids there are two types, (i) time independent and (ii) time dependent fluids. The viscoelastic systems are classified as linear viscoelastic and nonlinear viscoelastic materials depending on the response to the applied stress.

Table 2.1 Classification of materials on the basis of rheological behaviour.

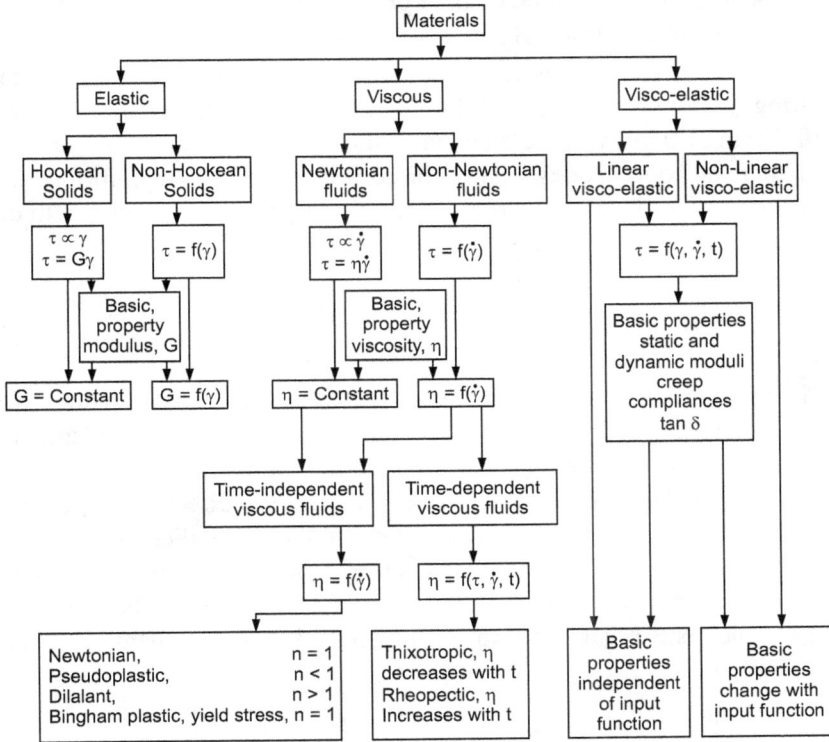

Materials

- **Elastic**
  - **Hookean Solids**: $\tau \propto \gamma$, $\tau = G\gamma$ → Basic, property modulus, G → $G = $ Constant
  - **Non-Hookean Solids**: $\tau = f(\gamma)$ → $G = f(\gamma)$
- **Viscous**
  - **Newtonian fluids**: $\tau \propto \dot{\gamma}$, $\tau = \eta\dot{\gamma}$ → Basic, property viscosity, η → $\eta = $ Constant
  - **Non-Newtonian fluids**: $\tau = f(\dot{\gamma})$ → $\eta = f(\dot{\gamma})$
    - **Time-independent viscous fluids**: $\eta = f(\dot{\gamma})$ → Newtonian, $n = 1$; Pseudoplastic, $n < 1$; Dilalant, $n > 1$; Bingham plastic, yield stress, $n = 1$
    - **Time-dependent viscous fluids**: $\eta = f(\tau, \dot{\gamma}, t)$ → Thixotropic, η decreases with t; Rheopectic, η Increases with t
- **Visco-elastic**
  - **Linear visco-elastic**: $\tau = f(\gamma, \dot{\gamma}, t)$ → Basic properties static and dynamic moduli creep compliances tan δ → Basic properties independent of input function
  - **Non-Linear visco-elastic** → Basic properties change with input function

2.2 RHEOLOGICAL PARAMETERS

Rheological investigations of any material involve the application of a force to the system and study of its response either in the form of strain (in case of solids), or rate of shear (in case of liquids) or stress relaxation and creep compliance in case of viscoelastic systems. In this section different rheological parameters are defined and discussed in brief.

2.2.1 Stresses

Stress is simply a constraining force or influence. It is defined as the intensity of surface force acting on the system[1, 2] that is, the forces on a member caused by loads: consisting of torsion or twisting, compression or pushing, tension or pulling, and shear or cutting. In continuum mechanics stress is a physical quantity that expresses the internal forces that the neighbouring particles of a continuous material exert on each other.

Two main types of stresses are; (a) direct or normal stress (tensile stress and compressive stress), and (b) shear stress. Another stress terms in use are principal stresses, which are the extreme value of the normal stresses possible in the material. These are maximum normal stress and minimum normal stress. Maximum normal stress is called major principal stress while minimum normal stress is called minor principal stress.

Any system being subjected to a deformation may be considered to be consisting of large number of small volume elements. If we arbitrarily choose an infinitesimal small volume element in the neighbourhood of a point in the system then the point in question may be considered to be in equilibrium due to the balance of forces acting on the surface of the volume element due to other body parts (Fig. 2.4). The limit of the ratio of force, ΔF, to the area, ΔS, as $\Delta S \to 0$, is termed as stress, σ, such that

$$\sigma = \lim_{\Delta S \to 0} \frac{\Delta F}{\Delta S} = \frac{dF}{dS} \qquad \qquad ...(2.1)$$

The forces acting on an infinitesimal small area (defined by the direction of its normal determined by the unit vector, n of the chosen volume element can be resolved in three mutually perpendicular directions X, Y and Z. Thus each area element will be subjected to a force component in a direction perpendicular to it i.e. along the unit vector, n and two mutually perpendicular components parallel to the surface (Fig. 2.5). This results in the stress system consisting of three stress components of each X, Y and Z stresses acting in these directions giving the elements of stress tensor acting on the volume element. This can be represented as,

$$\{\sigma\} \equiv \begin{vmatrix} \sigma_{xx} & \sigma_{xy} & \sigma_{xz} \\ \sigma_{yx} & \sigma_{yy} & \sigma_{yz} \\ \sigma_{zx} & \sigma_{zy} & \sigma_{zz} \end{vmatrix}$$

In the nomenclature of stresses, first subscript, say x in σ_{xy}, refers to stress acting on the plane perpendicular to x-axis and second subscript indicates the direction of stress. The stresses with the same subscript repeated i.e. σ_{xx}, σ_{yy} and σ_{zz} are normal stresses and those with different subscripts i.e. σ_{xy}, σ_{xz} etc. are the shear stresses. The normal stresses acting perpendicular to the surface are compressive when act inwards (considered negative) and elongational, extensional or tensile when act outwards (considered positive) and shear stress results when the forces act along the parallel surfaces of the system Fig. 2.6.

Shear stresses on the outer cylindrical surface of a cylindrical element (Fig. 2.7) subject the system to torsion or torsional stresses and the element in question deforms (twists) by an angle, Θ(say) whereas, the displacement is a function of radius. The application of the cyclic force results in cyclic stresses and the corresponding cyclic deformation of the system.

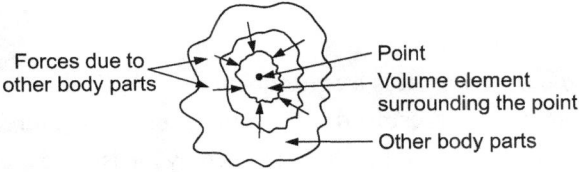

Fig. 2.4   System of stresses acting on a point.

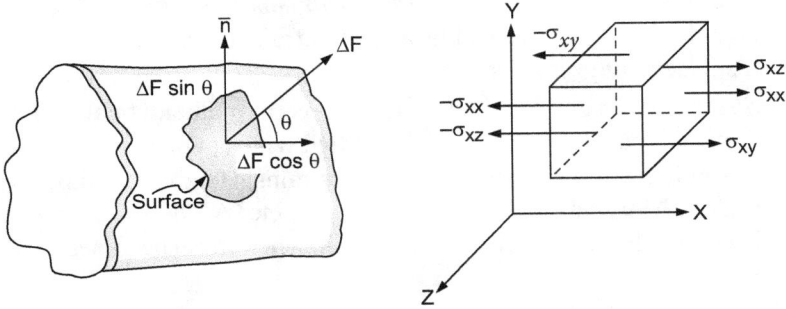

Fig. 2.5   System of Stress components acting on the surface of the volume element

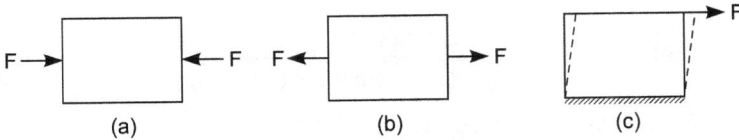

Fig. 2.6   System of stresses (a) compressive, (b) tensile and (c) shear stress

2.2.2 Strains

Strain is a measure of deformation of a flexible body. The application of a force to a material subjects it to the deformation. The extent of deformation depends on the magnitude of stress developed and the resistance to the deformation of material. The direction of deformation normally corresponds to the direction of force for example a tensile force will elongate the system in the direction of the force. Different types of strains are:

- *Average strain:* It is strain measured over a finite portion of the body, for example by using a strain gauge.
- *Point strain:* It is obtained by a limit process in which the dimension (s) of the gauged portion is made to approach zero.

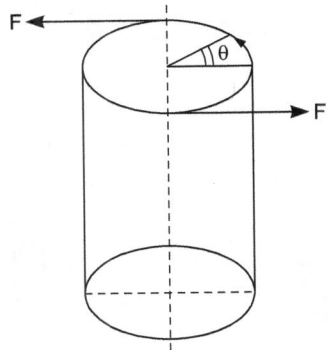

Fig. 2.7   Tortional deformation under tangential force

- *Normal strain or extensional or dimensional strain:* It measures changes in length along one specific direction.

- *Shear strain:* It measures the changes in angles with respect to two specific directions.
- *Mechanical strain:* It is a consequence of application of stress.
- *Thermal strain:* It results due to changes in temperature of the system.
- *Finite strain:* Finite strains are obtained by measuring the exact changes in dimensions or angle.
- *Infinitesimal strains or linearized or small strains:* These strains are obtained by linearizing a finite strain measure with respect to displacement gradients.
- *Strain measures:* For finite strains several mathematical expressions are in application particularly identified with a person name in front for example Langranian strain, Eulerian strains, Henckey strains, Almansi strains, Murnaghan strains, Biot strains, etc. All theses strains tend to become infinitesimal (linearized) strain when strain becomes very small i.e. $\ll 1$.

The strain experienced by the system:

$$\text{Normal Strain} = \frac{\text{(Extent of deformation in a particular direction)}}{\text{Original dimension of the test piece in that direction}}$$

Consider the application of tensile force to a metallic bar as shown in Fig. 2.8. If the initial length of the bar is L_0, and the final length is L_1, i.e. the change in the length after the application of the force is $\Delta L = (L_1 - L_0)$, then the linear strain or engineering strain is given as:

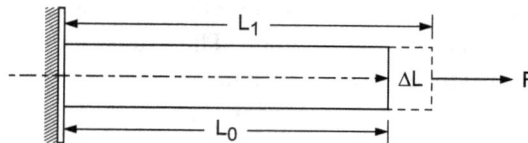

Fig. 2.8 Tensile deformation

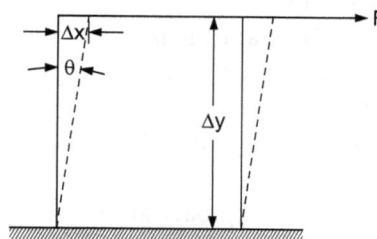

Fig.2.9 Shear deformation

$$\varepsilon_L = \frac{\Delta l}{L_0} = \frac{(L_1 - L_0)}{L_0} \qquad \qquad ...(2.2)$$

This strain is also known as Langrangian strain.

If L_0 in denominator is replaced by L_1 i.e., the final length then it is known as Eulerian strain.

The Lagrangian strain is preferred in solid and structural mechanics whereas the Eulerian strain is generally used in fluid mechanics.

Due to the application of the force at the end of the bar all points inside the bar are subjected to uniform differential strain, therefore the true strain, which is also known as Hencky strain will be obtained by the integration of the differential strain over the entire length. The differential Hencky strain is

$$d\varepsilon_H = \frac{dl}{L} \qquad \qquad ...(2.3)$$

and the Hencky strain is

$$\varepsilon_H = \int d\varepsilon_H = \int_{L_0}^{L_1} \frac{dl}{L} = \ln\left(\frac{L_1}{L_0}\right) \qquad \qquad ...(2.4)$$

Under the application of the shear stress the system undergoes an angular deformation (simple shear) by a certain degree (Fig. 2.9) say, θ. The resultant strain is known as shear strain, which is given by

$$\gamma = \frac{\delta x}{\Delta y} = \tan \theta \qquad \qquad ...(2.5)$$

If θ is very small,

then, $\qquad \qquad \tan \theta \cong \theta = \gamma \qquad \qquad ...(2.6)$

2.2.3 Rate of Strain or Rate of Shear

The system may be deformed at different rate of deformations. The rate of deformation has profound influence on the response of the material. A purely elastic material when deformed at very slow rate may flow or a purely viscous fluid deformed with very high rate may show elastic response. For tensile or compressive deformations the linear strain rate and Hencky strain rate are given respectively by

$$\dot{\varepsilon}_L = \lim_{\Delta t \to 0} \pm \frac{\varepsilon_L}{\Delta t} = \pm \frac{d\varepsilon_L}{dt} = \pm \frac{1}{L_0}\frac{dL}{dt} \qquad \qquad ...(2.7)$$

and $\qquad \qquad \dot{\varepsilon}_H = \pm d \ln\frac{L}{dt} \qquad \qquad ...(2.7(a))$

In the above equations the positive sign indicates tensile rate of strain and negative sign the compressive. For shear deformation the rate of shear or shear rate is given by

$$\dot{\gamma} = \lim_{\Delta t \to 0} \frac{\Delta \theta}{\Delta t} = \frac{d\theta}{dt} \qquad \qquad ...(2.8)$$

2.3 RELATIONSHIP BETWEEN THE RHEOLOGICAL PARAMETERS

The mathematical relationship between stress, strain and the rate of strain is known as constitutive equation of a body or the rheological equation of state. A mathematical function of the form

$$f(\sigma, \gamma, \dot{\gamma}) = 0 \qquad \qquad ...(2.9)$$

represents a general function between different rheological parameters. The science of rheology has developed from two different directions, (i) Theoretical analysis of response of the system to applied stimuli, and (ii) analysis of the experimental data to develop the empirical correlations. The theoretical analysis involves the relationships between stress, strain and rate of strain tensors resulting in highly complex generalized equations and only their simplified forms find the practical applications. However these equations give better insight into the combined effects of different types of stresses and the corresponding response. The experimental approach on the other hand results in simple applicable equations containing few material parameters or constants. Different rheological systems are discussed next.

2.4 RHEOLOGICAL SYSTEMS

2.4.1 Purely Elastic Behaviour

Purely elastic response of materials is characterized by their stress-strain behaviour. Different materials give different stress-strain curves for example brittle materials and, metals and alloys are characterized by high stress and low strain but plastics in general suffer brittle failure whereas the metals undergo yield (Fig. 2.10 curves 1 & 2) i.e. a ductile failure. The rubbers and rubber like materials show low stress and very high elongation and their stress-strain curves pass through a number of stages (curve 2.10.3). The cross-linking or vulcanization increases the stress value but reduces the elongation (curve 2.10.5) whereas the fillers give the similar effect but to a lesser extent (curve 2.10.4). It is of considerable interest to discuss the curve 2.10.3 which shows four distinct regions i.e. sections (i) OA, (ii) AB, (iii) BC and (iv) CD.

(i) *Section OA:* In this section the stress varies linearly with strain up to a small value of strain. The resistance to the deformation comes from the physical structure of the material. The structure deforms reversibly under low strain conditions and the system follows Hooke's law. This region is known as the first Hookean region.

(i) **Section AB:** In this section the nominal stress (stress based on initial cross section area) increases very slowly as compared to the strain and passes through a maximum but the true stress (stress based on actual cross section area) only shows a hump (curve 2.10.6). The resistance to the deformation comes from the positional rearrangement of particles, or molecular segments or that of macromolecule as a whole, resulting in an irreversible flow of material.

(iii) **Section BC:** In this section the nominal stress remains more or less constant or increases very slowly for a considerable increase in strain. It is because of disentanglement and uncoiling of the long molecular chains of rubber molecules resulting in negligible consumption of energy for large strain values. As the long chain molecules they get oriented in the direction of deformation. The true stress increases more or less linearly with strain albeit very slowly hence this region is known as the second Hookean region.

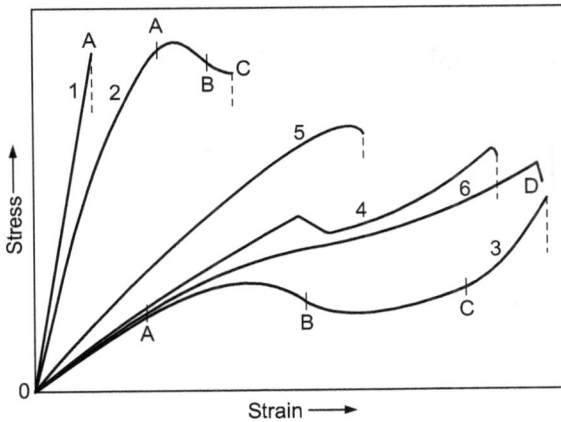

Fig. 2.10   The stress strain curves of different materials

(iii) **Section CD:** In this section the stress increases relatively sharply with small strain showing considerable enhancement in the capacity of rubber to resist the deformation or a build up of internal strength of the rubber matrix. This additional matrix strength arises from, (i) what is known as strain induced crystallization due to the orientation of chains in the direction of deformation thereby bringing the chains so close to each other that they unite to form a crystallite and (ii) the stretching of the main chain -C-C- bonds before failure.

The actual magnitude of the stress i.e. the true stress experienced by the deforming test piece is always higher than the nominal or conventional stress as the area continuously decreases as the elongation increases. The change in area may take place in different ways (Fig. 2.11) i.e.

(i) Uniform reduction over the entire length. Such reduction takes place only for small strain values (Fig. 2.11(a)),

(ii) Gradual reduction from both ends to a minimum at some point in the middle subjecting the test piece to a uniform necking (Fig. 2.11(b)) and

(iii) Sudden reduction at the middle of the test piece giving deep necking and have uniform value up to the ends (Fig. 2.11(c)).

Under uniform necking conditions the new cross section area may be assumed to be constant over the length. If A and A_0 are the areas of elongated and original test pieces and F is the force applied then the true stress σ and nominal stress (also known as Engineering Stress) σ_0 are related as

$$F = A_0\sigma_0 = A\sigma \qquad \qquad ...(2.10)$$

As the volume of test piece is constant

$$A_0 l_0 = A l \qquad \qquad ...(2.11)$$

$$\therefore \qquad \sigma = \frac{A_0\,\sigma_0}{A} = \frac{\sigma_0\,l}{l_0} = \sigma_0\lambda \qquad \qquad ...(2.12)$$

where l_0 and l are the initial and final length of the test piece and λ is the elongation ratio.

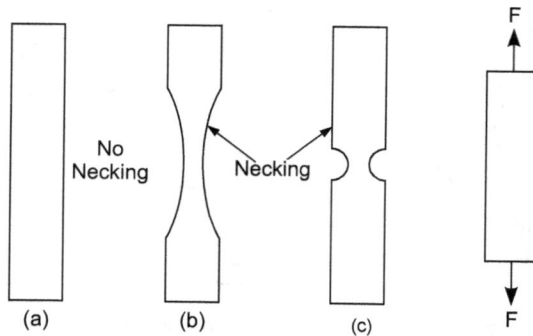

No Necking

Necking

F

F

(a)   (b)   (c)   F

Fig. 2.11   Extensional deformation with necking

The above relations hold up to the point of necking. The phenomenon of necking is the behaviour of materials, which arises due to the non-uniformity of the structural strength and existence of faults in the test pieces. The stress concentration will develop around these faults resulting in the reduction in the area. Necking may be defined as a shape discontinuity characterized by a change of specimen boundary surface from the condition of parallelism with the direction of the force[3]. It therefore arises as a result of instabilities in material flow due to uneven stress distribution in the test piece. The necking as shown in the Fig. 2.11(c), results when the material flows on further elongation from shoulders of the piece towards the neck due to cold drawing. The cold drawing is the consequence of ability of the polymer chains to disentangle and slip past each other easily. The neck under these conditions remains stabilized till the material keeps flowing and thereafter reduces on further elongation and ultimately fails. Other important properties that can be obtained from stress-strain curves are yield energy and toughness.

2.4.2 Yield Energy

Yield energy is defined as the work per unit volume, required to make a polymer yield. The yield energy, Ey is given as, $Ey = \int(\text{tau})\, d\gamma$. In case of metals this behaviour is called resilience. For polyethylene at room temperature $Ey = 2.5$ J/cc. The polymers being the viscoelastic in nature, their response to the mechanical loading is time dependent and they display delayed elasticity in general. For this reason, the magnitude of the yield stress depends on the rate of strain.

2.4.3 Toughness

During elongational deformation the specimen may rupture at stress higher than σ_y and the strength of the polymer is defined as the highest stress it can support. σ_y may or may not be strength of polymer depending on its behaviour after yielding. The ultimate properties i.e. those at rupture, are used to define toughness of the polymer as the total work per unit volume required to produce rupture. Toughness is given by the area under the stress-strain curve from zero strain up to rupture (γ_r) i.e. Toughness $= \int \tau\, d\gamma$.

It is important to point out that the tensile and shear deformational behaviours of polymers depend on type of loading, the experimental procedure, the temperature and the pressure.

EXAMPLE 2.1. An elongational force of 160 N is applied to a piece of vulcanized rubber of dimensions $4 \times 4 \times 30$ mm. Calculate the true and nominal stresses at 50% elongation.

Solution: Nominal Stress is based on the original cross section area,

$$A_0 = 4 \times 4 = 16 \text{ mm}^2 = 16 \times 10^{-6} \text{ m}^2$$

$$\sigma_{\text{nominal}} = \frac{F}{A_0} = \frac{160}{16 \times 10} = 10 \text{ MPa.}$$

True Stress $\sigma_{\text{true}} = $ Force / True cross section area

The true cross section area

$$A_1 = \frac{A_0 L_0}{L_1}$$

where L_1 is the length after extension $= 1.5 \times 30 = 45$ mm

Therefore $\qquad A_1 = \dfrac{4 \times 4 \times 30}{45} = 10.67 \text{ mm}^2$

$$\sigma_{true} = \dfrac{160}{10.67 \times 10^{-6}} = 15 \text{ MPa}$$

2.5 PURELY VISCOUS BEHAVIOUR

2.5.1 Newtonian and Non-Newtonian Fluids

Purely viscous behaviour is characterized by the dependence of shear stress on the rate of shear (also sometimes written as the rate of deformation or the shear rate). The liquids are capable of sustaining infinite deformation and the linear relationship between shear stress, τ, and the rate of shear, $\dot{\gamma}$ is the well known Newton's law of viscosity representing the Newtonian fluid behaviour. The behaviour of such fluids can be best demonstrated by shearing a fluid between two parallel plates as shown in Fig. 2.2. If the distance between the plates is y and a force F is applied to the top plate of surface area, A, so that it moves with a velocity of V_x with the lower plate being fixed, the shear stress and the rate of shear can be calculated and the relationship between them can be written as

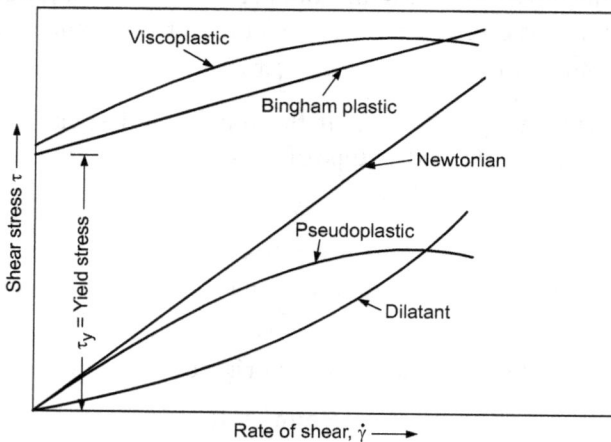

Fig. 2.12 Plots of shear stress versus rate of shear for different fluids

$$\tau = \frac{dF}{dA} = \eta \left(\frac{dV_x}{dy} \right) \qquad \qquad ...(2.13)$$

where η is the viscosity of the fluid.

Most of the low molecular weight liquids, organic solvents and dilute solutions of polymers follow this equation. The viscosity for these liquids is constant over the entire range of rate of shear at a particular temperature. A plot of shear stress and rate of shear for these fluids gives a straight line passing through the origin (Fig. 2.12) and the slope of the line is the viscosity of the fluid. Some of the examples of Newtonian fluids are, blood plasma, custard and water.

2.5.2 Time Independent non-Newtonian Fluids

A large number of liquids like polymer and rubber melts, concentrated polymer solutions, solids suspensions and slurries invariably do not follow equation 2.13 but show the viscosity to be a function of rate of shear, $\dot{\gamma}$ only. Such fluids are known as the time independent non-Newtonian fluids and if the viscosity changes also with time then they are known as time dependent fluids. The Table 2.1 gives the classification of materials based on their rheological behaviours. For these fluids the viscosity is either found to decrease with rate of shear as in the case of pseudoplastic fluids or increase as in the case of dilatant liquids. The Bingham plastic fluids on the other hand demonstrate the existence of yield stress i.e. to say that these fluids require a definite amount of energy before they start flowing and then flow as Newtonian fluids. Some polymer melts may flow as pseudoplastic liquids after yielding, these are known as plastoviscous or viscoplastic fluids (Fig. 2.12). The physical reasons for different types of fluid behaviors are discussed next.

2.5.3 Pseudoplastic Fluids

The viscosity of these fluids reduces with the shear rate and the curve of shear stress versus rate of shear is concave downwards. A log-log plot (Fig. 2.13) of shear stress and rate of shear clearly shows three different regions. These are, (i) a low shear region of a small range with the slope of the line = 1, known as the first Newtonian region, (ii) an intermediate shear region with the line of slope of <1 known as pseudoplastic or shear thinning region and (iii) a high shear region with slope of the line = 1 known as second Newtonian region. Depending on the type of polymer the slope of the line in the intermediate region may vary from just <1 to very close to zero. The extent of pseudoplasticity or the shear thinning behaviour increases with the decrease in the value of slope below 1. The viscosity and shear stress or rate of shear plot for such fluids shows a general nature with the viscosity being constant at low shear region, continuously falling in the intermediate region and then again constant in the high shear region particularly for the polymer solutions and low MW polymer melts (Fig. 2.14). Such a behaviour can be explained by assuming that in the polymer melts and solutions, which show shear thinning nature, the macromolecules or the particles are completely random in the spatial orientation when at rest (Fig. 2.15(a)) and are bound by weak intermolecular forces. When the applied stress is relatively low the molecules maintain their randomness and the molecular forces do not allow them to be oriented in the direction of the force resulting in constant viscosity in this region. As the shear rate increases the polymer molecules start orienting in the direction of flow. As the polymer system contains a molecular weight distribution the smaller molecules, which require less energy to change the direction start orienting, initiating the reduction in the viscosity. The high molecular weight molecules have very long coiled chains, which are entangled

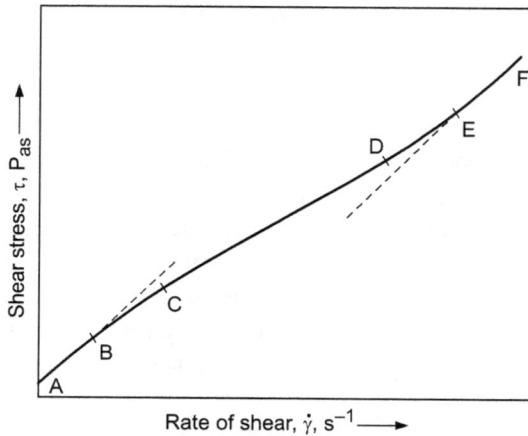

Fig. 2.13   Log-log plot of shear stress versus rate of shear for pseudoplastic fluids[6]

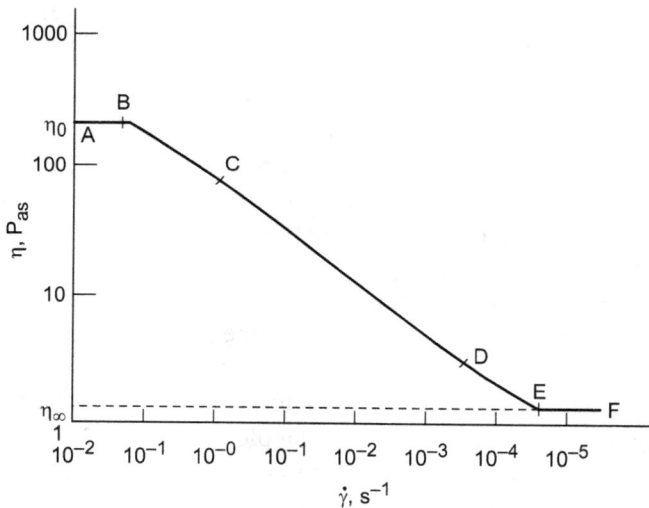

Fig. 2.14   Variation of viscosity with rate of shear[6]

with each other. These molecules require higher energy for disentanglement, uncoiling and then orientation. As the shear stress and shear rate increase further the system is subjected to increasingly higher energy for deformation resulting in more and more molecules being oriented. The resistance to the flow of the oriented molecules is reduced due to the fact that the oriented molecules can slip pass each other with relative ease. This gives a continuous decrease in the viscosity till all the molecules are oriented (Fig.2.15(b)) at some higher value of rate of shear. Beyond this value of rate of shear there is no further increase in the extent of orientation of molecules and the systems show a constant viscosity or the second Newtonian region. Some of the examples of materials, which show shear thinning nature are nail polish, whipped cream, ketchup, molasses, syrups, paper pulp in water, latex paint, ice, blood, some silicone oils, some silicone coatings, sand in water, polymer solutions, starch solutions etc.

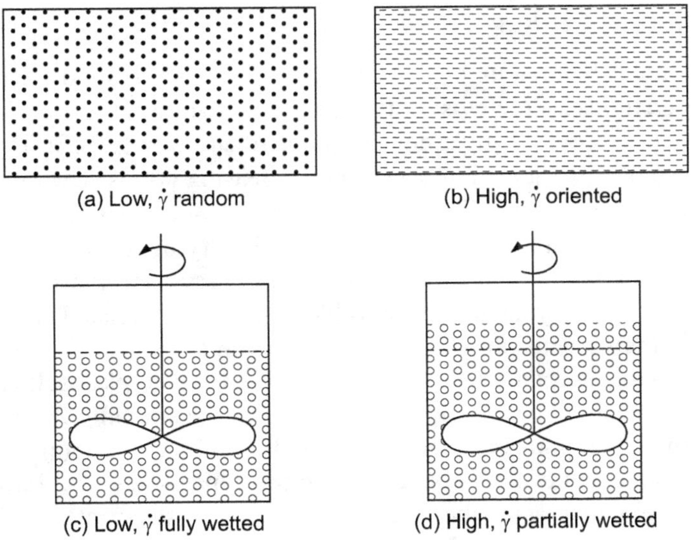

(a) Low, $\dot{\gamma}$ random   (b) High, $\dot{\gamma}$ oriented

(c) Low, $\dot{\gamma}$ fully wetted   (d) High, $\dot{\gamma}$ partially wetted

Fig. 2.15 Pseudoplastic (a, b) and dilatant (c, d) fluids under deformation

EXAMPLE 2.2. The following shear stress- shear rate data was obtained from an Instron high-pressure capillary viscometer for poly (Vinyl Chloride), PVC melts at 200°C. From the plot of shear stress versus rate of shear identify the rheological behaviour of the melt.

Rate of Shear, $\dot{\gamma}$, s^{-1}	Shear Stress, τ, Pa
3.06	80,000
6.12	110,000
12.23	170,000
30.59	330,000
61.17	590,000
122.34	1,020,000
305.85	2,000,000
611.70	3,200,000
1223.40	4,900,000
3058.50	7,800,000

Solution: The curve (I) (Figure E-2.1) obtained by plotting shear stress versus rate of shear is concave downwards indicating a continuously decreasing slope with increasing rate of shear representing the pseudoplastic nature. Thus the PVC melt is a pseudoplastic fluid.

2.5.4 Dilatant Fluids

The rheological behaviour of dilatant fluids is exactly opposite to that of pseudoplastic fluids i.e. their viscosity increases with rate of shear or shear stress and the plot of τ versus $\dot{\gamma}$ gives concave upward curve (Fig. 2.12). These fluids show a constant viscosity at low shear rate, increasing viscosity with increasing rate of shear in the intermediate region but at high shear region the behaviour is not well established due to experimental difficulty in collecting the reliable rheological data. Examples of such fluids are not very common and only some highly filled polymer propellant systems like hydroxyl terminated polybutadiene propellants, the suspensions of beach sand, starch suspensions in water, gum solutions, aqueous suspensions of titanium dioxide, the PVC plastisols etc. display this nature. The shearing of suspensions in which the amount of solids present is so high that the liquid is just enough to fill the voids (Fig. 2.15(c)) will tend to release the solids out of fluid contact due to the expansion (Fig.2.15(d)) during shearing thereby increasing the viscosity. At very low shear rate the liquid still fills the voids causing lubrication effect and giving constant viscosity.

EXAMPLE 2.3. A certain polymer melt gave the following rheological data. From the rheograph (a plot of shear stress vs. rate of shear) characterize the melt rheologically.

Shear rate, $\dot{\gamma}$, s^{-1}	Shear stress, τ, Pa	Shear rate, $\dot{\gamma}$, s^{-1}	Shear stress, τ, Pa
100	100,000	1400	3,500,000
400	560,000	1800	5,600,000
700	1,200,000	2000	7,150,000
1000	2,000,000	—	—

Solution: The rheograph (II) (Figure E-2.1) is concave upwards giving continuously increasing slope i.e. proportionately higher shear stress for the same increase in the rate of shear. The melt in question therefore is a dilatant fluid.

EXAMPLE 2.4. A certain polymer melt gave the following rheological data using a high-pressure capillary viscometer. Determine the rheological nature of melt.

Shear rate, $\dot{\gamma}$, s^{-1}	Shear stress τ, k Pa	Shear rate, $\dot{\gamma}$, s^{-1}	Shear stress, τ, kPa
50	480	1200	2320
100	540	1800	3240

240	800	2200	3910
800	1670	—	—

Solution: The rheogram (III) (Figure E-2.1) gives a linear variation of shear stress with rate of shear with an intercept at the stress axis at rate of shear equal to zero. Thus it is a fluid having yield with Newtonian behaviour under deformation i.e. the Bingham plastic melt, which is discussed next.

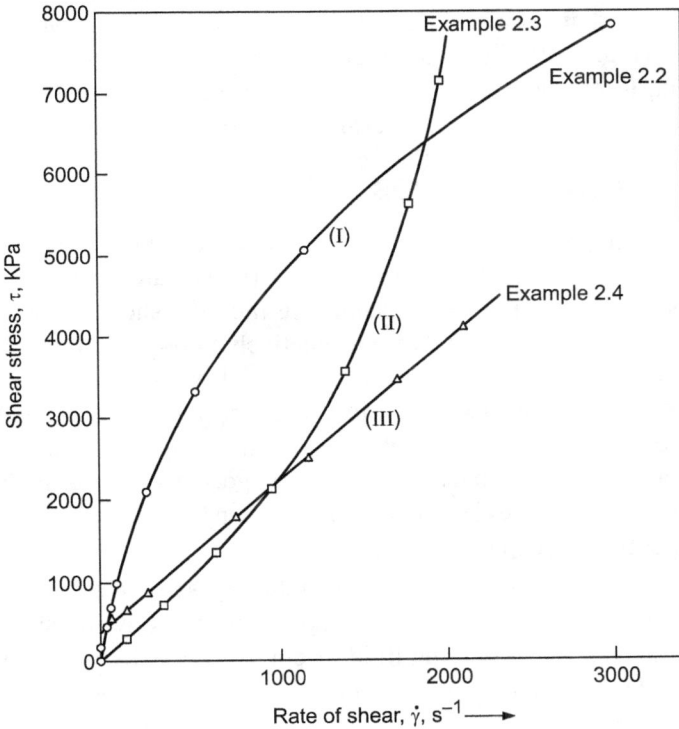

Fig. E-2.1   Plots of shear stress versus rate of shear for examples 2.2, 2.3 and 2.4.

2.5.5 Bingham Plastic Fluids

Bingham plastic fluids require a certain force to develop what is known as yield stress, τ_y before they start flowing and then flow like Newtonian liquids (Fig. 2.12) i.e. the plot of τ, versus $\dot{\gamma}$ is a straight line and it intercepts the stress axis at some finite value equal to the yield stress. The viscosity of these fluids is constant during the flow at shear stresses higher than yield stress, $\tau > \tau_y$ and increases sharply as the stress approaches yield stress. These fluids, when at rest, are believed to develop a three-dimensional structure due to the presence of intermolecular or interparticle forces. This structure resists the deformation to such an extent that it does not allow the fluid to deform till the applied force gives enough energy to break it down and once the fluid begins to deform it flows like a Newtonian liquid. After the deforming force is removed the structure forms again when at rest. The systems which show such a behaviour are thickened hydrocarbon greases, some asphalts, bitumen, suspensions of clay, fly ash or metal oxides in water, sewage sludges, jellies, tomato ketchups, tooth pastes etc.

2.5.6 Viscoplastic Fluid

These fluids also show yield stress but during flow do not follow the Newton's law of viscosity but show a shear rate dependent viscosity beyond the shear stress $\tau > \tau_y$ (Fig. 2.12). Filled polymeric systems, heavy crude oils with high wax content, tooth paste, fine particle suspensions, paint slurries, pastes and certain food stuffs are some of the examples of such behaviour.

2.6 TIME DEPENDENT FLUIDS

The time dependent fluids are those liquids, which show either a decrease or increase in the viscosity with time at a particular rate of shear i.e. if these fluids are subjected to a constant shear rate then the shear stress will either decrease or increase. These fluids are respectively known as **thixotropic fluids** and **rheopectic fluids** (Fig. 2.16). Examples of thixotropic systems are filled polymers, bentonite clay suspended in water, drilling muds, cement slurries, crude oils, coal-water slurries, yoghurt, salad dressing, mayonnaise, xanthan gum solutions, aqueous iron oxide gels, gelatin gels, pectin gels, hydrogenated castor oil, montmorillonite clay suspensions, carbon black suspensions in molten tire rubber, colloidal suspensions etc.

The rheopectic behaviour is demonstrated by some clay suspensions in water, specific gypsum paste, printer ink etc. In the body the synovial fluid (a viscous transparent lubricating fluid, occurring typically within the tendon sheaths and the capsular ligaments surrounding movable joints) exhibit the property of rheopexy[3a].

Thixotropic fluids are essentially pseudoplastic types where the process of orientation of molecules at a particular shear rate is a time dependent as compared to being instantaneous for the pseudoplastic liquids. These fluids therefore can be described as time dependent pseudoplastic fluids. The rheopectic fluids are dilatant type where the dilatant nature is time dependent and can be described as time dependent dilatant fluids. Both these fluids when deformed first with increasing and then with decreasing shear stress do not follow the same path back but show hysteresis (Fig. 2.17). The area under the hysteresis curves represents the loss of energy during the cycle of deformation. This energy is consumed in the changes in the molecular configuration and other structural alterations these liquids might have undergone as a result of shearing.

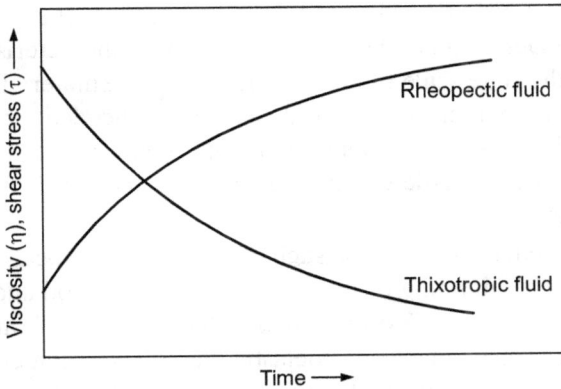

Fig. 2.16 η and τ versus time plots for time dependent fluids[6]

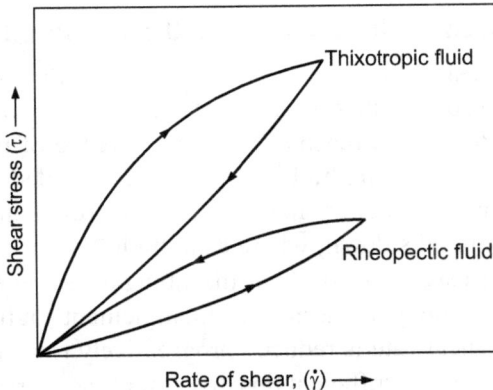

Fig. 2.17   Hysteresis curves for (i) thixotropic and (ii) rheopectic fluids[6]

2.7 VISCOELASTIC FLUIDS

As the name implies these fluids posses both purely elastic and purely viscous characteristics i.e. they store a part of the deformational energy and dissipate the remaining as heat due to viscous drag. Both these behaviours influence the polymer processing. Slow process takes long time to complete and the molecules get sufficient time to uncoil, disentangle, orient and relax as the viscous nature dominates. Whereas when the process is faster the time available for the molecules to respond is much less and the elastic response dominates. In case of flow process for example if the flow is slow then if the time scale for the flow is much less than the relaxation time of the material then the elastic effect are prevalent. Whereas on the other hand if the relaxation time is larger than the time scale, and the system is relaxed completely then the viscous effects are pronounced. The ratio of time scale of flow to the relaxation time is defined as the Deborah number. It is often used to characterize the fluidity of the materials under specific flow condition. It is based on the thinking that given enough time even the solids can flow. Reiner[3b] defined a dimensionless number namely Deborah number as a ratio of some characteristic time of the material, λ_c to the time scale of deformation, λ_s of the material and is written as $D_e = \lambda_c/\lambda_s$. The characteristic time, λ_c, is the time needed for the material to acquire 63.2% i.e. $[1 - (1/e)]$ of its ultimate retarded elastic response to a step change. If $D_e > 1$ the elastic effects prevail and if $D_e < 0.5$ then the viscous effects dominate.

The experimental evidence of such a behaviour has been shown in Fig. 2.3(a), and (b). The climbing of fluid along the rotating rod and the extrudate swell are the consequence of normal stress difference, which is also responsible for many other effects typical to viscoelastic fluids. The polymeric systems like filled systems, polymer melts and their solutions and blends, bread dough and egg white in general show this kind of behaviour.

2.7.1 Normal Stress Difference[4] and the Weissenberg Effect[5]

Under the simple shear flow it can be shown that the only nonzero component (using the Cartesian coordinate system) are shear stresses τ_{xy} and τ_{yx}, and normal stresses τ_{xx}, τ_{yy} and τ_{zz}. The reason for the occurrence of normal stresses is that the polymer molecules are highly entangled and coiled in nature and this entanglement forms a temporary network due to Brownian motion. At high strain this network breaks down, where as at higher shear rate the molecules experience large stresses due to chain disentanglement eventually reaching a balance between entanglement and disentanglement giving uniform value of stresses. As the shear rate is reduced progressively to a very low value the disentangled molecules get sufficient time to entangle again resulting in reduced magnitude of stresses and second normal stress may acquire even a negative value (tainstruments.com/AN007).The shear stress terms τ_{xy} and τ_{yx} are equal. The normal stress components as such are not responsible for the deformation of an incompressible fluid but the their differences are responsible for a number of effects. The primary or the first normal stress difference, $(\tau_{xx} - \tau_{yy})$, the secondary or the second normal stress difference, $(\tau_{yy} - \tau_{zz})$ and shear stresses can cause material to deform. When axes x, y and z are expressed as 1, 2 and 3 respectively these differences are expressed as

The primary or the first normal stress difference,

$$N_1 = (\tau_{xx} - \tau_{yy}) = (\tau_{11} - \tau_{22}) \qquad \qquad ...(2.14)$$

The secondary or the second normal stress difference,

$$N_2 = (\tau_{yy} - \tau_{zz}) = (\tau_{22} - \tau_{33}) \qquad \qquad ...(2.15)$$

For Newtonian fluids both N_1 and N_2 vanish under simple shear deformation, however for polymeric systems these have non-zero values. Lodge[4] reported that the shear stress is proportional the rate of shear and the normal stress differences are proportional to the square of the rate of shear. These are expressed as

$$\tau_{12} = \tau_{21} = \eta\dot{\gamma} \qquad\qquad ...(2.16)$$

$$N_1 = \psi_1\dot{\gamma}^2 \qquad\qquad ...(2.17)$$

$$N_2 = \psi_1\dot{\gamma}^2 \qquad\qquad ...(2.18)$$

Where η, ψ_1 and ψ_2 are the viscosity, first normal stress coefficient and second normal stress coefficient respectively. The different flow phenomena attributed to the viscoelastic nature of the polymers are discussed next.

1. Weissenberg Effect[5]

The phenomenon of viscoelastic fluid rising on the rotating rod was first explained by Weissenberg in 1947 as a consequence of normal stress differences. He demonstrated this effects by shearing such fluid in a cone and plate rheogoniometer, (Fig. 2.18), which is fitted with a set of piezometric tubes. The stored elastic energy manifests itself in the form of normal stresses resulting in the climbing of fluid in the tubes. The intensity of normal stresses is maximum at the center and progressively reduces to be minimum at the periphery. Lodge[7] has reported that the polymeric liquids will exhibit both positive and negative Weissenberg effects (i.e. the liquid will move either up or down the immersed rotating rod, under the conditions where the contact angle and centrifugal force effects are negligible), if the value of the factor $R: = 1 + 2\left(\dfrac{\tau}{N_1}\right)\dfrac{dN_2}{d\tau}$ is positive or negative. This equation can not be solved directly but needs a trial and error approach to estimate the normal stress differences. Although the rising of the liquid up the rotating rod is a common phenomenon with polymeric fluids the climbing down has not been observed although the solution to the above equation predicts it. The reasons for this discrepancy are not clear. A similar rod climbing effect of an interface between two Newtonian fluids has been demonstrated by Bonn et al. and explained to occur when a Taylor-Couette instability happens in the less viscous of the two liquids but not in the more viscous one[8].

Weissenberg Number: A dimensionless number known as the Weissenberg number has been defined by Weissenberg, as $Wi = \dfrac{\text{Elastic forces}}{\text{Viscous forces}} = \dfrac{\tau_{xx} - \tau_{yy}}{\tau_{xy}}$

$$= \frac{\lambda\mu\gamma^2}{\mu\gamma} = \gamma\lambda$$ [R.J. Poole, Deborah and Weissenberg numbers, Rheology Bulletin., 53(2) 32-39 (2012).]. This number is used to study the viscoelastic flows. It relates stress relaxation time of the fluid and a specific process time for example, in a simple steady shear, the Weissenberg number is defined as shear rate, times the relaxation time. The Weissenberg number helps to explain the effects caused due to normal stress difference.

2. Die Swell

The extrudate from the extruder die rapidly increases in the cross section area as soon as it emerges from the die. This phenomenon is known as the extrudate swell or the die swell. The melt while flowing from the die experiences extensional stresses causing in the generation of the normal stress differences, which are responsible for this effect (Fig. 2.3(a)). The die swell, α is defined as the ratio of extrudate diameter to the die diameter. It is generally agreed that the die swell is a consequence of elastic stress relaxation behaviour, however there is no single theory that can explain the die swell behaviour of all the fluids like Newtonian fluids, polymer solutions and polymer melts. For Newtonian fluids it has been reported by Middleman and Gavis[9] that in the laminar flow region the die swell, α, values vary from 0.87 for Reynolds number (defined as the ratio of inertia force to viscous force) $N_{Re} >$ 100 to 1.12 for $N_{Re} < 2$. At the low N_{Re} values it should be noted that the inertia forces are almost negligible in comparison to the viscous forces however at higher N_{Re} values both these forces are important. It is important to understand the role of the viscous forces at low N_{Re} value at and near the die exit. The velocity distribution inside the capillary is a fully developed laminar flow profile, which gets disturbed at the die exit due to the viscous effects.

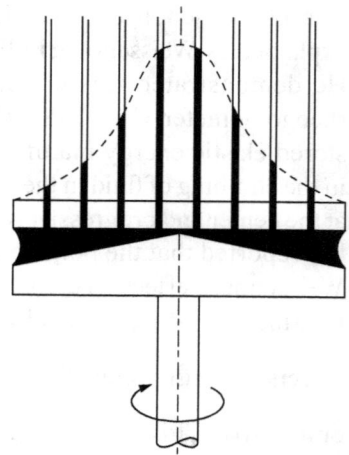

Fig. 2.18   Normal stresses as observed in a rheogoniometer[1]

3. Draw Resonance

When the polymer melt is drawn during polymer processing like film extrusion, coating, fiber spinning, film casting, film blowing etc. and drawn into the quenching medium, a phenomenon called draw resonance takes place. Draw resonance is an instability phenomenon associated only with stretching motions. This instability was first observed by Christensen[10] in the extrusion coating process of polypropylene. Hyun[11] has forwarded a new criterion for draw resonance based on the interaction of the travelling times of some kinetic waves propagating along the spin-line from the die exit to the take up position. As soon as one wave crosses the take up, another similar wave immediately appears at the spinneret and begins traveling toward the take-up. This process repeats itself to result in the perpetuation of draw resonance phenomenon. The instability manifests itself in the occurrence of periodic fluctuations in the extrudate dimensions (fiber diameter or film thickness) when a certain critical value of take-up speed is exceeded. It may result in the fluctuations in the extrudate diameter up to 5–30%. As the draw resonance can limit the productivity or deteriorate the product uniformity its understanding and control is essential during polymer processing.

4. Melt Fracture

The occurrence of the die swell has been accepted to be a consequence of elastic recovery or the elastic stress relaxation. Another phenomenon believed to be associated with the elastic response of the polymer melts is melt fracture. As the melt is extruded through the dies with progressively increasing flow rate the extrudate surface undergoes a number of transformations[12]. These include, (i) "Matte or sharkskin", a fine scale surface roughness, (ii) "Ripple", a fairly regular wavy appearance on the surface, (iii) "Bamboo or palm tree", extrudate containing alternatively relatively smooth longer section followed by a smaller section having appearance of bamboo knot, hence the name, (iv) "Screw or helix", having the appearance of helical screw or a rope and (v) "Gross melt fracture", a totally random surface distortion having absolutely no regularity in appearance (Fig. 8.21). Melt fracture therefore represents a real problem in polymer processing affecting the product quality in almost all extrusion and film blowing industries. The melt fracture phenomenon has been reviewed comprehensively by Tordella[13], Petrie and Denn[14] and White[15]. The parameters, which cause the melt fracture include (i) die characteristics like L/D ratio and entrance geometry (ii) material of construction, which may cause the wall slip and stick and slip phenomenon (iii) processing parameters like temperature, rate of shear and shear stress and (iv) polymer characteristics like chemical structure, molecular weight, molecular weight distribution, branching and relaxation and creep behaviour etc. It has been more or less accepted that the melt fracture originates at the die entry and the die exit only aggravates it.

5. Stress Relaxation

Stress relaxation is the phenomenon associated with almost all polymer melts and solutions. It is the response of the system subjected to a finite input strain. As soon as a finite strain is applied to a system, a pure elastic material gets subjected to finite stress corresponding to its modulus G_i, and retain it till the strain is applied upon, a pure viscous liquid experience an infinite rise in stress, which instantly drops to zero following a spike or delta function. Whereas the viscoelastic materials which have both viscous and elastic characteristics to varying degree store a part of elastic energy giving a finite rise in stress instantly. The stress developed in these systems will be less than that stored in purely elastic solid as some energy is dissipated as heat due to the viscous nature. The stored elastic energy decays exponentially over a period of time[12,13] to an equilibrium stress known as the residual stress. Stress relaxation is expressed as (a) Static stress relaxation and (b) Dynamic stress relaxation. These will be discussed separately in detail.

6. Creep Compliance

Creep is a slow deformation under applied stress. The test piece is subjected to the step change in the stress at a particular time interval and the change in the strain or the extent of deformation is measured over the period of time. The response of different materials under the applied stress considerably differs due to the variation in their rheological behaviour. Purely elastic materials under the creep deformation get instantly subjected to finite value of the strain, at the time of application of step change in stress. The magnitude of strain depends on the modulus of the system. Purely viscous fluids start flowing and the strain continuously increases with time at a constant rate of strain giving linear increase for Newtonian liquids. The viscoelastic systems show a non-linear increase in the strain with time and acquire a finite maximum value depending on the extent of energy dissipation due to the viscous nature of the system. The ratio of output strain, to the input stress is defined as the creep compliance[12,13].

7. Flow Behaviour at Capillary Entry

When the melt enters the capillary from a larger cross section area of extruder, the flow instability causes the fluid to circulate, which form vortices at the entry point. The circulation and vortex formation is a consequence of the viscoelastic behaviour of the melts[16, 20]. Experimental observations for polyethylenes have been presented by Tordella[21]. There is a distinct behavioral difference between linear and branched polymers at the entry of die. The vortex formation and the melt circulation are prominent for branched polymers, whereas the linear polymers fill the space completely at the die entry. Also Cogswell[20] reported that the polymers having higher extensional viscosity form larger vortices and vice versa. The vortex formation and associated circulation are the main reasons for the melt fracture of the extrudate due to consequent instability with increasing flow rate. Fig. 1.2(e) clearly shows this effect for IUPAC-LDPE melt where both larger vortex formation and extrudate swell are visible with increasing flow rate[22].

8. Pressure Hole Error Effect

The hole pressure error (pH) effect was first discovered by Broadbent, Kae, lodge and Vale in 1968[23] during the use of pressure tap to measure the normal stresses exerted on the boundary of a non-Newtonian material. Their measurement led to a significant error. Thereafter Han[24] studied the hole pressure effect using HDPE and PP melts and observed no pH error. Higashitani and Pritchard[25] developed a kinematical theory to explain the phenomenon and to estimate the hole pressure. Han and Kim[26] carried out experimental measurements with Newtonian fluids using slit dies of various aspect rations having different sizes of pressure holes. They also studied using Indipol polybutenes of different grades (L-50 and L-100) having different viscosities. These systems also gave pH, which could be correlated with viscous effects alone. They also investigated with viscoelastic polymeric systems like (i) 2 wt.% polybutelene in decalene, (ii) 2 wt.% CMC in water and (iii) aqueous solutions of polyacrylamide (ET-597) having different concentrations ie 0.6, 2.0, 4.0, 6.0 and 8.0 wt%. The result of these studies concluded that substantial pH error was obtained for 0.6 % solution and the error progressively reduced with increasing concentration up to 6.0%. Beyond 6.0% only negligible errors were observed which were within the experimental error.

The concentration dependence of the pressure hole error indirectly corroborates the earlier finding by Han[24] that polymer melts gave no discernible pressure hole error. Baired[27] reported that pressure hole error is sensitive to the changes in the molecular weight for different solutions of PS in dioctyl phthalate (DOP) and that this error could be used to monitor the change in molecular structure occurring during the mechanical degradation of Polyethylene oxide (PEO) solution. Higashitani and Lodge[28] experimented with dilute polyacrylamide (Separan AP30 at 23C) solutions flowing through the slit dies. They found that for low Reynolds number, Re ($10^{-4} < Re < 20$) the scatter in pH value was within ± 35 dyn/cm^2. However for each of the aqueous solutions the data lie within $\pm 20\%$ of the values calculated using the equation pH $= -0.18\, N_1$ (σ_w) for $350 \le N_1 \le 3600$. If N_2 is smaller than N_1, the pH data are compatible with the predictions of Higashitani and Pritchard applied to the holes of circular cross section. However they reported that their data for Newtonian and non-Newtonian luiquids did not agree with those of Han and Kim[26]. Many other workers[29,34] also have experimented with different systems using slit dies and reported similar behaviour of the pH loss. It has been suggested[28] that use of pH is essential in the estimation of normal stress differences but for viscosity and shear stress calculations pH can be safely ignored. It is possible that due to lower viscosity dilute polymer solutions are capable of circulating inside the pressure hole as shown in Fig. 2.19, (concentrated solutions and melts can not circulate due to high viscosity) resulting in continuous entry and exit of entangled long chain polymer molecular solution giving upward thrust resulting in reduced pressure reading.

9. Viscoelastic Effects in Calenders

The viscoelastic material during calendering forms a fringe pattern which is a function of roll speed and nip gap. These patterns appear on the roll surface at regular intervals as reported by Shenoy[35]. The exact reason for the appearance of the fringe patterns are not understood but are supposed to be due to viscoelasticity of material being calendered. The probable reason may be that while the melt is passing through nip, the polymer molecules will get disentangled and stretched. After exiting from the nip area theses molecules tend to relax causing fringe pattern to appear.

10. Parallel Plate Separation

Viscoelastic polymer melts and solutions when placed in between two parallel plates where one plate is rotating and other is fixed then the plates are exerted upon by a nonzero pressure created due to the viscoelastic nature of the material. It tries to separate the two plates where as such an effect is absent in Newtonian fluids[36, 37] (Fig. 2.20).

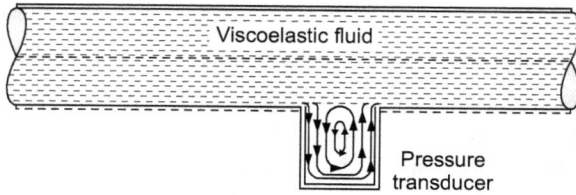

Fig. 2.19   Circulation inside the pressure hole

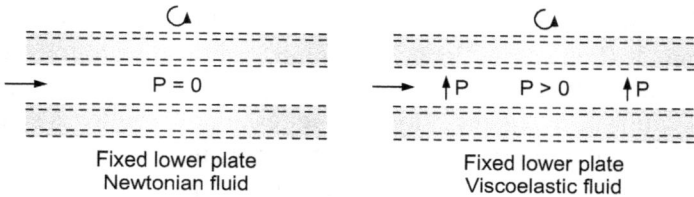

Fig. 2.20   Parallel plate separation in a viscoelastic fluid[36, 37]

11. Tubeless Siphon

Tubeless siphoning is another typical effect produced by the viscoelastic fluids, where during the siphoning process these fluid continue to siphon out the liquid even after the tube is lifted out of the fluid. The Newtonian fluids on the other hand will stop flowing as soon as the tube is lifted out[38, 39]. This effect is possibly due to the fact that the polymer molecules being coiled and entangled with each other act as if these are hooked together. When the siphoning tube is removed this entanglement seems to establish a virtual tube making the melt or solutions continue to flow (Fig. 2.21).

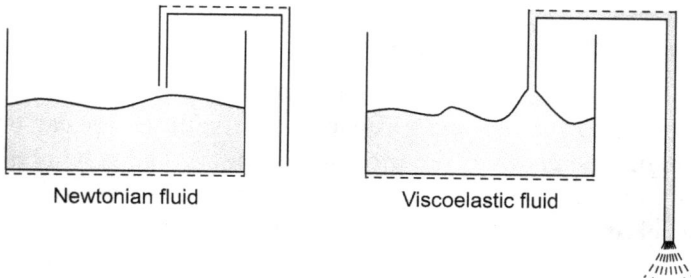

Fig. 2.21   Tubeless siphoning of a viscoelastic fluid[38, 39]

12. Ubler Effect

Ubler effect is reported to occur during the production of plastic foams wherein a gas like nitrogen is used for foaming purpose. The polymer melts along with the gas bubbles when pass through a sudden contraction, the gas bubbles tend to stop suddenly for a short period at the entrance of the contraction. The bubble size reported[35, 38, 39] is of the order of 1/6 to 1/8 of the smaller tube diameter. These bubbles move approximately one minute after stopping.

2.8 RHEOLOGICAL OR CONSTITUTIVE EQUATIONS

The development of mathematical relationship between the input parameters and the corresponding response variables, which will describe all the factors of material deformation, is an impossible proposition. Theoretical rheologists however, have developed some generalized constitutive equations like Criminale-Ericksen-Filbey (CEF)[40] and Goddard-Miller (G-M)[41] equations. The CEF equation is a second order fluid constitutive equation for steady shear flow containing constant viscosity and first and second normal stress coefficient functions. It could be reduced to generalized Newtonian fluid (GNF) equation if the normal stress coefficients are ignored. The G-M equation yields for small deformations the general linear viscoelastic (LVE) fluid equation representing the stress relaxation behaviour. These equations however fall seriously short of actual practical applications. Practical rubber and plastic scientists, engineers and technologist therefore rely mostly on developing the empirical correlations based on the experimental data and there are a large number of such relationships available in the literature. The following sections discuss some of the important empirical correlations, their applicability and limitations.

The flow of the polymer melts and solutions invariably show shear thinning or pseudoplastic nature having three distinct regions, (i) low shear rate or the lower or initial Newtonian region of constant viscosity (section AB in Fig. 2.14), (ii) intermediate rate of shear or shear-thinning region of falling viscosity (section CD) and (iii) high shear rate or the upper Newtonian region of constant viscosity (section EF). The small sections in the curve i.e. BC and DE represent the inflection regions where the slope of the curve varies in the nonlinear fashion. Polymer solutions and low MW melts show all the three regions but high MW polymer melts develop melt fracture in the high shear region and may not show the upper Newtonian region. The development of rheological empirical correlations has centered on this basic nature of flow with additional parameters to describe other variations like occurrence of yield stress or time and temperature dependency. Both the constant viscosity regions are well described by one parameter Newton's law of viscosity (Equation 2.13) for incompressible fluids. To describe the behavior of non-Newtonian fluids a number of empirical correlations have been proposed by different workers. The complexity of these equations varies depending on the range of rate of shear data covered. These correlations can be broadly categorized as (i) two parameter models, (ii) three parameter models, (iii) four parameter models, (iv) generalized correlations and (v) polynomial equations. Some of the important models in each category are discussed in the next sections.

2.8.1 Two-parameter Models

These models contain two adjustable material parameters, which depend on temperature, concentration (for solutions) and the filler loading (for filled systems). Some of the commonly used models of this type are:

(i) Oswald-de Waelle[42] or Power Law Model[43]

$$\tau = k(\dot{\gamma})^n \qquad ...(2.19)$$

where k is the consistency index and n is the non-Newtonian index or the flow behavior index, τ is shear stress and $\dot{\gamma}$ is the rate of shear. For $n < 1$, the fluid is pseudoplastic, $n = 1$, the fluid is Newtonian, and $n > 1$, the fluid is dilatant. Equation 2.19 when plotted on a double log scale gives a straight line with a slope $= n$ and an intercept $= k$. The apparent viscosity η_a, can be calculated as

$$\eta_a = \frac{\tau}{\dot{\gamma}} = k(\dot{\gamma})^{n-1} \qquad ...(2.20)$$

This equation shows the viscosity to be a function of $\dot{\gamma}$ such that it will increase with $\dot{\gamma}$ for $n > 1$ and decrease for $n < 1$. The equation 2.19 gives a good data fit for a large number polymer and rubber melts, polymer solutions, low to moderately filled rubber in the intermediate range of rate of shear. Depending on the material the intermediate range may lie between 10^{-1} to $10^5 s^{-1}$[44, 45]. In the low and high shear region the power law model deviates considerably from the experimental data and should not be extended to these ranges at all. To describe the behavior in these zones other equations have been developed.

EXAMPLE 2.5. For the experimental data given in (a) examples 2.2 and (b) 2.3 apply the power law model and calculate the model constants by using (i) the graphical method and (ii) Linear regression method.

Solution: (a) **Data of Example 2.2**

(i) *Graphical method:* A plot of ln τ versus ln $\dot{\gamma}$ -Rheogram (I) Fig. E-2.2 is a straight line giving the non-Newtonian Index, $n = 0.73$ and consistency index, $k = 31.5$ kPa s$^{0.73}$

(ii) *Linear Regression (LR) method:* For an equation of straight line of the form $y = A + Bx$, the LR method is applicable and gives the intercept

$$A = \frac{\Sigma y - B\Sigma x}{N}$$

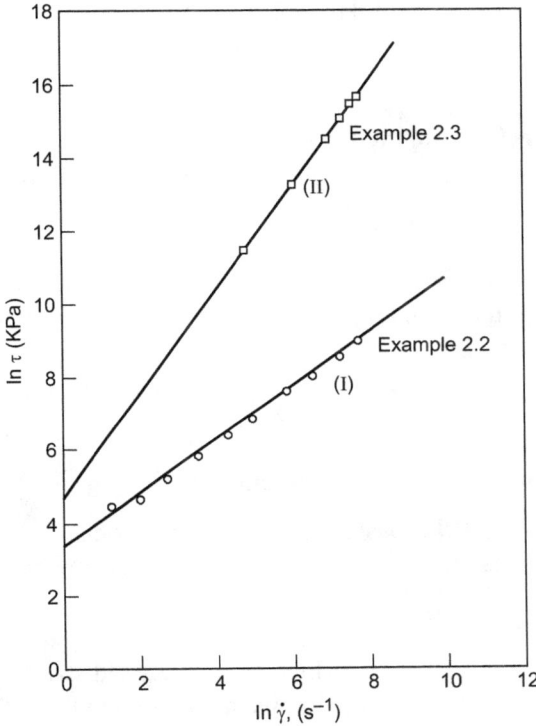

Fig. E 2.2   Log-Log plots of shear stress and rate of shear

The slope, $B = \dfrac{N\Sigma x.y - \Sigma x \cdot \Sigma y}{N\Sigma x^2 \cdot (\Sigma x)^2}$

And the correlation coefficient, r is calculated as

$$r = \dfrac{N\Sigma x.y - \Sigma x \cdot \Sigma y}{\sqrt{\left[N\Sigma x^2 - (\Sigma x)^2\right]\left[N\Sigma y^2 - (\Sigma y)^2\right]}}$$

where N is the number of data points
For the power law equation the linear form is log form i.e.

$$\ln \tau = \ln k + n \ln \dot{\gamma}$$

The values of n and K are: $n = 0.7$

$$k = 32.94 \text{ kPa } s^{0.7}$$

and $r = 0.9983$

As $n < 1$ the melt is pseudoplastic in nature

(b) Data of Example 2.3

Graphical solution- Rheogram (II) (Figure E 2.2)

$$n = 1.4$$

$$k = 127.7 \text{ k Pa } s^{1.4}$$

LR solution $n = 1.41$

$$k = 133.4 \text{ k Pa } s^{1.41}$$

$$r = 0.9964$$

As $n > 1$ therefore the melt is dilatant in nature

The solutions obtained by both the approaches are similar.

(ii) Eyring-Prandtl Model[46, 47]

This model has been derived from the Eyring kinetic theory of liquids and is applicable only to the pseudoplastic fluids in the low and intermediate shear region. The equation is

$$\tau = A \sinh^{-1}\left(-\frac{1}{B}\dot{\gamma}\right) \qquad \qquad ...(2.21)$$

where A and B are two adjustable parameters.

As $\dot{\gamma} \to 0$, $\eta_a \to A/B$ (constant), giving the Newtonian behavior at low shear rate. This equation is unable to predict the high shear rate behaviour

(iii) Bingham Model[44, 48, 49]

This model describes the fluids, which show the yield stress and then flow as Newtonian fluids. Such liquids are known as Bingham plastic fluids and follow the equations 2.22 and 2.23. In these equations

$$\tau - \tau_Y = \beta\dot{\gamma} \quad \text{for } \tau \geq \tau_y \qquad \qquad ...(2.22)$$

$$\dot{\gamma} = 0 \quad \text{for } \tau < \tau_y \qquad \qquad ...(2.23)$$

τ_y is the yield stress and β is the consistency index or the coefficient of rigidity or the coefficient of plastic viscosity.

EXAMPLE **2.6.** For the Bingham plastic fluid of example 2.4 calculate the consistency index and yield stress.

Solution: (a) Graphical method – From the plot of shear stress and shear rate, (Rheogram (iii) Fig. E-2.1)

The yield stress = 400 k Pa.

Consistency index =1.6 kPa s

Linear regression method gives yield stress

= 399.5 kPa

Consistency index =1.6 k Pa s and $r = 0.99993$.

2.8.2 Three-Parameter Models

(i) Viscoplastic Fluids Model[1, 45]

The fluids having both yield stress as well pseudoplasticity are known as the viscoelastic materials. These fluids show the shear thinning flow behavior after the shear stress exceeds the yield stress value. The corresponding equations are

$$\tau - \tau_y = \beta(\dot{\gamma})^n \quad \text{for } \tau \geq \tau_y \qquad \qquad ...(2.24)$$

$$\dot{\gamma} = 0 \quad \text{for } \tau < \tau_y \qquad \qquad ...(2.25)$$

where n is the flow behaviour or non-Newtonian index.

(ii) Ellis Model[49,50]

This empirical model describes the rheological behaviour of a number of polymer and rubber melts in the low and intermediate shear region. The relationship is of the form

$$\tau = \left[\frac{\eta_0 \dot{\gamma}}{1 + |\tau/\tau_{1/2}|^{\alpha-1}} \right] = \frac{\dot{\gamma}}{\phi_0 + \phi_1 |tau|^{\alpha-1}} \qquad ...(2.26)$$

Both these forms are reported in literature. The parameters η_0, $\tau_{1/2}$, α and ϕ_0, ϕ_1 are interrelated and can be calculated from the experimental data (Fig. 2.22). These variables have some physical significance, i.e.

$\eta_0 = \underset{\tau \to 0}{\text{Lim}}\, \eta_a$, which, is known as zero shear viscosity,

$\tau_{1/2} = \tau|_{at\, \eta = \eta_0/2}$,

$\alpha \approx \dfrac{1}{n}$,

$\phi_0 = 1/\eta_0 = $ the fluidity of the system at very low shear rate and

$$\phi_1 = \left[\frac{1}{\eta_0 |\tau_{1/2}|^{\alpha-1}} \right] \qquad ...(2.27)$$

Also if $\alpha > 1$, the fluid is pseudoplastic, if $\alpha < 1$ it is dilatant and if $\alpha = 1$, then it is Newtonian.

As $\tau \to 0$, the model reduces to a Newtonian model with a constant viscosity of η_0.

(iii) Eyring-Powell Model[49]

This model is a modification of two-parameter Eyring-Prandtl model to extend its application to the high shear region. It is expressed as

$$\tau = A\dot{\gamma} + \left(\frac{1}{B} \right) \sinh^{-1} \left(\frac{\dot{\gamma}}{C} \right) \qquad ...(2.28)$$

where, A, B and C are the material parameter. For both lower and upper Newtonian regions this equation reduces to Newton's law of viscosity such that

as $\dot{\gamma} \to 0$, $\tau \to \left[A + \dfrac{1}{BC} \right] \dot{\gamma}$ and $\eta_a \to \eta_0 = \left[A + \dfrac{1}{BC} \right]$, and as $\dot{\gamma} \to \infty$, $\tau \to A\dot{\gamma}$

and $\eta_a \to \eta_\infty = A$

At the intermediate region the equation gives the shear thinning behavior. This equation cannot be used for the dilatant fluids unlike the power law model or the Ellis model

(iv) Carreau Model[51]

It is a three-parameter model applicable for the low and intermediate shear region only. It does not cover the high shear region. The equation is written as

$$\tau = \left[\frac{A}{(1 + B\dot{\gamma})^c} \right] \dot{\gamma} \qquad ...(2.29)$$

A, B and C are the three material parameters. For a number of polymers this model has been found to give a better representation of the data as compared to the Ellis model. The data for a large number of systems under different temperatures has been used to compare these models and it has been found that this model predicts the viscosity within a close range of experimental data.

EXAMPLE 2.7. The high-pressure capillary viscometer data for Styrene Acrylonitrile rubber (Luran 368 R from BASF Germany) at 190°C is available. Apply (a) Power law model, (b) Ellis model and (c) Carreau Model to the data and estimate the model constants. (Data from Rheologie, Vol. 2, Kendaten fuer die Verarbeitung themoplastischer Kunststoffe, Carl Hanser Verlag, Munich, Page-381, (1980).

Rate of shear, $\dot{\gamma}$, s^{-1}	Shear Stress, τ, kPa	Rate of shear, $\dot{\gamma}$, s^{-1}	Shear Stress, τ, kPa
1.33	32.6	117.0	187.0
2.76	48.6	252.0	231.0
7.88	79.8	510.0	266.0
18.20	110.0	1220.0	3000.0
42.30	147.0	—	—

Solution: The apparent viscosity is calculated as $\eta_a = \left(\frac{\tau}{\dot{\gamma}} \right)$ and the log-log plots of viscosity versus shear stress and rate of shear are used for the estimation of the Ellis model parameters. The parameter α is obtained as an average value. The Carreau model parameters are estimated by using the limiting cases. Power law model constants are obtained by following the method discussed earlier.

(a) *Power law model:* The plot of ln τ versus ln $\dot{\gamma}$ does not give a straight line but only can be approximated if first two experimental points are not considered (Graph-I Fig. E-2.3). For this system the $n = n\,(\dot{\gamma})$. The values of the constants calculated are (i) graphically $n = 0.258$ and

$k = 51.5$ k Pa $s^{0.258}$ and (ii) by LR method these are $n = 0.262$, $k = 51.28$ kPa $s^{0.262}$ and $r = 0.989$.

If the total range of rate of shear is taken and an approximate straight line equation is obtained then the model constants by LR method are $n = 0.322$, $k = 37.18$ kPa $s^{0.322}$ and $r = 0.9821$. It may be noted however that the LR method does not indicate the range of rate of shear for which the power law model may not be applicable. The graphical approach can give the idea about the experimental data, which may be considered out of the range of application of the power law model

(b) *Ellis Model*: From the log-log plots of viscosity vs. rate of shear (Graph-II) and from the log-log plot of viscosity vs. shear stress (Graph-III) (Fig. E-2.3) the values of zero shear viscosity, η_0 and $\tau_{1/2}$ are calculated as

$$\eta_0 = 28.28 \text{ kPa s}$$

and

$$\tau_{1/2} = 59.87 \text{ k Pa}$$

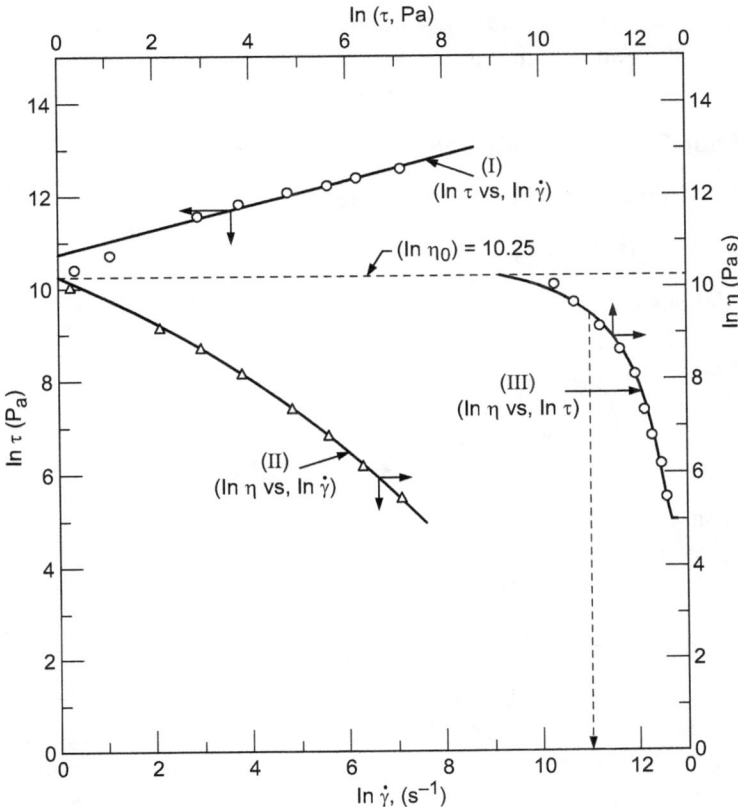

Fig. E-2.3 Plots for data of example 2.7

The value of α has been estimated by trial and error from the experimental data for four different data points and an average value taken. These are

For $\qquad\qquad\qquad\quad \dot{\gamma} = 1.33 \qquad\qquad\qquad\qquad \alpha = 4.074$

$\qquad\qquad\qquad\qquad\quad\;\; \dot{\gamma} = 18.2 \qquad\qquad\qquad\qquad\; \alpha = 3.144$

$\qquad\qquad\qquad\qquad\quad\;\; \dot{\gamma} = 117.0 \qquad\qquad\qquad\qquad \alpha = 3.47$

$\qquad\qquad\qquad\qquad\quad\;\; \dot{\gamma} = 1220.0 \qquad\qquad\qquad\quad \alpha = 3.94$

Average value of $\alpha = 3.66$

(c) *Carreau Model*: The model is written as $\tau = \dfrac{A\dot{\gamma}}{(1+\beta\dot{\gamma})^{C}}$

As $\dot{\gamma} \to 0$, $\tau \to A\dot{\gamma}$ therefore $A = \eta_0 = 28.28$ k Pa s

and $\qquad$ As $\dot{\gamma} \to \infty$, $\tau \to \dfrac{A\dot{\gamma}}{(\beta\dot{\gamma})^{C}}$

Taking last two values of rate of shear and shear stress as these are the highest in the experimental range and substituting in the above equation and solving the two simultaneous equations for B and C gives

$$B = 0.2 \text{ and } C = 0.86$$

2.8.3 Four-Parameter Models

Some of the important models in this category are:

(i) Meter Model[49, 52]

This model is similar to the three-parameter Ellis model but can be applied to the entire range of rate of shear due to an additional term corresponding to the upper Newtonian region. The relationship is as follows

$$\tau = \left[\eta_\infty + \frac{\eta_0 - \eta_\infty}{1 + \left| \tau/\tau_m \right|^{\alpha - 1}} \right] \dot{\gamma} \qquad\qquad ...(2.30)$$

The constants η_0, η_∞ and α have the similar significance as discussed earlier in the case of Ellis model. τ_m is the shear stress corresponding to the viscosity value equal to $(\eta_0 + \eta_\infty)/2$ (Fig.2.22).

(ii) Modified Power Law Model[52]

The two-parameter power law model has been modified by Gupta[41] to cover the entire range of rate of shear. The experimental data of a large number of polymer melts has been fitted to calculate the constants. This equation predicts the viscosity within 15% of the experimental data. The expression is as follows

$$\tau = a\dot{\gamma}^{(b + c/\dot{\gamma}^d)} \qquad\qquad ...(2.31)$$

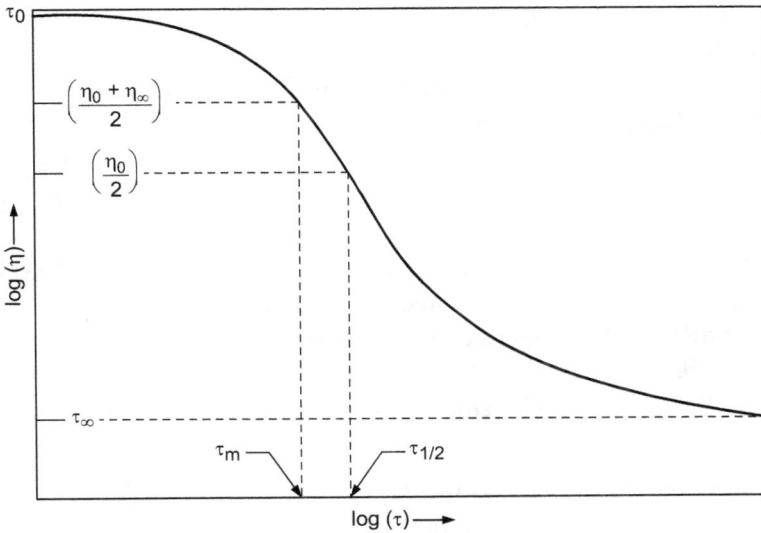

Fig. 2.22 Ellis model and Meter model parameters[49]

The values of a, b, c and d are given in Annexure-I(a).

The range of applicability of the power law model has been extended by applying it over two rate of shear ranges. The power law constants n and K have been evaluated for these ranges. The rate of shear ranges and the corresponding values of power law constants are given in Annexure-I(b).

2.8.4 Generalized Rheological Correlations[1]

A generalized model applicable for a number of plastics and rubber melts is

$$\frac{\eta_a}{\eta_0} = \left[1 + 0.4\left[\frac{\dot{\gamma}\eta_0}{1.32 \times 10^4}\right]^{0.355} + \left[\frac{\dot{\gamma}\eta_0}{1.32 \times 10^4}\right]^{0.71}\right]^{-1} \qquad ...(2.32)$$

2.8.5 Polynomial Rheological Relations[54, 55]

A number of equations developed by the statistical curve fitting of the experimental data are available in the literature. One example of such a relation is Münstedt equation, which is a fourth degree polynomial:

$$\log \eta_a = A_0 + A_1 \log \dot{\gamma}_a + A_2 (\log \dot{\gamma}_a)^2 + A_3 (\log \dot{\gamma}_a)^3 + A_4 (\log \dot{\gamma}_a)^4$$
$$...(2.33)$$

The constants A_0, A_1, A_2, A_3 and A_4 are the temperature dependent material parameters and, $\dot{\gamma}_a$ is the apparent rate of shear.

2.8.6 Equation for Filled Styrene Butadiene Rubber[45, 56]

An empirical correlation developed by this author for filled SBR system gives viscosity as a function of unit shear viscosity, $\eta_1 = \kappa$ (the consistency index in the power law model), t, the temperature in °C, filler loading in parts per hundred of rubber, P and the apparent rate of shear, $\dot{\gamma}_a$. This equation is valid for $\dot{\gamma}_a > 20 \text{ s}^{-1}$. The relationship is as follows

$$\left[\frac{\eta_a}{\kappa}\right] = 9.42 \left[\frac{\dot{\gamma}_a \kappa t}{P}\right]^{-0.74} \qquad \text{...(2.34)}$$

EXAMPLE 2.8. For the experimental data in example 2.7 compare the values of shear stress or rate of shear calculated by power law, Ellis and Carreau models using four different data points and on the basis of the % deviation suggest the best model for the system.

Solution: Power law model equation used is based on all the experimental points with the equation

$$\tau = 37180 \, \dot{\gamma}^{0.322}$$

Ellis Model is $\qquad \tau = \dfrac{28282.5 \, \dot{\gamma}}{1 + (\tau/59874)^{2.7}}$

Carreau Model is $\qquad \tau = \dfrac{28282.5 \, \dot{\gamma}}{(1 + 0.2 \dot{\gamma})^{0.862}}$

S. No.	$\dot{\gamma}_{exp}, s^{-1}$	$\tau_{exp} \text{ } kPa$	Power law		Ellis Model		Carreau model	
			$\tau_{cal} \text{ } kPa$	%δ	$\dot{\gamma}_{exp}, s^{-1}$	%δ	$\tau_{cal} \text{ } kPa$	%δ
1	1.33	32.6	40.76	+ 25.0	1.38	+ 3.8	30.5	− 5.8
2	18.2	110.0	94.6	− 14.0	24.33	+ 33.7	137.0	+ 24.5
3	117.0	187.0	172.3	− 7.9	152.2	+ 30.0	211.0	+ 12.8
4	1220.0	300.0	366.5	+ 22.2	1364	+11.8	300.0	0.0

The error is the highest for the Ellis Model, intermediate for Power law model and the least for the Carreau model. Therefore the Carreau model is most suitable.

2.9 RHEOLOGY SOFTWARE

A computer program in Java language has been developed[57] for (a) solving four different models, (b) the estimation of minimum cumulative error, (c) comparison of the errors of different models (d) deciding the most appropriate model for a given rheological data of a melt and (e) plotting the experimental and calculated values of the viscosities based on different models. The models used are (i) power law model, (ii) Ellis model, (iii) Carreau model and (iv) Meter model. The program is given in Annexure-II. In developing the program all the models were transformed to the form of a straight line and appropriate statistical equations were used to calculate the slope and intercept of the straight line and the standard error from the linear regression analysis. The constants of different models have been calculated for a number of polymeric systems and the most appropriate model for the systems has been worked out using the rheology program. The calculated results are given in Annexure-III.

EXAMPLE 2.9. For the experimental data given in example 2.7 apply the computer program given in Annexure-II and suggest the best model for the system.

Solution: The results obtained from the computer software are given below

S.No.	Model used	Model Parameters	Standard Error
1	Power law model	$k = 37174$ Pas^n $n = 0.323$	0.15
2	Ellis model	$\phi_0 = 3.96 \times 10^{-5}$ (Pa s)$^{-1}$ $\phi_1 = 4.65 \times 10^{-21}$ (Pa$^{-\alpha}$s^{-1}) $\eta_0 = 25252.5$ Pa s $\tau_{1/2} = 78519$ Pa $\alpha = 4.26$	0.17
3	Carreau model	$A = 30416$ Pa s $B = 0.34$ s $C = 0.8$	0.06
4	Meter model	$\eta_0 = 25212$ Pa s $\eta_\infty = 2.72$ Pa s $\tau_m = 78495$ Pa $\alpha = 4.26$	0.18

The Carreau model gives the least error amongst these models therefore Carreau model is the best model for the given melt.

PROBLEMS

1. Consider the equilibrium state of the stresses acting on x-y, y-z and x-z planes of the stress system shown in Fig. 2.5. Show that for the shear stress the following conditions are valid

$$\left| \sigma_{xy} \right| = \left| \sigma_{yx} \right|, \left| \sigma_{xz} \right| = \left| \sigma_{zx} \right| \text{ and } \left| \sigma_{yz} \right| = \left| \sigma_{zy} \right|$$

2. Show that for a stress system acting on a two dimensional (biaxial) plane x-y there always exists a condition irrespective of the coordinate system that (a) the sum of the normal stresses σ_{xx} and σ_{yy} and (b) their product are always constant. Specify the assumptions involved.

3. For a longitudinal elongation the Poisson's ratio, μ, is defined as the ratio of lateral contraction $\varepsilon^{\perp}$ to the longitudinal extension ε i.e. $\mu = \varepsilon^{\perp}/\varepsilon$. Show that for an incompressible fluid under small deformation $\mu = \dfrac{1}{2} - \dfrac{3}{8}\varepsilon$.

4. Under a simple shear deformation of a fluid contained between two parallel plates, at a distance y apart, with the lower plate fixed and the upper plate moving with a constant velocity due to the application of a force, F_x, in x direction, show that (a) the shear stresses σ_{xz}, σ_{zx}, σ_{yz} and σ_{zy} are zero and (b) the system is acted upon by a set of normal stresses σ_{xx}, σ_{yy} and σ_{zz}.

5. Show that for a steady simple deformation of an incompressible Newtonian fluid the normal stresses are all zero.

6. Discuss why for an incompressible non-Newtonian purely viscous and viscoelastic fluids the normal stresses have different magnitudes?

7. Derive the Eyring model from the concept of kinetic theory of liquids proposed by Eyring and coworkers.

8. The data on the shear stress and rate of shear obtained with Instron capillary viscometer for hytrel melt is given below

Shear rate $\dot{\gamma}$, 1/s	TEMPERATURE, °C			
	Shear Stress, $\tau/10000$, Pa.			
	170	180	190	200
3.06	42.0	26.0	22.5	13.5
6.12	76.0	50.0	40.0	25.0
12.23	142.0	100.0	71.0	50.0
30.60	300.0	220.0	157.0	120.0
61.17	528.0	385.0	284.0	225.0
122.34	840.0	650.0	483.0	415.0
305.85	1700.0	1300.0	970.0	830.0
611.70	2500.0	1950.0	1625.0	1450.0
1223.40	3500.0	2890.0	2415.0	2000.0
3058.50	4680.0	4270.0	3630.0	3250.0

(a) Characterize the fluid behavior and calculate the flow behavior index, n
(b) Plot the variation of n with temperature and explain the variation

9. For an unknown polymer melt τ, versus $\dot{\gamma}$ data is given below. Apply the power law, Ellis and Meter models to the data and calculate the constants of these models.

$\tau/10000$, N/m²	$\dot{\gamma}$, 1/s	$\tau/10000$, N/m²	$\dot{\gamma}$, 1/s
2.5	0.75	490.0	1354.8
5.2	2.00	780.0	3387.0
7.4	3.10	1120.0	5000.0
17.0	13.50	1550.0	7000.0
33.0	33.90	1780.0	8000.0
59.0	67.70	2000.0	9000.0
200.0	338.70	2200.0	10000.0

10. A Bingham plastic fluid gave the following shear stress- rate of shear data. Plot the rheogram and calculate β and τ_y. Also plot η versus τ and η versus $\dot{\gamma}$ and explain the reasons for sudden rise of viscosity at low values of $\dot{\gamma}$ and τ.

$\tau/10000$, N/m²	$\dot{\gamma}$, 1/s	$\tau/10000$, N/m²	$\dot{\gamma}$, 1/s
10.0	0.1	210.0	20.0
19.6	1.0	700.0	70.0
31.0	2.0	1005.0	100.0
49.0	4.0	2100.0	200.0
92.0	8.0	4010.0	400.0
164.0	15.0	—	—

11. A gum natural rubber melt (SMR-L) gave the following shear stress- rate of shear data on a Davenport high-pressure capillary viscometer. Apply the power law model and calculate its constants and also establish the type of fluid. Calculate the apparent viscosity using the power law constants and determine its % deviation with the experimental values.

$\tau/10000$, N/m²	$\dot{\gamma}$, 1/s	$\tau/10000$, N/m²	$\dot{\gamma}$, 1/s
34.5	8.64	56.5	60.5
40.0	17.30	58.6	69.1
43.4	25.90	60.7	77.1
46.9	34.60	62.1	86.4
50.3	43.20	68.9	130.0
55.2	51.80	75.8	173.0

12. EPDM and BIIR and their 50/50 blends gave the following rheological data at 120°C. Plot the rheograms and calculate the power law constants. Comment how the values of constants for blend compare with those for pure elastomers.

	$\tau/10000$, N/m²						
$\dot{\gamma}$, 1/s	EPDM	Blend	BIIR	$\dot{\gamma}$, 1/s	EPDM	Blend	BIIR
3.0	740.0	470.0	668.0	122.24	4000.0	2520.0	2150.0
6.12	970.0	682.0	1030.0	305.85	4450.0	3150.0	2300.0
12.23	1280.0	1070.0	1370.0	611.70	5080.0	3900.0	2550.0
30.58	2070.0	1880.0	1740.0	1223.40	5150.0	4300.0	2700.0
61.17	3000.0	2290.0	1940.0	—	—	—	—

13. The natural rubber (SMR CV-70) was masticated on a two-roll mill and extruded through a 2.5 mm dia. High-pressure capillary viscometer (Davenport Viscometer). The rheological data obtained is given below. Estimate the effect of mastication on the rheological behaviour of the melt.

$\dot{\gamma}$, 1/s	Time, min./$\tau/10000$, N/m²				
	0	5	10	15	20
1.73	20.7	13.8	—	—	—
3.46	24.1	—	—	—	—
5.18	26.2	—	—	—	—
6.48	—	—	17.2	—	—
6.98	28.3	—	—	—	—
8.64	29.3	24.1	19.0	—	—
17.30	33.8	26.9	22.8	22.4	13.8
34.60	38.6	31.7	25.9	24.9	17.9
51.80	42.1	34.8	28.3	27.6	20.7
69.10	44.8	36.2	30.3	29.3	23.4
86.40	47.6	39.6	32.4	31.0	24.5
130.00	51.7	42.1	34.5	33.1	26.5
173.00	56.9	46.5	36.5	35.2	28.3
216.00	60.3	47.6	38.6	37.9	30.3
259.00	—	—	39.3	36.5	31.0

14. Blends of natural rubber and HDPE gave the following shear stress versus shear rate data. Plot the viscosity versus shear stress and rate of shear and comment on the behavior of these systems.

$\dot{\gamma}$, 1/s	Blend composition (NR/HDPE)/τ/10000, Pa		
	70/30	50/50	30/70
3.06	81	170	370
6.12	130	250	500
12.23	210	370	660
30.60	390	620	1000
61.20	600	880	1320
122.30	900	1240	1650
305.80	1450	1900	2150
611.70	1950	2250	2400
1223.40	2400	2500	2600
3058.50	2900	3000	3000

15. The high-pressure capillary data for LDPE and LLDPE at 170 deg C using different L/D ratio capillaries is given below.

$\dot{\gamma}$, 1/s	L/D Ratio / System /Shear Stress,τ (k Pa)					
	12.5		37.1		75.1	
	LDPE	LLDPE	LDPE	LLDPE	LDPE	LLDPE
37.5	66.0	140.8	48.2	129.7	54.9	142.8
75.0	88.0	198.0	63.8	166.8	64.1	195.9
112.5	106.2	239.8	74.1	203.9	73.2	238.1
225.0	140.8	316.8	96.4	265.0	95.2	274.7
375.0	176.0	378.5	115.8	322.5	120.9	307.6
750.0	242.0	453.3	174.2	381.8	157.5	338.8
1125.1	268.4	488.5	179.8	415.1	175.8	402.9
2250.0	391.7	558.9	244.6	481.9	236.2	446.8
3750.4	440.0	734.9	291.9	539.3	272.8	496.2

Calculate the power law and Ellis parameters for both these systems. Which of the polyethylenes give higher viscosity and why? Explain your answer from the molecular structure point of view. [Ref.- D.G.Baired and D.I.Collias, Polymer Processing – Principles and design., 1998, pp311-314]

16. Using the model constants calculated from the data in problem number 15 the viscosity for the above systems. Estimate the % deviation in the calculated values as compared to the experimental values.

17. Apply the linear regression analysis to the data of problem 15 and calculate the power law constants. Estimate the % deviation in the viscosity so obtained and those obtained by graphical method and discuss the results.

18. Compare the % deviation obtained by using the Ellis model (prob.15) and by power law model (prob.17) and suggest the better model out of these two.

19. For the data of problem 15 calculate the three parameter Carreau model constants and % deviation in viscosity and suggest the best model for these data amongst power law, Ellis and Carreau models.

20. Use the two parameter Eyring-Prandtl model for the data in problem 15 and calculate the % deviation and compare with those obtained by other models.

21. Apply the Meter model to the data (prob.15) and compare it with other two and three parameter models.

REFERENCES

1. G.V. Vinogradov and A. Ya. Malkin, Rheology of Polymers, Mir Publishers, Moscow, 1977. (a) p 2, (b) pp 105-116, (c) p 120, (d) p 176.

2. S. Timoshenko, G. H. Mac Cullongh, Elements of Strength of Materials, 3rd Ed., Von Nostrand, New York, 1949.

3. R .S. Lenk, Polymer Rheology, Applied Science Pub. Ltd., London, 1978, p 19.
   (a) O'Neill, " The inverse Thixotropic Behaviour of Synovial Fluid., PP1, (1996).
   (b) Reiner M., *Deformation and Flow,* Lewis, London, 1949.

4. A.S. Lodge, Elastic Liquids, Academic Press, New York, 1964, p-62.

5. K. Weissenberg, Nature, **159**, 310 (1947).

6. S. Ramachandran and E.B. Christiansen, J. Non-Newt. Fl. Mech., **13**, 21 (1983).

7. Lodge A. S, J. Chem. Phys., **88**, 4001 (1988)

8. Daniel Bonn , Mathias Kobylko , Steffen Bohn, Jacques Meunier , Alexander Morozov, and Wim van Saarloos , Physical Rev. Letters. **93**, Number 21, November 2004 (214503-1)

9. S. Middleman and J. Gavis; Phys. Fluids, **4**, 355 (1961).

10. R. E. Christensen , " Extrusion Coating of polypropelene" SPE J**18**, 751 (1962)

11. Jay Chun Hyun, Korea Australia Rheology Journal, **11(4)**, 279-285, (1999).

12. S Middleman, Fundamentals of polymer processing, McGraw Hill Book Co. New York (1979).

13. J. P. Tordella, Rheology , Vol. 5, F. R. Eirich ed. Academic Press New York(1969).

14. C.J.S. Petrie and M. M. Dean, A. I. CH. E. J., **22**, 209 (1976).

15. J.L. White, Appl. Polym. Symp. **20**,155 (1973).

16. Y. Oyonagi, Appl. Polym. Symp., **20**, 123-36 (1973).

17. J.H. Southern, and Paul, D. R., J. Polm. Engg. Sci., **14**, 560 (1974).

18. J.L. White, Appl. Polym. Sci., **20**, 155-74 (1975).

19. E. Bourdreaux, Jr., Cucuto, J. A., J. Macromol. Sci., Rev., Macromol. Chem., C-16, 39-77(1977).

20. F.N. Cogswell, Polym. Engg. Sci.,12, 64-73 (1972).

21. J. P.Tordella, J. Appl. Phys. 27, 454-8 (1956).

22. Barakos, G. and Mitsoulis, E. "Numerical Simulation of Extrusion through Orifice Dies and prediction of Bagley correction for an IUPAC-LDPE Melt" J. Rheology, 39, 193-209 (1995).

23. Broadbent J. M., Kaye A., Lodge A.S., and Vale D. G., Nature, 217 (5123), 55-56 (1968).

24. Chang Dae Han, AICHEJ, 18(1), 116-12 (1972).

25. Higashitani, K. and Pritchard, W. G.,Transactions Society of Rheology, 6, 687 (1972).

26. Chang Dae Han and Kwang Ung Kim, Trans. Soc. Rheology, 17 (1), 1512 (1973).

27. Donald Baird, J. Appl. Polym. Sci., 20, 3155-3173 (1976), and J. Non-Newtonian Fluid Mechanics, 148,13-23 (2008)

28. K. Higashtani, and Lodge A. S., Trans. Soc. Rheol., 19, 307-35 (1975).

29. Berman, Neil S. and Gurney George B., AICHEJ, 20(2), 393-394 (1974).

30. P. F. Lobo and H. R. Osmers, Rheologica Acta, 13(3), 457-462 (1974).

31. Malkus, D. S.,Pritchard and M. Yao, Rheologica Acta, 31, 521 (1992).

32. E.A. Jensen and J. deC. Christiansen, Non Newtonian Fluid Mechanics, 148 (91-3),1-46 (2008).

33. Townsend P., PDF URL ADA068865.

34. Paulo, F. Teixeira, Loic Hilliou, Jose A. Covas and Joao, M. Maia, Rheologica Acta, 52 (7),661-672 (2013).

35. A.V. Shenoy, Rheology of filled polymer Systems, Kluwer Academic Publishers, Dordrecht, The Netherlands, (1999).

36. Chung, J. T.,Fundamentals of Polymer Materials, in Encyclopedia of Fluid Mechanics, Gulf Publishing, T X,Vol. 7, Chapt. 33,1061-80 (1988).

37. Bird R. B., Armstrong R. C. and Hasager, O., Dynamics of Polymeric Liquids, Vol. 1, and 2,Wiley, Newyork, (1977).

38. A.B., Metzner, A. I. Ch. E. J., 13, 316-18, (1967).

39. A.B. Metzner, Ubler, E. A., and Chan C. F., A. I. Ch. E.J. 15, 750-8, (1969).

40. W. O. Chriminale, Jr., J. L. Ericsen, and G. L. Filbey, J. Arch. Rat. Mech. Anal., 1, 410 (1958).

41. J.D. Goddard and C. Miller, Rheol. Acta, 5, 177 (1966).

42. W. Ostwald, Kolloid-Z 36, 99 (1925).

43. A. de Waelle, oil and Colour Chem. Assoc. J., 6, 33 (1923)

44. J.M. Coulson and Richardson, Chemical Engineering, Vol. 3, The English Language Book Society and Pergamon Press Ltd., Oxford 1971.

45. B. R. Gupta, Rubber Processing on a Two-Roll Mill, Allied Pub. New Delhi, 1998, pp 71-77.

46. H. Eyring, J. Chem. Phys., 4, 283 (1963).

47. F.H. Ree, T. Ree and H. Eyring, Ind. Chem. Engg., 50, 1036 (1958).

48. R. B. Bird, W. E. Stewart and E. N. Lightfoot, Transport Phenomena, Wiley International Ed., New York, 1960.

49. P. A. Longwell, Mechanics of Fluid Flow, McGraw Hill Book Co., New York, 1966.

50. M. Reiner, Deformation Strain and Flow, Interscience Pub. Inc., New York, 1960.

51. Rheologie, Verband Deutscher Machinen-und Anlagenbau e.V. (VDMA) and Fachgemeinschaft Gummi- und Kunststoffmaschinen, Vol. 2, Carl Hanser Verlag, Muenchen Wien, 1982.

52. D. M. Meter and R. B. Bird, A. I. Ch. E. J., **10**, 878 (1964).

53. B. R. Gupta, Proc. International Rubber Conference, Vol. II, Calcutta, (1997), pp 94-98

54. N.S. Rao, Design Formulas for Plastics Engineers, Hanser Pub. Munich, 1991, p 13

55. H. Muenstedt, Kunststoffe., **68**, 92 (1978).

56. B.R. Gupta and M. Haeberlein, Kautsch. + Gummi Kunstst., **43(8)**, 673 (1990).

57. S.S. Srikant, Rheology Software, M. Tech. Thesis, Rubber Technology Centre, I.I.T. Kharagpur, India, January, 2000.

❑   ❑   ❑

Fluid Flow Analysis   **3**

The polymer melt processing gives rise to different flow situations in different processing equipment. The rate of shear of melt in mixers and extruders is relatively smaller as compared to that in injection moulding. As the viscosity of the polymer melt being very high the flow regime is invariably laminar but the flow through the runner and the gate of the injection moulding may become turbulent as the flow cross sections of these flow channels are small. The non-Newtonian nature of these materials associated with the elastic behaviour, occurrence of the normal stresses, and extensional viscosity etc. present additional problems in the flow analysis. These melts give non-linear relation between pressure drop and volumetric flow rate, and exhibit the die swell, wall slip, melt fracture, stress relaxation and, uncoiling and orientation of the highly coiled long chain high molecular weight molecules. The distribution of molecular weight, chain lengths and coiled and entangled structure of the polymer molecules causes the flow behaviour of theses melts to be quite different than that of Newtonian fluids. The application of the Newtonian fluid mechanics is completely inadequate for the flow analysis of such complex fluids although its basic principles form the basis for the treatment of the non-Newtonian fluid flow.

During the processing operations the rubbers and polymers have to be first heated and melted before these can flow through different processing equipment. Therefore both fluid flow and heat transfer analysis need to be discussed. The melt has to flow through different cross section areas for example in a two roll mill the material is forced through the nip area of the mill, which is essentially a rectangular flow area. In molding machines like compression and blow molding the melt is forced in to the die through a circular cross sections having flow channels of varying lengths. In the film blowing process the melt and the air are forced through an annular channel to flow into a spiral mandrel die. During the production of polymer sheets the melt flows through a rectangular long slot die i.e. it flows between two parallel plates with a very small gap between the plates. Thus it is clear that during processing the polymer melts passes through either a circular or a rectangular die which may have a constant or varying cross section areas like tapered dies. Thus it is important to analyse the melt flow through different flow cross section areas as mentioned above. As the polymer melts are non-Newtonian in nature different non-Newtonian rheological models need to be used in the melt flow analysis. It is relatively straightforward to carryout flow analysis of Newtonian fluids and non-Newtonian fluids, but analysis becomes increasingly complex as the number of constants in the rheological model increase. Mathematical analysis has been presented by Pearson and others[1, 2].

This chapter contains the laminar flow analysis of different non-Newtonian fluids including the velocity distribution, shear stress and rate of shear distribution, true shear stress and rate of shear for melt flow, energy loss correction factors, flow through the non-circular flow channels, slip velocity, transition between laminar and turbulent flow, turbulent flow analysis, modified Reynolds number for non-Newtonian fluids, friction factor Reynolds number relations and extensional flow analysis.

3.1 LAMINAR FLOW THROUGH CIRCULAR CROSS SECTION

The flow through a circular cross section is the most frequently encountered flow situation in industrial practices and due to high melt viscosity the flow is invariably laminar in nature. For simplicity of the analysis it is assumed that the isothermal and steady flow conditions are maintained. Based on the Newtonian fluid mechanics analysis some generalized relationships applicable to all types of fluids[3, 4] are given below.

For the flow through a circular cross section the shear stress is linearly related to the radius and the relationship between shear stress, τ at any radius, r is expressed in terms of shear stress at the wall, τ_0 and wall radius, r_0 as

$$\frac{\tau}{\tau_0} = \frac{r}{r_0} \qquad \qquad ...(3.1)$$

where the shear stress at the wall for a Newtonian fluid is $\tau_0 = \dfrac{r_0 \Delta P}{2L}$, ΔP is the

pressure drop and L is the length of the flow channel.

The rate of shear can be expressed in terms of shear stress and shear dependent viscosity as

$$\dot{\gamma} = \left(-\frac{du}{dr}\right) = f(\tau) = \frac{\tau}{\eta(\tau)} = \frac{\tau_0 \, r}{r_0 \, \eta(\tau)} \qquad ...(3.2)$$

The velocity distribution for one dimensional, isothermal, steady, and fully developed laminar flow under no wall slip i.e. for fully wetting fluid can be obtained by integrating the expression for rate of shear as

$$u = \int_u^0 du = \int_r^{r_0} \left(-\frac{du}{dr}\right) dr = \frac{\tau_0}{r_0} \int_r^{r_0} \frac{r \, dr}{\eta(\tau)} \qquad ...(3.3)$$

The estimation of velocity distribution, u, using the above expression requires the information about the variation of rate of shear with radius, r, or the viscosity, η, as a function of shear stress, τ. Application of different rheological models suitable for a specific fluid will be needed for this purpose.

The expression for the bulk velocity or the average velocity, U, in terms of radius of the tube, r_0, and shear stress at the wall, τ_0, is derived as follows

$$U = \frac{2}{r_0^2}\int_0^{r_0} ur\, dr = \frac{1}{r_0^2}\int_0^{r_0}\left(-\frac{du}{dr}\right)r^2\, dr = \frac{\tau_0}{r_0^3}\int_0^{r_0}\frac{r^3\, dr}{\eta(\tau)} \quad ...(3.4)$$

where $(-du/dr)$ is the rate of shear and η is the viscosity.

The rate of shear at the wall for purely viscous Newtonian fluids, $\dot\gamma_{wa}$, flowing through the circular cross section area, say a tube is given in terms of tube diameter, D_0, and volumetric flow rate, Q as

$$\dot\gamma_{wa} = \left(-\frac{du}{dr}\right)_{0,N} = \frac{8\,U}{D_0} = \frac{32\,Q}{\pi\,D_0^3} \qquad\qquad ...(3.5)$$

This quantity $\left(\dfrac{8\,U}{D_0}\right)$ is a very important parameter in terms of which the

rate of shear at the wall, $\left(-\dfrac{du}{dr}\right)_0$ for the non-Newtonian fluids can be expressed

as

$$\left(-\frac{du}{dr}\right)_0 = f_1\left(\frac{8\,U}{D_0}\right) \qquad\qquad ...(3.6)$$

The rate of shear at any point in the flow system also can be expressed as some other function of $\left(\dfrac{8\,U}{D_0}\right)$ as

$$\left(-\frac{du}{dr}\right) = f_2\left(\frac{8\,U}{D_0}\right) \qquad\qquad ...(3.7)$$

Using different rheological models suitable for a specific polymer melt the functions f_1 and f_2 can be estimated.

EXAMPLE 3.1. Water flows through a pipe of diameter 25 mm at a rate of 144×10^{-3} m³/hr at 20°C. Calculate (a) velocities at the center, at 6 mm and 10 mm from the center, (b) wall rate of shear, (c) wall shear stress and (d) pressure gradient under the given flow conditions.

Solution: At 20°C the density of the water, $\rho = 1$ gm/cc and the viscosity, $\eta = 1$ cp.

For the estimation of velocity at a point it is important to know if the flow is laminar or turbulent by calculating the Reynolds number, $N_{Re} = DU\rho/\eta$. From the given data,

$$N_{Re} = \frac{D\left(\dfrac{Q}{A}\right)\rho}{\eta}$$

$$= \frac{\dfrac{25\times10^{-3}\times144\times10^{-3}}{3600\times10^3}}{\left(\dfrac{\pi}{4}\right)(25\times10^{-3})^2\times10^{-3}} \quad \frac{\left[m\left(\dfrac{m^3}{s}\right)\left(\dfrac{kg}{m^3}\right)\right]}{\left[(m^2)\left(\dfrac{kg}{ms}\right)\right]}$$

$$= 2037 < 2100$$

therefore the flow is laminar and the velocity distribution will be parabolic as the water is a Newtonian fluid. Hence,

(a) $u_r = 2U\left[1-\left(\dfrac{r}{r_0}\right)^2\right]$ where the average velocity

$$U = \frac{Q}{A} = \left[\frac{\left(\dfrac{144\times10^{-3}}{3600}\right)}{\left(\dfrac{\pi}{4}\right)(25\times10^{-3})^2}\right] = 0.0815 \text{ m/s}$$

Therefore, $u_{r=0} = 2 \times 0.0815 = 0.163$ m/s;

$$u_{r=6} = 2 \times 0.0815\left[\left(\frac{1-6}{12.5}\right)^2\right] = 0.125 \text{ m/s}$$

and $u_{r=10} = 2 \times 0.0815\left[\left(\dfrac{1-6}{12.5}\right)^2\right] = 0.0587$ m/s

(b) The rate of shear at the wall,

$$\dot{\gamma}_{wa} = \frac{8U}{D_0} = \frac{8\times0.0815}{25\times10^{-3}} = 26.08 \text{ s}^{-1}$$

(c) The shear stress at the wall,

$$\tau_w = \eta\dot{\gamma}_{wa} = 10^{-3} \times 26.08 = 26.08 \times 10^{-3} \text{ Pa}$$

(d) The pressure gradient,

$$\left(\frac{\Delta P}{L}\right) = \frac{4\tau_w}{D_0} = \frac{4\times26.08\times10^{-3}}{25\times10^{-3}} = 4.173 \text{ Pa/m}$$

3.2 FLOW ANALYSIS USING RHEOLOGICAL MODELS

3.2.1 Power Law Model[5, 6]

The power law model when applied to equations 3.3 and 3.4 the corresponding expressions for the velocity distribution and $\left(\dfrac{8U}{D_0}\right)$ i.e. apparent rate of shear can be derived as follows

$$u = \int_u^0 du = \int_r^{r_0}\left(-\frac{du}{dr}\right)dr \qquad \text{...(3.7a)}$$

and the power law equation

$$\tau = \kappa\left(-\frac{du}{dr}\right)^n = \kappa\dot\gamma^n \qquad \text{...(3.8)}$$

Substitution of equation 3.8 in equation 3.7a and subsequent integration and simplification results in velocity distribution for power law fluids as

$$u = \frac{nr_0}{n+1}\left(\frac{\tau_0}{\kappa}\right)^{1/n}\left(1 - \lambda^{1+n/n}\right) \qquad \text{...(3.9)}$$

where $$\lambda = \frac{r}{r_0} \qquad \text{...(3.10)}$$

Similarly the expressions for average velocity, U, ratio of point velocity to the average velocity, $\dfrac{u}{U}$ and apparent rate of shear, $\left(\dfrac{8U}{D_0}\right)$ are obtained as given below,

$$U = \frac{nr_0}{3n+1}\left(\frac{\tau_0}{\kappa}\right)^{1/n} \qquad \text{...(3.11)}$$

$$\therefore \qquad \frac{u}{U} = \frac{3n+1}{n+1}\left(1 - \lambda^{n+1/n}\right) \qquad \text{...(3.12)}$$

and $$\frac{8U}{D_0} = \frac{4n}{3n+1}\left(\frac{\tau_0}{\kappa}\right)^{1/n} \qquad \text{...(3.13)}$$

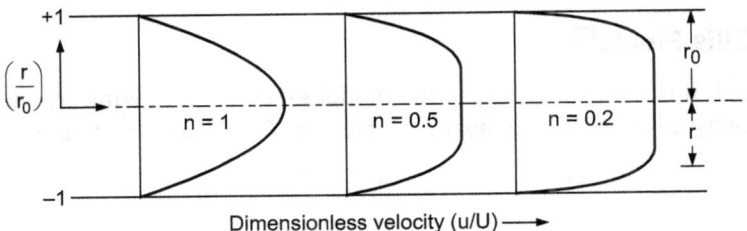

Fig. 3.1   Velocity distribution of power law fluids

It can be seen from the above analysis that the velocity distribution (Equation 3.9) for the power law fluid is a function of power law parameters and the radius of the flow channel. Fig. 3.1 gives its graphical representation. As the value of n reduces the velocity profile becomes more flat ultimately reaching a plug flow profile as $n \to 0$.

EXAMPLE 3.2. A 3% aqueous solution of carboxymethylcellulose (CMC) shows a non-Newtonian fluid behaviour following a power law model with the constants $n = 0.566$ and $k = 9.29$ (Pa $s^{0.566}$). If the solution flows through a 15 mm diameter tube at a flow rate of 150×10^{-3} m³/hr. under laminar flow condition calculate the velocities at the center of the tube and at 5 mm from the center, wall shear stress, wall shear rate if the pressure gradient for the flow is 38.25 k Pa/m.

Solution: (a) The expression for the velocity distribution for the power law model is Equation 3.9. Substituting the values of $n = 0.566$, $r_0 = 7.5 \times 10^{-3}$ m, $\left(\dfrac{\tau_0}{k}\right) = \dot{\gamma}^{0.566}$, $k = 9.29$ Pa $s^{0.566}$ and $\lambda = 0$ and (5/7.5) for the center and at 5 mm radius respectively the flow velocities can be calculated.

The wall shear stress,

$$\tau_w = \left(\frac{D}{4}\right)\frac{\Delta P}{L} = \left(\frac{15\times10^{-3}}{4}\right)\times 38.25\times10^3$$

$$= 143.44 \text{ Pa}$$

The rate of shear at the wall,

$$\dot{\gamma}_w = \left(\frac{\tau_w}{k}\right)^{1/0.566} = 125.9 \text{ s}^{-1}$$

The velocity at the center,

$$u_{r=0} = \left[\frac{nr_0}{n+1}\right]\left[\left(\frac{\tau_w}{k}\right)^{1/0.566}\right] = 0.34 \text{ m/s}$$

The velocity at $r = 5$ mm will be,

$$u_{r=5\text{ mm}} = u_{r=0}\left[1-\left(\frac{r}{r_0}\right)^{(1+n)/n}\right] = 0.228 \text{ m/s}$$

3.2.2 Ellis Fluids[7]

For Ellis fluids the application of the model gives the following equations for the velocity distribution, u, average velocity, U and apparent rate of shear, $\left(\dfrac{8U}{D_0}\right)$

$$u = \frac{\phi_0 \tau_0 r_0}{2}(1 - \lambda^2) + \frac{\tau_0^{\alpha} \phi_1 r_0}{\alpha + 1}(1 - \lambda^{a+1})$$

$$= \frac{\tau_0 r_0}{\eta_0}\left[\frac{1 + \lambda^2}{2} + \left(\frac{\tau}{\tau_{1/2}}\right)^{\alpha-1}\left(\frac{1 - \lambda^{\alpha+1}}{\alpha + 1}\right)\right] \quad ...(3.14)$$

$$U = \tau_0 r_0\left[\frac{\varphi_0}{4} + \frac{\varphi_1 \tau_0^{\alpha-1}}{\alpha + 3}\right]$$

$$= \frac{\tau_0 r_0}{\eta_0}\left[\frac{1}{4} + \left(\frac{1}{\alpha + 3}\right)\left(\frac{\tau_0}{\tau_{1/2}}\right)^{\alpha-1}\right] \quad \text{--- 3.15}$$

and

$$\frac{8U}{D_0} = \tau_0\left[\varphi_0 + \frac{4\varphi_1 \tau_0^{\alpha-1}}{\alpha + 3}\right]$$

$$= \frac{\tau_0}{\eta_0}\left[1 + \left(\frac{4}{\alpha + 3}\right)\left(\frac{\tau_0}{\tau_{1/2}}\right)^{\alpha-1}\right] \quad ...(3.16)$$

The velocity distribution for Ellis fluid depends on $\left(\frac{\tau_0}{\tau_{1/2}}\right)$ as well as α.

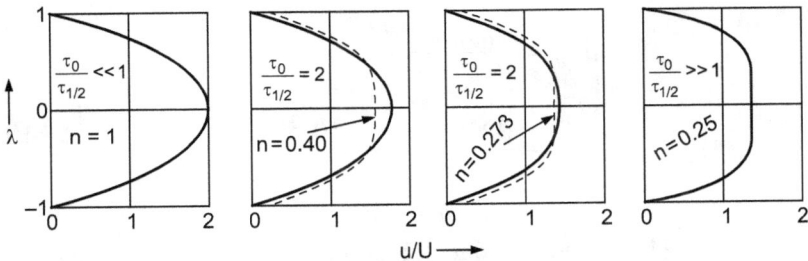

Fig. 3.2 Velocity distribution for Ellis fluid[4]

Figure 3.2 gives the velocity distribution for different values of $\left(\frac{\tau_0}{\tau_{1/2}}\right)$ for a particular value of α. As $\left(\frac{\tau_0}{\tau_{1/2}}\right)$ increases the velocity distribution becomes increasingly flat.

However as the $\left(\frac{\tau_0}{\tau_{1/2}}\right) \rightarrow 0$ the velocity profile tends to become parabolic i.e. the fluid approaches the Newtonian behaviour.

Example 3.3. The rheological behaviour of 0.6% aqueous solution of high molecular weight CMC at 30°C is described by the Ellis model. For a wall shear stress of 20 N/m^2, calculate the velocity at the center for the flow through a 10 mm diameter tube, zero shear viscosity and $\tau_{1/2}$ values. The Ellis parameters are, $\alpha = 1.707$, $\phi_0 = 2.89$ m^2/Ns and $\phi_1 = 0.28$ m$^{2\alpha}$/Ns$^\alpha$

Solution: Using Equation 3.15 for the average velocity for the Ellis fluids and substituting the values of required parameters we get

$$U = 20(5 \times 10^{-3}) \left[\left(\frac{2.89}{4} \right) + \frac{0.28 \times 20^{0.707}}{4.707} \right]$$

$$= 0.122 \text{ m/s}$$

The rate of shear at the wall

$$\dot{\gamma}_w = [\phi_0 + \phi_1(\tau_0)^{\alpha-1}] \, \tau_0$$

$$= [2.89 + 0.28(20)^{0.707}] \times 20 = 104.36 \text{ s}^{-1}$$

$$\eta_0 = \frac{1}{\phi_0} = 0.346 \text{ Pa s}$$

$$(\tau_{1/2})^{\alpha-1} = \frac{1}{(\eta_0 \times \phi_1)} = \frac{\phi_0}{\phi_1}$$

Substituting the required parameters we get

$$\tau_{1/2} = 27.2 \text{ Pa}$$

The velocity at the center of the tube can be obtained using Equation 3.14 and substituting $\lambda = 0$. We get

$$u_{r=0} = \frac{\phi_0 \, \tau_0 \, r_0}{2} + \frac{\tau_0^\alpha \, \phi_1 \, r_0}{\alpha + 1}$$

$$= \frac{2.89 \times 20 \times 5 \times 10^{-3}}{2} + \frac{20^{1.707} \times 0.28 \times 5 \times 10^{-3}}{2.707}$$

$$= 0.23 \text{ m/s}$$

EXAMPLE 3.4. The rheological data for natural rubber (SMR-L) masticated on a two-roll mill for 5 min. has been obtained from a Davenport high-pressure capillary extrusion rheometer at 120°C. The Rheology Software (Appendix-II) was used to calculate the constants of power law, Ellis, Carreau and Meter models from the experimental data. The model constants are given below

   (i) Power law constants

$$n = 0.19, \, k = 166091 \text{ Pa s}^n$$

  (ii) Ellis model constants

$$\phi_0 = 2.31 \times 10^{-5} (1/\text{Pa s})$$
$$\phi_1 = 7.88 \times 10^{-37} \text{ Pa}^{-\alpha} s^{-1}$$
and $\quad\quad\quad\quad \alpha = 6.81$

(iii) Carreau model constants

$$A = 72476 \text{ Pa s}$$
$$B = 0.23 \text{ s}$$

and

$$C = 0.92$$

(iv) Meter model constants

$$\eta_0 = 43204 \text{ Pa s}$$
$$\eta_\infty = 2.72 \text{ Pa s}$$
$$\alpha = 6.81$$

and

$$\tau_m = 265908 \text{ Pa}$$

For a given value of apparent shear stress at the wall = 414 k Pa, calculate the rate of shear at the wall using these models and calculate the % deviation from the experimental value of 130 s^{-1}

Solution: (i) *Power law model*

$$\dot{\gamma}_{wa} = \left(\frac{\tau}{k}\right)^{1/n} = \left(\frac{414000}{166091}\right)^{1/0.19} = 122.4 \text{ s}^{-1}$$

$$\% \text{ deviation} = \frac{122.4 - 130.0}{130.0} = -5.88\%$$

(ii) *Ellis Model*

$$\dot{\gamma}_{wa} = \tau_0[\phi_0 + \phi_1 \tau^{\alpha-1}]$$
$$= 414000[2.31 \times 10^{-5} + 7.88 \times 10^{-37}(414000)^{5.81}]$$
$$= 150.2 \text{ s}^{-1}$$

$$\% \text{ deviation} = +15.56$$

(iii) *Carreau model*

$$\tau = \frac{A\dot{\gamma}}{(1 + B\dot{\gamma})^C} = \frac{72476\dot{\gamma}}{(1 + 0.23\dot{\gamma})^{0.92}}$$

Solving for rate of shear by trial and error gives $\dot{\gamma} = 175 \text{ s}^{-1}$ and % deviation = +34.6%

(iv) *Meter model:* Substituting the shear stress and Meter model constants in the Meter equation we get rate of shear, $\dot{\gamma} = 135.0 \text{ s}^{-1}$ and the % deviation = 3.8 %

3.2.3 Binghamplastic Fluids[4]

The Binghamplastic fluids show the yield stress, and beyond yield stress they flow like Newtonian fluids. For a fully developed laminar flow under steady and isothermal flow conditions these fluids give a central core of finite diameter where the stress acting on the fluid is constant and equal to yield stress, τ_y. The radius of this core is $\left(r_y = \dfrac{\tau_y r_0}{\tau_0}\right)$. The shear stress, rate of shear and velocity will all be the function of radius for $r \geq r_y$. The velocity of the fluid in the central core is maximum and the rate of shear zero. The expressions for the velocity distribution and average velocity can be easily derived and are given below. The velocity distribution is

$$u = \frac{\tau_0 r_0}{2\beta}\left[1 - \lambda^2 - \frac{2\tau_y}{\tau_0}(1-\lambda)\right] \text{ for } r \geq r_y > \left(\frac{\tau_y r_0}{\tau_0}\right) \quad ...(3.17)$$

The maximum velocity at the central core is

$$u = u_m = \frac{\tau_0 r_0}{2\beta}\left[1 - \frac{2\tau_y}{\tau_0} + \left(\frac{\tau_y}{\tau_0}\right)^2\right] \text{ for } r \leq r_y \leq \left(\frac{\tau_y r_0}{\tau_0}\right) \quad ...(3.18)$$

The average velocity is given by

$$U = \frac{\tau_0 r_0}{4\beta}\left[1 - \frac{4\tau_y}{3\tau_0} - \frac{1}{3}\left(\frac{\tau_y}{\tau_0}\right)^4\right] \quad ...(3.19)$$

The velocity distribution is a function of radius as well as the ratio of yield stress to wall shear stress, $\left(\frac{\tau_y}{\tau_0}\right)$. Figure 3.3 gives the velocity distribution for these fluids at different values of $\left(\frac{\tau_y}{\tau_0}\right)$. As this ratio increases the velocity profile becomes increasingly flat and at very high value of this ratio the flow tends to become the plug flow. However as $\left(\frac{\tau_y}{\tau_0}\right) \to 0$ the systems becomes Newtonian and gives the parabolic velocity profile. It can be seen that the magnitude of the yield stress has a dominating effect on the flow behaviour of theses fluids.

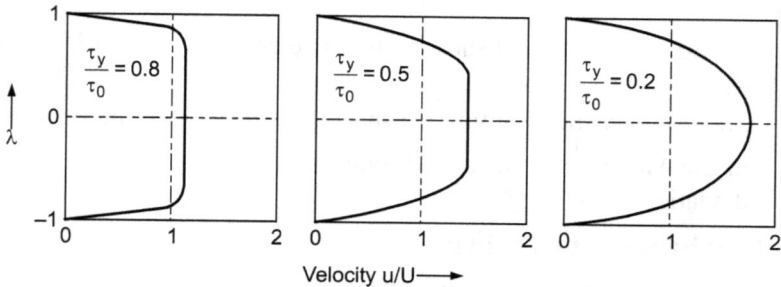

Fig. 3.3   Velocity distribution of Bingham plastic fluid[4]

3.2.4 Viscoplastic Fluids[8]

The viscoplastic fluids are the polymer melts and solutions, which show the shear thinning behaviour after yielding i.e. to say that these fluids possess both, yield stress and pseudoplastic characters. The expressions for the velocity distribution and apparent rate of shear are derived following the similar approach described for other fluid models. These fluids also give a central core of constant velocity. The diameter of the core is a function of model constants and the yield stress. The expression for the velocity distribution is given next assuming that the power law model is applicable to the material beyond the yield stress.

for $\qquad r \geq r_y \geq \left(\dfrac{\tau_y r_0}{\tau_0} \right)$

$$u = \left(\frac{n r_0}{n+1} \right) \left(\frac{\tau_0}{\beta} \right)^{1/n} \left[\left(1 - \frac{\tau_y}{\tau_0} \right)^{\frac{1+n}{n}} - \left(\lambda - \frac{\tau_y}{\tau_0} \right)^{\frac{1+n}{n}} \right] \qquad ...(3.20)$$

and, for $r \leq r_y \leq \left(\dfrac{\tau_y r_0}{\tau_0} \right)$

$$u = u_m = \frac{n r_0}{n+1} \left(\frac{\tau_0}{\beta} \right)^{\frac{1}{n}} \left(1 - \frac{\tau_y}{\tau_0} \right)^{\frac{1+n}{n}} \qquad ...(3.21)$$

The Average velocity expression (for $n \neq -1, -2$ and -3, which is always true) is

$$U = \left(\frac{r_0}{\tau_0^3 \beta^{1/n}} \right) \left[(\tau_0 - \tau_y)^{\frac{n+1}{n}} \left\{ \frac{n(\tau_0 - \tau_y)^2}{1+3n} + \frac{2n\tau_y(\tau_0 - \tau_y)}{1+2n} + \frac{n\tau_y^2}{1+n} \right\} \right]$$

$$...(3.22)$$

It can be seen from the above equations that velocity distribution is affected both by n and yield stress of the melt. For a polymer melt with a given value of yield stress the reduction in n values will tend to further flatten the velocity profile.

3.3 PRESSURE DROP-VOLUMETRIC FLOW RATE RELATIONSHIP

In most industrial applications the polymer melts and solutions often flow through long circular and non-circular flow channels. In other applications however the fluids may be subjected to tangential or circular flow. In case of flow through the long channels the energy is consumed in overcoming the resistance to the flow due to viscous drag. The energy required for the flow can be calculated from the knowledge of pressure drop, which increases with increase in the volumetric flow rate. Also in all laboratory measurements one measures only the pressure drop and volumetric flow rate. This data is to be used to calculate the shear stress and shear rates acting on the fluid during flow for its rheological characterization. It is therefore important to establish the relationship between the experimental data and the parameters required for analysis.

For Newtonian fluid flow through a circular cross section under steady, isothermal, and fully developed laminar flow conditions the volumetric flow rate and pressure drop relation is given by the Hagen-Poiseuille equation. This equation forms the basis for the similar relationship for the non-Newtonian fluids as it is derived from the basic fluid mechanics principles like equation of motion and equation of continuity. The Hagen-Poiseuille equation relates the parameters like, volumetric flow rate, Q, Pressure drop, ΔP, viscosity, η, diameter, D_0, and length, ΔL, of the channel as

$$Q = \frac{\pi D_0^4}{128 \eta}\left(-\frac{\Delta P}{\Delta L}\right)$$ ...(3.23)

The shear stress and rate of shear are given as

$$\text{Shear stress, } \tau = \frac{D_0}{4}\left(\frac{\Delta P}{\Delta L}\right)$$ ...(3.24)

and $\quad$ Rate of shear, $\dot{\gamma}_{w,a} = \left(-\frac{du}{dr}\right) = \frac{8U}{D_0} = \frac{32Q}{\pi D_0^3}$ ...(Ref. eq. 3.5)

Also from the generalized equation 3.4 for average velocity the expression for the apparent rate of shear is

$$\frac{8U}{D_0} = \frac{4}{r_0^3}\int_0^{r_0}\left(-\frac{du}{dr}\right)r^2 dr = \frac{4}{\tau_0^3}\int_0^{\tau_0}\left(-\frac{du}{dr}\right)\tau^2 d\tau$$ ...(3.25)

Equation 3.25 is a very important relation relating rate shear at any point in the flow system with apparent rate of shear at the wall, radius and shear stress. It suggests that for a specific fluid, as the rate of shear, $\left(-\frac{du}{dr}\right)$ is a function only of shear stress, τ, the right hand side of the equation depends only on the upper limit, τ_0 as it vanishes at the lower limit. It therefore implies that for a given fluid at a particular wall shear stress τ_0, flow channels of different cross sections will give same apparent rate of shear, $\left(\frac{8U}{D_0}\right)$. Thus the rheological data obtained in the laboratory using a capillary can be applied to the flow through the bigger cross sections provided the flow conditions in both the cases are similar.

Differentiating equation 3.25 with respect to τ_0 one gets

$$\frac{d(8U/D_0)}{d\tau_0} = -\frac{12}{\tau_0^4}\int_0^{\tau_0}\tau^2\left(-\frac{du}{dr}\right)d\tau + \frac{4}{\tau_0^3}\left[\tau_0^2\left(-\frac{du}{dr}\right)_0\right]$$ ...(3.26)

Substituting for the integral on the R.H.S. of the above equation from equation 3.25 and rearranging gives

$$\left(-\frac{du}{dr}\right)_0 = \frac{\tau_0}{4}\frac{d\left(8\,U/D_0\right)}{d\tau_0} + \frac{3}{4}\left(\frac{8\,U}{D_0}\right) \qquad \text{...(3.27)}$$

This equation was developed by Rabinowitsch[9] and Mooney[10] and is known as Mooney-Rabinowitsch equation. Metzner and Reed[11] further modified it as

$$\left(-\frac{du}{dr}\right)_0 = \frac{\left(\dfrac{8U}{D_0}\right)}{4}\frac{d\left(\dfrac{8U}{D_0}\right)/\left(\dfrac{8U}{D_0}\right)}{d\tau_0/\tau_0} + \frac{3}{4}\left(\frac{8U}{D_0}\right)$$

$$= \frac{\left(8U/D_0\right)}{4}\left(\frac{d\ln\left(8U/D_0\right)}{d\ln\tau_0} + 3\right) \qquad \text{...(3.28)}$$

The term $\dfrac{d\ln\left(8U/D_0\right)}{d\ln\tau_0}$ is the inverse of the slope of the straight line obtained by plotting $\ln\tau_0$ versus $\ln\left(8\,U/D_0\right)$. Denoting this as $1/n'$ and simplifying one gets

$$\left(-\frac{du}{dr}\right)_0 = \frac{3n'+1}{4n'}\left(\frac{8U}{D_0}\right) \qquad \text{...(3.29)}$$

The equation 3.29 is an important relation as it gives the true rate of shear at the wall in terms of apparent rate of shear. The term $\dfrac{3n'+1}{4n'}$ is the Rabinowitsch correction factor for the true wall shear stress.

The slope n' is the slope of the straight line plotted between $\ln\tau_0$ versus $\ln(8\,U/D_0)$. Both τ_0 and $(8\,U/D_0)$ are the quantities calculated from the experimental data based on the Newtonian fluid analysis hence do not represent the true shear stress and shear rate for the non-Newtonian fluids. The plots based on the apparent rheological parameters are referred as the pseudo-shear diagrams and those based on the corrected parameters as the true-shear diagrams. The pseudo-shear diagrams of polymer melts and solutions may be either (i) a straight line or (ii) a curve. The relationship below

$$\tau_0 = \kappa'\left(\frac{8\,U}{D_0}\right)^{n'} \qquad \text{...(3.30)}$$

represents an equation to a straight line. When the shear diagram is a straight line over the experimental range of apparent rate of shear the n' value is constant and the equation 3.30 represents this line, however when it is a curve this equation represents the equation to the tangent to the curve at the given value of the rate of shear. In the first case n' is equal to the true non-Newtonian index, n for the fluid and in the second case it varies with the rate of shear. An interrelation between n and n' is developed next.

3.3.1 Estimation of True Power Law Constants

Case 1: When pseudo-shear diagrams are straight line and $n = n'$ holds.

Applying the power law equation to the wall condition we can write

$$\tau_0 = \kappa \left(-\frac{du}{dr} \right)_0^n \qquad \qquad ...(3.31)$$

EXAMPLE 3.5. The apparent shear stress-shear rate data for unmasticated natural rubber (SMR CV-70) at 120°C as given below was obtained using a Davenport extrusion rheometer. Apply the power law model and calculate the Rabinowitsch correction factor and true wall rate of shear.

Apparent Rate of Shear, $\dot{\gamma}_{wa}$, s^{-1}	Shear Stress, τ_a, k Pa	Calculated True Rate of Shear, $\dot{\gamma}_w$, s^{-1}
1.73	207.0	3.3
3.46	241.0	6.6
5.18	262.0	9.9
6.91	283.0	13.18
8.64	293.0	16.48
17.3	338.0	33.0
34.6	386.0	66.0
51.8	421.0	98.8
69.1	448.0	131.8
86.4	476.0	164.0
130.0	517.0	248.0
173.0	569.0	330.0
216.0	603.0	412.0

Solution: The plot (not shown) of ln τ versus ln $\dot{\gamma}$ gives a straight line with slope,

$$n' = 0.23 \text{ and the intercept}$$
$$k' = 174.5 \text{ k Pa.}$$

By using the Rheology software (Appendix-II) the values are
$$n' = 0.216 \text{ and}$$

$$k' = 183.6 \text{ k Pa.}$$

The Rabinowitsch correction factor is calculated based on the results obtained from software calculations.

$$\text{The correction factor} = \frac{3n' + 1}{4n'} = 1.907$$

and the true wall rate of shear is calculated by using Equation 3.29.

The calculated values of the rate of shear are given in the table.

Comparing equations 3.30 and 3.31 the true consistency index can be written as

$$\kappa = \kappa' \left(\frac{4n}{3n + 1} \right)^n \qquad \qquad ...(3.32)$$

Case 2: When $n' \neq n$, using equation 3.29 and expressing it in the log differential form as

$$d \ln \left(-\frac{du}{dr} \right)_0 = d \ln \left(\frac{3n' + 1}{4n'} \right) + d \ln \left(\frac{8U}{D_0} \right) \qquad ...(3.33)$$

and dividing both sides by $d \ln \tau_0$ one gets

$$\left[\frac{d \ln \left(-\dfrac{du}{dr} \right)_0}{d \ln \tau_0} \right] = \left[\frac{d \ln \left(\dfrac{3n' + 1}{4n'} \right)}{d \ln \tau_0} \right] + \left[\frac{d \ln \left(\dfrac{8U}{D_0} \right)}{d \ln \tau_0} \right] \qquad ...(3.34)$$

From the definitions of n and n' it can be written as

$$\frac{1}{n} = \frac{1}{n'} + \left[\frac{d \ln \left(\dfrac{3n' + 1}{4n'} \right)}{d \ln \tau_0} \right] \qquad ...(3.35)$$

The value of n can be calculated from equation 3.35 after estimating the second term on the right hand side from the log-log plot between $\left(\dfrac{3n' + 1}{4n'} \right)$ and τ_0. The expression for the consistency index k can be written as

$$k = k' \left(\frac{8U}{D_0} \right)^{n' - n} \Big/ \left(\frac{3n' + 1}{4n'} \right)^n \qquad ...(3.36)$$

Thus the power law fluids can be completely characterized.

EXAMPLE 3.6. For the data given in Example 3.5 calculate the true power law constants.

Solution: As the log-log plot of shear stress and rate of shear (not shown) is a straight line over the entire experimental range of rate of shear we can assume that the true flow behaviour index, $n = n' = 0.216$.

The true consistency index, k can be calculated from the Equation 3.32

$$k = k'\left[\frac{4n}{(3n+1)}\right]^n = 183593\left(\frac{1}{1.907}\right)^{0.216}$$

$$= 159.7 \text{ k Pa.}$$

3.3.2 Energy Losses and Bagley Correction[12]

The polymer and rubber melts invariably flow through varying cross sections during processing in a equipment and are subjected to the change in the direction of flow or undergo sudden expansion or contraction due to the design of the equipment as shown in Fig. 3.4. This results in additional energy loss as compared to that for fully developed flow. Under these situations the fully developed flow condition is not likely to be always maintained and due to additional energy loss high pressure drop results, which if not accounted for, will give erroneous result in volumetric flow rate and other design calculations. The shear stress estimated based on the fully developed flow pressure drop will be higher in magnitude than the true shear stress. It is therefore important to calculate the energy loss and the corresponding correction factor. The correction factor applied to estimate the true shear stress for non-Newtonian fluids is known as the Bagley correction factor.

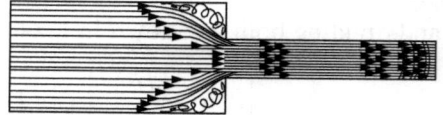

Fig. 3.4   Schematic diagram showing sudden contraction of flow

When the polymer melts flow from a larger cross section channel to a smaller cross section and then come out as extrudate for example in a ram extruder the total energy consumed is always higher than that in the case of an uniform cross section flow of similar length. The energy losses correspond to various reasons like, (i) change in direction of the fluid streamlines due to sudden reduction in the flow cross section area, (ii) circulation of melt in the stagnation zone at the confluence of two flow channels, (iii) the flow development in the initial region of smaller cross section channel, (iv) fluid flow in the fully developed region i.e. in the major part of the flow channel and (v) the sudden expansion of the fluid as it comes out of the die. The total energy loss is measured in terms of the pressure drop, ΔP_T between die entry and exit by introducing a pressure transducer at the die entry. The first three losses are termed as the entry losses, ΔP_{entry}, the fourth as frictional flow loss due to the viscous drag, ΔP_{fd}, and the last as the sudden expansion loss or the exit loss, ΔP_{exit}. Hence

$$\Delta P_T = \Delta P_{entry} + \Delta P_{fd} + \Delta P_{exit} \qquad \qquad ...(3.37)$$

The third term in the RHS is relatively small as compared to first two terms and can be neglected.

3.3.3 Estimation of Bagley Correction Factor

To estimate the Bagley correction factor the experiments are performed using capillaries of different L/D ratios but having the same diameter so as to ensure same entry losses for all capillaries at a particular flow rate. The pressure drop is plotted against L/D ratio for different flow rates or apparent rates of shear as shown in Fig. 3.5. The plots are straight lines not passing through the origin but giving intercepts at both the axes. The intercept at the pressure axis at $L/D = 0$ is proportional to the entry losses. The intercept at the L/D axis at $\Delta P = 0$ represents the length of the capillary which will give pressure loss equal to the entry loss if the same fluid was flowing under fully developed flow condition. The apparent shear stress now can be corrected to give true value. Bagley suggested the following equation for the true shear stress

$$\tau = \frac{\Delta P_T}{4\left[\left(\dfrac{L}{D}\right)+\left(\dfrac{L}{D}\right)'\right]} \qquad ...(3.38)$$

The factor $(L/D)'$ is known as the Bagley correction factor that estimated graphically. This factor represents the extra capillary length required to give the pressure drop equivalent to ΔP_{entry}. Whereas the intercept $\Delta P'$ at $L/D = 0$ represents the pressure drop due to, (i) change in direction of the fluid streamlines due to sudden reduction in the flow cross section area and (ii) circulation of melt in the stagnation zone at the confluence of two flow channels. Assuming the pressure drop due to the flow development in the initial region of smaller cross section channel to be negligible one may apply the correction in pressure drop and write the shear stress equation as

$$\tau = \frac{\Delta P_T - \Delta P'}{4\,(L/D)} \qquad ...(3.39)$$

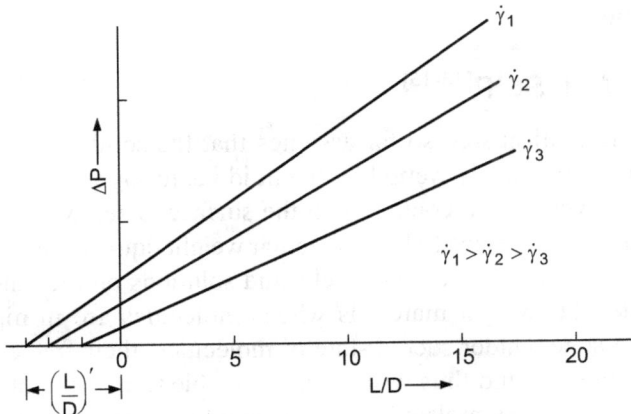

Fig. 3.5   Plot of pressure drop versus L/D ratio.

Although both the equations 3.38 and 3.39 can be used to estimate the entry losses but the later may give slightly higher shear stress as compared to the former. The true shear stress is always smaller than the apparent shear stress and for smaller dies it is important to calculate it from the point of view of the die design.

The entry losses can be reduced, (i) by making the die entry in the form of a cone with single or compound cone angle so as to smoothen the flow disturbance due to sudden change in the direction of flow and (ii) by using a long flow channel of a large L/D ratio. By using the conical design of die entry, the entry loss becomes negligible in comparison to the fully developed flow loss and the applicable minimum L/D ratio of the capillary can be reduced to 40 as compared to 100 for a flat entry die.

EXAMPLE 3.7. A natural rubber based compound gave the following pressure drop-volumetric flow rate data when extruded through a 1.5 mm dia. capillary. The pressure drops for different L/D ratio capillaries for a flow rate of 330.7×10^{-9} m³/s are available. Calculate the Bagley correction.

L/D Ratio of the capillaries	20.0	10.0	5.0	1.0
Pressure drop, ΔP, MPa	64.3	37.0	24.8	14.5

Solution: A plot of ΔP versus L/D ratio gives a straight line (not shown), which does not pass through the origin but gives an intercept on the L/D axis (at ΔP = 0) equal to − 4.5. Thus the Bagley correction for the melt at the given flow rate is, $(L/D)' = 4.5$. The true shear stress will be given as

$$\tau = \frac{\Delta P}{4[L/D + (L/D)']} = \frac{\Delta P}{4[L/D + 4.5]}$$

The intercept at ΔP axis at $L/D = 0$ which is 12 M Pa represents the pressure losses associated with melt circulation and change of direction of the stream lines of the melt.

3.4 THE WALL SLIP[13-16]

The flow analysis discussed so far assumes that the solid surface in contact with the fluid is completely wetted by the fluid i.e. to say that there is no wall slip or the fluid velocity in contact with the surface is zero. This assumption holds good for the low viscosity low molecular weight liquids, which completely wet the solid surface. The polymer melts and solutions on the other hand are the high molecular weight materials whose molecules are in highly coiled and entangled state. Under such a state of molecules, their free ends may lie any where in the coil and thus will not be available to make contact with the surface. Hence only few molecules wet the solid surface and others only roll on while the melt is flowing. This results in a finite velocity of the fluid at the interface between the solid surface and the fluid, which is known as the slip velocity. It is important to have a reasonably good estimate of the slip velocity for calculating the correct volumetric flow rate of the melt. The actual flow rate may be much higher than the design flow rate calculated without the wall slip correction resulting in considerable loss of material particularly in injection moulding operations where only the exact quantity of the material flow at a predetermined rate is most desired.

3.4.1 Estimation of the Wall Slip Velocity[17-20]

The generalized equation for the point velocity can be written to include the slip velocity at the wall, u_{r_0} as

$$\int_{u_r}^{u_{r_0}} du = u_{r_0} - u_r = \int_r^{r_0} \left(\frac{du}{dr}\right) dr \qquad \text{...(3.40)}$$

or

$$u_r = u_{r_0} + \int_r^{r_0} \left(-\frac{du}{dr}\right) dr \qquad \text{...(3.41)}$$

Using equations 3.2, 3.40 and 3.41 in the generalized equation for volumetric flow rate and simplifying we get

$$Q = 2\pi \int_0^{r_0} u_r r \, dr$$

$$= \pi r_0^2 u_{r_0} + \pi \left(\frac{r_0}{\tau_0}\right)^3 \int_\tau^{\tau_0} \left(-\frac{du}{dr}\right) \tau^2 d\tau \qquad \text{...(3.42)}$$

As the apparent rate of shear at the wall is related to the volumetric flow rate it can be written as

$$\dot{\gamma}_{wa} = \frac{4Q}{\pi r_0^3} = \frac{4u_{r_0}}{r_0} + \frac{4}{\tau_0^3} \int_\tau^{\tau_0} \left(-\frac{du}{dr}\right) \tau^2 d\tau \qquad \text{...(3.43)}$$

The second term in the right hand side of the above equation depends only on the shear stress, therefore if the rate of shear data using different capillaries is collected at the constant shear stress it will vanish, reducing it to a simple equation as

$$\dot{\gamma}_{wa} = \frac{4Q}{\pi r_0^3} = \frac{4u_{r_0}}{r_0} \qquad \text{...(3.44)}$$

This equation is an equation of straight line of $\dot{\gamma}_{wa}$ versus $1/r_0$ plot, giving slope equal to $4u_{r0}$. Thus by performing rheological experiments with the capillaries of different diameter but the same L/D ratio and collecting the rate of shear data at a particular shear stress the wall slip velocity can be calculated. Figures 3.6, 3.7 and 3.8 give respectively shear stress-apparent rate of shear;

apparent rate of shear-radius inverse and wall slip velocity-shear stress plots for natural rubber melts. The wall slip velocity increases sharply in the high shear stress region where as in the low shear stress values its magnitude is rather small. There appears to be a transition value of shear stress beyond which it becomes a dominating parameter and therefore will influence the volumetric flow rate considerably. The mastication of rubber gives higher slip velocity as compared to the unmasticated sample.

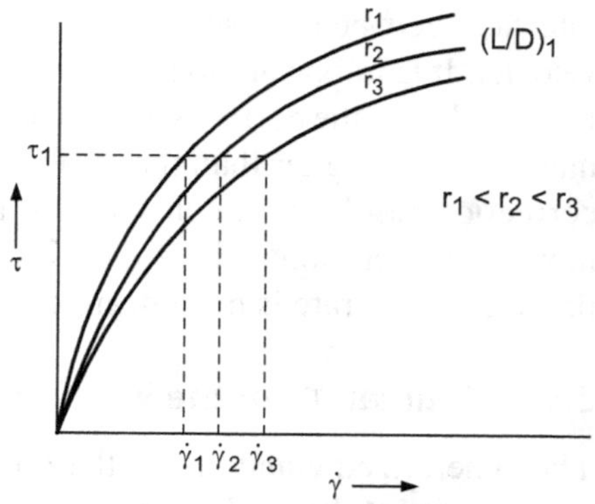

Fig. 3.6 Shear stress versus rate of shear plots for different capillaries

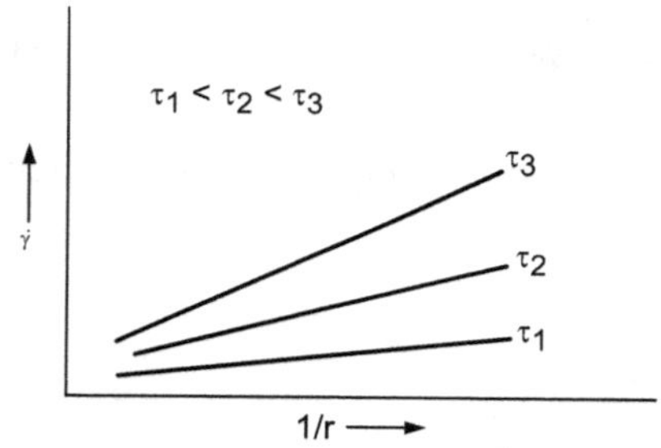

Fig. 3.7 Apparent rate of shear versus radius inverse plots at different shear stress

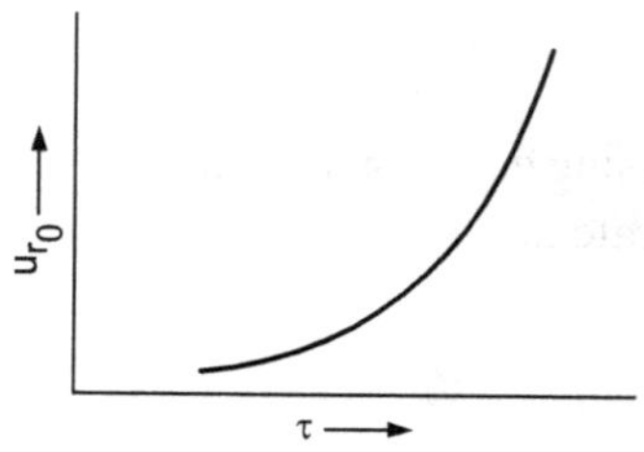

Fig. 3.8 Wall slip velocity versus shear stress plot

3.5 FLOW BETWEEN THE PARALLEL PLATES[21a, 22]

The flow of polymer melts in the flat molds, slit dies etc. can be considered as the flow through the parallel plates. The analysis of such a flow can be carried out following the procedure used to analyze the flow through the circular cross sections. Figure 3.9 gives the schematic of the velocity distribution and the forces acting on a fluid element in the flow through the parallel plates under steady, isothermal and fully developed laminar flow conditions.

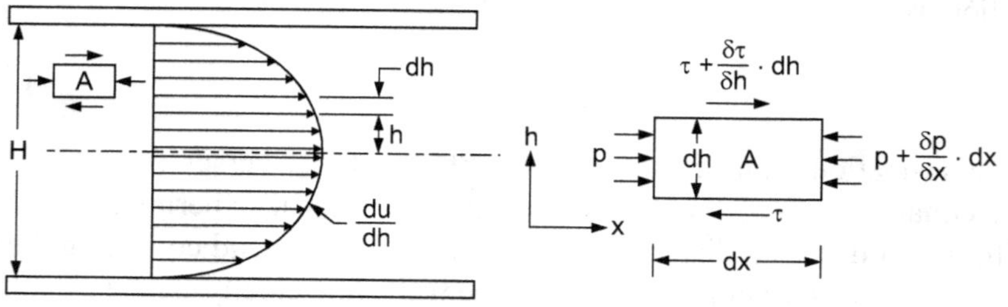

Fig. 3.9 Schematic of flow through parallel plates

The expressions for shear stress and rate of shear are given below

The wall shear stress,

$$\tau_0 = \frac{H}{2}\left(\frac{d\,p}{d\,x}\right) = \frac{H}{2}\frac{\Delta P}{L} \qquad \qquad ...(3.45)$$

The wall shear rate,

$$\dot{\gamma}_w = \left(-\frac{du}{dh}\right)_0 = \frac{2}{wH^2}\left(2Q + \Delta P\frac{dQ}{d\,\Delta P}\right) \qquad ...(3.46)$$

An equation (Equation 3.47) similar to the Rabinowitsch equation for the flow through circular cross sections can be derived for this flow situation also as

$$\left(-\frac{du}{dh}\right)_0 = \frac{2n'' + 1}{3n''}\left(\frac{6Q}{wH^2}\right) \qquad \qquad ...(3.47)$$

where

$$\frac{1''}{n} = \frac{d\ln\left(6Q/wH^2\right)}{d\ln\tau_0} \qquad \qquad ...(3.48)$$

In these equations Q is the volumetric flow rate, H is the gap between the plates and w is the width of the plates. Integrating the equation 3.46, one can derive the velocity distribution expression for the power law fluid flowing between the parallel plates. The final equation is

$$u = \frac{n}{n+1}\left[\frac{1}{\kappa}\left(\frac{dP}{dx}\right)\right]^{1/n}\left[\frac{H}{2}\right]^{(n+1)/n}\left[1 - \left(\frac{2h}{H}\right)^{(n+1)/n}\right]$$

$$...(3.49)$$

The maximum velocity will occur at the middle of the two plates i.e., $u = u_m$ at $h = 0$ and the corresponding expression is

$$u_m = \frac{n}{n+1}\left[\frac{1}{\kappa}\left(\frac{dP}{dx}\right)\right]^{1/n}\left[\frac{H}{2}\right]^{(n+1)/n} \qquad ...(3.50)$$

The average velocity can be expressed as

$$U = \frac{nH}{2(2n+1)}\left[\frac{H}{2k}\frac{dp}{dx}\right]^{1/n} \qquad \qquad ...(3.51)$$

3.6 FLOW THROUGH AN ANNULUS[21b, 23]

The flow of polymer melts through an annular space is encountered in a number of practical applications like hose making, wire coating and blow moulding. Figure 3.10 gives the force balance diagram on a fluid element flowing through the annular space. The analysis for the annular cross section flow is carried out following similar procedure as done for a circular section flow. Consider the polymer melt flowing through an annular space (formed by two concentric cylinders) having inner radius of outer cylinder, r_o and outer radius of inner cylinder, r_i. Assume that the steady, isothermal and fully developed laminar flow conditions exist. The forces acting in the direction of flow (x-direction) on the control volume of radius, r, thickness, δr, and the length, δx are

1. Pressure force acting at $x = x$: $(2\pi r \delta r)P$

2. Pressure force acting at $x = x + \delta x$: $(2\pi r \delta r)\left[P + \left(\dfrac{dp}{dx}\right)\delta x\right]$

3. Shear force at $r = r$: $(2\pi r \delta x)\tau_{rx}$

4. Shear force at $r = r + \delta r$: $2\pi r \delta x \tau_{rx} + 2\pi \delta x \left[d\left(\dfrac{r\tau_{rx}}{dr}\right)\delta r\right]$

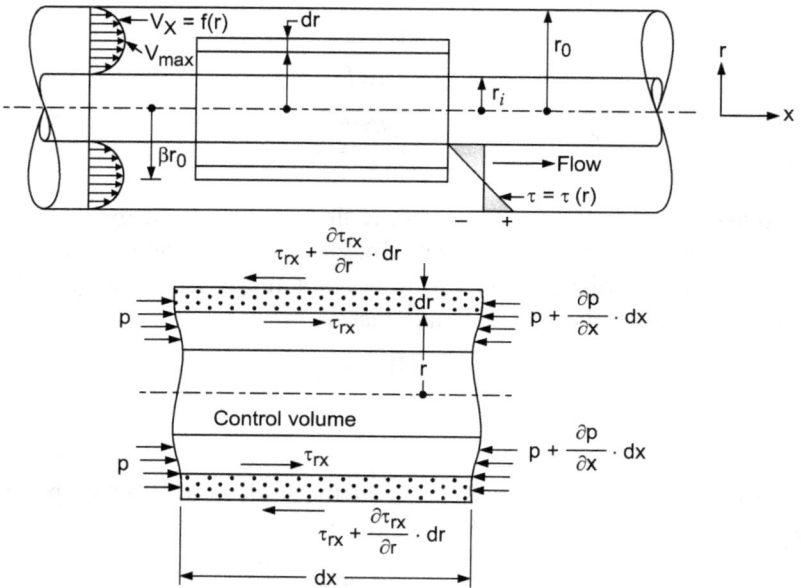

Fig. 3.10   The force balance on control volume in an annular flow

For a horizontal flow the net force due to gravity will be zero. Thus the control volume moves with a constant velocity (incompressible fluid) due to the balance between the shear forces and the pressure forces. Taking the sum of the above forces and equating it to zero and considering that the volume of the fluid element ($\delta V = 2\pi r \delta r \delta x$) is finite, one gets after simplification

$$\frac{1}{r}\frac{d(r\tau_{rx})}{dr} = -\frac{dP}{dx} \qquad\qquad ...(3.52)$$

Separating the variables and integrating the equation 3.52 gives the following set of equations

$$\int_{(r\tau_{rx})_i}^{(r\tau_{rx})_r} d(r\tau_{rx}) = -\frac{dP}{dx}\int_{r_i}^{r} r\,dr \qquad\qquad ...(3.53)$$

$$(\tau_{rx})_r = \frac{r_i(\tau_{rx})_i}{r} - \frac{1}{2}\frac{dP}{dx}\left(r - \frac{r_i^2}{r}\right) \qquad ...(3.54)$$

The equation 3.54 gives a non-linear variation of shear stress with radius. This equation will reduce to the equation for the flow through circular cross section if $r_i = 0$ showing a linear variation.

For the power law fluids the shear stress is given by

$$\tau_{rx} = \kappa\left(-\frac{du}{dr}\right)^n \qquad ...(3.55)$$

Substituting this equation in equation 3.54 the rate of shear expression can be written as

$$\left(-\frac{du}{dr}\right) = \left[\frac{r_i}{r}\left(-\frac{du}{dr}\right)_i^n - \frac{1}{2\kappa}\left(\frac{dP}{dx}\right)\left(r - \frac{r_i^2}{r}\right)\right]^{1/n} \qquad ...(3.56)$$

Separating the variables in the equation 3.56 and integrating it, one can derive the expression for the velocity distribution as

$$u = -\int\left[\frac{r_i}{r}\left(-\frac{du}{dr}\right)_i^n - \frac{1}{2\kappa}\left(\frac{dP}{dx}\right)\left(r - \frac{r_i^2}{r}\right)\right]^{1/n} dr \qquad ...(3.57)$$

The velocity of the fluid in contact with both the tube surfaces will be zero as no slip condition has been assumed hence it is expected that the velocity will be maximum at some point in between the annular space. Thus the velocity will increase from zero at $r = r_i$ to a maximum value, u_{max} at some radius say $r = \beta r_0$ and then decrease again to zero at $r = r_0$. Such a velocity distribution will result in negative shear stress and rate of shear in the region between $r = r_i$ and $r = \beta r_0$ and positive in the region between $r = \beta r_0$ and $r = r_0$. When $1/n$ is a whole number (i.e., 1, 2, 3, etc.) then the right hand side of the equation 3.57 can be integrated to directly obtain the expression for the velocity, u. However for the polymers the value of n is usually a fraction (shear thinning systems), inverse of which may not be an integer, in that case the integral must be evaluated numerically. The solution for a Newtonian fluid can be easily worked out to give the following equation for the velocity distribution

$$u = \frac{r_0^2}{4\eta}\left(\frac{dP}{dx}\right)\left[1 - \left(\frac{r}{r_0}\right)^2 + \frac{1 - \alpha^2}{\ln(1/\alpha)}\ln\left(\frac{r}{r_0}\right)\right] \qquad ...(3.58)$$

where η is the viscosity of the Newtonian fluid and α is the ratio r_i/r_0 and r is the radius in the annular space such that $r_i < r < r_0$.

For a small value of $(r_0 - r_i)$ and large r_0 the annulus may be assumed to be a bent slit and the problem of flow through such a geometry may be approximated as flow through parallel plates. Under these assumptions we have

$$w = \pi(r_0 + r_i) \quad \text{and} \quad H = (r_0 - r_i)$$

The expressions for the shear stress and the rate of shear at the wall are given below

The shear stress at the wall,

$$\tau_0 = \frac{(r_0 - r_i)}{2}\left(\frac{dP}{dx}\right) = \frac{(r_0 - r_i)}{2}\frac{\Delta P}{L} \qquad \qquad ...(3.59)$$

Wall rate of shear,

$$\left(-\frac{du}{dr}\right)_o = \left[\frac{2}{WH^2}\right]\left(2Q + \Delta P\frac{dQ}{d\,\Delta P}\right) \qquad ...(3.60)$$

or

$$\left(-\frac{du}{dr}\right)_o = \frac{2(2n''+1)}{n''}\left[\frac{Q}{WH^2}\right] \qquad \qquad ...(3.61)$$

In the above equations Q is the volumetric flow rate and, n'' is defined by equation 3.48.

EXAMPLE 3.8. Low-density polyethylene (LDPE) melt at 190°C flows through an annulus of outside radius $r_o = 44.0$ mm, an inside radius $r_i = 42.0$ mm and length, $l = 500.0$ mm. For a given mass flow rate of 72.0 kg/hr calculate (i) pressure gradient and (ii) the pressure drop for the flow. The density of the melt is 720 kg/m³. The melt follows the power law model equation, $\tau = 7276\,\dot{\gamma}^{0.3}$.

Solution: The mass flow rate, $m = 72.0$ kg/hr $= 72/3600$ kg/s $= 0.02$ kg/s

The volumetric flow rate, $Q = m$/density $= 0.02/720 = 2.8 \times 10^{-5}$ m³/s

For an annulus with large difference in the inside and outside radii the shear stress acting on the two surfaces differ considerably however for small gaps an average value of the shear stress at the wall may be calculated by assuming the annulus to be a slit having following dimensions.

$$\text{The height, } H = (r_o - r_i) = (44.0 - 42.0)$$
$$= 2.0 \text{ mm} = 2 \times 10^{-3} \text{ m}$$
$$\text{And width, } w = \pi(r_o + r_i) = \pi(44.0 + 42.0)$$
$$= 270.18 \text{ mm} = 0.2702 \text{ m}$$

The apparent wall rate of shear,

$$\dot{\gamma}_{wa} = \frac{6Q}{\pi(r_o + r_i)(r_o - r_i)^2}$$

$$= \frac{6 \times 2.8 \times 10^{-5}}{270.2 \times 4 \times 10^{-9}} = 155.0\,s^{-1}$$

The shear stress can be obtained from the power law equation as,

$$\tau_0 = 7276.0 \times (155.0)^{0.3} = 33036.4 \text{ Pa}$$

The pressure gradient,

$$\frac{\Delta P}{L} = \frac{2\tau_a}{H} = \frac{(2 \times 33036.4)}{2 \times 10^{-3}} = 33.04 \text{ M Pa/m}$$

The pressure drop over a 500 mm length = 16.52 M Pa.

3.7 MELT FLOW IN A WIRE COATING DIE[21a, 23]

The electrical wires and cables are coated with insulating polymer layer to protect the conductors and prevent persons from electrical shocks. The wire coating die is used for this purpose. The flow analysis in this case is carried out following the analysis for the flow through the annulus where the inner tube i.e. the wire moves with a finite velocity. The following assumptions are made

1. The melt flow is steady, isothermal, incompressible and fully developed laminar in nature.
2. The power law applies to describe the rheological behaviour of the fluid and the elastic effects i.e. the die swell can be neglected.
3. The converging part of the die is neglected.
4. The pressure drop through the die is negligible (although in actual practice a small pressure is applied to control the coating thickness).

The fluid elements moving through the die assembly will be subjected only to the shear force exerted by one layer of the fluid on the other as the pressure force is negligible. Hence the force balance will give the following differential equation (3.62) for the shear stress. The wire velocity is kept relatively low so that the melt is not subjected to the extensional deformation.

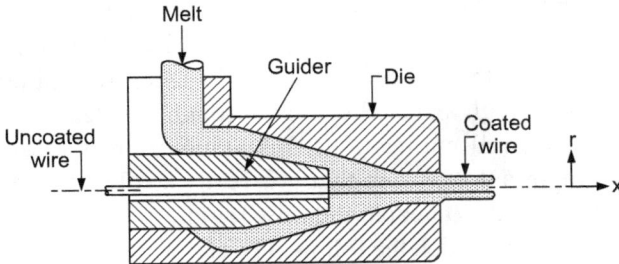

Fig. 3.11 The flow through a wire coating die[23]

$$-\left(\frac{d\left(r\tau_{rx}\right)}{dr}\right) = 0 \qquad\qquad ...(3.62)$$

Equation 3.62 on integration gives

$$\tau_{rx} = C_1/r \qquad\qquad ...(3.63)$$

Using the power law model and integrating the resulting expression to get the velocity distribution as

$$u = V\left[\frac{r^\varphi - r_o^\varphi}{r_i^\varphi - r_o^\varphi}\right] \qquad\qquad ...(3.64)$$

where V is the velocity of the wire and $\varphi = (n-1)/n$. In actual practice it is seen that the wire velocity V and the die radius, r_0 depend on the thickness of the polymer to be coated. A mass balance on the region of the coated wire starting at the exit of the die and ending where the fluid takes same velocity as the wire, results in the following expression to calculate the thickness of the coating.

$$\frac{2r_o^2}{\alpha^{1-s}-1}\left[\frac{1-\alpha^{3-s}}{3-s} - \frac{1-\alpha^2}{2}\right] = \delta(\delta + d_i) \qquad\qquad ...(3.65)$$

where δ is the coating thickness, $\alpha = r_i/(r_i + \delta)$, d_i is the wire diameter, $s = 1/n$, $r_0 = r_i + \delta$ and r_i is the wire radius. Thus the wire-coating die can be completely designed.

There are many other complicated flow situations, which the rubber and plastic process engineers and technologist have to face for example two-roll mill (Fig. 1.2d and Fig. 1.6) and calendars (Fig. 1.11) where, the melt flows through a continuously decreasing cross section area. Internal mixers (Fig. 1.7) where the rotors have complex design (Fig. 1.8) and continuous mixers with special screw and barrel designs, (Figs. 1.9 and 1.10) for the melt to have cross flow streams to get efficient mixing. Single screw and twin screw extruders, (Figs. 1.12 and 1.14), injection blow moulding systems (Fig. 1.15), film blowing, melt spinning, blow moulding, thermoforming and pultrusion processes all have complex flow paths. The actual flow analysis for all these processes is highly complicated and only the numerical solutions are possible. Some of these processes are discussed in chapter-8.

In the flow analysis in previous sections laminar flow assumptions has been always made. As the viscosity of high molecular weight polymer melts and solutions is very high and as they flow through smaller cross section area in majority of industrial applications this assumption is reasonably valid. However, it is important to ensure that it is really so. The approach followed for the Newtonian fluids to decide the flow regime is based on the Reynolds number, R_e according to which the following criteria apply for the fluid flowing through a circular cross section

If, $R_e \leq 2100$, the flow is laminar and

If $R_e > 2100$, the flow is turbulent.

The definition of Reynolds number is based on the constant viscosity for the Newtonian fluids, which is not valid for the non-Newtonian fluids flow. At the same time it is not possible to develop independent theoretical analysis for the non-Newtonian fluids, which are highly complex in rheological behaviours. Therefore the approach to identifying the transition between the laminar and turbulent flow for non-Newtonian fluids and the subsequent turbulent flow analyses are based on the modification of the Newtonian fluid approach. These approaches are discussed in the following sections.

3.8 TRANSITION BETWEEN LAMINAR AND TURBULENT FLOW[24-27]

The approach to find the criteria for the transition between laminar and turbulent flow is based on the modification of Reynolds number so as to incorporate the variable viscosity effects. Ryan and Johnson[24], and Hanks[25] have proposed two such approaches. The first approach was proposed essentially for the power law fluid flowing through a circular cross section and is not applicable to other non-Newtonian fluids flowing through non-circular channels. The Hanks method on the other hand is very general and can be applied to all types of fluids and cross sections.

3.8.1 Ryan-Johnson Criterion[4,24]

Ryan and Johnson defined a dimensionless stability parameter similar to the Reynolds number as

$$Z = \frac{r_o \rho u}{\tau_o / (-du/dr)} = \frac{r_o \rho u}{\tau_o} (-du/dr) \qquad ...(3.66)$$

A comparison of Z parameter with the Reynolds number $(D_0 U \rho / \eta)$ reveals that this definition of stability parameter retains the basic characteristics of Reynolds number that is to say that it is also a ratio of inertia force to the viscous force and therefore is dimensionless. Here diameter has been replaced with radius of the tube, r_0, average velocity, U, by point velocity, u, and the viscosity, η, by a ratio of shear stress at the wall to rate of shear at an arbitrary point in the flow system, $(\tau_o / (-du/dr))$. It can be seen that Z satisfies the following boundary conditions

(i) At $r = 0$, $(-du/dr) = 0$ $\quad \therefore \quad Z = 0$ $\qquad ...(3.67)$

(ii) At $r = r_0$, $u = 0$ $\qquad \therefore \quad Z = 0$ $\qquad ...(3.68)$

Thus Z vanishes both at the center of the pipe and the pipe wall indicating that a maximum in Z will occur at some intermediate value of radius (Fig. 3.12). The position of the maxima in Z can be found by differentiating Eq. 3.66 with radius and equating the first derivative to zero. For this purpose one must have the velocity distribution for the given fluid. Thus the need for the use of rheological models is essential. The Newtonian fluid model is first applied and then the condition of Reynolds number equal to 2100 is superimposed to identify the critical Z value for the transition between laminar and turbulent flow. Using the velocity distribution and rate of shear expressions for a Newtonian fluid flow Z can be written as

$$Z = \frac{r_o \rho}{\tau_o}(u)\left(-\frac{du}{dr}\right) = \frac{r_o \rho}{\tau_o}\left[\frac{\tau_o r_o}{2\eta}\left(1-\left(\frac{r}{r_0}\right)^2\right)\right]\left(\frac{\tau_o r}{r_0 \eta}\right)$$

...(3.69)

where
$$\lambda = \frac{r}{r_0}$$
...(Ref. eq. 3.10)

Z can be written as

$$Z = \frac{r_o \rho}{\tau_o}\left(\frac{\tau_o r_o}{2\eta}(1-\lambda^2)\right)\left(\frac{\tau_o \lambda}{\eta}\right)$$
...(3.70)

Simplifying equation 3.70, differentiating it with respect to **r** and equating the differential to zero results in the value of radius where the maxima of Z will occur i.e.

$$Z = Z_{max} \text{ at } r = \frac{1}{\sqrt{3}}r_o = 0.577\, r_0$$
...(3.71)

Substituting this in equation 3.70 one gets

$$Z_{max} = \frac{2}{3\sqrt{3}}\left(\frac{\tau_o r_o}{4\eta}\right)\left(\frac{2r_o \rho}{\eta}\right)$$

$$= \frac{2}{3\sqrt{3}}(U)\left(\frac{D_o \rho}{\eta}\right) = \frac{2}{3\sqrt{3}}R_e$$
...(3.72)

Using the critical value of Reynolds number, $R_{e,C} = 2100$ for the transition between laminar and turbulent flow of Newtonian fluids in the above equation one gets

$$Z_{max} = Z_C = 808$$
...(3.73)

Assuming that this stability criterion is applicable to the power law fluids it can be said that

(i) For $Z \leq 808$ the flow is laminar and

(ii) For $Z > 808$ the flow is turbulent

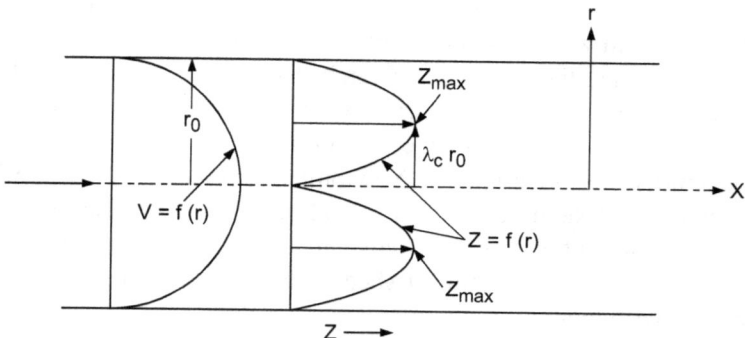

Fig. 3.12 Variation of Z with radius

Application of Ryan-Johnson Criterion to the Power Law Fluids

The velocity distribution and rate of shear for the power law fluid model can be substituted in the Ryan-Johnson stability parameter expression to further develop the analysis. For the power law model equation 3.9 gives the velocity distribution

$$u = \frac{n r_0}{n+1}\left(\frac{\tau_0}{\kappa}\right)^{1/n}\left(1 - \lambda^{\frac{1+n}{n}}\right) \qquad ...(3.73a)$$

Differentiation of above equation with radius results

$$\left(-\frac{du}{dr}\right) = \left(\frac{\tau_0}{\kappa}\right)^{1/n}\lambda^{1/n} \qquad ...(3.74)$$

Substituting these equations in the expression for Z and simplifying one gets

$$Z = \frac{n r_o^2 \rho}{(n+1)\tau_o}\left(\frac{\tau_0}{\kappa}\right)^{2/n}\left(\lambda^{1/n} - \lambda^{(n+2)/n}\right) \qquad ...(3.75)$$

Differentiating the equation 3.75 w.r.t. r and equating it to zero and simplifying one gets an expression for the critical value of $\lambda = \lambda_c$ as

$$\lambda_c = \left(\frac{1}{n+2}\right)^{n/(n+1)} \qquad ...(3.76)$$

This equation suggests that as the pseudoplasticity of the system increases i.e. to say that as the value of n decreases and approaches zero, the maxima in Z value progressively shifts to the wall and ultimately occurs at the wall at $n = 0$ (Fig. 3.13). The velocity distribution corresponding to the power law fluids changes from being marginally flat at the value of n just less than 1, to highly flat at $n \ll 1$, to plug flow at $n = 0$ (Fig. 3.1).

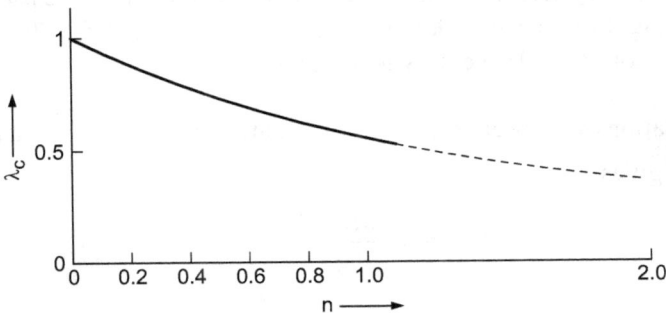

Fig. 3.13 Variation of λ_c with n

Substitution of equations 3.76, and 3.11 (for average velocity), and critical value of $Z = Z_C = 808$ in equation 3.75 and using $U = U_C$ and $\tau_0 = \tau_{0,C}$ under the critical conditions and simplifying we get

$$\tau_{0,C} = \frac{\rho U_C^2}{808} \frac{(3n+1)^2}{n} \left(\frac{1}{n+2}\right)^{(n+2)/(n+1)} \qquad ...(3.77)$$

and
$$\frac{\tau_{o,C}}{(\frac{1}{2}\rho U_C^2)} = f_C = \frac{\varphi(n)}{404} \qquad ...(3.78)$$

where f_C is the critical Fanning's friction factor for the power law fluid and

$$\phi(n) = \frac{(3n+1)^2}{n} \left(\frac{1}{n+2}\right)^{(n+2)/(n+1)} \qquad ...(3.79)$$

The transition from the laminar to the turbulent flow for the Newtonian fluids is described by

$$f_C = \frac{16}{R_{e,C}} \qquad ...(3.80)$$

If some suitable Reynolds number say a modified Reynolds number, R'_e could be defined for power law fluids then it is expected to follow equation 3.80 in terms of new parameter. Hence an expression for $R'_{e,C}$ could be written from equations 3.78 and 3.80 as

$$R'_{e,C} = \frac{6464}{\phi(n)} \qquad ...(3.81)$$

Thus the critical value of modified Reynolds number can be calculated, as the non-Newtonian index, n is known for the fluid. The critical Reynolds number for different values of n and the % deviation with respect to a value of 2100 has been tabulated in Table 3.1. It is seen from this table that the % deviation lies within ±15% for $0.2 < n < 1.5$ and only for highly pseudoplastic fluids $n < 0.2$ the deviation is large. Thus it can be said that the Ryan-Johnson criterion gives a satisfactory prediction of critical Reynolds number for the pseudoplastic fluids flowing through a circular cross section and that, for $Z \le 808$ the flow is laminar and for, $Z > 808$ the flow is turbulent.

Application of the pseudo-power law equation, $\tau_0 = \kappa' \left(\frac{8U}{D_o}\right)^{n'}$ at the critical conditions gives

$$\tau_{0,C} = \kappa' \left(\frac{8U_C}{D_o}\right)^{n'} \qquad ...(3.82)$$

Dividing both sides of equation 3.82 by $\left(\frac{\rho U_C^2}{2}\right)$ and simplifying and then comparing it with equations 3.78, 3.80 and 3.81 one gets an expression for critical modified Reynolds number[10, 26] as

$$R'_{e,C} = \frac{6464}{\varphi(n)} = \frac{D_o^{n'} \rho U_C^{2-n'}}{8^{n'-1} \kappa'} \qquad ...(3.83)$$

Also from the pseudo-power law equation, the pseudo-apparent viscosity is given as

$$\eta_{a,p} = \kappa' \left(\frac{8U}{D_o} \right)^{n'-1} \qquad ...(3.84)$$

Using this viscosity term in general definition of Reynolds number, we can write an expression for modified Reynolds number as

$$R'_e = \frac{D_o^{n'} \rho U^{2-n'}}{8^{n'-1} \kappa'} \qquad ...(3.85)$$

Comparing equations 3.83 and 3.85 it can be said that both the equations are same and equation 3.85 can be obtained by dropping the suffix C from U_C in equation 3.83. For the case when $n = n'$, equation 3.85 can be expressed in terms of true flow behaviour index, n as

Table 3.1   Critical Reynolds number for different values of n

S.No.	n	$\phi(n)$	$R'_{e,C}$	%δ
1	0.1	4.100	1577.0	−25.0
2	0.2	3.016	2143.0	+02.0
3	0.3	2.757	2345.0	+11.7
4	0.4	2.698	2396.0	+14.0
5	0.5	2.714	2382.0	+13.4
6	0.6	2.766	2337.0	+11.3
7	0.7	2.836	2279.0	+08.5
8	0.8	2.913	2219.0	+05.7
9	0.9	2.995	2158.0	+02.8
10	1.0	3.076	2100.0	00.0
11	1.1	3.164	2043.0	−02.7
12	1.5	3.491	1852.0	−11.8
13	2.0	3.859	1675.0	−20.2

$$R'_e = \frac{D_o^{n} \rho U^{2-n}}{8^{n-1} \kappa} \qquad ...(3.86)$$

For the case when $n \neq n'$, n has to be estimated from Eq. 3.35. Thus the modified Reynolds number for power law fluids is completely defined.

The generalized friction factor versus Reynolds number charts (Moody's charts) developed for Newtonian fluids can be used for non-Newtonian fluids in the laminar region by using the R'_e in place of R_e. However the R-J criterion has been reported to be unsatisfactory for the Bingham plastic fluids[27].

EXAMPLE 3.9. Polypropylene melt at 190°C gave the following rheological data.

Calculate (i) the value of $\lambda = r/r_0$ at which the Ryan-Johnson stability parameter will be maximum and (ii) the critical Reynolds number for the transition between laminar and turbulent flow.

S.No.	Rate of shear, s^{-1}	Shear stress, k Pa
1	3.67	22.4
2	7.73	32.7
3	22.8	53.6
4	51.4	61.8
5	114.0	87.1
6	306.0	111.0
7	682.0	148.0
8	1380.0	173.0
9	3340.0	202.0
10	6390.0	221.0

Solution: A plot of log shear stress versus log rate of shear (not shown) gives a straight line over the experimental range of the shear rate. Hence the power law is applicable and the slope of the line i.e. the non-Newtonian index, $n = 0.26$ and $k = 25.79$ k Pa s^n

The λ_c is given as,

$$\lambda_c = \left(\frac{1}{n+2}\right)^{n/(n+1)} = \left(\frac{1}{0.26+2}\right)^{0.26/(1.26)} = 0.845$$

The critical Reynolds number $R_{eC} = 6464/\phi(n)$

where

$$\phi(n) = \frac{(3n+1)^2}{n}\left(\frac{1}{n+2}\right)^{(n+2)/(n+1)}$$

$$= \frac{(3\times 0.26+1)^2}{0.26}\left(\frac{1}{2.26}\right)^{(2.26)/(1.26)} = 2.823$$

The critical Reynolds number for the transition between laminar and turbulent flow of the given melt will be, $R_{eC} = 6464/2.823 = 2289.7$

The calculated Reynolds number is slightly higher than 2100, the critical R_e for the Newtonian fluids.

3.8.2 Hanks Criterion[25]

The stability parameter defined by Hanks is similar to the Ryan-Johnson parameter but more general in the application. In its definition an attempt has been made to remove the limitations of Ryan-Johnson parameter by defining it to be the ratio of rate of change of kinetic energy (KE) with radius of the flow conduit to the pressure gradient. As the kinetic energy is proportional to the inertial energy and the pressure gradient is the consequence of the viscous dissipation of energy this parameter is identical in meaning to the Reynolds number. The Hanks stability parameter, k is defined as

$$k = \frac{\left(d(\rho u^2/2)/dr\right)}{(dP/dx)} = \frac{1}{2}\frac{\rho\left(du^2/dr\right)}{(dP/dx)} \qquad ...(3.87)$$

The Hanks parameter uses the viscous dissipation of energy calculated from the pressure gradient as against the Ryan-Johnson (R-J) parameter, which uses an arbitrary viscosity term as the ratio of wall shear stress to the rate of shear at any point in the flow system. Also the use of K.E. term as against the velocity and density in the R-J criterion makes the Hanks parameter more general in nature and therefore finds much wider applications. Following the similar approach of analysis as followed in case of R-J criterion for the flow of a Newtonian fluid flowing through a circular cross section and applying the criterion for transition between laminar to turbulent flow as $R_{e,C} = 2100$, the critical value of k can be worked out as

$$k_c = 404 \qquad\qquad ...(3.88)$$

It is applicable for different non-Newtonian fluids flowing through both circular and noncircular cross sections. For the analysis of the flow through the noncircular cross sections the use of some form of equivalent in terms of circular cross section is recommended. For this purpose the use of equivalent diameter[28] and substitute radius[29] have been reported. These are discussed next.

3.8.3 Equivalent Diameter[28]

Rubber and polymer melts quite often flow through a noncircular channels or dies and to be able to apply the Hanks criterion for analysis it is important to define an equivalent diameter. The equivalent diameter, D_e is defined in terms of mean hydraulic radius, R_H as

$$D_e = 4R_H = \frac{4(\text{Flow cross section area})}{\text{wetted perimeter}} \qquad ...(3.89)$$

The values of mean hydraulic radius for different cross sections are given in Table 3.2.

3.8.4 Substitute Radius[29]

Schenkel defined an equivalent radius namely the substitute radius, R_{th}, for
different cross sections applicable for the power law fluids. The expression for
R_{th} is as

$$R_{th} = \left[\frac{(2^{(1+n)/n}) A^{(1+2n)/n}}{\pi B^{(1+n)/n}} \right]^{n/(1+3n)} \quad ...(3.90)$$

where A is the cross section area, B is the circumference and n is the non-
Newtonian index. This formula has been applied to calculate the equivalent
radius for design calculations of pressure drop, volumetric flow rate and energy
required for the flow of power law fluids flowing through the square, rectangular,
annulus, elliptical or any other complex cross section flow channels. Equation
3.90 reduces to $R_{th} = d/2$ for a Newtonian fluid flowing through a circular cross
section of diameter d.

Table 3.2   Values of R_H for different cross sections [28]

Cross section	R_H
1. Circle of diameter, D	$D/4$
2. Annulus of inner dia.= d and outer dia. = D	$(D-d)/4$
3. Square of side = D	$D/4$
4. Rectangle of sides a and b	$ab/2(a+b)$
5. Ellipse of major axis = 2a and minor axis = 2b	$ab/K(a+b)*$

* Values of K, If S = $(a-b)/(a+b)$

S =	0.2	0.3	0.4	0.5	0.6	0.7	0.8	0.9	1.0
K =	1.010	1.023	1.040	1.064	1.092	1.127	1.168	1.216	1.273

3.8.5 Analysis of Bingham Plastic Fluid Flow[27]

The Hanks criterion can be applied to describe the flow behaviour of the
Bingham plastic fluids. Following the procedure adopted for the pseudoplastic
fluids using the R-J criterion the critical modified Reynolds number expression
for the Bingham plastic fluids has been derived as

$$R'_{e,C} = \frac{D_o U_C \rho}{\beta}$$

$$= k_C \sqrt{3} \frac{\{3-4(\tau_y/\tau_{o,C})+(\tau_y/\tau_{o,C})^4\}}{(1-(\tau_y/\tau_{o,C}))^3} \quad ...(3.91)$$

where $(\tau_y/\tau_{o,C})$ is calculated from

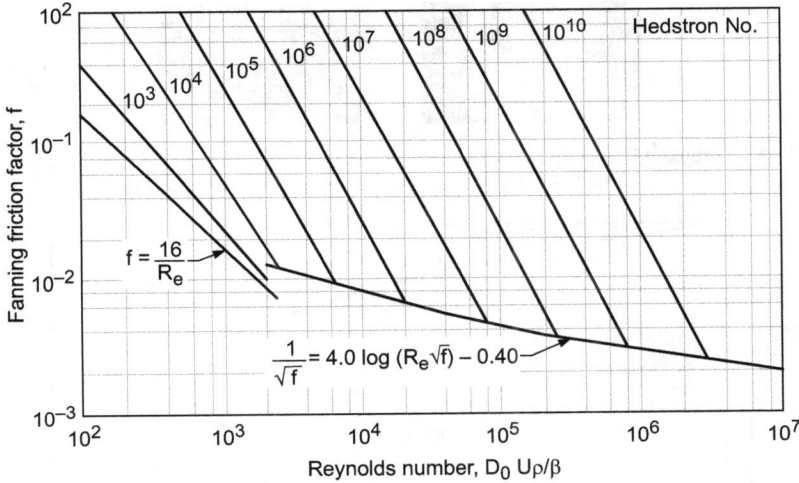

Fig. 3.14 Friction factor Reynolds number plots fir Bingham fluid[4]

$$\frac{(\tau_y/\tau_{o,C})}{(1-\tau_y/\tau_{o,C})^3} = \frac{1}{16800}\left(\frac{\rho D_o^2 \tau_y}{\beta^2}\right) = \frac{H_e}{16800} \qquad ...(3.92)$$

where H_e is the Hedstrom number.

The experimental data agrees well with equation 3.92 for $H_e < 10^5$ (Fig. 3.14) but there is a considerable deviation for higher values of the Hedstrom number. Hanks, after the analysis of the Bingham plastic fluids concluded that this model was adequate to determine the transition correctly when it predicts the radius of the unsheared plug as less than half the pipe radius. Under this condition the laminar to turbulent transition occurs progressively at higher Reynolds number as the Headstrom number increases. Hanks also applied his criterion to the Eyring-Powell model fluids for a case with larger Hedstrom number and predicted the transition accurately.

3.9 TURBULENT FLOW ANALYSIS

The turbulent flow analysis for the Newtonian fluids has been developed based on the experimental observations by Reynolds and mostly by Prandtl and Tietjens. Von Karman derived the relationship between friction factor and Reynolds number. When the fluid flow is turbulent the visual experiments show the presence of secondary motions, eddy currents and irregular fluctuations in velocity (Fig. 3.15(a) and (b)). The fluid elements near the center of the flow channel are subjected to the component of velocity or momentum in the direction perpendicular to the main flow direction. This momentum is lost due to the viscous drag between the adjacent fluid elements producing the superimposition of eddying motion, velocity fluctuations and intermixing effects on the main fluid motion. The velocity distribution changes from the parabolic profile for

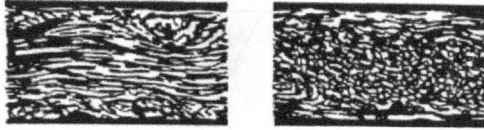

Fig. 3.15 Turbulent flow in open channel (a) camera speed same as fluid speed at the wall, (b) camera speed equal to the fluid speed at the center[4]

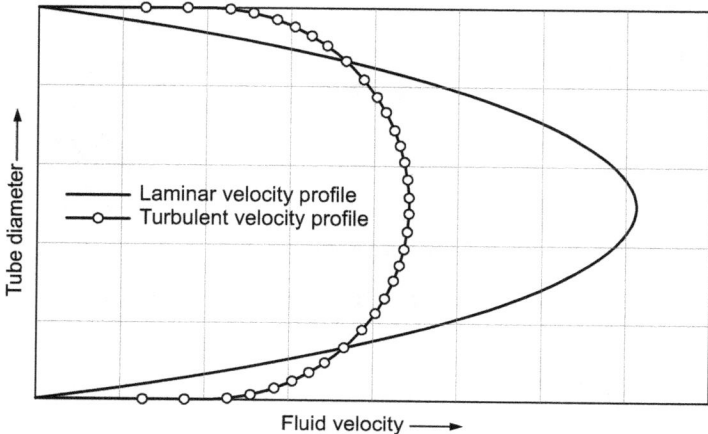

Fig. 3.16 Velocity profiles for flow of water through a circular tube[4]

laminar flow to the flat profile for turbulent flow (Fig. 3.16)[31]. The turbulent flow analysis of non-Newtonian fluids is carried out following (i) von Karman's approach and (ii) universal velocity profile approach. These approaches are discussed next.

3.9.1 Von Karman's Approach

Von Karman developed the empirical friction factor-Reynolds number correlation for turbulent flow of Newtonian fluids, which has been extended to the non-Newtonian fluids by Dodge and Metzner[32]. For Newtonian fluids the correlation is

$$\frac{1}{\sqrt{f}} = 4.0 \log\left[R_e\sqrt{f}\right] - 0.4 \qquad ...(3.93)$$

For non-Newtonian fluids the equation has been modified as

$$\frac{1}{\sqrt{f}} = \frac{4.0}{(n)^{0.75}} \log\left[R'_e\,(f)^{(1-n/2)}\right] - \frac{0.4}{(n)^{1.2}} \qquad ...(3.94)$$

where n is the non-Newtonian index, R'_e is the modified Reynolds number and f is the Fanning's friction factor for the fluid. This equation predicts the friction factor to be a function of Reynolds number and n. The variation of friction factor has been shown in Figure 3.17. It can be seen from the curves that (i) friction factor reduces with increase in Reynolds number and (ii) it also reduces with decrease in n values.

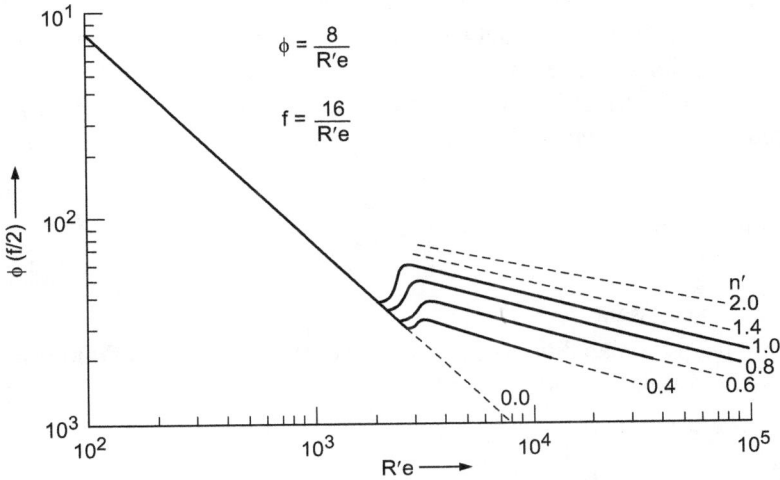

Fig. 3.17 Friction factor versus Reynolds number plot for non-Newtonian fluids[3]

3.9.2 Universal Velocity Profile Approach[3, 4, 33, 34]

The universal velocity profile approach uses a dimensionless velocity, which is the ratio of point velocity to the friction velocity and a dimensionless radius. These parameters are defined below

Friction velocity or shearing stress velocity, u^* is defined as

$$u^* = \sqrt{\frac{\tau_o}{\rho}} = \sqrt{\left(\frac{U^2\tau_o/2}{\rho U^2/2}\right)}$$

$$= \sqrt{f(U^2/2)} = U\sqrt{f/2} = U\varphi^{1/2} \qquad \text{...(3.95)}$$

where U is the average velocity, $\varphi = f/2$, and $f = (\tau_0/(\rho U^2/2))$ is the Fanning's friction factor.

The universal velocity or the dimensionless velocity, u^+ is defined as

$$u^+ = \frac{u}{u^*} = \frac{u}{\varphi^{1/2}U} = \frac{u}{U}\varphi^{-1/2} \qquad \text{...(3.96)}$$

The dimensionless distance, r^+ is defined as

$$r^+ = \frac{r_d}{(\eta/\rho u^*)} = \frac{\rho u^* r_d}{\eta} \qquad \text{...(3.97)}$$

where $r_d = (r_0 - r)$, i.e. the distance from the wall of a circular flow channel.

The application of this approach is based on the fact that any fluid flowing under turbulent flow region has two distinct zones within the system. These are (i) turbulent central core where the liquid elements flow in eddies and

(ii) a laminar zone known as viscous sub layer adjacent to the wall where the fluid flows in laminas (Fig. 3.18). As the extent of the turbulence increases, the thickness of the viscous sub layer decreases. The universal velocity profile clearly differentiates between the two zones through the nature of the velocity variation. Bogue and Metzner[34] analyzed the velocity distribution for Newtonian fluids and developed the following relations, which have been extended to the power law fluids also. For Newtonian fluids in viscous sub layer:

$$u^+ = r^+ \qquad\qquad ...(3.98)$$

in turbulent zone:

$$u^+ = 2.5 \ln r^+ + C \qquad\qquad ...(3.99)$$

where C is a constant

For non-Newtonian fluids (power law fluids) the definitions of dimensionless velocity and dimensionless distance are

$$u^+ = \frac{u}{u^*} = \left(\frac{u}{U}\right)\varphi^{-1/2} \qquad\qquad ...(3.99a)$$

and

$$r^+ = \frac{(r_d^n)\,(\rho u^{*(2n-1)})}{8^{n-1}\,\kappa} \qquad\qquad ...(3.100)$$

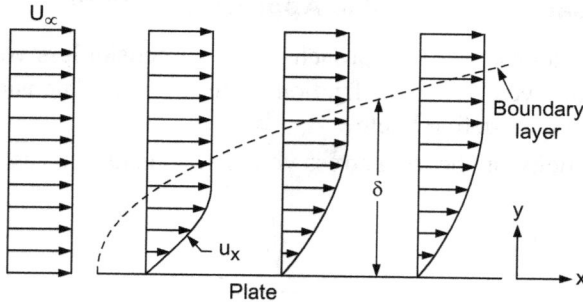

Fig. 3.18 Schematic of viscous sub layer and turbulent zones

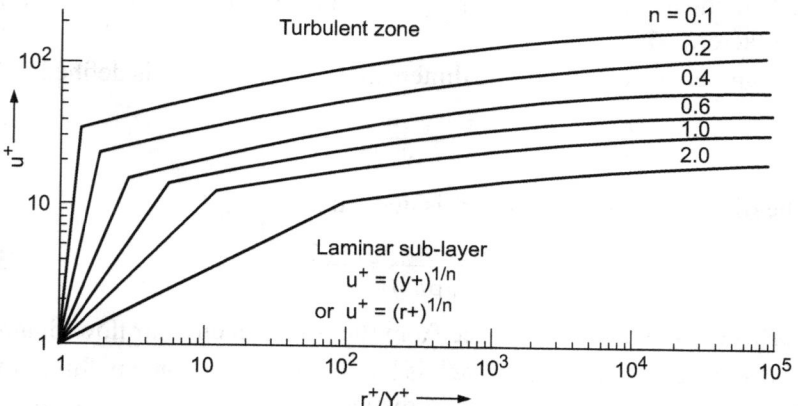

Fig. 3.19 Plots of universal velocity versus dimensionless distance as a function of n[3]

Multiplying and dividing with R'_e on the right hand side of Equation 3.100 and simplifying gives

$$r^+ = \left(\frac{r_d}{D_o}\right)^n \left(\frac{u^*}{U}\right)^{2n-1} R'_e \qquad ...(3.101)$$

The corresponding velocity distributions are

For laminar viscous sub layer:

$$(u^+)^n = r^+ \qquad ...(3.102)$$

For turbulent zone:

$$u^+ = \left(\frac{C_1}{n}\right) \ln (r^+) + C_2 \qquad ...(3.103)$$

The constants C_1 and C_2 have been evaluated for power law fluids. Using these constants in equation 3.103 it becomes[34]

$$u^+ = 2.42 \ln(r^+)^{1/n} - 1.21 \left[\frac{D_o^n u^{*(2n-1)}}{\kappa} \rho\right]^{1/n} + 0.984\sqrt{\varphi} + 3.63$$

$$...(3.104)$$

The plots of u^+ versus r^+ are shown in Figure 3.19. It can be seen that the region of the intersection of turbulent and laminar sub-layer progressively shifts closer to the wall of the flow channel as the value of the non-Newtonian index, n, decreases i.e. as the fluid becomes more and more shear thinning. Thus for very low value of $n \to 0$ the thickness of the laminar sub-layer may approach zero resulting in a plug flow situation.

3.10 FLOW OF VISCOELASTIC FLUIDS

Fluids showing the viscoelastic effects show lower pressure drop for the same volumetric flow rate as compared to the purely viscous non-Newtonian fluids. The reduction in pressure gradient is more pronounced in the turbulent flow condition. A number of experimental results by Shaver and Merrill[35] using aqueous solutions of CMC-70, ammonium alginate and carboxypolymethylene, and solutions of polyisobutylene in cychlohexane show such a reduction in the pressure gradient. Based on the data of these systems following friction factor-Reynolds number relationship was developed.

$$f = \frac{0.079}{n^5 (R'_e)^{(2.63/10.5^n)}} \qquad ...(3.105)$$

This equation is valid for $n > 0.4$.

The investigations by Meter[34] using viscoelastic solutions, which follow the Ellis model showed that the friction factor decreases with increase in

$(\tau_0/\tau_{1/2})$, where τ_0 is the wall shear stress and $\tau_{1/2}$ is the value of shear stress at the apparent viscosity equal to the half of the zero shear viscosity (Fig. 3.20).

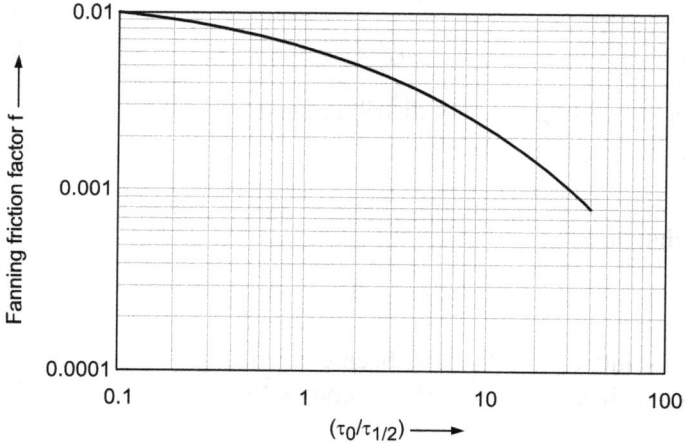

Fig. 3.20   A plot of friction factor versus $(\tau_0/\tau_{1/2})$[4]

3.11 TURBULENCE DAMPING (TOMS'S EFFECT)[37]

Dilute solutions of polymers have been found to exhibit an interesting effect wherein they show considerably lower friction factor as compared to the pure solvent.

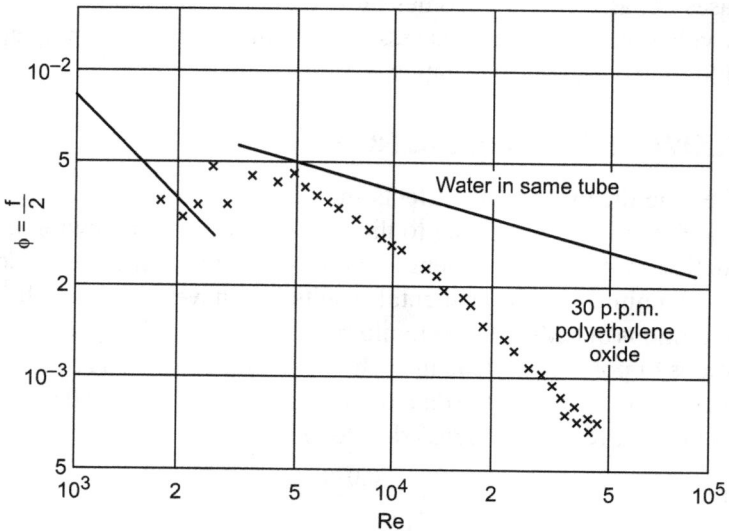

Fig. 3.21   Friction factor versus Reynolds number plots for very dilute polymer solutions[3]

This is known as the Toms' effect. The concentration of the polymer required for this effect to be produced is in ppm (parts per million) level.

The solutions behave as Newtonian fluids and conventional friction factor - Reynolds number plots can be drawn. Figure 3.21 shows such a plot for 30 ppm solution of polyethylene oxide in water. It appears that the extent of reduction in friction factor increases with increasing Reynolds number and that the transition region is relatively unstable. The reasons for such a behaviour are not yet fully understood. However it has been postulated that the random rotational motion (characteristic of free turbulence) of the fluid elements tend to repeatedly deform the long chain molecules resulting in the dampening of the turbulence. The deformed molecules release the stored energy during the relaxation and thus tend to accelerate the adjacent solvent molecules thereby giving higher flow rate at a particular pressure drop. It is also likely that the deformation of long chain molecules may result in their migration to near the wall and stabilization of the laminar boundary sub-layer and subsequent reduction in the fluid drag. These propositions are very difficult to verify experimentally but the drag reduction effect is a real one. This property has considerable industrial application in reducing the energy consumption in pumping of liquids by using the drag reducing agents.

It was observed by Tom[37] that the friction resistance in turbulent flow decreased by 50-80% in the dilute solutions of macromolecular polymers with the linear structure of molecules. His experiments revealed that the presence of microorganisms in sea weeds excrete the micromolecular compounds (biopolymers), which, when present in water even at very low concentrations, reduce the resistance to a body moving at high speed. Similar effect was also observed with the presence of poly-oxyethelene and a number of other water-soluble synthetic polymers.

Myagchenkov and Chichkanov[37a] have pointed out that application of Toms effect in the industries that transport large volume of fluids like, in oil production, pumping of water, oils and oil emulsions etc. are of great advantage due to reduction in the resistance to flow.

Takahiro and Kawaguchi[37b] have reported on the application of turbulent kinetic energy (k) and its dissipation rate (ε) and developed a model known as K-epsilon (K-ε) model. This model has been found to be applicable to the viscoelastic fluids to predict the frictional drag reduction and turbulence modification in a wall-bounded turbulent flow.

3.12 EXTENSIONAL FLOW

During polymer processing and shaping operations the polymer and rubber melts invariably flow from a bigger cross section channel to a smaller cross section resulting in change in the direction of the flow lines undergoing predominantly extensional deformation. Extensional flow in its simplest form is defined as a deformation, which involves stretching of the fluid elements along the streamlines. The extensional flows are of different types i.e. (i) simple extension, (ii) biaxial extension and (iii) planar extension[38]. The response of the melt under any of these extensional deformations depends on (i) rate of extension (ii) extent of extension and (iii) time after the application of the extensional force. For a very slow rate of strain or a small total strain the linear viscoelastic theory provides an unifying approach to the responses of different types of deformations, for example it can be shown that the stress growth coefficient under extensional deformation is three times that under of shear deformation. However for large and rapid deformation this theory is totally inadequate and the response depends on the rate, extent and the kinematics of deformation. Specific characteristics of different extensional flows are given next.

Simple Extension: In the simple extension or the uniaxial extension the distance between any two points on a streamline increases due to stretching in the direction of flow or the axis of symmetry. The uniaxial compression results in exactly opposite effect. The simple extensional flows are axisymmetric, uniform and shear free flows in which the strain rate is constant for every fluid element. The fluid elements do not undergo any rotation during such a deformation.

Biaxial Extension: The biaxial extension is also an axisymmetric, shear free flow involving compression along the axis of symmetry and stretching in the radial direction.

Planar Extension: The planar extension is a non-axisymmetric shear free flow in which the stretching takes place in the direction of applied force say in x-direction associated with compression in only one of the perpendicular direction say in y-direction but having no displacement in z-direction. An example of such a deformation is the stretching of a thin walled tube in its axial direction while pressurizing it so as to keep the diameter constant.

Some kind of extensional flow is always present in almost all polymer processing steps. Examples of such flows are:

1. *Flow through the internal mixers and extruders:* The flow of the melt through the gap between rotor tip and mixer body and that through the gap between screw flight and barrel involves both planar and uniaxial extension.

2. *The flow through the nip of the two roll mills and calenders:* The nip area provides the smallest flow area for the melt. The melt is pushed in the nip area due to the rotation of the rolls and undergoes continuous contraction, resulting in the planar extension of the melt.

3. *Flow through the extruder dies:* The melt flow from a bigger cross section to a smaller cross section area involves a simple extension where as the flow through a divergent die involves the flow in between the planar and the biaxial extension.

4. **Film blowing:** This operation involves the flow of melt in the downstream of the die where the melt is stretched along the flow direction due to the force exerted by the pinch rollers. Simultaneously there is an increase in the diameter due to air pressure involving the biaxial extension.

5. **Film casting:** The flow here is intermediate between planar and uniaxial extensions.

6. **Parison sag:** Essentially the gravity force is involved in the parison sag and thus it is a case of simple extensional flow.

7. **Parison inflation:** Here the melt flow is intermediate between the planar and biaxial extensions.

8. **Melt spinning:** It is an example of the uniaxial extensional flow where the strain rate changes along the flow path.

3.13 EXTENSIONAL VISCOSITY

3.13.1 Simple Extension

In the uniaxial extension when a system at rest is subjected to a steady extensional force resulting in a Hencky strain rate of $\dot{\varepsilon}_H$, the extensional stress, τ_E, in the beginning (i.e. for small extensions during the start up flow) are functions of both time and strain rate. The relationship between these parameters is

$$\tau_E(\dot{\varepsilon}_H, t) = \eta_E^+(\dot{\varepsilon}_H, t)\dot{\varepsilon}_H \qquad \qquad ...(3.106)$$

where $\eta_E^+(\dot{\varepsilon}_H, t)$ is the tensile stress growth coefficient, which is also the function of strain and time.

For a longer period of time when the stress is assumed to approach a limiting constant value, the tensile stress growth coefficient becomes the function of rate of strain only such that

$$\underset{t \to \infty}{\mathrm{Lim}}\left[\eta_E^+(\dot{\varepsilon}_H, t)\right] = \eta_E^+(\dot{\varepsilon}_H) \qquad \qquad ...(3.107)$$

For a linear viscoelastic material under a very low strain rate the tensile stress growth coefficient becomes independent of the rate of strain and is only a function of time. Numerically it approaches the value of three times the shear stress growth coefficient, $\eta^+(t)$, i.e.

$$\underset{\dot{\varepsilon} \to 0}{\mathrm{Lim}}\left[\eta_E^+(\dot{\varepsilon}_H, t)\right] = \eta_E^+(t) = 3\eta^+(t) \qquad \qquad ...(3.108)$$

For Newtonian fluids both shear viscosity and extensional viscosity are constant and are independent of time and rate of deformation. These are related by the Trouton[39] equation

$$\eta_E = 3\eta \qquad \qquad ...(3.109)$$

The investigations on the simple extensional behaviour of low-density polyethylene (LDPE)[40-54], high-density polyethylene (HDPE) and other linear polymers[43, 50, 56-63] and filled and unfilled acrylonitrile-butadiene-styrene (ABS) copolymer[64-71] have been reported. The response of the LDPE is different from those of other linear polymers because of the branching in its main chain. Some of the important observations regarding the extensional flow behaviour of these systems are reported next.

Extensional Behaviour of LDPE: A number of authors have reported on tensile stress growth function, $\eta_E^+(t, \dot{\varepsilon})$ [40, 41, 48, 51-54], and recoverable strain or the tensile recoil[44, 47]. Some of the important observations are listed below

  (a) For very low strain rates i.e. $\dot{\varepsilon}_H < 0.001 \, s^{-1}$ the tensile stress growth function, $\eta_E^+(t, \dot{\varepsilon})$, is independent of the strain rate and is equal to three times the shear stress growth function, $3\eta^+(t)$. This is more or less true for other polymers also.

  (b) As the strain rate is increased, $\eta_E^+(t, \dot{\varepsilon})$ follows the $3\eta^+(t)$ curve for progressively smaller times and starts rising sharply giving the onset of extension thickening or the strain hardening behaviour as the system starts to deviate from the linear behaviour and behaves as non-linear one (Fig. 3.22)[48].

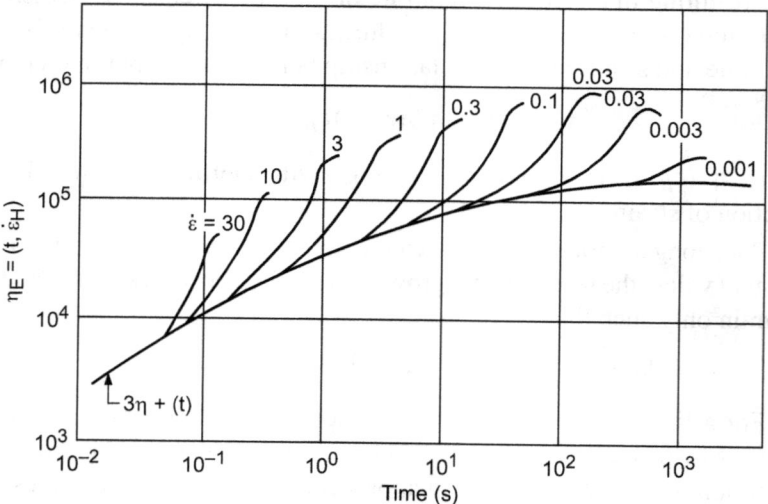

Fig. 3.22 Extensional viscosities of polyethylenes[54]

  (c) The deviation from the linear behaviour appears to start at the Hencky strain of about 0.5 for strain rates between 0.01 and 3 s^{-1}. For higher and lower strain rates it occurs at higher strain values[48].

  (d) For different strain rates the stress approaches a steady state value at some strain value above 4 (LDPE IUPAC A at 150°C) thereby enabling one to calculate the extensional viscosity of the melt[41]. However for some other LDPE melt samples no steady state regime has been reported[49] even up to Hencky strain of 7.

(e) The recoverable strain or the strain recoil[44, 47]: It has been reported that (i) at short times or the small elongation all the strain is recoverable for high strain rates, (ii) for longer times or high elongation the recoverable strain reaches a constant value, which increases with strain rate and (iii) at very slow strain rates the recoverable strain approaches zero indicating a purely viscous flow.

(f) The ultimate tensile recoil i.e. the recoverable strain increases almost linearly with tensile stress up to stress of about 1000 Pa beyond which it increases slowly and finally levels off[41].

(g) The LDPE samples having high degree of long chain branching in general give a smooth extensional flow with a broad constant stress region facilitating the estimation of the tensile viscosity[50].

(h) The increase in polydispersity index (PDI) i.e., $\bar{M}_w/\bar{M}_n$ increases the tensile viscosity[50]. Higher PDI materials contain higher amount of high molecular weight chains hence it can be said that the extension viscosity is quite sensitive to the presence of high molecular weight segments. The ratio of the maximum value of extensional viscosity to three times the zero shear viscosity i.e. ($\eta_{E,max}/3\eta_0$) has been suggested to be the useful measure of the extension thickening. It has been found to increase with increasing $\bar{M}_w/\bar{M}_n$ value[50].

(i) The ratio of ($\eta_{E,max}/3\eta_0$) has been reported to decrease with increasing molecular weight for the LDPE samples having similar molecular weight distribution[50].

(j) Increase in the degree of branching of the LDPE samples having similar $\bar{M}_w$ increases the extent of extension thickening giving increase in the ratio of $\eta_{E,max}/3\eta_0$.

(k) It has been reported[51-53] that both extensional viscosity and strain hardening behaviour for LDPE melt decrease with increasing temperature in the range of 140-180°C. It has been argued that the increase in temperature increases the inter-chain distance thereby reducing the interaction between the long chain branches and the main chain and thus reducing the strain hardening capacity of the melt.

Extensional Behaviour of Linear Polymers

A number of authors[43, 50, 54-63] have carried out investigations on the extensional behaviour of linear polymers. Both the extension thickening[50, 56-60] and the extension thinning[43, 53-56, 61-63] characteristics have been reported. The reasons for this discrepancy is that these polymers are relatively less extension thickening as compared to the branched polymers hence it is difficult to maintain the uniform sample diameter throughout the experiment. As a result of this these materials cannot be stretched isothermally as much as the branched ones[57] without the occurrence of the necking.

High-density polyethylene melts with relatively high molecular weight ($\bar{M}_W$ in the range of 150,000 to 220,000) have shown extension thickening[50,57]. Whereas those with relatively low $\bar{M}_W$ (in the range of 90,00 to 150,000) value gave no extension thickening but demonstrated a gradual reduction in the extensional viscosity showing a tendency of extension thinning behaviour. The extensional properties of these systems have been found to be very sensitive to the high-end molecular weight distribution but less dependent on the polydispersity index[50].

Linear low-density polyethylene (LLDPE) melts have been reported to give decrease in the extensional viscosity with increase in temperature in the range of (140–180°C) in the linear region (low strain region)[51-53]. However the strain hardening nature increases with increase in temperature particularly above 170°C at times corresponding to the strain higher than 3.5. It has been suggested that under the elastic deformation condition the increase in temperature increases the thermodynamic flexibility of the chain backbone resulting in the greater resistance to the extensional deformation hence giving the extension hardening behaviour.

The extensional behaviour of polystyrenes (PS) of different molecular weight distribution containing modest amount of high molecular weight fraction demonstrated enhanced non-linearity, high extension thickening and high recoverable strain[58-60]. The tensile viscosity approached $3\eta_0$ value at very low rates of extension.

Polymethyl methacrylate (PMMA) melts also show the extension thickening behaviour under simple extensional flow[60].

Extensional Behaviour of Copolymers and Polymer Blends: Styrene-butadiene-styrene (S-B-S) linear and star (four arms) block copolymer melts have been reported to show the extension softening[61] at temperature where the samples contain the lamellar structure. The lamellar structure is assumed to under go mostly the plastic orientation without any elastic recovery giving strain softening. The acrylonitrile styrene (AS) copolymer melts show the extension thickening behaviour at different strain rates (in the range of 0.005 to 0.5 s^{-1}) at 150°C[64].

The extensional flow behaviour of the blends of LDPE and LLDPE in different proportion has been published in the literature[51-53]. It has been reported that

(a) In the early stages of elongation the increase in temperature decreases the extensional viscosity of the blends much in the same way as the shear viscosity.

(b) These blends show the strain hardening behaviour, which increases with increase in the temperature for a single phase LLDPE rich blends essentially due to the dominant elastic response of the LLDPE main chain.

(c) LDPE rich two-phase blends show decrease in strain hardening with increase in temperature. It is related to the reduction in the elastic response of the long chain branching of the LDPE with increasing temperature. The response of the polystyrene blends of different molecular weight does not show any extension thickening characteristics even for the systems containing some very high molecular weight fraction[43]. It has been suggested that the extension thickening behaviour is essentially associated with long chain branching[43, 62, 63].

Extensional Behaviour of Filled Polymer Melts

The extensional flow behaviour of acrylonitrile – butadiene – styrene (ABS) two phase polymer melt containing cross linked butadiene rubber particles grafted with acrylonitrile styrene (AS) matrix has been reported by a number of workers[64-69]. Some of the observations of this system and other filled systems[69-71] are:

(a) ABS containing agglomerated butadiene particles up to 42.3% (wt.) demonstrated yielding when tested using a vertical elongational rheometer[64].

(b) The blends of these melts show strain hardening characteristics (using a vertical elongational rheometer), which depend on $\bar{M}_w$ of the AS matrix, particle size and the dispersion characteristics of the particles[65-67].

(c) For the AS matrix containing the butadiene particles below 20%(wt.) the strain hardening behaviour did not change appreciably when measured with a horizontal elongational viscometer[68].

(d) The hardness of the butadiene particles has a considerable influence on the strain hardening of the ABS melts. The presence of soft deformable particles (with very low cross linking) gave relatively higher extension thickening, which increased with the increase in strain rate as well as particle concentration[69]. This has been attributed to the capacity of the soft particles to deform in the direction of strain and thus consume more energy giving higher extensional viscosity.

(e) The hard butadiene (highly cross linked, nondeformable) particles filled ABS melt on the other hand gave much reduced extension thickening, which further reduced and changed to extension softening at higher strain rates[69].

(f) Similar observations of reduction in strain hardening behaviour have been reported for HDPE and PS melts containing short glass fibers[70] and for PS melts containing $CaCO_3$[74] and other short fibrous fillers[72].

Extensional Behaviour of Polymer Solutions[73-76]

The extensional flow behaviour of polyisobutylene solution in mixed solvent[73-75] and other solutions[76] has been reported using converging or contraction flow method. These solutions have demonstrated the extension thinning behaviour.

Mixing and the Extensional Flow

The role of extensional flow in mixing of polymer melts and the solid fillers has been amply emphasized by a number of workers[77-79]. The melt under goes the extensional flow in two-roll mill, extruders and intermixes. The twin-screw extruder has been reported to generate a stronger extensional flow field and provide better mixing[77]. Similarly a combination of single screw extruder and an elongational flow mixer gives improved mixing. The dispersion of the solid fillers in the rubber matrix has been reported to improve considerably in presence of the extensional flow field[78, 79].

3.13.2 Biaxial Extension

A biaxial extension results in an axisymmetric flow, where the net stretching stress, τ_B, acts in the direction perpendicular to the axis of symmetry. For a biaxial start-up flow when a constant strain rate, $\dot{\varepsilon}_B$, is suddenly applied at time $t = 0$, the biaxial stress and the rate of strain are related as

$$\tau_B(t, \dot{\varepsilon}_B) = \eta_B^+(t, \dot{\varepsilon}_B)\dot{\varepsilon}_B \qquad \qquad ...(3.110)$$

where, $\eta_B^+(t, \dot{\varepsilon}_B)$ is the biaxial stress growth coefficient.

Under the steady state condition i.e. after sufficiently long time interval when the stress is no more a function of time, the stress growth coefficient depends only on the rate of strain and is called the biaxial extensional viscosity i.e.,

$$\eta_B(\dot{\varepsilon}_B) = \lim_{t \to \infty} \left[\eta_B^+(t, \dot{\varepsilon}_B) \right] \qquad \qquad ...(3.111)$$

For a very slow rate of deformation i.e. at $\dot{\varepsilon}_B \to 0$ the linear viscoelastic behaviour is observed giving

$$\lim_{\dot{\varepsilon}_B \to 0} \left[\eta_B^+(t, \dot{\varepsilon}_B) \right] = \eta_B^+(t) = 6\eta^+(t) \qquad \qquad ...(3.112)$$

and
$$\lim_{\dot{\varepsilon}_B \to 0} \left[\eta_B(\dot{\varepsilon}_B) \right] = 6\eta_0 \qquad \qquad ...(3.113)$$

The biaxial extensional flow in the polymer melts can be generated by using sheet inflation[80], lubricated squeeze flow[81,82], diverging pressure flow[83] and rotary clamp[84,85], techniques. There is practically no reliable experimental data available for any polymer melts under this mode of deformation possibly because of the fact that all these methods require elaborate equipment and exceptional experimental skills.

3.13.3 Planar Extension

Planar extension is a uniform deformation in which the melt is stretched in one direction say in x-direction and compressed in y-direction but there is no flow in z-direction. Thus it implies that to generate a planar extensional flow the tensile forces of different magnitude must be applied in both x and y directions. These forces will give rise to the normal stresses in all the three directions i.e. τ_{xx}, τ_{yy}, and τ_{zz}. However from the rheological considerations only the normal stress differences are of significance. The three normal stress differences for this flow are,

$$N_1 = \tau_{xx} - \tau_{yy} \qquad\qquad \text{...(Ref. eq. 2.14)}$$
$$N_2 = \tau_{yy} - \tau_{zz} \qquad\qquad \text{...(Ref. eq. 2.15)}$$
$$N_3 = \tau_{xx} - \tau_{zz} \qquad\qquad \text{...(3.114)}$$

For an incompressible fluid only two of the above stress differences are independent and any two can be used to define the viscosity, for example Meissener et al.[89] and Laun and Schuch[43] have used N_2 and N_3. For small or slow deformations the planar stress growth coefficients are related to the shear stress growth coefficients as

$$\eta_{p1}^+ = \frac{N_3}{\dot{\varepsilon}} = 4\eta^+(t) \qquad\qquad \text{...(3.115)}$$

$$\eta_{p2}^+ = \frac{N_2}{\dot{\varepsilon}} = 2\eta^+(t) \qquad\qquad \text{...(3.116)}$$

The planar extensional flow can be generated using sheet inflation technique[86,87], lubricated planar stagnation flow[88], rotary clamp method of Meissner[89] and tube stretching technique of Laun and Schuch[43]. Some investigations on the planar extension for LDPE have been reported[43], which suggest that the first planar stress growth function $\eta_{p1}^+/2$, exhibited the extension thickening as in the case of simple extension. The second planar stress growth function however did not give any extension thickening tendency.

3.14 EXPERIMENTAL TECHNIQUES

Experimental measurement of the extensional behaviour of the polymer melts is relatively difficult in comparison to the shear measurements. The difficulties arise because of[38]

(i) Absence of undeformed material surface that can be used to support the sample. As it cannot be supported on any other solid surface the sample must be suspended or floated in an oil bath.

(ii) Clamping of the sample at the ends is difficult as the sample is soft. The use of rotary clamps and application of a suitable adhesive has been used with some success.

(iii) Requirement of a very long oil bath, as the final length of the sample may be very large (extension ratio may be as high as 200 under a strain of more than 5) making the temperature control very difficult.

(iv) For materials, which show extension thinning or very little thickening, the sample will undergo necking leading to nonuniform deformation and ductile failure.

The experimental techniques, which have been widely used and are reported to give a reasonably reliable data up to the strain rates of $< 1 \ s^{-1}$ are (i) the "Extensional creep meter", originally proposed by Cogswell[90] and modified by Muenstedt[56], (ii) the "Rotary clamp rheometer" of Meissner[47] and (iii) the" Universal extensional rheometer" of Muenstedt[42]. For the higher strain rates the uniaxial extension is generated using, (i) extrudate drawing or melt spinning to estimate the melt strength[91] and (ii) pressure driven flow in converging channel[63, 92]. These two approaches have been used to estimate an apparent extensional viscosity. It has been pointed out that the melt strength test does not yield a well-defined rheological material function because of continuous variation of strain rate acting on the filament as it moves ahead[93] during drawing process. On the other hand in the pressure driven converging flow the deformation is due to, both the shear stress as well as the extensional stress and it is difficult to separate these. Because of this reason only the apparent extensional viscosity can be evaluated from the entrance pressure drop for capillary flow as reported by Cogswell[63, 92]. Based on the entrance pressure drop measurements a number of empirical models have been proposed. These are discussed next.

3.15 RHEOLOGICAL MODELS FOR EXTENSIONAL VISCOSITY

Recently Rides and Chakravorty[94], Binding[95] and Gibson[96] have reviewed the converging flow methods for extensional behaviour of the polymer melts. A number of authors have published empirical models for the extensional viscosity from entrance pressure drop measurements. These models are based on two main considerations

1. Constrained flow[92, 97-102], where the entry to the die is profiled, which prevents the polymer from forming its own "free convergence" flow pattern in to the die (Fig. 3.23). These models can be used to describe the flow behaviour when the die profile is lubricated with low viscosity polymer or an oil to reduce or eliminate the effect of shear deformation and

2. Free convergence flow[92, 103-106] with die having either 90° entry (Fig. 3.24) or conical or profiled entry geometry wherein the melt's converging flow boundary is contained within that of the profile boundary. These models are useful in predicting the position of the converging flow boundary for comparison with the experimental observation.

3. It has been suggested by a number of authors that the high-pressure drop at the die entry is caused by the recoverable shear effects[107]. However the experimental and theoretical (numerical modelling) evidence by other workers[108-111] suggest that the pressure drop at the entrance is

dominated by extensional flow behaviour of the melt rather than its elastic behaviour

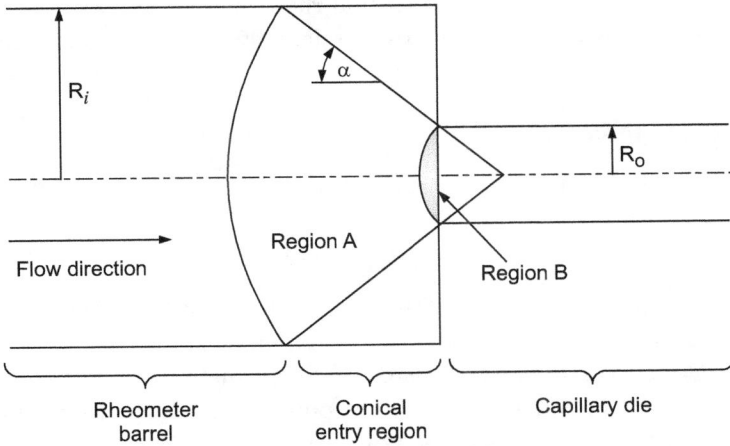

Fig 3.23 Constrained extensional flow model (Gibson model) geometry[98]
(Region A- Pressure drop due to both shear and extensional flows and Region B-that due only to extensional flow)

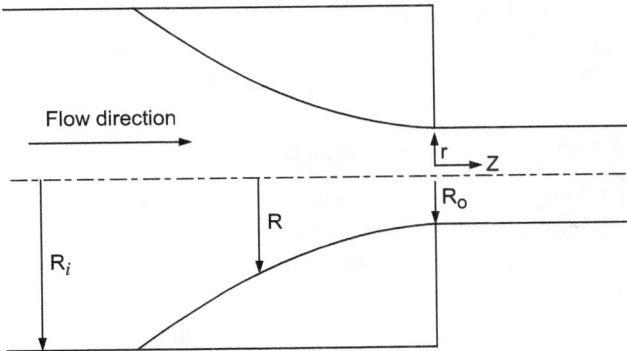

Fig. 3.24 Free convergence flow model (Binding model) geometry[104]

Out of the numerous empirical models available in literature for the estimation of extensional flow behaviour by entrance pressure drop measurements the most widely used are those proposed by Cogswell[92], Binding[104] and Gibson[98]. These models are discussed next.

3.15.1 COGSWELL MODEL

Cogswell in 1972[92] proposed a strain rate independent extensional viscosity model analogous to the Newtonian model for the shear viscosity. It is

$$\tau_E = \eta_E \dot{\varepsilon} \qquad \qquad ...(3.117)$$

where η_E is the extensional viscosity, which is assumed constant. For shear viscosity the power law model for pseudoplastic systems has been used.

This model has been applied for LDPE, HDPE, LDPE/HDPE blends PP, and Acrylic polymers using conical die entries of 90°. The extensional flow data obtained from constant stress extensional rheometer (Cam type)[90] has been analyzed using this model. It has given relatively poor agreement for LDPE melts[94].

3.15.2 BINDING MODEL

Binding in 1988[95] proposed a power law type model for the extensional viscosity as

$$\tau_E = \eta_E \dot{\varepsilon}^m \qquad \qquad ...(3.118)$$

It has been applied to predict free convergence profile for 90° conical dies based on the energy balance for a number of polymer solutions and PP melts. It has been reported to give satisfactory results for estimating extensional viscosity of shear thinning polymer solutions but poor results for constant viscosity solutions[94]. For shear viscosity the power law model has been used.

3.15.3 GIBSON MODEL

This model[107] is same as the Binding model. It has been used to predict the free convergence (90° dies) as well as constricted entry flow profiles and pressure drops for nylons, PP and unsaturated polyesters bulk-moulding compounds (BMC) melts. The best data fit has been reported for die entry angle < 30°.

PROBLEMS

1. Equation 3.9 gives the velocity distribution for a power law fluid. Calculate and plot the velocity distribution for special cases of $n = 0$ and $n = \infty$. Discuss the results with specific reference to the volumetric flow rate and pressure drop.

2. Derive a relationship for the fully developed velocity profile of a Newtonian fluid flowing down an open channel as a function of angle of inclination, thickness of the fluid film and physical properties of the fluid. Discuss the special cases when the angles of inclination of the channel are 0° and 90°.

3. Derive the expression for the velocity distribution as in the case of Pr. 2 when the fluid is a power law fluid.

4. Equation 3.9 gives the velocity distribution for a power law fluid flowing through a circular cross section. How will this expression be modified if the flow channel was inclined by an angle α to the vertical? Will the velocity distribution change if the fluid flows down or up?

5. Two immiscible liquids with different density and viscosity flow together through a horizontal circular flow channel. Assuming that the flow is fully developed laminar flow derive the expression for the velocity distribution, and shear rate and shear stress profiles. Assume that (a) both fluids are Newtonian and (b) the fluid with lower density is a power law fluid and that with higher density is a Newtonian fluid.

6. In Pr. 5 how will the velocity profile change if both fluid are power law fluids, one with high density has $n = 0.8$ and that with lower density has $n = 0.2$.

7. Equation 3.23 is the well-known Hagen-Poiseuille equation for a Newtonian fluid flowing through a horizontal circular channel under fully developed laminar flow conditions. Develop a similar relation for a power law fluid. Include both Rabinowitch and Bagley corrections.

8. Derive the volumetric-flow rate equation for a Bingham plastic fluid as derived by Buckingham and Reiner. [Buckingham E., proc. ASTM, 21, 1154 (1921); Reiner M. Deformation and Flow, Lewis, London (1949).]

9. Derive an expression for the volumetric flow rate for an Ellis fluid flowing down an inclined circular tube.

10. A Bingham plastic fluid is allowed to flow down an inclined circular tube by changing the inclination of the tube to the horizontal. Derive an expression for the maximum angle of inclination, which will allow the fluid to start flowing down under gravity.

11. For a flow of a Newtonian fluid through a capillary tube show that the viscosity of the fluid can be expressed as $\eta = At - B/t$, where A and B are constant. Explain the significance of the two terms in the right hand side of the equation.

12. Apply the Hagen-Poiseuille equation to the capillary flow (as in Pr. 11) of high viscosity polymer solution and compare the two equations for the viscosity of the solution. Explain if the constant A can be calculated from these two equations. Specify the assumptions involved.

13. Calculate and plot the velocity distribution for 4% CMC solution flowing through a circular cross section of 2.5 cm. dia. at 30°C. The Ellis parameter calculated for this fluid are $\phi_0 = 0.137$ (cm²/s.dynes), $\phi_1 = 0.3211$ (cm²$^\alpha$/s.dynes²$^\alpha$), $\alpha = 1.17$.

14. Derive the expressions for velocity distribution for the power law fluid flowing through (i) Parallel plates, (ii) A square section die and (iii) A hemispherical die

15. For a given annular die designed for extruding a pipe at 180°C with an outer dia. $D_0 = 0.075$ m and inner dia. $D_i = 0.0635$ m for HDPE for the extrusion rates corresponding to the maximum wall shear stress, $\tau_0 = 1.0 \times 10^5$ Pa. Calculate the volumetric flow rate and pressure drop for the melts having following power law parameters (i) $n = 0.56$, $\kappa = 6.19 \times 10^3$(Pa s^n), (ii) $n = 0.25$, $\kappa = 6.19 \times 10^3$ (Pa s^n) and (iii) $n = 1.0$, $\kappa = 6.19 \times 10^3$ (Pa s^n). Compare and comment on the volumetric flow rate pressure drop variations for different fluid.

16. PVC melt at 200°C has the following rheological data, which shows the Binghamplastic behaviour

τ, Pa	$(8U/D_0)$, 1/s
450	0.15
500	0.78
800	5.00
1000	9.35
2000	33.33
10000	285.70
100000	2857.00

If the melt flows through 2.0 cm dia. pipe, calculate and plot the velocity distribution at $\tau_y/\tau_0 = 0.5$. Calculate the radius and velocity of the central core. Assuming the density of PVC to be 620 Kg./m³,calculate the Hedstrom number and critical Reynolds number for the melt.

17. A pseudoplastic fluid flowing through a 2.5 cm. Diameter pipe with a pressure gradient of 16.51×10^3 N/m², gave the following shear stress rate of shear data

$(-du/dr)$, 1/s	τ, N/m²	$(-du/dr)$, 1/s	τ, N/m²
10.0	7.85	150.0	60.90
25.0	18.43	230.0	74.64
50.0	31.84	300.0	84.22
75.0	41.65	400.0	95.85
100.0	49.12	500.0	105.05
		600.0	114.34

From the above data calculate (i) Maximum velocity, (ii) Average velocity at $r = 0.6$ cm. and the ratio of Maxm./Avg. velocity, by the graphical integration of the data.

18. For the fluid in the problem No. 17 calculate and plot f vs. Re, if the density of the fluid is 0.98 gm/cc.

19. Following experimental data is reported for a natural rubber melt flow with a viscometer having barrel dia. = 22.5 mm., L/D of capillary = 5.0 at 120°C.

Ram speed, mm/min.	[ΔP, psi]		Ram speed, mm/min.	[ΔP, psi]
	$D_0 = 2.5$ mm	2.0 mm		$D_0 = 1.0$ mm
2.0	900	1030	0.4	1120
4.0	1000	1200	0.6	1220
8.0	1140	1400	0.8	1320
12.0	1260	1540	1.0	1380
16.0	1360	1700	1.5	1500
20.0	1450	1800	2.0	1580
30.0	1600	2000	4.0	1910
40.0	1775	2200	6.0	2060
50.0	1910	2400	8.0	2330
60.0	2050	2600	12.0	2600
70.0	2150	—	—	—
80.0	2350	—	—	—
90.0	2500	—	—	—

Calculate the slip velocity for the system and plot it as a function of shear stress.

20. Apply Ryan-Johnson stability theory to an Ellis fluid flowing through a pipe to derive an expression for λ_c at, (a) $\left(\dfrac{\tau_o}{\tau_{1/2}}\right)^{\alpha-1} \to 0$ and (b) $\left(\dfrac{\tau_o}{\tau_{1/2}}\right)^{\alpha-1} \to \infty$.

21. How does the velocity distribution under laminar flow condition inside a circular cross section for a power law fluid depend on n? Plot u/U vs. r/r_0 for $n = 0.2, 0.5, 0.8$. and 1.0

22. Engineers working in process plant require Q vs. ΔP plots for a non-Newtonian pseudoplastic fluid flowing through a 0.1 m dia. pipe for the control of the process operation. The laboratory experimental rheological data gave a power law equation to fit the data as, $\tau_0 = 0.6055(8U/D_0)^{0.47}$. If the specific weight of the fluid is 1025.3 kg/m^3, develop the required plot.

    [*Hint:* Test for the flow region, laminar or turbulent, then transform the power law equation in terms ΔP and Q]. Comment how the plot will be useful in process control.

23. An Ellis fluid flows down a wide vertical surface in the form of a thin film of thickness 2 mm. Calculate the fully developed volumetric flow rate of the fluid. The Ellis model constants and the density of the fluid are $\rho = 1.2$ gm/cc, $\phi_0 = 0.3$ cm.s.gm, $\phi_1 = 0.2$ (cm$^{1.3}$ s$^{1.6}$/g$^{1.3}$) and $\alpha = 3$.

24. Apply the Hanks criterion to the Eyring–Powell model fluid to predict the laminar-turbulent transition.

25. Derive Eq. 3.102 and 3.103 for an Ellis fluid.

26. Define free converging flow, constricted converging flow, primary flow, secondary flow and primary-secondary flow boundary with reference to the converging flow. Will there exist any primary-secondary flow boundary for a diverging flow as in the case of converging flow?

27. Use the Cogswell model to derive the expressions for (i) the extensional flow pressure drop and (ii) primary-secondary flow boundary profile.

28. Use the Gibson model to derive the expression for extensional pressure drop using the constricted converging flow.

29. Use the Binding model for estimating the extensional pressure drop in a free convergence flow field.

30. Modify the extensional flow pressure drop expression derived in Prob. 27 (1), by using a power law extensional viscosity function and the Rabinowitsch correction for rate of shear.

REFERENCES

1. *http-//investigadores.ciqa.mx/maestria/docs/reologia3a.pdf. Polymer melt flow.pdf*
2. Pearson J.R.A., Mechanics of polymer Processing, Elsevier Applied Science, London (1985).
3. J.M. Coulson and J.F. Richardson, Chemical Eng. Vol. 3, The English Language Book Society & Pargaman Press, Oxford, 1971.
4. P.A Longwell, Mechanics of Fluid Flow, McGraw Hill Book Inc., New York, 1966.
5. W. Ostwald, Kolloid-Z, **36**, 99 (1925).

6. A. De Waelle, Oil & Colour Chemistry Assoc. J., **6**, 33 (1923)

7. M. Reiner, Deformation Strain and Flow, Inter Science Publishers Inc., Inc., New York 1960.

8. G.V. Vinogradov and A. Ya.Malkin, Rheology of Polymers, Mir Publishers, Moscow, 1980.

9. B. Rabinowitsch, Z. Physik Chem. **A 145**, (1929)

10. M. Mooney, J. Rheol; **2**, 210 (1931).

11. A.B. Metzner and J. C. Reed, A.I.Ch.E. J; **1**, 434 (1955).

12. E. B. Bagley, End corrections in the capillary flow of polyethylenes, J. Appl. Phys; **28**, 624 (1957).

13. J.J. Benbow, R. V. Charley and P. Lamb, Nature, **192**, 223 (1961)

14. B. Maxwell and J. C. Galt; J. Polym. Sci., **62**, 850 (1962).

15. J.L. den Otter, Rheol. Acta, **10**, 200 (1971)

16. L. L. Blyler Jr. and A. C. Hart Jr., Polym. Eng. Sci.,**10**, 193 (1970)

17. A. V. Ramamurthy, J. Rheol., **30**, 337 (1986)

18. D. S. Kalika and M. M. Denn, J. Rheol., **31**, 815 (1987)

19. B.R. Gupta , Ind. J. Nat. Rubber Res., **2**, 134 (1989)

20. B.R. Gupta and M. Haeberlein, Kautsch. Gummi Kunstst., **43**(8), 673-676 (1990)

21. Z. Tadmor and G.G. Gogos, Principles of Polymer Processing, John Wiley & Sons, New York 1979 (a) pp 509-512 , (b) pp 560-564.

22. J. A. Brydson, Flow Properties of Polymer Melts, Butter worth & Co. (Plastic Institute), London, 1970 p-30.

23. D.G. Baird and D.I. Collias, polymer Processing- Principles and Design, John Wiley & Sons, Inc., New York, 1998, pp-14-18.

24. N. W. Ryan and M. M. Johnson, A. I. Ch. E. J., **5**, 433 (1959).

25. R. W. Hanks , A. I. Ch. E. J., **9**, 1 (1963).

26. R. B. Bird, A. I. Ch. E. J., **2**, 428 (1956).

27. R. W. Hanks and E. B. Christiansen, A. I. Ch. E.J., **8**, 467 (1962).

28. J.H. Perry (Ed.), Chemical Engineering Hand book, McGraw Hill Co. Inc., New York, 3rd Ed, 1950, p 378

29. N.S. Rao, Design Formulas for Plastic Engineers, Hanser Publishers, Munich, 1991.

30. D. G. Thomas, A. I. Ch. E. J., **6**, 631 (1960).

31. J. Nikuradse, V D I- Forschungs., **356**, 1 (1932).

32. D. W. Dodge and A. B. Metzner, A. I. Ch. E. J., **5**, 189 (1959).

33. A. H. P. Skelland, Non- Newtonian Flow and Heat Transfer, Wiley, 1967.

34. D.C. Bogue and A. B. Metzner , Ind. Eng. Chem. Fundamentals, **2**, 143 (1963).

35. R.G. Shaver & E. W. Merril, A. I. Ch. E. J., **5**, 81 (1959).

36. D. M. Meter , A. I. Ch. E. J. **10**, 881 (1964).

37. B. A. Toms, International Congress on Rheology, Part II, North Holland Publishing Co., Amsterdam, 1949, pp.135-141.

    (a) V. A. Myagchenkov and S. V. Chichkanov, Russian J. Appl. Hem., **78(4)**, 521-537 (2005).

    (b) T. Takahiro and Y. Kawaguchi, J. Appl. Maths., 2013, Article ID 197628, 15 pages (2013)

38. J. M. Dealy & K. F. Wissbrun, Melt Rheology & its role in Plastic Processing-Theory and Application, Chapman & Hall, London, 1995.
39. J. Dealy, Polym. Eng. Sci., **11**, 433-45 (1971).
40. H. M. Laun and H. Muenstedt, Rheol. Acta, **15**, 517 (1976).
41. H. M. Laun and H. Muenstedt, Rheol. Acta, **17**, 415 (1978).
42. H. Muenstedt, J. Rheol. **23**, 421 (1979).
43. H. M. Laun and H. Schuch, J. Rheol., **33**, 119 (1989)
44. J. Meissner, Rheol. Acta, **8**, 78 (1969).
45. J. Meissner, Pure and Appl. Chem., **42**, 553 (1975)
46. M. H. Wagner, J. Non-Newt. Fl. Mech., **4**, 39 (1978).
47. J. Meissner, Rheol. Acta, **10**, 230 (1971).
48. H. Muenstedt and H. M. Laun, Rheol. Acta, **18**, 492 (1979).
49. T Raible, A. Desmarmels and J. Meissner, Polym. Bull.,**1**, 379 (1979).
50. H. Muenstedt and H. M. Laun, Rheol. Acta, **20**, 211 (1981).
51. P. Micic, S. N. Bhattacharya, and G. Field, Intern. Polym. Processing **XIII (1)**, 50 (1998).
52. P. Micic, S. N. Bhattacharya and G. Field, Polym. Eng. Sci. **38** (10), 1685 (1998).
53. P. Micic, S. N. Bhattacharya, and G. Field, Intern. Polym. Processing **XII (2)**, 110 (1997).
54. M. Padmanabhan, C. W. Macosko, Rheol. Acta, **36** (2), 144 (1997)
55. D. J. Fleming, Annl. Trans. Nord. Rheol. Soc., **4**, 50 (1996).
56. H. Muenstedt, Rheol. Acta, **14**, 1077 (1975).
57. J. J. Linster and J. Meissner, Polym. Bull., **16**, 187 (1986).
58. H. Muenstedt, J. Rheol., **24**, 847 (1980).
59. A. Frank and J. Meissner, Rheol. Acta, **23**, 117 (1984).
60. J. J. Linster and J. Meissner, Macromol. Chem., **190**, 599 (1989).
61. T. Takahashi, H. Toda, K. Minagawa and K. Koyama, Proc. Pacific Conf. On Rheology and Polymer Processing, 1994, P. 134.
62. Y. Ide and J. L. White, J. Appl. Plym. Sci., **22**, 1061 (1978).
63. F. N. Cogswell, Trans. Soc. Rheol., **16**, 383 (1972).
64. H. Muenstedt, Polym. Eng. Sci., **21**(5), 259 (1981).
65. Y. Sato, J. Soc. Rheol. Jpn., **10**, 123 (1982).
66. Y. Sato, J. Soc. Rheol. Jpn., **10**, 128 (1982).
67. Y. Sato, J. Soc. Rheol. Jpn., **10**, 135 (1982).
68. L. Li, T. Masuda and M. Takahashi, J. Rheol., **34** (1), 103 (1990).
69. T. Takahashi, W. Wu, H. Toda, J. Takimoto, T. Akatsuka and K. Koyama, J. Non-Newtonian Fluid Mech., **68**, 259 (1996).
70. Y. Chen, J. L. White and Y. Oyanagi, J. Rheol., **22**(5), 507 (1978).
71. L. Li, T. Masuda, J. Soc. Rheol. Jpn., **18**, 190 (1990).
72. M. Kobayashi, T. Takahashi, J. Takimoto and K. Koyama, Polymer, **36**(20), 3927 (1995).
73. V. Tirtaatmadja, and T. Sridhar, J. Rheol., **37**, 1081 (1993).
74. N. E. Hudson and T. E. R. Jones, J. Non-Newt. Fl. Mech., **46**, 69 (1993).
75. J. M. Maia and D. Binding, Rheolo. Acta, **38**, 160 (1999).

76. T. Hasegawa and H. Nakamura, J. Non-Newt. Fl. Mech., **38**,159 (1991).
77. L. A. Utracki and A. Luciani, Intern. Plastics Eng. Tech., **2**, 37 (1996).
78. N. Nakajima, Rubber Chem. Tech., **53**, 1088-1105 (1980).
79. G. R. Cotton, Plastics and Rubber Processing, **4**, 89-95 (1979)
80. M. C. Yang and J. M. Dealy, J. Rheol., **31**,113 (1987).
81. Sh. Chatraei, C. W. Macosko and H. H. Winter, J. Rheol., **25**, 433 (1981).
82. P. R. Soskey and H. H. Winter, J. Rheol., **29**, 493 (1985).
83. H. H. Winter, C. W. Macosko and K. E. Bennet, Rheol. Acta, **18**, 323 (1979).
84. J. Meissner, T Raible, and S. E. Stephenson, J. Rheol., **25**, 1 (1981).
85. J. Meissner, Polym. Eng. Sci., **27**, 537 (1987).
86. C. D. Denson and D. L. Crady, J. Appl. Polym. Sci., **18**, 1611 (1974).
87. C. D. Denson and D. C. Hilton, Polym. Eng. Sci., **20**, 535 (1980)
88. R. B. Secor, C. W. Macosko and L.E. Scriven, J. Non-Newt. Fl. Mech., **23**, 355 (1987).
89. J. Meissner, S. E. Stephenson, A. Desmarmels and P. Portman, J. Non-Newt. Fl. Mech., **11**, 221 (1982).
90. F. N. Cogswell, Plast. Polym., **36**,109 (1968).
91. W. F. Busse, J. Polym. Sci., **A-2** (5), 1249 (1967).
92. F. N. Cogswell, Polym. Eng. Sci., **12** (1), 64 (1972).
93. K. F. Wissbrun, Polym. Eng. Sci., **13**, 342 (1973).
94. M. Rides and S. Chakravorty, NPL Report CMMT (A) **80**, August 1997
95. D. M. Binding, Contraction flows and new theories for estimating extensional viscosity, in Techniques in rheological measurements, Ed. Collyer, A. A., Chapman and Hall, London, 1993.
96. A. G. Gibson, Converging Dies in Rheological Measurements, A. A. Collyer and D.W. Clegg Ed. Elsevier Applied Science, London, 1988.
97. A. G. Gibson, and G. A. Williamson, Polym. Eng. Sci., **25** (15), 968(1985).
98. A. G. Gibson, Composites, **20** (1), 57 (1989).
99. H. C. Kim, A. Pendse and J. R. Collier, J. Rheol., **38**, 831 (1994).
100. D. M. Binding and D. M. Jones, Rgeol. Acta, **28**, 215 (1989)
101. A. E. Average and R. L. Ballman, Nature, **273**, 213 (1978)
102. S. Zahorski, J. Non-Newt. Fl. Mech., **41**, 309 (1992).
103. A. B. Metzner and A. P. Metzner, Rheol. Acta, **9**, 174 (1970).
104. D. M. Binding, J. Non-Newt. Fl. Mech., **27**, 173 (1988).
105. D. M. Binding, J. Non-Newt. Fl. Mech., **41**, 27 (1991).
106. B. H. Bersted, Polym. Eng. Sci., **33** (16), 1079 (1993).
107. W. Philipoff and F. H. Gaskin, Trans. Soc. Rheol., **2**, 263 (1985).
108. S. Okubo and H. Hori, J. Rheol., **25**(1),153 (1981).
109. J. L. White and A. Kondo, J. Non-Newt. Fl. Mech., **3**, 41 (1977/78).
110. S. A. White and D. G. Baird, J. Non-Newt. Fl. Mech., **20**, 93 (1986).
111. S. A. White and D. G. Baird, J. Non-Newt. Fl. Mech., **30**, 47 (1988).

Polymer Rheology – Effect of Various Parameters

4

The rheological behaviour of the polymers is influenced by a number of variables viz. temperature, pressure, shear rate, shear stress, shear induced crystallization, and Polymer molecular parameters like molecular weight, molecular weight distribution, molecular structure, extent of branching and grafting, extent and quality of solid fillers, plastisizers, blend ratio and the extent of mastication in case of rubbers. The parameters which influence the rheological behaviour can be grouped as: (a) Material variables like physical and chemical properties of polymers. The physical characteristics are evident as, extensional stress, strain and yield stress, shear thining, shear thickening etc. (b) Processing variables like, temperature, pressure, rate of shear, shear stress, extensional stress etc. (c) Time as the rheological response depends on the strain history of the material particularly on the previous loads. (d) Other factors like pH value and magnetic and electric field strengths. Effect of these parameters has a profound influence on the processing characteristics of the polymers and rubbers in different processing equipment. The effect of rate of shear and shear stress has been dealt with in previous chapters. Next sections will present a discussion on the effect of other parameters on the rheological behaviour of the polymers.

4.1 EFFECT ON EXTENSIONAL BEHAVIOUR

4.1.1 Effect of Temperature

The variation of temperature has considerable effect on stress-strain behaviour of different polymers. With increase in temperature, the yield stress, Young's modulus and yield energy decrease, while the yield strain increases. This behaviour is observed over the entire temperature range from the brittle-ductile transition to the melting point. Table 4.1 shows the variation of stress with strain at different temperatures for polyethylene[1].

Table 4.1   Tensile properties of polyethylene at different temperatures[1]

Temperature, (K)	Yield stress, (M Pa), σ_y	Yield strain γ_y	Young's modulus (Gpa)	Yield energy (J/cc)
294	29.3	0.102	1.39	2.47
318	21.2	0.132	0.70	2.31

Temperature, (K)	Yield stress, (M Pa), σ_y	Yield strain γ_y	Young's modulus (Gpa)	Yield energy (J/cc)
350	11.7	0.150	0.30	1.42
390	3.7	0.260	0.06	0.63

This data is at strain rate = 2/min, Yield energy, $E_y = \int\limits_0^{\varepsilon_y} \sigma d\varepsilon$ [1]

4.1.2 Effect of Strain Rate

The yield stress, Young's modulus and yield energy are functions of strain rate. As the strain rate increases all these properties increase whereas the yield strain remains more or less constant. However the temperature has the similar effect on yield stain as on other properties.

4.1.3 Effect of Environment

Exposer to different gaseous surroundings also affects tensile properties particularly at low temperature, which causes the surface to craze. Except helium, which is inert, other gases like nitrogen, argon, carbon dioxide and water vapour have been found to produce crazes depending on test temperature and partial pressure[2]

4.1.4 Effect of Pressure

Pressure also influences the extensional behaviour particularly yield stress and Young's modulus, which have been reported to increase with pressure. It has been reported by Imai and Brown[3] that yield stress gives a linear variation with pressure at a strain rate of 0.2/min. Both crystalline and amorphous polymers show pressure dependence of tensile properties. For comparison the room temperature data of tensile properties of polyethylene a crystalline polymer and polycarbonate an amorphous polymer are shown below[4, 5]:

Polyethylene	Polycarbonate
Yield stress, σ_y = 26.1 MPa,	64
$d\sigma_y/dP$ = 0.094,	0.13
Youngs modulus, Y = 1.24 GPa	2.35
dY/dP = 8.7	3.0

Polymers show relatively high pressure dependence as compared to metals. It is due to the fact that the yield for polymers occur at a higher stress as compared to metals. The ratio of the yield stress to bulk modulus for polycrystalline metals lies in the range of 0.001 to 0.01 whereas for polymers it is 0.02 or higher[6]. There are a few theoretical propositions available in literature. The most widely used theory is the rate theory to explain the yield in polymers. A short discussion about the rate theory is given below.

4.1.5 The Rate Theory

When the polymeric system is not under any applied external stress the polymer molecules are in a state of mutual equilibrium and posses a particular level of potential energy or enthalpy ΔH. The application of stress, say σ, subjects the system to strain rate of $\dot{\varepsilon}$ and disturbs the state of equilibrium. As a result of this deformational effect, the polymer molecules acquire additional segmental motion in the direction of applied stress and reduce the intermolecular distance. The potential of some set of molecules, which are relatively close to each other i.e. where the intermolecular distance is small rises by an amount $\sigma\Delta V$, where ΔV is known as the activation volume. As the system tends to get back to the equilibrium state the potential of another set of molecules, where intermolecular distance is relatively large, reduces by the same amount. This keeps the potential of the total system at initial level.

Eyring derived a relationship between for applied strain rate as,

$$\text{Strain rate, } \dot{\varepsilon} = \dot{\varepsilon}_0 \exp\left[\frac{(\Delta H - \sigma\Delta V)}{RT}\right] \qquad ...(4.1)$$

*Where $\dot{\varepsilon}_0$ is a constant (that includes the concentration of the segment, entropy effects etc.). **Solving for σ_y

$$\sigma_y = \frac{\Delta H}{\Delta V} + \left(\frac{RT}{\Delta V}\right)\ln\left(\frac{\dot{\varepsilon}}{\dot{\varepsilon}_0}\right) \qquad ...(4.2)$$

It gives σ_y as a linear function of log strain rate, at constant temperature. Such a variation is a common feature and is often observed and this fact is usually taken as the justification for the use of Eyring theory. Differentiating equation 4.2 gives

$$\left(\frac{\partial\sigma}{\partial\ln\dot{\varepsilon}}\right) = \frac{RT}{\Delta V} \qquad ...(4.3)$$

The activation volume can be determined from a plot of σ_y vs. log strain rate. Hartmann et al. have reported the values of activation volume for various polymers and are available in literature[1,2,2a,2b,2c]. These values for crystalline polymers are smaller by a factor of two or three than for amorphous polymers[5a], which vary from 5 to 20 nm^3. The volume of one polymer segment, v is given as

$$v = \frac{M}{\rho N_A} \qquad ...(4.3a)$$

where M is the molecular weight, ρ is the density and N_A is the Avogadro's number. The volume ratio, $\Delta V/v$ therefore must refer to number of segments, which are activated at the yield, ie yield involves the cooperative movement of many chain segments (in the range of 10-50 segment). However the number of chain segments taking part in coordinated movement are different for different polymers for example, for PE it is 48[1], Nylon 6,6 it is 11[7], and for PP it is 27[7a]. ΔV increases with increase in temperature, which suggests that with increasing temperature the correlated chain movement also increases. Such a behaviour is also observed with glassy polymers but not with metals[7b]. It has been reported by Bauwens-Crowet et al.[8] and Truss et al.[9] that the plot σ_y/T vs log strain rate, which is normally linear initially shows some curvature in the wide range indicating two processes acting in parallel and that the yield stress must consists of two additive terms of the form of Equ. 4.2. These have been inferred to be due to the molecular transition in the polymer.

The discussion so far about the tensile yield refers to at atmospheric pressure and as the high pressure has considerable influence this treatment has been extended to include the effect of hydrodynamic pressure[1]. The relationship for the rate of strain then is expressed as

$$\dot{\varepsilon} = \dot{\varepsilon}_0 \left[\frac{-(\Delta H - \tau\Delta V + P\Delta\Omega)}{RT} \right] \qquad \qquad ...(4.4)$$

where τ_y is the yield stress, P the hydrostatic pressure, and ΔV is the shear activation volume and $\Delta\Omega$ is the pressure or bulk activation volume. This analysis has been reported to fit the experimental data reasonably well.

4.2 EFFECT ON VISCOUS BEHAVIOUR

4.2.1 Effect of Temperature

The polymer viscosity is a strong function of temperature. The increase in temperature reduces the melt viscosity in general except in case of cross linking or thermosetting or vulcanization operations where the temperature helps in building a three dimensional molecular structure thereby increasing the viscosity due to the increase in the molecular weight. In general the dependence of viscosity of polymers on temperature has a pronounced influence on their processability as it is highly sensitive to the change in temperature and therefore is an extremely important consideration from the point of view of setting optimum processing conditions.

Eyring has extensively studied the dependence of viscosity of Newtonian fluids on temperature. His theoretical proposition is based on the theory of absolute reaction rate,[10] which envisages the crossing of potential energy barrier by a molecular kinetic unit and simultaneous generation of a "hole" in the space to accommodate these units. Such a movement of kinetic units thus becomes easy due to higher kinetic energy of molecules at higher temperature and availability of free space, requiring lower energy to bring about the transfer as the temperature increases. The theoretical analysis of Eyring resulted in the viscosity variation with temperature being related by an Arrhenius type equation

$$\eta_0 = Ae^{E/RT} \qquad \qquad ...(4.5)$$

where A is a constant, E is activation energy, R is the universal gas constant and T the absolute temperature. Kartsovnik and Pelekh[11] proposed flow mechanizm based on the conformational changes of the polymeric chains. The elasticity of straightened chains having an entropy origin (or a freely jointed polymer chain or an ideal chain maximizing its entropy due to reduction of the distance between its two ends[11a]) can be the reason of reduction of activation energy for the transition of the molecular-kinetic unit of the chains in to a new equilibrium state in flow direction during the thermal fluctuations.

For polymer melts and solutions Doolittle suggested a theoretical proposition based on free volume theory[12]. He carried out extensive rheological experiments with low molecular weight hydrocarbons like n-alkanes and found that the zero shear viscosity, η_0 can be expressed as

$$\eta_0 = A \exp\left(\frac{Bv}{\phi}\right) \qquad \qquad ...(4.6)$$

where A and B are constants and ϕ is the free volume of liquid at experimental temperature. The free volume, ϕ is given as

$$\phi = v - v_0 \qquad \qquad ...(4.7)$$

where v is the specific volume at any temperature and v_0 is the specific volume at zero free volume condition. The zero free volume condition refers to a condition of the molecular spacial arrangement where all molecules are in a state of

EXAMPLE 4.1. The shear stress and shear rate data for polystyrene at three different temperatures is available. Calculate the apparent viscosity at say 148.4 s^{-1} at these temperatures and discuss its variation with temperature.

S.No.	Temperature, °C					
	210		230		250	
	Rate of Shear, $\dot{\gamma}, s^{-1}$	Shear Stress, τ, Pa	Rate of Shear, $\dot{\gamma}, s^{-1}$	Shear Stress, τ, Pa	Rate of Shear, $\dot{\gamma}, s^{-1}$	Shear Stress, τ, Pa
1	7.96	34800	20.1	35200	18.4	21200
2	23.3	52600	45.0	48100	40.5	32900
3	54.3	67000	105.0	62700	91.3	45500
4	126.0	82400	319.0	83800	276.0	64700
5	369.0	105000	704.0	98700	617.0	79100

6	790.0	122000	1510.0	116000	1320.0	92600
7	1620.0	138000	3930.0	138000	3570.0	118000
8	4120.0	161000	7900.0	155000	7260.0	136000
9	8170.0	179000	15600.0	172000	14300.0	152000
10	16100.0	196000	—	—	—	—

Solution: The $\ln \tau$ versus $\ln \dot{\gamma}$ data is plotted for different temperatures and values of shear stress at $\dot{\gamma} = 148.4\ \text{s}^{-1}$ are read. The viscosity is calculated from $\eta_0 = \tau/\dot{\gamma}$.

Temperature, °C	210	230	250
Viscosity, Pa s	580.6	464.1	362.1

The magnitude of the viscosity progressively reduces as the temperature increases.

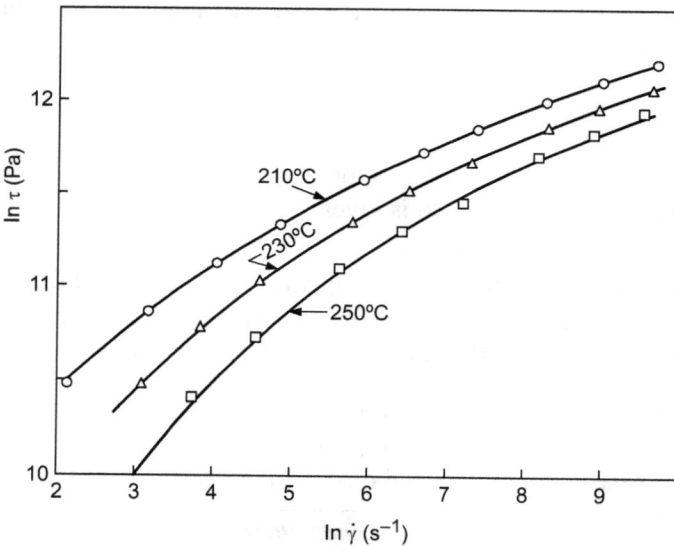

Fig. E 4.1   Log-log plots of shear stress versus rate of shear.

EXAMPLE 4.2: For the data in Example 4.1 calculate the activation energy at constant rate of shear and develop the Arrhenius equation of variation of viscosity with temperature at constant rate of shear.

Solution: At constant rate of shear the Arrhenius equation for viscosity is

$$\eta_{a,\dot{\gamma}} = Ae^{E_{\dot{\gamma}}/RT}$$

The activation energy at constant rate of shear, $E_{\dot{\gamma}}$ is calculated from the slope of the plot of $\ln \eta$ at constant rate of shear versus $1/T$. The plot (not shown) gave a straight line and the slope was calculated using the linear regression method.

$$\text{The slope, } \frac{E_{\dot\gamma}}{R} = 2.985 \times 10^3$$

and
$$E_{\dot\gamma} = R \times 2.985 \times 10^3 = 24.82 \text{ KJ/mole.}$$

And the intercept, $A = 1.211$ Pa s

Thus the desired equation is

$$\eta_{a,\dot\gamma} = 1.211 e^{24.82/RT}$$

EXAMPLE 4.3. For the data in Example 4.1 calculate the activation energy at constant shear stress and develop the Arrhenius equation of variation of viscosity with temperature at constant shear stress.

Solution: As in Example 4.2 the viscosity at constant shear stress is calculated at three temperatures and the activation energy at constant shear stress is calculated following the same procedure.

$$\text{The slope, } \frac{E_\tau}{R} = 11.312 \times 10^3$$

and
$$E_\tau = R \times 11.312 \times 10^3 = 94.24 \text{ KJ / mole.}$$

And the intercept, $A = 0.116 \times 10^{-6}$ Pa s

Thus the desired equation is

$$\eta_{a,\tau} = 0.116 \times 10^{-6} e^{94.24/RT}$$

closed packing without having any free space in between them i.e. to say that v_0 is the volume occupied by molecules of the material. It must be kept in mind that for rheological measurements at the state of zero free volume must not mean an un-deformable state but it is a just deformable condition. Such a state of a system may be assumed to occur at some temperature higher than the glass transition temperature, T_g. Let this temperature be referred to as T_a.

Glass transition temperature is the temperature at which, all long range-coordinated motions of the molecular chains are frozen. At this state, the system will not allow irreversible shearing and hence makes the shear viscosity measurements highly erroneous and inconsistent. For the rheological measurements to be reproducible the molecules must be reasonably mobile, should posses all types of molecular motions, be able to slip past each other and lose deformational energy in viscous drag. Hence the reference temperature T_a will represent almost identical physical state of the long chain molecules of all the polymers. It has been established from the experimental observations[13, 14] that indeed it is so and that this reference temperature bears an approximate relation with the glass transition temperature i.e.

$$T_a \approx Tg + 50(°C) \qquad \qquad ...(4.8)$$

Around this reference temperature the specific volume and the free volume of the liquid may be assumed to vary linearly with temperature. These variations are represented below

$$v = v_a + \alpha_v(T - T_a) \qquad \qquad ...(4.9)$$

and

$$\phi = \phi_a + \alpha_\phi(T - T_a) \qquad \qquad ...(4.10)$$

where v_a, α_v, ϕ_a and α_ϕ are specific volume at T_a, specific volume thermal expansion coefficient, free volume at T_a and free volume thermal expansion coefficient respectively.

Williams, Landle and Ferry developed the free volume theory further by applying the equation 4.6 at any temperature T and reference temperature T_a and using the definitions in equations 4.7, 4.9 and 4.10 derived the following relationship,

$$\log\left(\frac{\eta_{0,T}}{\eta_{0,T_a}}\right) = -\frac{B_1(T - T_a)}{B_2 + (T - T_a)} = \frac{-8.86(T - T_a)}{101.6 + (T - T_a)} \quad ...(4.11)$$

Equation 4.11 is the famous WLF equation. Williams et al.[13] also estimated the constants of equation 4.11. These constants are applicable for different polymers. These authors also suggested that as the temperature T_a was to be estimated from the experimental rheological data, it might be more useful to use T_g as the reference temperature. They calculated the constants B_1 and B_2 corresponding to T_g and gave the following correlation

$$\log\left(\frac{\eta_{0,T}}{\eta_{0,T_g}}\right) = \frac{B_1^g(T - T_g)}{B_2^g + (T - T_g)} = \frac{-17.44(T - T_g)}{51.6 + (T - T_g)} \quad ...(4.12)$$

The constants in the equation are known as the universal WLF constants. The same authors derived these constants theoretically in terms of free volume at glass transition temperature and free volume expansion coefficient. These expressions are

$$B_1^g = \left(\frac{1}{2.303\phi_g}\right) = -17.44 \qquad \qquad ...(4.13)$$

$$B_2^g = \left(\frac{\phi_g}{\alpha_\phi}\right) = \left(\frac{1}{2.303\,B_1^g\alpha_\phi}\right) = 51.6 \qquad ...(4.14)$$

While applying these equations, it should be kept in mind that although T_g can be estimated accurately for most of the polymers but the accurate estimation of viscosity at glass transition temperature is practically impossible as the polymers change to the glassy state. On the other hand the reference temperature, T_a is an arbitrary temperature, which has to be estimated from the rheological data and may be in considerable error. However the use of equation 4.11 gives more accurate prediction than equation 4.12.

Equations 4.11 and 4.12 are the generalized relationships, which give the generalized curves for different polymer melts. Such a curve based on T_g is shown in figure 4.1 for a large number of polymers. The curve based on reference temperature, T_a is reported to give least scatter in the temperature range of $T_g < T < (T_g + 100)$ where as that based on the glass transition temperature gives more scattering[15-17].

The estimation of reference temperature T_a is also important from the point of processing of polymers in the processing equipment. As this refers to the temperature at which the long chains of the polymer molecules have become completely mobile and thus are supposed to have acquired reasonable level of the fluidity to be processed without any difficulty. Thus this temperature may be considered to be the minimum temperature at which a polymer should be processed. Williams et al[13] have reported the values of reference temperature for a number of polymers.

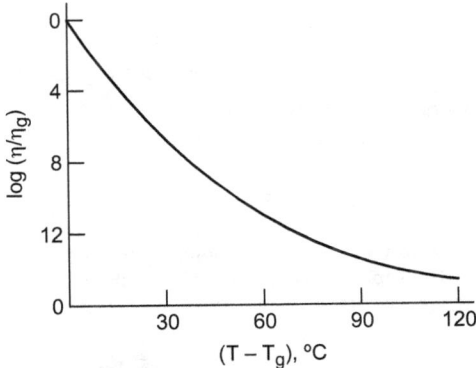

Fig. 4.1   Generalized curve based on WLF equation[10]

EXAMPLE 4.4. The zero shear viscosity of polyethylene (LDPE) at five different temperatures has been calculated using Carreau model. These values are given below.

Temperature, °C	150	170	190	210	230
η_0, Pa s	11842	8471	5327	4507	3239

Calculate the glass transition temperature using the WLF equation (Equation 4.8).

Solution: By using the two values of the zero shear viscosity the term η_{0,T_g} can be eliminated from Equation 4.12 and the T_g can be calculated by trial and error method.

Taking the η_0 at 150 and 210°C and eliminating the η_{0,T_g} one gets

$$\log\left(\frac{4507}{11842}\right) = -0.4195 = 17.44\left[\frac{423.2 - T_g}{474 - T_g} - \frac{483.2 - T_g}{534.8 - T_g}\right]$$

or $$0.02405 = \left[\frac{483.2 - T_g}{534.8 - T_g} - \frac{423.2 - T_g}{474.8 - T_g} \right]$$

Solving by trial and error one gets RHS = $0.0241 \approx$ LHS at temperature of 145 K

Hence, $T_g = 145$ K

EXAMPLE 4.5. Calculate the reference temperature T_a form the data given in above example using WLF equation (Equation 4.11) and establish equation $T_a = T_g + 50$.

Solution: Following the same procedure as in the above example one gets

$$0.04735 = \left[\frac{483.2 - T_a}{534.8 - T_a} - \frac{423.2 - T_a}{474.8 - T_a} \right]$$

and at a temperature of 195°C we get RHS

= 0.04742 which very close to LHS.

Hence the reference temperature

$T_a = 195$ K

Thus $T_a = T_g + 50$ is established for polyethylene melt.

Table 4.1 Glass transition and reference temperature of different polymers[13, 22]
(As reported in Wikipedia Encyclopedia)

Polymers	T_a, °K	T_g, °K	$(T_a - T_g)$, °K	$\%\delta = \dfrac{T_a - T_g}{T_g + 50} \times 100$
Polyvinyl chloroacetate	346	296	50	0
Polyisobutylene	243	202	41	−18
PMA	324	276	48	−4
PVA	349	301	48	−4
Polystyrene	408	373	35	−30
PMMA	433	378	55	+10
SBR	268	212	56	+12

Table 4.1   (a) Glass Transition Temperature and minimum processing temperature for other Polymeric systems (based on equn 4.8)

Systems	T_g K,	*Min. Proc. Temp., K*
Tire rubber	203	253[19]
Polyvinylidene fluoride (PVDF)	242	292[20]

Systems	T_g K,	Min. Proc. Temp., K
Polypropylene (PP atactic)	253	303[20]
Polyvinyl fluoride (PVF)	253	303[19]
Polypropylene (PP isotactic)	273	323[20]
Poly-3-hydroxybutyrate (PHB)	288	338[20]
Poly(vinyl acetate) (PVAc)	303	353[20]
Polychlorotrifluoroethylene (PCTFE)	318	368[20]
Polyamide (PA) (Nylone-6x)	320-333	370-383
Polylactic acid (PLA)	333-338	383-388
Polyethylene terephthalate (PET)	343	393[20]
Poly(vinyl chloride) (PVC)	353	403[20]
Poly(vinyl alcohol) (PVA)	358	408[20]
Polystyrene (PS)	368	418[19]
Poly(methyl methacrylate) (PMMA atactic)	378	428[20]
Acrylonitryl butadiene styrene (ABS)	378	428[21]
Polytetrafluoroethylene (PTFE)	388	438[21]
Poly (carbonate) (PC)	418	468[20]
Polysulfone	458	508
Polynorbornene	488	538[20]

Glass transition Temperature of some common biodegradable polymers*

Polygycolic acid (PGA)	35-40	LPLA	60-65
DLPLA	55-60	PCI	55-60
PDO	−10-0	85/15 DLPLG	50-55
65/35 DLPLG	45-50	50/50 DLPLG	45-50

* J.C. Middleton and A. J. Tipton, Biomaterials, 21, 2335-2346 (2000).

It can be seen from the Table 4.1 that the equation 4.8 is a poor approximation as it gives wide range of error when used with different melts. It is therefore of considerable importance that the reference temperature for different polymers be evaluated from the corresponding rheological data. However if the equation 4.8 is considered to be valid for a polymer molecule, which can be considered to have a specific chain stiffness then the molecules, which have smaller value of $(T_a - T_g)$ than 50 must have lower chain stiffness and vice-versa. It may be argued that this approximation may also be true for other polymers listed in Table 4.1(a).

The thermal sensitivity of viscosity of different polymers is quite different i.e. to say that the viscosity of some polymers falls much more rapidly than the others. Fig. 4.2 gives a plot of viscosity versus temperature for different polymer melts having different molecular weight. It can be seen that the viscosity of polystyrene and polyethylene is highly sensitive to the change in temperature as compared to other rubbers and polymers. Natural rubber, ethylene-propylene copolymer and polyisobutylene of lower molecular weight show higher sensitivity at lower temperature as compared to higher temperature, on the other hand high molecular weight polyisobutylene and plasticized butyl rubber show uniform rate of decrease of viscosity. The thermal sensitivity of viscosity of polymers is due to the combined effect of different factors such as molecular weight distribution, molecular entanglement, degree of coiling, extent of branching and presence of heavier chemical groups.

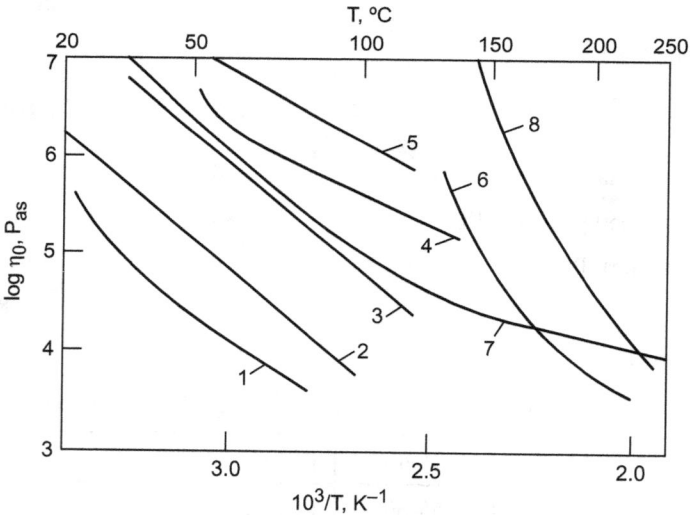

Fig. 4.2 Temperature dependence of viscosity[10]
1. Polyisobutylene (M = 8 × 10^4); 2. Same (1 × 10^5); 3. Butyl rubber (1 × 10^5) plasticized with 15% mineral oil; 4. Natural rubber; 5. Styrene-butadiene rubber (87:23) (3 × 10^3); 6. High-pressure polyethylene; 7. Ethylene-propylene copolymer at 25°C and 8. Polystyrene (3.6 × 10^5)

4.3 ACTIVATION ENERGY OF FLOW

The term activation energy was first introduced in 1889 by a Swedish scientist Swante Arrhenius as the minimum energy required to initiate a chemical reaction. This concept has been extended as the minimum energy required to start any process. Thus the activation energy of flow is of fundamental importance in understanding the relationship between the consumption of energy in the flow process and the viscosity of liquids. As the viscosity is defined as the resistance to flow, the activation energy refers to the minimum energy required to overcome this resistance. Equation 4.5 represents the variation of viscosity with temperature and can be rewritten as

$$\ln \eta_0 = \ln A + (E/RT) \qquad \qquad ...(4.15)$$

In this equation E is the activation energy of flow, which can be calculated as a slope to the plot of $\ln \eta_0$ versus $1/T$. Thus the activation energy of flow can be defined as the minimum amount of energy that must be supplied to start the flow at any temperature T. It is assumed here that E is a constant and does not depend on temperature. For simple low molecular weight liquids the activation energy is found to be constant but for the high molecular weight polymer melts, where the molecules are in a highly coiled and entangled state, E is a function of temperature. The WLF equation amply demonstrates it to be so. By comparing WLF equation and equation 4.5, the expression for the activation energy for the flow can be written as

$$E = \frac{3.85 \, RT(T - T_a)T_a}{(101.6 + T - T_a)} \qquad \qquad ...(4.16)$$

The equation 4.16 clearly indicates the complex dependence of E on temperature and for polymer melts the activation energy cannot be assumed to be constant.

For polymer melts and solutions the activation energy of flow depends on a number of molecular and other factors like,

   (i) Molecular weight of the polymer,

   (ii) Extent of molecular chain flexibility and interaction, which include
      (a) Chain microstructure, (b) Extent of polarity in the chain and
      (c) Extent of side branching and its distribution, and

(iii) Temperature.

The effect of molecular weight of polymers on the activation energy of flow is felt only in the low molecular weight region i.e. up to the segmental length equivalent to 25 to 50 carbon atoms in case of linear hydrocarbons, which show a limiting value of the activation energy of about 25 to 29 kJ/mole. Depending on the molecular chain structure and degree of flexibility, and molecular interaction this limiting value has been found to vary considerably for example 15 kJ/mole for polydimethyl siloxanes with high flexibility and weak intermolecular interaction to 145 kJ/mole[10, 23] for copolymer of tetrafluoroethylene and polyvinyl chloride, which has high degree of chain stiffness and molecular interactions. Table 4.2 gives the values of activation energy of flow for different polymers.

Table 4.2 Activation energy of flow for various polymers

System's characteristics	Activation Energy, kJ/mole
cis-Polybutadienes having different cis- 1,4-units content [Vinogradov et al., 24]	
40	33.6
60	33.6
80	31.0
92	22.5
96	19.5
Polydimethyl siloxan	20.0
Polybutadien	22.7
Polyethylene glycol ether	27.2
Linear polyethylene	31.0
Etylene-1 butene copolymer	35.6
Polydecamethylene sebacate	37.6
Styrene-butadiene rubber	37.6
Polypropylene	40.0
Poly-ε-capramide	41.0
Natural rubber	40.5
Polyisobutylene	50.0
Polyvinyl acetate	58.3
Polystyrene	87.5
Poly-4-metyl-1-pentene	116.7
Poly-α-methylstyrene	130.5

The chain flexibility parameter is defined as

$$\sigma = \sqrt{\frac{\overline{L}_0^2}{\overline{L}_f^2}} \qquad \qquad ...(4.17)$$

where $\overline{L}_0^2$ is mean square distance between the ends of unperturbed macromolecule in theta-solvent and $\overline{L}_f^2$ is the same for the freely jointed chain at fixed valance angle. The activation energy of flow increases with increasing σ or with increasing stiffness of the molecular chains. Polybutadienes give varying values of the activation energy of flow depending on their cis content[24]. Table 4.2 shows that with increasing cis content the E values keep decreasing.

Branching of the polymer chains influences the magnitude of E. It has been reported that, in going from linear to high-pressure polyethylenes, in which there are 5-10 side branches per 100 atoms of the backbone chain, the value of E increases from 25-29 to 38-50 kJ/mole.

The length of the side branch also has a considerable influence on the magnitude of E values as reported by Porter et al[25], Shirayame et al.[26], and Combs et al.[27]. It first increases with increasing length, attains a maximum value and then decreases. As the length of the side branch increases initially it adds to the stiffness of the chain however a further increase in the length beyond a certain value acts as a long segment thereby increasing the flexibility and reducing E.

4.4 ACTIVATION ENERGY AND NON-NEWTONIAN BEHAVIOUR

The polymer melts and solutions are non-Newtonian fluids and power law model is the most frequently applied rheological equation to describe this behaviour. The magnitude of the non-Newtonian index, n, describes the extent of non-Newtonianity of the system. This parameter was considered to be a purely arbitrary and empirical factor. The concept of activation energy of flow and the analysis of temperature dependence of viscosity brings out an important correlation between the activation energy and non-Newtonian index, n[28, 29] This development is discussed next.

For the variation of viscosity with shear stress and temperature one can write, $\eta = \eta(\tau, T)$ and its total differential gives

$$d\eta = \left(\frac{\delta\eta}{\delta\tau}\right)_T d\tau + \left(\frac{\delta\eta}{\delta T}\right)_\tau dT \qquad ...(4.18)$$

At constant $\dot{\gamma}$ it can be written as

$$d\eta = \left(\frac{\delta\eta}{\delta\tau}\right)_T \left(\frac{\delta\tau}{\delta T}\right)_{\dot{\gamma}} dT + \left(\frac{\delta\eta}{\delta T}\right)_\tau dT \qquad ...(4.19)$$

Using Newton's law of viscosity for derivative of shear stress with respect to viscosity and rearranging the equation 4.19, it can be shown that

$$\left(\frac{\delta\eta}{\delta T}\right)_\tau = \left[1 - \dot{\gamma}\left(\frac{\delta\eta}{\delta\tau}\right)_T\right]\left(\frac{\delta\eta}{\delta T}\right)_{\dot{\gamma}} \qquad ...(4.20)$$

From equation 4.20 it can be seen that the rate of change of viscosity with temperature at constant shear stress will be equal to that at constant rate of shear when the first term on the right hand side of the equation is equal to one. It can be possible under the following two conditions

1. When $\dot{\gamma} = 0$, i.e. when the system is at rest and

2. When $\left(\frac{\delta\eta}{\delta\tau}\right)_T = 0$, i.e. when viscosity is constant or for the Newtonian fluids only.

Thus it can be said that for the non-Newtonian fluids these rates of variation of viscosity will be constant only under very low rate of shear or very high rate of shear condition where the viscosity of the melt has limiting constant values. In the intermediate region of processing the viscosity falls with shear stress for shear thinning or pseudoplastic fluids and increases for the dilatant fluids.

Hence for the pseudoplastic fluids $\left(\dfrac{\delta\eta}{\delta\tau}\right)_T$ is negative and the first term of the equation 4.20 will be more than one whereas for the dilatant fluids it will be less than one. Accordingly for pseudoplastic fluids,

$$\left(\frac{\delta\eta}{\delta T}\right)_\tau \Big/ \left(\frac{\delta\eta}{\delta T}\right)_{\dot\gamma} > 1 \qquad\qquad ...(4.21)$$

for dilatant fluids,

$$\left(\frac{\delta\eta}{\delta T}\right)_\tau \Big/ \left(\frac{\delta\eta}{\delta T}\right)_{\dot\gamma} < 1 \qquad\qquad ...(4.22)$$

and for Newtonian fluids,

$$\left(\frac{\delta\eta}{\delta T}\right)_\tau \Big/ \left(\frac{\delta\eta}{\delta T}\right)_{\dot\gamma} = 1 \qquad\qquad ...(4.23$$

Thus the ratio on the left hand side of equations 4.21 to 4.23 must represent a material parameter whose numerical value indicates the rheological behaviour of the polymer melts or solutions.

The viscosity dependence on the temperature can also be expressed in terms of activation energy by using Arrhenius-type equation

$$\eta = A \exp\left(\frac{E_\tau}{RT}\right) = B \exp\left(\frac{E_{\dot\gamma}}{RT}\right) \qquad\qquad ...(4.24)$$

where E_τ and $E_{\dot\gamma}$ are the activation energies at constant shear stress and constant rate of shear respectively and A & B are constants. Taking the partial derivatives of viscosity with respect to temperature we get

$$\left(\frac{\delta\eta}{\delta T}\right)_\tau = \eta \frac{E_\tau}{RT^2} \qquad\qquad ...(4.25)$$

and

$$\left(\frac{\delta\eta}{\delta T}\right)_{\dot\gamma} = \eta \frac{E_{\dot\gamma}}{RT^2} \qquad\qquad ...(4.26)$$

Substituting these equations in equations 4.21, 4.22 and 4.23 one gets the following conditions in terms of the activation energy

For pseudoplastic fluids $(E_\tau/E_{\dot\gamma}) > 1,$

For dilatant fluids $(E_\tau/E_{\dot\gamma}) < 1$

and   For Newtonian fluids $(E_\tau/E_{\dot\gamma}) = 1$

Using the power law model the apparent viscosity can be expressed as,

$$\eta_a = k^{1/n}\tau^{(n-1)/n} \qquad \qquad ...(4.27)$$

Dropping the suffix a and differentiating with τ and simplifying gives

$$\left(\frac{\delta\eta}{\delta\tau}\right)_T = \frac{n-1}{n}k^{1/n}\tau^{-1/n}$$

$$= \frac{n-1}{n}\left(\frac{k}{\tau}\right)^{1/n} = \frac{n-1}{n}\frac{1}{\dot\gamma} \qquad \qquad ...(4.28)$$

Substituting equation 4.28 in equation 4.20 and combining with equations 4.21 through 4.27 and simplifying one gets

$$\left(\frac{E_\tau}{E_{\dot\gamma}}\right) = \frac{1}{n}$$

or

$$\left(\frac{E_{\dot\gamma}}{E_\tau}\right) = n \qquad \qquad ...(4.29)$$

Thus we can see that the non-Newtonian index or flow behaviour index, **n** is the ratio of the activation energy of flow at constant rate of shear to that at constant shear stress.

EXAMPLE 4.6. Calculate the flow behaviour index, n in the power law equation for polystyrene from the data of example 4.1.

Solution: The values of activation energy at constant rate of shear, $E_{\dot\gamma}$ and that at constant shear stress, E_τ are calculated in examples 4.2 and 4.3 as

$$E_{\dot\gamma} = 24.82 \text{ KJ/mole}$$

and

$$E_\tau = 94.24 \text{ KJ/mole}$$

Hence, from Equation 4.29,

$$n = \frac{E_{\dot\gamma}}{E_\tau} = \frac{24.82}{95.24} = 0.261$$

4.5 EFFECT OF PRESSURE ON VISCOSITY

The rubber and polymer melts encounter high pressure during processing in different equipment particularly in extruders and injection moulding operations.

Although the temperature has an overwhelming effect on the viscosity of the melts, the pressure effects at the pressure levels of 1 to 2 thousands atmosphere cannot be discounted. It may lead to different problems during processing as reported by Lenk[29] and others[30-42] like

(i) Vitrification and crystallization particularly of crystalline polymers[30],

(ii) Occurrence of pulsating flow or sometimes even stopping of flow[31].

These practical problems are associated with different effects the high pressure causes to the physical state of the melt. The possible changes in the physical states are,

(i) Increase of the glass transition temperature with pressure[10],

(ii) Deformation of the coiled macromolecule to an elongated state,

(iii) Increase in molecular contact with the solid surface of the flow channel giving increased adhesion,

(iv) High rate of uncoiling of molecules under sudden contraction,

(v) Decrease in the intermolecular distance due to the compressive force,

(vi) Decrease in the free volume of the macromolecules,

(vii) Increase in the elastic component of the viscoelastic behaviour of the melt and

(viii) Occurrence of the elongational flow at the die entrance.

Combinations of these parameters give rise to the processing problems. When the melt is flowing through a constant flow cross section, then depending on the rate of shear the polymer molecules will flow in a particular state of orientation and increase in the pressure will tend to bring the oriented molecules closer, thereby increasing the probability of pressure induced crystallization to take place. If the pressure applied is high enough the non-Newtonian pseudoplastic behaviour may tend to change to the dilatant nature.

During flow under sudden contraction as occurring at the die entrance where the flow cross section changes from a large value to a small one the melt undergoes the flow instability due to the transition from low velocity flow to the high velocity condition. As a result of this the macromolecules tend to uncoil, disentangle, orient and elongate in flow direction. Under high-pressure conditions the orientation and extensional flow results in the oriented molecules coming very close to each other and form pressure induced crystallites resulting in the sudden increase in the viscous as well elastic nature of the system. In the flow channel down stream further orientation takes place but at the same time the wettability of the melt to the channel surface also increases due to high pressure thereby increasing the adhesion of the melt to the surface of the channel. These factors are likely to induce the stick-slip phenomenon. A very high magnitude of the pressure will cause the stick mechanism to dominate causing pulsating flow and eventually at still higher pressure the flow may even stop[31].

In different practical applications the pressure encounterd is very high for example in lubricating layer it may be of the order of 1 GPa. It has been reported by Boza and Gallegos[32] that the viscosity almost doubles its magnitude when the pressure increases from 1 atm. to 100 MPa. Thus the viscosity of the lubricating liquid will increase by several decades. Such an increase in the viscosity and the corresponding change in rheological response at these pressures can not be ignored. In injection molding the pressure in the range of 0.1 GPa is encounterd, which may cause 10 to 100 times increase in polymer viscosity depending on the MW and molecular structure of the polymer being processed. Barness,[33] reported that the viscosity of water decreases with pressure at ambient temperature up to about 0.1 GPa and then increases. Operations at such a large pressure encounter equally large pressure drops as the melt flows from about 0.1 GPa at the starting of the process to an atmospheric pressure at the mould's end. Such high pressure drops causes non-equilibrium PVT effects as reported by Driscoll and Bougue[34].

The high pressure experimental data of polystyrene at 180°C using a capillary viscometer has been reported by Kamal and Nyun[35] to fit the WLF equation well after the thermal effect corrections were applied. It has been shown that the pressure effect on viscosity is quite significant and must not be ignored. The high pressure performance data obtained in the studies of extruder performance during extrusion of polymer melt by Aghazadeh[36] gave a satisfactory functional relationship between volumetric flow rate and pressure drop. Li et al.[37], studied the effect of pressure on different polymer melts like Poly (e-caprolactum) (PA6), Poly (ethylene terepthalate) (PET), Low density polyethylene (LDPE) and Polypropylene (PP) using a high pressure capillary viscometer. They used Barus equation, although not very accurate at high pressure to fit the experimental data, and calculated the pressure coefficients. The resulting pressure coefficients of the shear viscosity demonstrated that the degree of pressure dependence of the viscosity was in the order of PP < PET < LDPE < PA6.

The effect of dissolved gases like carbon dioxide and nitrogen in polymer melts at high pressure also has been reported by Smolinski et al.[38]. The dissolved gases reduce the flow resistance during processing in various equipments resulting in higher flow rates for same pressure drop. These are also used as blowing agents in plastic processing. The use of gases eliminate hydrocarbons as plasticizers. PS and PMMA have been used to study the effect of dissolved CO_2(5 wt%) using high pressure capillary viscometer.

Another investigation by Dealy and Park[39] using a newly designed high-pressure sliding plate rheometer (HPSPR) studied the variation of slip velocity and effect of pressure on viscosity. Under the applications at high pressure polymer processing, slip is of considerable importance as it often occures in linear polymers above a critical stress in melt rheometers and processing equipments. It has been reported that the increase in pressure reduces the slip considerably for polybutadiene. The HPSPR was also used to study the effect of high pressure on the viscosity of high density polyethylene (HDPE) using a pressure range of 0.1 to 52 MPa (Fig. 4.3). It has been reported that although the pressure increases the viscosity but dissolved CO_2 reduces it. The extent of reduction compensates more than the increase due to pressure effect.

Southern and Porter[40] have investigated the combined effect of pressure and orientation on the crystallizaton of HDPE in a Instron Capillary Rheometer. They reported that a transparent, high-modulus and highly oriented filament structure was produced. Niikuni and Porter[41] investigated the annealing characteristics of ultra-oriented HDPE fibers using DSC, thermomechanical analysis, and tensile testing. The annealing behaviour was found to be consistent with the thermal instability of the surface free energies of the ultra-oriented crystalline fibers as well as sources of melting-point reduction such as structural defects. Perkins and Porter[42] investigated on ultra high modulus by drawing single crystal mats of high molecular weight polyethylene.

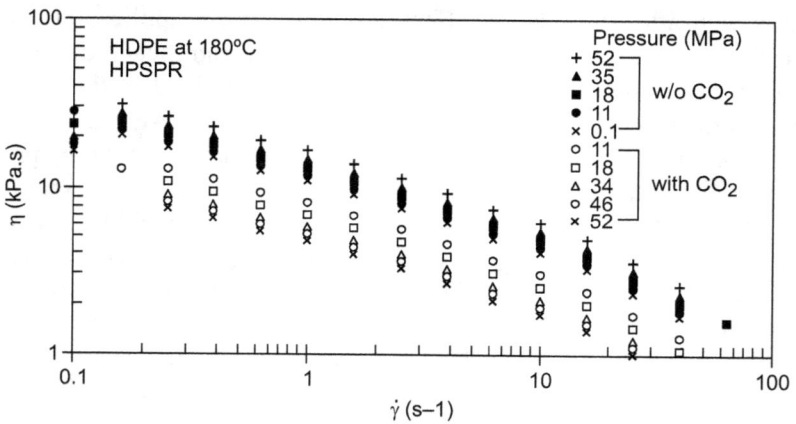

Fig. 4.3   Viscosity of HDPE measured at several pressures with and without dissolved CO_2 using HPSPR[39]

Increase in the glass transition temperature of the melt at high pressure is a direct consequence of change in the spatial configuration of the molecules, which tends to restrict their mobility by reducing molecular free volume and consequently causing the freezing of the long-range motions to occur relatively at higher temperature. Thus processing the polymers at high pressure and at temperatures close to the glass transition may cause unforeseen processing problems.

The free volume expression to give the combined effect of pressure and temperature may be written as[10]

$$\phi = \phi_a + \alpha_\phi (T - T_a) + \beta_\phi (P - P_v) \qquad ...(4.30)$$

where β_ϕ and P_v refer to the compressibility coefficient and the pressure at which the vitrification starts at the given processing conditions and the other terms are explained under equation 4.10. A similar expression for the specific volume of polymers may be written and used in the WLF equation to get the viscosity variation with both temperature and pressure. The increase in pressure reduces the free volume and therefore increases the viscosity of the polymer melts. Westover[43] reported the experimental observations to this effect for polystyrene melts, which are shown in the Fig. 4.4.

Example 4.7 below gives the variation of apparent viscosity of polystyrene melt which indicates the extent of increase in the viscosity for an increase of hydrostatic pressure from 13.79 MPa to 172.38 MPa, the shear stress increases from 6.9 kPa to 60.68 kPa, and the apparent viscosity increases from 1.38 kPas to 12.14 kPas, ie an almost 9 fold increase in viscosity showing that initially viscosity increases slowly and then with faster rate as the pressure increases. This trend gives some idea of the magnitude of viscosity in an injection moulding system where the pressure encountered are of the order of 0.1 Gpa.

4.6 EFFECT OF MOLECULAR PARAMETERS ON POLYMER RHEOLOGY

A number of molecular parameters influence the rheological behaviour of the polymer melts and solutions. These are (i) molecular weight, (ii) molecular weight distribution, (iii) branching and grafting, (iv) molecular entanglement, and (v) co-polymerization.

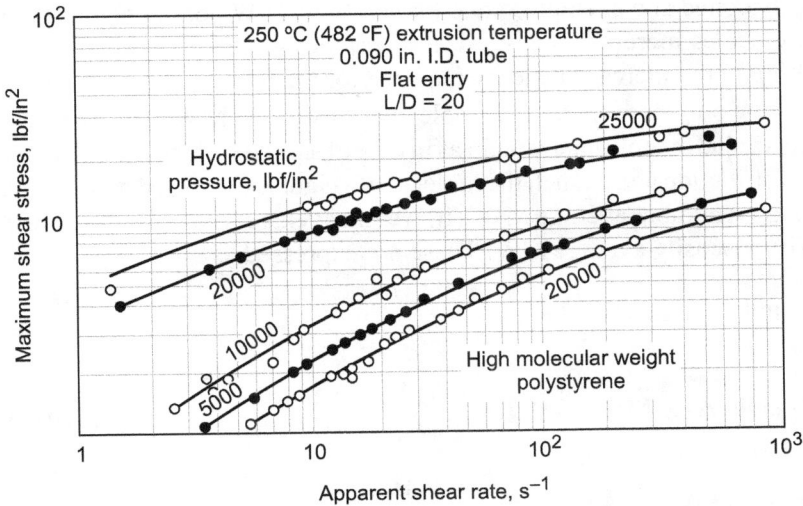

Fig. 4.4 Variation of shear stress with increase in pressure on the polystyrene melts[43]

EXAMPLE 4.7. For polystyrene melt the following data is available
Apparent rate of shear, $\dot{\gamma}_{w,a} = 5 \text{ s}^{-1}$

Hydrostatic Pressure, M Pa	13.79	34.48	68.95	137.90	172.38
Shear stress, K Pa	6.90	9.65	14.48	44.82	60.68

Calculate and plot the apparent viscosity and hydrostatic pressure and comment on the variation of viscosity at low and high pressures.

Solution: The calculated apparent viscosity data at the given pressures are

Apparent viscosity, η_a, k Pas,   1.38   1.93   2.90   8.96   12.14

The plot of viscosity versus hydrostatic pressure gives an exponentially increasing curve (not shown) giving relatively slow rise of viscosity with pressure at low pressure and exponential increase at high pressure.

4.7 SOLUTION VISCOSITY: EFFECT OF MOLECULAR WEIGHT

The rheological behaviour of the polymer melts and solutions is considerably influenced by the molecular weight (MW) of the polymers. As the molecular weight increases the melt viscosity also increases under the similar conditions of temperature and rate of shear. The resistance to the flow arises from the loss of energy in relative motion of molecules during flow where different molecular species slip past each other. For small molecules the resistance to slipping past each other is low as there is hardly any interaction between the individual molecules. However as the molecular weight increases the chain length also increases and the spatial structure of the molecules changes from being a straight chain for low MW to a mildly coiled for intermediate MW and then to a highly coiled and entangled state for very high MW polymers. As the size of the molecules increases the physical interactions between them also increase resulting in the increased energy required for the flow giving higher viscosity.

As the polymers contain the molecules of varying MW, it is not possible to assign a value of molecular weight as in case of simple molecules. Polymers have MW distribution and as such the molecular weight is expressed as some average value as (i) number average MW, (ii) weight average MW, (iii) Z-average MW and (iv) viscosity average MW. These are expressed as

Number Avg. MW $\quad \bar{M}_n = \dfrac{\Sigma N_i M_i}{\Sigma N_i}$ $\qquad$ ...(4.31)

Weight Avg. MW $\quad \bar{M}_w = \dfrac{\Sigma N_i M_i^2}{\Sigma M_i N_i}$ $\qquad$ ...(4.31(a))

Z-Avg. MW $\qquad M_z = \dfrac{\Sigma N_i M_i^3}{\Sigma M_i^2 N_i}$ $\qquad$ ...(4.31(b))

Viscosity Avg. MW $\quad M_v = \left[\dfrac{\Sigma N_i M_i (1+\alpha)}{\Sigma M_i N_i} \right]^{1/\alpha}$ $\qquad$ ...(4.31(c))

(In $-Z$ Avg. MW, Z stands for Zentrifuge means centrifugation)

N_i is the number of molecules of MW, M_i, α is exponent in Mark-Houwink equation that relates the intrinsic viscosity to MW.

A number of workers have tried to give theoretical propositions to correlate the viscosity and molecular weight. The theoretical developments are based on the analysis of two types of systems i.e. (i) polymer solutions (dilute and concentrated) and (ii) polymer melts.

4.7.1 Analysis Based on Dilute Solutions

This method is based on the estimation of the intrinsic viscosity of different polymers of known molecular weight and then developing an empirical relationship between molecular weight and the intrinsic viscosity of solutions of polymer in some suitable solvent. The intrinsic viscosity of the solvent is defined as the limiting value of the reduced specific viscosity, $\eta_{sp,rd}$ as the concentration of polymer in solutions tends to be zero. It is obtained by extrapolation of curve of reduced specific viscosity to zero concentration. The intrinsic viscosity is expressed as

$$[\eta] = \lim_{c \to 0} \eta_{sp,rd} = \lim_{c \to 0}\left[\frac{\eta - \eta_0}{\eta_0 c}\right] \qquad ...(4.31)$$

where η, η_0 and c are the solution viscosity, solvent viscosity and the concentration of the polymer in the solution. Flory and Fox[44] made use of the Einstein's[45,46] equation of viscosity for dilute colloidal suspensions (volume fraction of solids $\ll 0.01$) of incompressible, rigid, non-interacting, uncharged, solid, spherical particles in a liquid, which is

$$\eta = \eta_0(1 + 2.5\,\phi_v) \qquad ...(4.32)$$

where ϕ_v is the volume fraction of the solids in suspensions. In case of solutions of polymers in a solvent, the conditions necessary for equation 4.32 can be assumed to exist in the dilute solutions where the inter-molecular distances are large so that they have no interaction. Although the polymer molecules will not be exactly rigid and spherical but an approximation to this effect is made.

If n_2 is the number of polymer molecules in the solvent, $\phi_{e,m}$ is the effective volume of each molecule (assuming that each molecule in the system is of same molecular weight, M, i.e. mono-dispersed polymer) and V is the volume of the solution, then

$$\phi_v = \frac{n_2\phi_{e,m}}{V} = \frac{\phi_{e,m}\,cN_{av}}{100\,M} \qquad ...(4.33)$$

where c and N_{av} are the concentration of the polymer in solution in gm/dl and the Avogadro's number respectively. Substituting equation 4.33 in equation 4.32 and rearranging, it can be written as

$$\eta = \eta_0\left[\frac{1 + 0.025(\phi_{e,m}\,cN_{av})}{M}\right] \qquad ...(4.34)$$

The specific viscosity, η_{sp} is related to the relative viscosity, η_r as

$$\eta_{sp} = (\eta_r - 1) = \left(\frac{\eta}{\eta_0}\right) - 1 = \frac{(\eta - \eta_0)}{\eta_0} \qquad ...(4.35)$$

From equation 4.34 and 4.35

$$\eta_{sp} = \frac{0.025\, \phi_{e,m}\, cN_{av}}{M} \qquad \qquad ...(4.36)$$

The basic assumption in the Einstein equation is the non-interacting spherical particles. As the polymer molecules in the solution are highly mobile and are acted upon by van der Waals forces, there always exists some kind of interaction between them until and unless they are far apart. This condition of the molecules being far apart is achievable only under very low concentration of polymer in the solution where the condition $c \to 0$ applies. Thus for the dilute solution at $c \to 0$ condition the solution viscosity is expected to be independent of concentration as the inter-molecular distances will be large. Therefore defining the reduced specific viscosity, $\eta_{sp,rd}$ as the ratio of specific viscosity to the concentration one can write its expression from equation 4.36 as

$$\eta_{sp,rd} = \left(\frac{\eta_{sp}}{c}\right) = \left[\frac{(\eta - \eta_0)}{\eta_0 c}\right]$$

$$= \frac{0.025\, \phi_{e,m} N_{av}}{M} \qquad \qquad ...(4.37)$$

Flory and Fox[44] further assumed that the effective volume of a polymer molecule, $\phi_{e,m}$ can be expressed in terms of the mean square radius of gyration of the molecule, $\langle S^2 \rangle_0$ or the mean square end to end distance of the chain, $\langle L^2 \rangle_0$ as

$$\phi_{e,m} \propto \langle S^2 \rangle_0^{3/2} \propto \left[\frac{\langle L^2 \rangle_0}{6}\right]^{3/2} \qquad \qquad ...(4.38)$$

Substituting equations 4.37 and 4.38 in equation 4.31 and simplifying the intrinsic viscosity can be expressed as

$$[\eta] = K_1 \left[\frac{\langle L^2 \rangle_0}{M}\right]^{3/2} M^{1/2} \qquad \qquad ...(4.39)$$

Equation 4.39 predicts the intrinsic viscosity as a function of (i) flexibility parameter, which is proportional to $\left[\langle L^2 \rangle_0 / M\right]^{3/2}$ and (ii) molecular weight to the power 1/2. It must be noted that this equation has been derived on the basis of no interaction between polymer and solvent molecules. For theta solvents no interaction situation may be assumed to be true but for other solvents it is not always applicable.

The concept of theta solvent and the related terms is explained next for better understanding. A theta solvent is an organic solvent for polymers producing random coil conformation of individual chains. This means that the polymer coils act like ideal chains, assuming exactly their random walk coil dimensions. Although the conformation assumed by polymer chains in dilute solutions can be modelled as a random walk of monomer subunits using freely jointed chain model it does not account for steric effects. Real coils are more closely represented by a self avoiding walk because the conformation in which the different chain segments occupy the same space are not physically possible. This is the excluded volume effect, which causes the polymers to expand. The excluded volume concept was introduced by Warner Kuhn in 1934 and was applied to polymer molecules by Paul Flory.

Recently Jiang et al.[44a] have investigated the effects of chain flexibility on conformational behaviour as a function of temperature using coarse-grained molecular dynamics (MD) simulation. They presented the simulation results regarding the potential energies and structural parameter of a single polymer chain made of 100, 200 or 300 beads with various chain flexibilities. The simulation results show a conformational transition from a random coil to a folded structure. It shows that the main factor inducing the structural transition lies in non-bonded interaction. It clearly explains that the long-range interaction plays an important role in the polymer folding process.

An ideal chain (or freely jointed chain Fig. 4.5) is the simplest model to describe polymers such as nucleic acid and proteins [en.wikipedia.org, edited Feb. 2017]. A random walk is a mathematical object, known as a stochastic or random process, that describes a path that consists of a succession of random steps on some mathematical space such as integers. Self avoiding walk (SAW) is a sequence of moves on a lattice (a lattice path) that does not visit the same point more than once. The idea of excluded volume refers to the idea that one part of the long chain molecule can not occupy space that is already occupied by another part of the same molecule Fig. 4.5(a). The excluded volume causes the ends of a polymer chain in a solution to be further apart than they would be when there was no excluded volume (ie in case of ideal chain model).

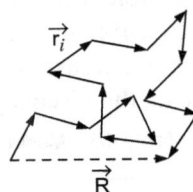

Fig. 4.5 A freely jointed chain model ($\vec{R}$ is total end-to-end vector and $\vec{r_i}$ are the vectors corresponding to individual sub limits in polymer)

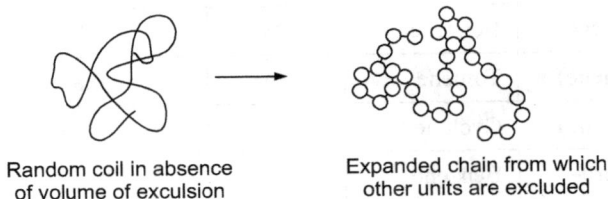

Random coil in absence of volume of exculsion

Expanded chain from which other units are excluded

Fig. 4.5(a) Excluded volume effect (Expansion of chain)

Other models proposed to explain the behaviour of the polymer systems are, Freely rotating chain, Hindered rotation model, Rotational isomeric state model, Worm-like chain model. The worm-like chain model is used to describe semi-flexible polymers.

The solubility is a temperature dependent phenomenon and hence to account for the molecular interaction a polymer-solvent-temperature interaction parameter known as the expansion factor α_s is introduced to give the expression for the intrinsic viscosity as

$$[\eta] = K_1 \left[\frac{\langle L^2 \rangle_0}{M} \right]^{3/2} M^{1/2} \alpha_s^3 \qquad ...(4.40)$$

For a theta-solvents (where the solvent polymer interaction is equivalent to the inter-molecular interaction of the polymer chains) α_s is always equal to unity and $[\eta]$ is proportional to $M^{0.5}$. However for other solvents α_s has been found to vary as molecular weight to the power 0.1 and for such systems intrinsic viscosity is expressed as

$$[\eta] \propto M^{0.8} \qquad ...(4.41)$$

On the basis of experimental results an empirical relationship of similar nature known as Mark-Kuhn-Houwink-Sakurada equation[47] has been proposed as

$$[\eta] = KM^a \qquad ...(4.42)$$

where K and 'a' are material parameters. For mono-dispersed or narrow molecular weight distribution polymers the value of a is $0.5 \leq a < 1$[10,48], but for rigid rod-like molecules it may be greater than 1. As the concentration of the polymer increases the intrinsic viscosity is no more relevant as a parameter and instead, the viscosity is correlated to the molecular weight of the polymer.

Table 4.3 Mark Howvink parameters for various polymers and solutions*

Polymer	Solvent	Tempera-ture, °C	$K / \left[\dfrac{cm^3}{gm} \times \left(\dfrac{gm}{mol} \right)^a \right]$	a
Atactic Polystyrene	Benzene	25	0.034	0.65
Atactic Polystyrene	Toluene	25	0.00848	0.748
Atactic polystyrene	Cyclohexane	25	0.108	0.479
Poly(isobutylene)	Benzene	25	0.083	0.53
Poly(isobutylene)	Toluene	25	0.87	0.56
Poly(isobutylene)	Cyclohexane	25	0.040	0.72
Atactic polypro-pylene	Benzene	25	0.027	0.71
Atactic polypro-pylene	Toluene	30	0.0218	0.725

Polymer	Solvent	Tempera-ture, °C	$K/\left[\dfrac{cm^3}{gm}\times\left(\dfrac{gm}{mol}\right)^a\right]$	a
Atactic polypro-pylene	Cyclohexane	25	0.016	0.80
Poly(vinyl alco-hol)[b]	Water	25	0.0002	0.76
PMMA #	Acetone	—	0.0078	0.70
PMMA #	Benzene	—	0.0052	0.76
PMMA #	Toluene	—	0.007	0.71
Polyvinyl acetate #	Acetone	—	0.0102	0.72
Polyvinyl acetate #	Benzene	—	0.0563	0.62
Polyvinyl acetate #	Acetonitrile	—	0.0415	0.725

* From J. Brandrup and E. H. Immergut, eds. "polymer Handbook, 2nd edition, Wiley Interscience, New york,1975.

b P. J. Flory and F. S. Leutner, J. Polym. Sci., 3, 880 (1948)

VALUE @Amrita (The temperature was not specified) "Determination of viscosity Avg. MW of Polymer" Source of data not mentioned.

The data given in Table 4.3 can be used to calculate molecular weight of polymers provided the intrinsic viscosity is available.

EXAMPLE 4.8. H. P. Schreiber, (J. Appl. Polym. Sci., 9, 2101(1965)) has reported the intrinsic viscosity values of polyethylene solution in tetralin at 120°C. For a given sample the intrinsic viscosity $[\eta] = 1.923$ dl/gm. Calculate the molecular weight of the sample using Equation 4.42. The constants of the equation for the system reported are $K = 5.24 \times 10^{-4}$ and $a = 0.71$.

Solution: Application of the Equation 4.42 using the given constants gives

$$[\eta] = 1.923 = 5.24 \times 10^{-4}\ \bar{M}_w^{0.71}$$

$$\therefore \qquad \bar{M}_w = \left(\frac{1.923}{5.24\times10^{-4}}\right)^{1/0.71} = 104,865.0$$

4.7.2 Analysis of Concentrated Solutions

As the concentration of polymers increases in the solution the polymer molecules come progressively closer to each other and the mutual interaction between them increases and hence the energy required for the flow of the solutions also increases. The scope of concentrated solution extends from a critical concentration below which equation 4.42 is applicable to very high concentrations encountered in the plasticization of the polymers. The chemical nature of the solvents also bears a considerable influence on the flow properties of these solutions. As a result of the combined effects of polymer and solvent molecular characteristics, molecular weight of polymers and the concentration of the polymers, the initial or the zero shear viscosity of concentrated solutions is a complex function of these parameters. An empirical equation similar to the equation 4.42 for initial viscosity has been proposed[10] for the concentrated solutions as

$$\eta_0 = K'c^\alpha M^\beta \qquad\qquad ...(4.43)$$

where K', α and β are constants depending on the system's characteristics.

The viscosity of the solutions of flexible polymer molecules, (i.e. those having non-polar or weakly polar nature) in poor solvents is higher at low concentration than that in good solvents, but at the higher concentration the trend reverses[10]. The polar and highly polar polymers molecules show increasing chain stiffness reaching to the rod-like behaviour and the nature of the solvent has a very strong influence on the viscosity of their solutions.

The material parameters α and β in general depend on the polymer concentration. For linear flexible-chain polymers α has been reported to vary in the range of 5 to 7 and, β to have a constant value of 3.5. For these polymers up to a reasonably high concentration both these parameters change in the similar fashion and to the similar extent giving the ratio β/α to remain fairly constant being approximately equal to the value of exponent a in the Mark-Houwink equation[36] (equation 4.42) within the range of variation of 10 to 15%. A value of $\beta/\alpha = 5/8$ has been frequently used for a number of polymers[41]. On the contrary for the stiff polymer chains the value of α increase rapidly with concentration and extent of stiffness of the chains reaching a value of up to 15-17 whereas the β is not affected. This clearly indicates that α represents the effect of physical interaction between the polymer-polymer molecules, which is likely to increase with increasing molecular number density and the chain stiffness. The constant β, which does not change with the concentration represents the solute-solvent and solvent-solvent interaction.

The variation of viscosity of solution with both molecular weight of polymer and its concentration show dual nature i.e. below a particular value known as the critical value of molecular weight and concentration the viscosity increases slowly and beyond this value it rises with a much faster rate. Fig. 4.6 gives the variation of viscosity of solutions of (a) polyvinyl acetate in methyl-ethyl ketone[50] versus molecular weight and (b) polyvinyl alcohol in water[51] versus concentration. It can be seen that the critical molecular weight is a function of concentration such that for increasing concentration the criticality occurs progressively at lower molecular weigh. Similarly the critical concentration reduces with increasing molecular weight.

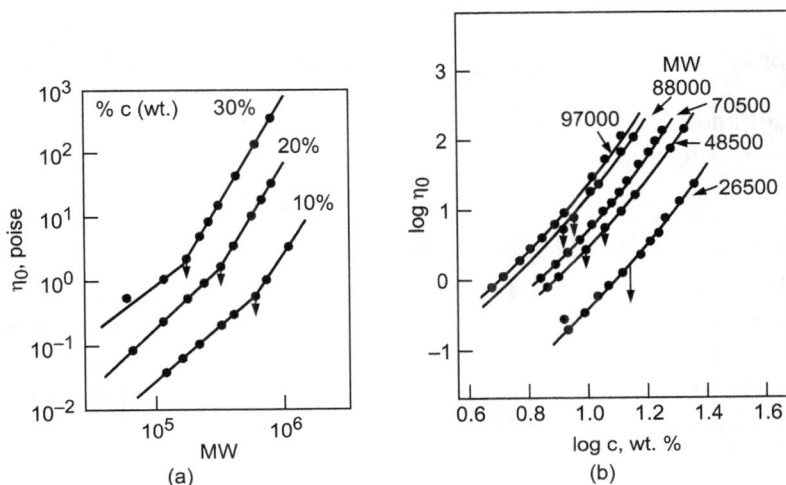

Fig. 4.6 Variation of solution viscosity versus (a) molecular weight[50] and (b) concentration[51]

4.8 EFFECTS OF MOLECULAR WEIGHT ON POLYMER MELT RHEOLOGY

The flow of polymer melts takes place at much higher temperature as against the solutions and the resistance to flow and the consequent pressure drop is the result of the viscous dissipation of energy due to the relative motion between the molecules. Increase in the molecular weight brings in the additional complications in the chains i.e. coiling and entanglement, which hinder the flow of molecules to a greater extent and require progressively higher energy for the flow. The first attempt to explain the variation of viscosity from the theoretical point of view was based on the statistical treatment of the flow system by Debye[52] and Bueche[53]. They developed the theory based on what is known as "Free-draining" molecule, as reported by Kumar and Gupta[54]. The relationship developed by them for the zero shear viscosity is as follows

$$\eta_0 = \frac{c}{6}\left(\frac{\xi_0}{M_s}\right)N_{av}\left(\frac{\langle s^2 \rangle_0}{\overline{M}_w}\right)\overline{M}_w \qquad ...(4.44)$$

where η_0 is the zero shear viscosity,

ξ_0 is the drag coefficient for the statistical segment

M_s is molecular weight of the repeat unit

N_{av} is Avogadro's number

$\overline{M}_w$ is the weight average molecular weight of polymer

$\langle s^2 \rangle_0$ is the mean square radius of gyration of the molecule

c is an empirical const.

For low molecular weight systems the quantities $\left(\dfrac{\xi_0}{M_s}\right)$ and $\left(\dfrac{\langle S^2\rangle_0}{\bar{M}_w}\right)$ are constant and therefore for such systems the viscosity varies linearly with molecular weight i.e.,

$$\eta_0 \propto \bar{M}_w \quad \text{or} \quad \eta_0 = K_1\bar{M}_w \qquad \qquad ...(4.45)$$

where K_1 is the proportionality constant.

This relationship is valid up to a molecular weight known as the critical molecular weight, beyond which the viscosity increases rapidly with molecular weight. Fox and coworkers[22, 55, 56] have modified the equation to account for the physical hydrodynamic interaction between the large molecules to show that the viscosity varies as 3.5th power to the molecular weight i.e.,

$$\eta_0 = K_2\bar{M}_w^{3.5} \qquad \qquad ...(4.46)$$

The constant K_2 depends on the molecular features like structure complexity, number and length of statistical segments, drag coefficient and the radius of gyration of the molecule etc. A log-log plot of viscosity versus molecular weight results in two straight lines, [Fox, Toosi et al., Allen and Fox, and Otter][22, 58-60] corresponding to equations 4.45 and 4.46 having slope equal to 1 and 3.5 respectively as shown in Figure 4.7. It can be seen that the viscosity variation passes through a transition at the critical molecular weight. Although the transition point is not a sharp one but it can be estimated experimentally at the intersection of the two extrapolated lines. For different polymer melts the onset of transition occurs at a molecular weight lower than the critical value and continues beyond it. As the molecular weight increases the chain length also increases and the molecular weight distribution becomes progressively loaded with large molecular weight chains. The long molecules tend to acquire coiled structure and also the average diameter of the coils increases with increase in the chain length. This results in progressive decrease in the intermolecular distance bringing the molecular segments very close to each other. The action of van der Waals forces gives rise to the Brownian motion, which causes the penetration of segments of one molecule in to the bulk of other and vice-versa resulting in the onset of the molecular entanglement at some molecular weight value lower than M_c. The critical point at which the polymer chains become large enough to entangle is called the molecular weight of entanglement M_c, which is the point where the physical properties of the polymer change. Most commonly the molecular weight of entanglement is determined by plotting the log of melt viscosity versus the log of molecular weight (Fig. 4.7). With increase in the molecular weight as the entanglement progressively increases the slope of the viscosity molecular weight curve becomes 3.5 or higher as reported by Tobolski et al., Bagley and West, and Berry[60-62]. The region of transition has been found to occupy the molecular weight range from $M_c/1.7$ to $1.7\,M_c$[63]. After this point the molecules deform in a totally entangled state. In fact during the flow both the processes of entanglement and disentanglement go on within the bulk of the polymer and at a particular molecular weight level, system attains equilibrium in entanglement.

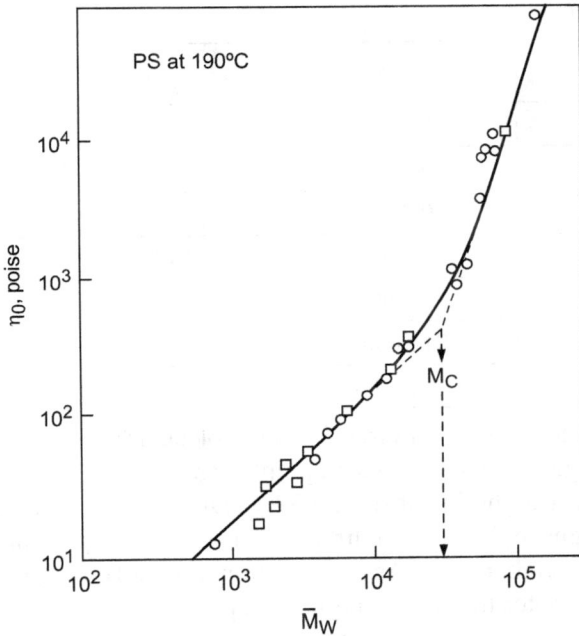

Fig. 4.7 Viscosity dependence on molecular weight[59]

The critical molecular weight is specific to each polymer and is controlled by the molecular characteristics of that polymer including the chain flexibility, pendent groups and intermolecular forces[10,54]. Table 4.2 gives the critical molecular weight values for some of the polymers.

Table 4.2 Critical molecular weight values for the polymers

Polymer melts	$M_c^{[54]}$	$M_c^{[10]}$
Polyethylene	4000	4000
Polystyrene	38000	35000
Polyvinylacetate	24000	22500
Polymethylmethacrylate	10500	27500
Polyisobutylene	17000	17000
Polybutadien	6000	5600
Nylon-6	5500	—
Polydimethyl siloxane	35000	29000

EXAMPLE 4.9. The data on the variation of zero shear viscosity with molecular weight of polybutadiene melt at $300°$ K has been published by J. T. Gruver and G. Kraus (J. Polym. Sci., 2, 797 (1964). From the data establish the relationship between η_0 and $\bar{M}_w$ and discuss the result in the light of reported relationship for other systems.

$\bar{M}_w \times 10^{-3}$	η_0, Pa s	$\bar{M}_w \times 10^{-3}$	η_0, Pa s
2.3	0.053	238.0	53040.0
3.9	0.105	278.0	99960.0
6.8	0.377	298.0	127500.0
9.7	0.980	334.0	185640.0
20.6	11.790	447.0	469200.0
61.0	714.000	524.0	775200.0
145.0	10,200.000	—	—

Solution: The log-log plot of viscosity and molecular weight (not shown) gives two distinct regions (i) first two data points give slope of 1.3 and (ii) rest of 11 points fall on a straight line of slope 3.39. Clearly the second region is the high molecular weight region and the first region seems to represent the end of the low molecular weight region. As the data points are not enough the relationship is developed only for the second region. It is

$$\eta_0 = 3.32 \times 10^{-14} \bar{M}_w^{3.39} \qquad \qquad ...(4.46(a))$$

The critical molecular weight, $M_c = 5623.0$. These results are in conformity with the results of other polymers at the high molecular weight region. It may be noted that the first three data points may be considered to be part of the region in which the molecular entanglement may be assumed to have grown to saturation.

Equation 4.12 gives the variation of viscosity with temperature and equation 4.46 gives that with molecular weight. The combination of these equations gives a combined correlation as

$$\log \eta_0 = 3.5 \log \bar{M}_w - \frac{B_1^g (T - T_g)}{B_2^g + T - T_g} + C \qquad ...(4.47)$$

where C is a constant to be evaluated for a particular system from the experimental data.

This equation gives a moderate fit with WLF universal constants. By using the experimentally determined constants a good fit of the data can be achieved as reported by Allen and Fox and coworkers and Ferry[16, 22, 58, 64].

4.8.1 The Mechanism of Entanglement

The entanglement as mentioned in the previous section is caused by the mutual interpenetration of the long chain polymer molecules. Graessley[65,66] while proposing his entanglement theory defined it as the process in which one of the segments of a particular molecule crosses the segments of other molecule and both these molecules undergo a coordinated motion in accordance with the applied rate of shear. The mechanism of entanglement can be understood by considering a system of polymer solution at steady state where, say any two long chain molecules are placed at a distance r^* and held there for a long period. After the lapse of certain time period these molecules will tend to interpenetrate each other's bulk without substantially changing their spatial configurations. But as the Brownian motions cause the molecular chains to continuously wreath and wriggle, the said two chains are bound to establish a coordinated motion of some sort between the segments, which have interpenetrated resulting in the formation of physical entanglement. Entanglement in concentrated random coil, flexible polymer molecule is considered as a network of bridges (Fig. 4.8(b)). As there are a large number of molecules present in the system it is likely that such a phenomenon will occur simultaneously in the bulk of solution. It is proposed that the rate of entanglement formation is very slow initially but rises sharply after some time interval and results in the system to attain equilibrium entanglement density. Figure. 4.8 shows this schematically. The rise in the rate of entanglement is so sharp that it may be considered as a step change at some time interval θ without introducing much error in the analysis. θ is a characteristic time required to attain equilibrium entanglement density. It depends on the polymer concentration for the solution and on the molecular structure for the melts. The characteristic time, θ, is given as

$$\theta = \frac{6\eta M}{\pi^2 cRT} \qquad ...(4.48)$$

where c is the concentration and M is the molecular weight of the polymer.

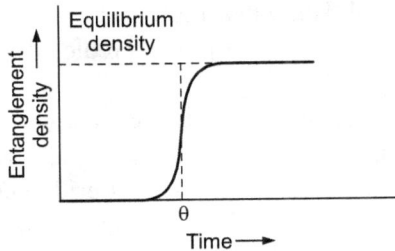

Fig. 4.8   Schematic of growth of entanglement[54]

In a flow system however different molecules move relative to each other with some velocity. During flow when a faster moving molecule approaches

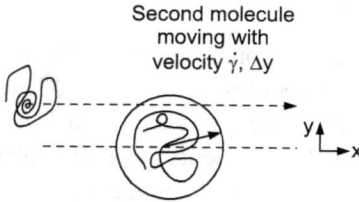

Fig. 4.8(a) Interaction between polymer
molecules in flow[54]

Fig. 4.8(b) The entangled state of polymer
chains or network of bridges
[Macromolecules, 26, 1564-1569 (1993)]

a relatively slower molecule and their spheres of influence happen to overlap each other there will occur a segmental interpenetration as shown in Fig. 4.8a. If there is no entanglement the molecules will rotate and dissipate viscous energy, however in presence of entanglement the molecules will tend to be tugged in different direction by each other in such a way that net force acting on the molecules will be in the direction of flow. Consider two molecules moving with velocities u and $[u + (du/dy) \times \Delta y)$ or $u + \dot{\gamma}\Delta y]$, where Δy is the distance between the centers of the two molecule masses. A sphere of influence may be considered to be present around each molecule and the entanglement is supposed to occur when these spheres of influence superimpose partially during the process of sliding past each other. The time of contact, θ_1 of faster molecule with the slower molecule will depend on the value of Δy and the relative velocity. If $\theta_1 < \theta$ the characteristic time then there will be no entanglement and if $\theta_1 \geq \theta$ then the entanglement formation will take place.

Graessley[65] successfully applied his entanglement theory to calculate the apparent viscosity as a function shear rate and an approximate estimation of normal-stress differences. It was shown that the viscosity did depend on rate of shear (as was established from experimental rheological measurements), and at sufficiently high shear rates it approached the variation as viscosity $\propto$ (shear rate)$^{-3/4}$. Also it was established that the onset of non-Newtonian behaviour occurs in the same region of shear rate as that predicted by coil distortion theory of steady flow.

Wool[66a] proposed the entanglement to be a network of bridges. A segment of polymer chain is considered to be long enough to form a loop on itself. On the basis of random-walk analysis the onset of bridge network is proposed to occur at critical entanglement MW, $M_c = 30.89\, C_0 M_0 / \alpha^2 j$, where C_0, M_0, j are the characteristic ratio, monomer molecular weight and the number of backbone bonds per monomer, respectively. The α is determined from the knowledge of C-axis dimension of the unit cell, bond length and number of monomers per C-axis length. The theoretical M_c expression, which contains no fitting parameters was found to be in excellent agreement with experimental M_c values.

The understanding of the molecular entanglement has been applied to explain some physical behaviours of polymers. Some of the observations are given below.

(a) Phenomenon like "coffee-ring" patterns of polymer droplets is reported by Biswas and Dutta[66b]. It has been reported that dried droplets of polymer solutions of different *MW* and concentration leave various types of coffee-ring patterns, which result from contact line motion. The patterns formed by very low *MW* systems contains spheroidal structure at the periphery whereas the central part contains continuous layer. For the high *MW* systems the contact line exibit the 'stick-slip' motion and very high *MW* polymers form a continuous layer. It has been explained in terms of chain entanglement, due to which, the chains can exhibit 'granular and collective' behaviour.

(b) Riggleman et al.[66c] have conceived an ideal polymer nanocomposite model in which the nanoparticles are dispersed throughout the polymer matrix. It has been explained that the improved properties of the polymer nanocomposites are due to the fact that the these composites are capable of considerably enhancing the entanglement network.

(c) It has been reported by Barzic et al.[66d] that the polymer flow behaviour, chain flexibility and entanglements are very important parameters to be monitored for establishing the optimal processing conditions from solution phase to film state for the purpose of liquid crystal orientation.

(d) The electrospinning process is used to manufacture nano-scale fibers of biopolymers like starch and pullulan. It has been reported by Kong and Ziegler[66e], that the molecular entanglement of constituent polymers in the spinning dope and the knowledge of their rheological behaviour are the essential prerequesites for successful electrospinning.

(e) The molecular entanglement plays an important roll in polyethylene film blowing process as reported by A. M. Sukhadia in his 1998 best paper, "The effect of molecular structure, rheology, morphology and orientation on polyethylene blown film properties".[66f] The polymer melt as it emerges from the die has some degree of molecular alignment or orientation along the machine direction (MD). The bubble stalk provides a region of stress relaxation prior to additional biaxial deformation during the bubble inflation stage. The degree to which the molecular relaxation can occur in the stalk region is likely to be influenced considerably by the degree of molecular entanglement present. A melt with higher molecular entanglements will exhibit longer relaxation time or in other words a slower relaxing melt. This results in greater degree of molecular alignment and orientation being preserved during subsequent cooling and non-isothermal crystallization under biaxial stress beyond the bubble neck region.

A theoretical mathematical analysis of the entanglement formation has been reported by Alison MacArthur[66g.] using the knots theory where the polymer entanglements are assumed to be knots like structures, which interact with each other during the flow. In this approach the author reported that it was possible to compute topological entropy of one-ring melt. The topology of entanglement has been related to the probability of one-ring melt not being in the state of being knotted or being un-knotted. Such a "unknot probability' will be unity for short rings and approach zero for long rings. The change of probability from zero to some value higher than zero is expected to occur at some characteristic length of polymer chain. Further it was possible to explore how far it was possible for a chain to wiggle before it cuts through another chain and "change the knot" made by all other chains in the system. The availability of the space to wiggle depends on the extent of entanglement of the chains. The chain stiffness, flexibility and its bulk all play a roll in probability of wiggle of rings. Figure 4.9 shows three different knot configurations, which are (i) a trefoil knot, (ii) an overhead knot and (iii) a molecular knot with eight crossings (an ensemble with 192 atoms coiled around a triple loop). The trefoil and overhead knots are most probable spacial configurations in polymer melts. The knot configuration with eight crossings is envisaged to user in super strong material in future.

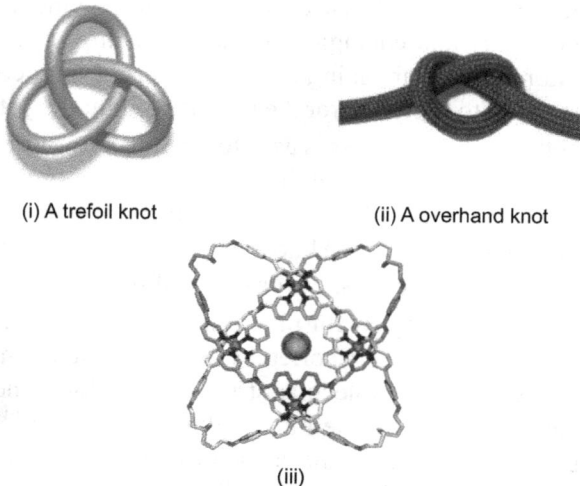

(i) A trefoil knot   (ii) A overhand knot

(iii)

Figs. 4.9   Different types of knots design to show the entanglements
[ZME SCIENCE, JAN. 2017]

The brief mention of knots above was felt necessary to get the idea of spacial configuration of molecules in the melt as the spacial configuration of polymer molecules has a direct bearing on the volume occupied by them and thus has direct consequence on the polymer rheology and their flow behaviour.

4.8.2   Effect of Shear Rate on the Viscosity-Molecular Weight Relationship

The effect of rate of shear on the viscosity of the polymer melts has been dealt in detail in Chapters 2 and 3 for different types of non-Newtonian fluids. However the viscosity-molecular weight relationship particularly in the entangled region is considerably influenced by the variation of rate of shear. For a constant molecular weight the shear thinning effect starts at a particular lower rate of shear (i.e. the end of the first Newtonian region) and continues up to a certain higher value (i.e. the onset of second Newtonian region). With increasing MW of the polymer the onset of shear thinning nature occurs progressively at lower rate of shear. Figure 4.10 gives the viscosity-shear rate plot for polydimethyl siloxane melts[67] at different molecular weights. At high M_w of 19×10^4 the onset of shear thinning appears at around $1.0\ s^{-1}$ and continues up to about $10^6\ s^{-1}$ whereas for lower M_w of 2.1×10^4 these transitions occur respectively at $10^4\ s^{-1}$ and $10^6\ s^{-1}$. Samples of very low M_w (below 0.54×10^4) do not show any shear thinning behavior. Thus the low MW polymers show the first Newtonian region over a large rate of shear range and this range shrinks as the MW increases. In this region the viscosity is a linear function of MW up to the critical MW, M_c. However, beyond M_c the slope of the η_0 vs. M_w line progressively falls from 3.5 corresponding to zero shear rate to 1 at high rate of shear as shown in the Fig. 4.11[68].

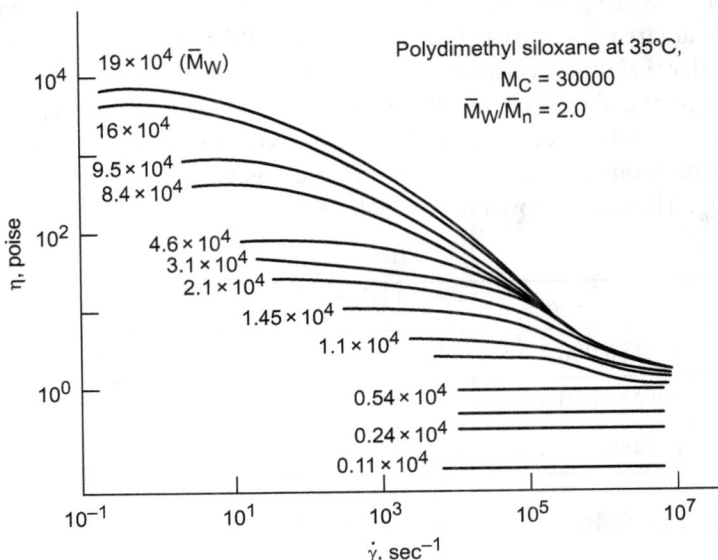

Fig. 4.10   Viscosity vs rate of shear plots at different MW[69].

The flow behaviour in the low molecular weight region is essentially Newtonian and therefore the viscosity remains unchanged but in the high MW region where the entanglement dominates, an increase in the rate of shear will cause uncoiling, disentanglement and alignment of the molecules thereby giving the shear thinning effect and reducing the viscosity. At high rate of shear the flow behaviour again changes to the Newtonian behaviour due to complete orientation of the molecular chains giving constant viscosity and following linear relation with the *MW*.

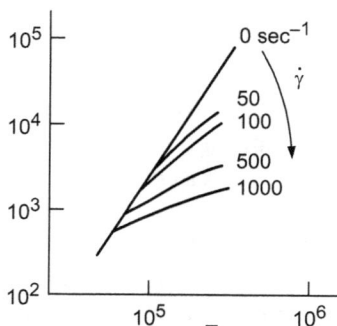

Fig. 4.11   Viscosity vs MW for PS at 227°C at different rate shear[70].

Similar investigations for polypropylene melts of different molecular weight [(i) HG265FB *MW*=180,000, (ii) JHG365FB *MW* = 230,000 and (iii) HB600TF *MW* = 460,000, used as received] has been reported recently by Njoroge[68a]. The experimental data after Bagley correction for shear stress and rate of shear data after correcting for Weissenberg-Robinowitch correction has been used to plot shear stress vs rate of shear for different *PP* samples to show the effect of molecular weight on viscosity. The results show that *PP* melt is also a shear thining type and the viscosity increases with *MW* as for other systems. The experimental results were fitted to linear approximation to give $y = mx + c$ type empirical relations for all *PP* systems, where y is viscosity η and x is true rate of shear $\dot{\gamma}_{tr}$. These are shown in Table 4.3 below

Table 4.3

PP Type	*Model*
PP 180,000 MW	$y = -0.39x + 2.8$
PP 230,000 MW	$y = -0.58x + 3.6$
PP 460,000 MW	$y = -0.70x + 4.2$

4.8.3 Effect of Molecular Branching[69, 70]

The branching in the polymer molecules is the consequence of replacement of a substituent e.g. hydrogen atoms on a monomer subunit, by another covalently bonded chain of that polymer or in the case of graft copolymer by the chain of another type. IUPAC definition of branched chain is "A chain with atleast one branch point intermediate between the boundary units." Figs. 4.12(*a* and *b*) show schematics of branching point and various random branching architectures in polymer molecules. Gotsis and Zeevenhoven[70b] have investigated the effect of branching on the viscosity of the *PP* by introducing the long chain branching on linear *PP* precursors in varying amount using reactive modification by peroxydicarbonates. The branched samples show distinct strain hardening, which was absent in the linear polymer melt. The shear viscosity at high shear rates is not affected much by the modifications. The zero-shear viscosity, however, increases considerably with the addition of a few branches, and the increase may be related to the degree of branching.

Many polymers like polyethylene, polystyrenes, polyvinyl acetate, polyisoprene, polyisobutadien, etc. have been reported to have side branches in the main chain. An example of (a) polybutadiene molecular structure is shown to give an idea of the complexity of (a) polybutadiene polymer molecule and (b) general molecular structure of polysiloxane

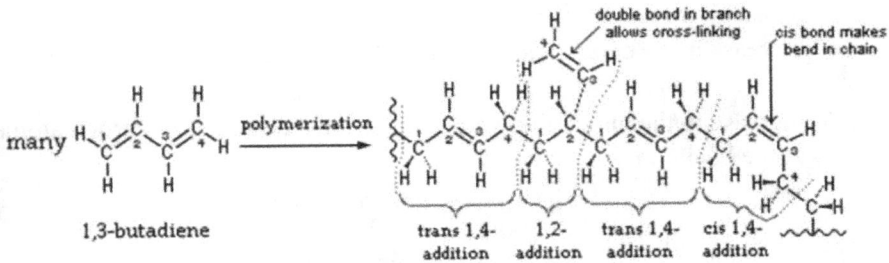

(a) Polybutadiene molecular structure
[https://en.wikipedia.org/wiki/Polybutadiene]

Structure of polysiloxane is also shown for the better perception of the complex molecular structure in polymer molecules and also to have an idea about degree of polymerization (*DP*), branch content (*Q*), branch length (in this case, alkyl length (*L*) and pendant group (*J*)[70a]. All the quantities mentioned above will change from one polymer molecules to another.

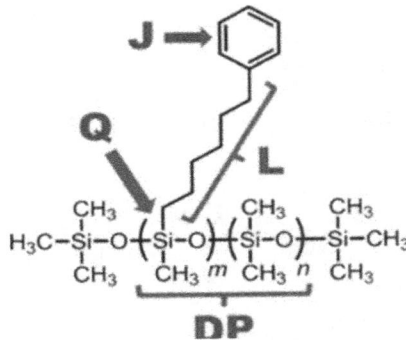

(b) Polysiloxane

The number and the length of the side branches have a considerable influence on the rheological behaviour of the polymers. In many cases the branching is the consequence of side reaction during polymerization due to variations in the reaction condition, quality of the catalyst and the presence of impurities. However these may be introduced deliberately by co-polymerization with a multifunctional monomer to achieve certain properties. Polymer molecules can have different types of branching like, block, star, comb, brush, dumbbell, star etc. The presence of the branching has the following effects on the molecular structure

(i) Increase in the free volume of the macromolecule,

(ii) Broadening of molecular weight distribution: it gives improved processing and lower die swell,

(iii) Radius of gyration is decreased compared to that of a linear chain of the same molecular weight therefore gives lower viscosity,

(iv) Increase in the entanglement density for long branching therefore gives high viscosity.

(v) Increase in the molecular weight for the same number of carbon atoms in the backbone.

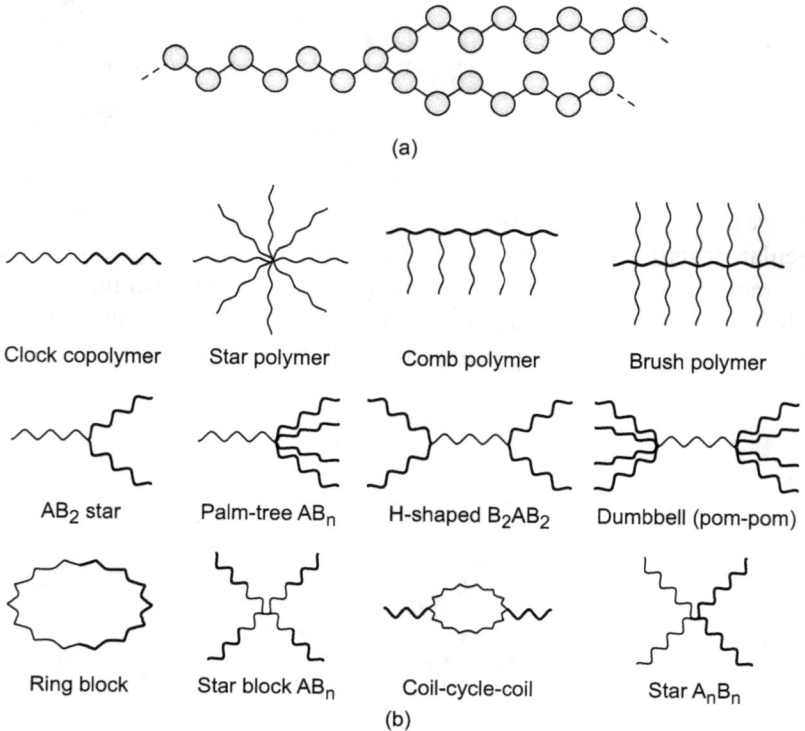

(a)

| Clock copolymer | Star polymer | Comb polymer | Brush polymer |

AB_2 star   Palm-tree AB_n H-shaped B_2AB_2 Dumbbell (pom-pom)

Ring block   Star block AB_n Coil-cycle-coil   Star A_nB_n

(b)

Fig. 4.12 Random Branching in Polymer Chains, giving various chain architectures. (a) Branch point in a polymer (b) Various architectures, [https://en.wikipedia.org/wiki/Polymer_architectur][70a].

HDPE (linear)   LDPE (branched)

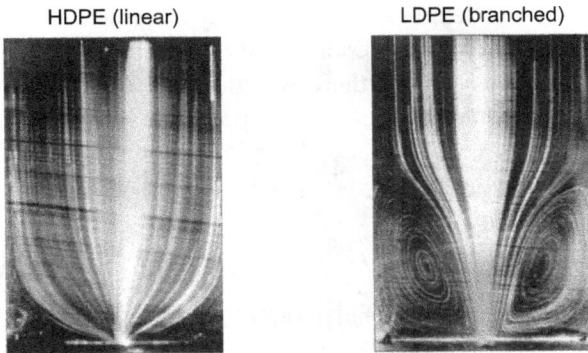

Fig. 4.13 Effect of molecular branching in the flow patterns and die swell of LDPE and HDPE, [Albena Lederer, Leibnitz Institute of Polymer Research, Dresden, 70c].

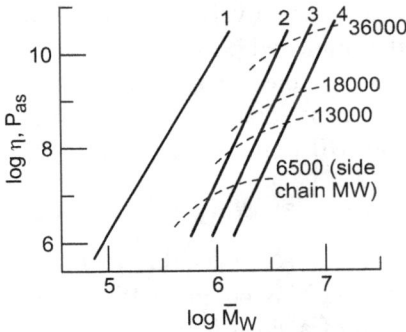

Fig. 4.13(a) Schematic of viscosity variation with MW of branched narrow MWD polystyrene. (1) Linear, (2-4) number of side chains: 2-10, 3-20, 4-40.[10] [71 (line-1), 72]

As a consequence of these changes in the microstructure the rheological response of the polymer changes considerably in comparison to the un-branched linear polymer. Increase in the free volume is due to the higher inter-chain distance in the polymer and it results in different flow patterns and decrease in the shear viscosity as compared to the linear polymer (Fig. 4.13 and 4.13(a)).

Length of the side chain is a critical parameter, which determines the extent of variation in the rheological properties. Initially at smaller branch lengths the branches have little or no flexibility and only help in increasing the inter-chain distance, increase the free volume and therefore reduce the viscosity. At still higher length but smaller than corresponding to the critical molecular weight of the linear chain the branched segments tend to make the spacial configuration more compact and reduce the radius of gyration therefore reduce the viscosity. But at much longer branch lengths the branches add to the overall entanglement density and the viscosity increases with a much faster rate than that for a linear chain.

The magnitude of the viscosity change varies with the branch chain length i.e. at smaller branching the reduction in viscosity may vary even by a factor of ten and at higher branch length the viscosity may be orders of magnitude higher than that of the linear chain as reported by Schreiber et al.,[73], Graessley[74] and Kraus and Gruver[75]. Thus there will be cross over point in the plot of viscosity versus molecular weight. The slope of the line at higher MW levels may be as high as 4.5.

A conformational parameter, $g^{1/2}$ has been defined to help to understand the variation of viscosity as

$$g^{1/2} = \langle S \rangle_{br} / \langle S \rangle_{lin} \qquad \text{...(4.49)}$$

and the square of the conformational parameter is known as branching index, g.

i.e.
$$g = \langle S^2 \rangle_{br} / \langle S^2 \rangle_{lin} \qquad \text{...(4.49(a))}$$

where $\langle S \rangle_{br}$ and $\langle S \rangle_{lin}$ are the root mean square radii of gyration of branched and linear macromolecules respectively[76,77]. In the region below critical MW radius of gyration of the branched polymer is smaller than that of the linear polymer (as the branched molecules are more compact as compared to the linear molecules) hence the conformational parameter is reduced thereby giving the lower viscosity. In the high MW region the entanglement is the controlling factor.

4.8.4 Effect of the Molecular Weight Distribution (MWD)[78,79]

The polymers have a distribution of molecular weight i.e. to say that they contain different numbers of varying MW chain lengths of polymer. Thus the MW of the polymer is some form of a statistical average. The number average MW, $\bar{M}_n$, and the weight average MW, $\bar{M}_w$ are the most commonly used MW averages. The IUPAC definition of polydispersity is $D_M = \bar{M}_W / \bar{M}_N$ [80]. MWD is generally expressed as, polydispersity index, PDI i.e.

$$D_M = PDI = \frac{\bar{M}_w}{\bar{M}_n} \qquad \text{...(4.50)}$$

The polymers have been broadly classified as (i) narrow MWD for $PDI <$ 1.8, (ii) medium MWD for $3.5 > PDI > 1.8$ and (iii) broad MWD with $PDI > 3.5$. It is a very arbitrary classification but helps in better understanding the polymer rheological and processing behaviour. A narrow MWD is characterized by the presence of similar chain length molecule in a very narrow range; medium distribution is likely to contain both high MW segments and low MW segments in an even distribution and broad MWD is dominated by presence of molecules of various lengths. A higher value of PDI contains more of high MW polymer chains where as a smaller value has more of low MW chains.

The polydispersity index increases with increase in the number of high MW species and always has a value > 1. A value of 1 refers to monodispersed polymers containing all molecules of the same chain length. Increase in PDI value considerably changes both the rheological and processing behaviours of the polymers. From the previous discussion on the effect of *MW* on the viscosity it is clear that as the *MW* increases the viscosity also increases, and as *PDI* value increases with increasing *MW*, viscosity must increase with increasing *PDI*. It has been reported that different polymer melt samples having similar $\bar{M}_w$ but different *PDI* show different viscosity. There is no single general formula available, which can describe the effect of *MWD* on the viscosity and therefore the discussion must remain only qualitative in nature. The polymers having different *MWD* can be tailor made by melt mixing two or more polymers having different *PDI* values. The rheological behaviour of the blend may be different than that of the base compounds. Ito and Shishido[81] reported the variation of polymer viscosity having different *PDI* as shown in Fig. 4.14.

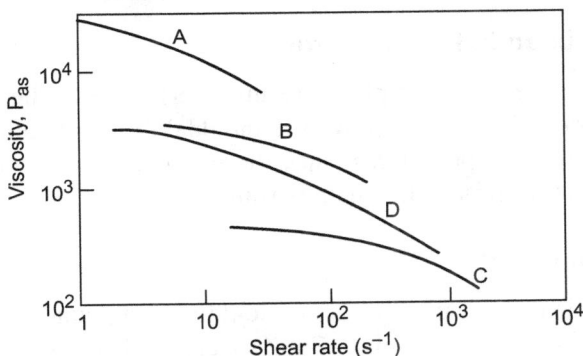

Fig. 4.14 Variation of viscosity of polymers having different PDI values A, B and C (pure polymers having PDI 2.0), D is the blend of A and C and has broad MWD, [Ito and Shishido, 81].

4.8.5 Effect of Fillers

In a number of applications solid particulate fillers are essential constituents to be used in the filled polymer composites. Particulate fillers, from the point of view of their size are of two types (a) Micro-fillers and (b) nano-fillers. The micro-fillers have been used for a long time to improve the mechanical properties and other properties of the composites. The nano-fillers find increasing applications due to a high improvement in the properties with small filler loading. The blends of micro and nanofillers also have been used to study the effects in enhancement of properties[82].

Basically three types of fillers are used: extender fillers being used to reduce the material cost and reinforcing and conductive fillers, which are used to improve mechanical properties, and thermal and electrical conductivity of the composites.

4.8.6 Fillers used in Rubber Compounds

Almost all virgin rubber compounds (99%) utilize either carbon black or inorganic fillers, like clay, ground or precipitated calcium carbonate, titanium dioxide, zinc oxide and several others as reinforcing agents and/or extenders for rubber compounds. Fillers like precipitated or fumed silica, have 0.02 micron average particle size and 150-175 square meter surface area per gram. Silica is added to improve modulus and tear strength of tire compounds.

The fillers in general fulfil a number of functions like colouring, cost reduction of the products (non-reinforcing fillers or extenders particularly in rubbers), strengthening of matrix (reinforcing fillers like carbon blacks and silica in rubbers), cross-linking (vulcanizing agents as sulphur) and many others used in small doses for the protection of the polymer matrixes like antioxidants, antiozonants, accelerators etc.

4.8.7 Fillers used in Plastic Compounds

Apart from extenders used in plastic resins to bring down the cost, properly selected mineral fillers can improve the mouldability and stability, increase the heat deflection temperature, reduce thermal expansion and impart other performance characteristics like flame-retardency.

4.8.8 Mineral Fillers

Mineral fillers commonly used in plastic moulding compounds include calcium carbonate, talc, silica, wollastonite ($CaSiO_3$), clay, calcium sulfate fibers (also known as Franklin fibre), mica, glass beads, and alumina trihydrate (to impart flame retardancy). Clay fillers improve electrical properties and processability of most resins. Mica in the form of small brown flakes boosts mechanical properties. Calcium sulfate fibres and wollastonite are bothersome fibrous fillers as these are needle-like materials and either marginally improve the mechanical properties or even reduce them. Other mineral fillers like glass beads, silica, calcium carbonate and talc generally weaken plastics. [machinedesign.com].

In polymers other than elastomers, many systems like ZnO-polypropylene[83], nano-calcium carbonet-polystyrene[84], nano-clay-filled glass fibre epoxy composites[85], nano-clay filled epoxy systems[86], graphite filled epoxy[87], ceramic particle filled glass/epoxy composites[88], Al_2O_3-filled glass fibre reinforced epoxy composites[89] have been reported. Many other fillers like rice husk flour[90], oil-palm ash[91], rice bran[92], cocoa pod[93] etc. have also been reported in literature. The list of the fillers-polymer system is exhaustive and some of these will be reported in chapter-7 on the rheology of filled polymer composites.

The solid fillers are dispersed in the continuous polymer matrix and are suspended in a high viscosity liquids. The rheological behaviour of such suspensions is totally different than that of the dilute suspensions and is influenced by a variety of factors relating to fillers like the loading, particle size, surface characteristics i.e. active or inert, shape factor and structure, surface area, the polymer filler interaction and the rheological behaviour of the polymer matrix. The rheological behaviour of such systems can be dealt with under following categories

(i) Suspensions in low viscosity Newtonian fluids,
(ii) Suspensions of inert solids in high viscosity fluids and polymer melts,
(iii) Suspensions of interactive or reinforcing fillers.

These can be further subdivided as dilute and concentrated suspensions.

Under the first category we have suspensions of minerals in water, fine coal powder in petroleum oil, suspensions of catalysts in reaction medium, suspensions of fillers and other ingredients in rubber latex, suspensions of reactants in monomers for reaction injection moulding systems, polymeric paints suspensions etc.

The second category of suspensions, include suspensions in glycerine, sugar syrups, liquid rubbers and polymers etc. In the third category are the blends of carbon blacks and silica in different rubber matrixes. A number of parameters influence the rheology of such suspensions and it is not possible to develop any theoretical correlation to describe their behaviour. As such a large number of empirical correlations have been developed to deal with different systems and specific variables.

The theoretical analysis carried out by Einstein for the rheology of very dilute colloidal suspensions of spherical, non-interacting particles in laminar flow condition forms the basis of numerous empirical correlations developed in this area. The Einstein's[45,46] formula is of the form

$$\eta_s = \eta_0(1 + 2.5\,\phi) \qquad \qquad ...(\text{Ref. eq. } 4.32)$$

where η_s, η_0 and ϕ are the suspension viscosity, medium viscosity and volume fraction of solids respectively.

The above assumptions imply that (i) there should be no hydrodynamic interaction between solid particles i.e. the inter particle distance in the suspension must be large and hence the dilute suspensions and (ii) no physico-chemical interaction between solid particles and medium i.e. to say that the particles must be absolutely inert. The equation 4.32 implies that the intrinsic viscosity of suspension

$$[\eta] = \frac{(\eta_s - \eta_0)}{\eta_0\phi} = \underset{\phi \to 0}{\text{Lim}}\ \eta_{inh} = 2.5 \qquad \qquad ...(4.52)$$

where η_{inh} is the inherent viscosity of the suspensions.

The Einstein's equation does not give any idea about different rheological behaviours the polymer melts and solutions may show as a consequence of deformable non-spherical nature of the molecules. However the theoretical mathematical analysis presented by Jeffery[94] and others[95] [96] for the suspensions of ellipsoidal particles gave insight to additional rheological phenomena like (i) dependence of viscosity on shear rate, (ii) development of normal stresses in shear flow and (iii) dependence of viscosity on the nature of flow i.e. shear or extensional flow. However similar conclusions can be drawn from the analysis of shearing flow of polymer melts, which is associated with the changes in spatial configurations of the molecules. In the first analysis the rotation of the ellipsoids and corresponding rotational diffusion coefficient and the rate of shear are important factors and in the second case the conformational entropy and deformational energy are important factors. Detailed discussion of these approaches is beyond the scope of this book.

4.8.9 Concentrated Suspension of Inert Fillers

The presence of solid particles in high concentration in a liquid causes the rheological behaviour to be entirely different than that corresponding to the Einstein equation. The effects of particle size, size distribution, shape factor, shear thinning and thickening, Bingham plasticity, shear rate dependence, time dependency and viscoelasticity are the prominent features appearing with increase in the concentration of solids in the suspension. It is practically impossible to visualize all these effects and correlate them in a single correlation. However attempts have been made to represent the effect of concentration of inert spherical particles by modifying the argument of concentration dependence in the Einstein equation to accommodate the maximum possible filler loading as[10]

$$\eta_s = \eta_0(1 + \Phi) = \eta_0 \left[1 + \frac{2.5\phi}{1 - (\phi/\phi^*)} \right] \qquad ...(4.52a)$$

where ϕ^* is a constant and Φ is the argument of the Eq. 4.52 which is given as

$$\Phi = \frac{2.5\phi}{1 - (\phi/\phi^*)} \qquad ...(4.53)$$

At low concentration as $\phi \to 0$ then $\Phi = 2.5\phi$, which is same as the argument in the Einstein equation whereas at high filler loadings corresponding to a limiting value of $\phi \to \phi^*$, the $\Phi \to \infty$. This implies that the maximum filler loading cannot exceed ϕ^* and the suspension will no longer be able to flow beyond this concentration.

Equation 4.52 is of limited application and quite a few of its modified forms are available in the literature. Out of these the Eiler's and Mooney's modification are of importance and are given below

Eiler's equation is

$$\eta_s = \eta_0 \left(1 + \frac{\Phi}{2} \right)^2 = \eta_0 \left[1 + \frac{1.25\phi}{1 - (\phi/\phi^*)} \right]^2 \qquad ...(4.54)$$

and Mooney's equation is

$$\eta_s = \eta_0 \exp(1 + \Phi)$$

$$= \eta_0 \exp\left[1 + \frac{2.5\phi}{1 - (\phi/\phi^*)}\right] \qquad ...(4.55)$$

Both these equations cover the wide range of concentration of solids in suspensions[97]. However it must be kept in mind that these equations refer to the mono-dispersed spherical particles. To take care of the effect of polydispersity, particle size and shape an equation similar to equation 4.54 is used as a second-degree polynomial as

$$\eta_s = \eta_0(1 + 2.5\phi_{eff} + K\phi_{eff}^2) \qquad ...(4.56)$$

where ϕ_{eff} is the effective volume fraction of solids, K is the material constant. The concept of effective volume fraction comes from the fact that when the particles are suspended in a fluid, a liquid layer known as the lyosphere having a specific thickness surrounds each particle. When the particle rotates (particularly in low viscosity liquids) during flow there is always a certain thickness of this layer, which strictly adheres to the particle and is therefore immobile. This layer neither deforms nor is available for the fluidity of the suspension and therefore effective volume of the particle is equal to its actual volume + the volume of the liquid in the lyosphere (Fig. 4.15)[98]. The lyosphere thickness depends on the wettability of the liquid and the flow Reynolds number. If thickness of the lyosphere is δ then the effective volume of the particle (spherical) will be

$$v_{eff} = \frac{\pi}{6}(d + 2\delta)^3 \approx \frac{\pi d^3}{6}\left(1 + \frac{6\delta}{d}\right)$$

$$= v_{actl}\left(1 + \frac{6\delta}{d}\right) \qquad ...(4.57)$$

The effective volume fraction ϕ_{eff} can then be calculated if the lyosphere thickness is known.

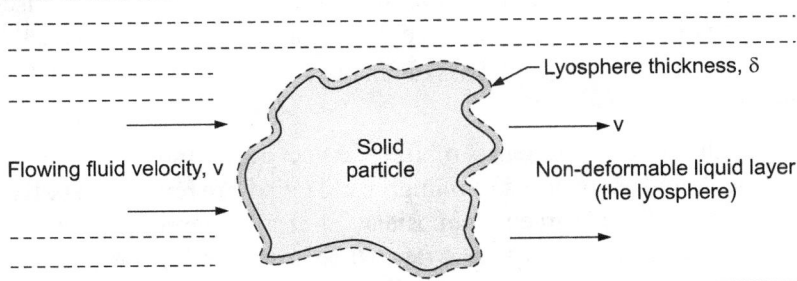

Fig. 4.15   Lyosphere around the filler particles and its thickness

4.8.10 Concentrated Suspensions of Active Fillers[99, 100]

Presence of different free radicals on the surface of the fillers makes them capable of chemically interacting with the polymer melts. The best examples of the active fillers are the carbon blacks and silica fillers used in rubber industry to enhance the strength of the green compound as well as the rubber vulcanizates. This is achieved because of the chemical interaction between the filler particle and the rubber molecules.

The polymer melts also form the lyosphere around the filler particles and the thickness of the layer consists of two parts, (i) chemically bound or the grafted layer[101] and (ii) physically bound or adsorbed layer or the occluded rubber[102, 103, 103a] and as such the thickness of the lyosphere is invariably higher than that on the inert particles. The equation 4.56 can be used to correlate the viscosity of these systems also but the value of K will be much higher to account for the additional energy required for the deformation of the system in which additional fluid has been made immobile due to its being bound on the surface of the filler. For natural rubber carbon black compounds, the value of K has been reported to be equal to 14.1[104] as compared to its value of 1.56 for the inert fillers, which is almost 10 times as high.

4.8.11 Fiber Filled Polymers Melts[105-109]

The short fibers are extensively used as fillers in the polymer and rubber melts to enhance the mechanical properties of the matrix and also to develop the electrically conductive polymers by using the carbon fibres. The rheological behaviour of these compounds is no different from those filled with other particulate fillers but the additional complications arise due to specific nature of the fibers like high l/d ratio, its poor dispersion, agglomerate formation and stretching of the fibers. Increasing the rate of shear can increase the extent of orientation of the fibers and it has been found that the viscosity sometimes decreases with the fiber loading essentially due to the fiber orientation particularly at higher rate of shear. Figure 4.16 gives the variation of viscosity with loading of short carbon fibers in a thermoplastic elastomer blends[104].

Other investigation on the effects of fibre loading on the rheology of different polymer systems are;

(a) Evaluation of influence of fibre aspect ratio and fibre content on rheologial characteristics of high yield pulp fibre reinforced polyamide 11 'HYP/PA11"green composites[105]. It was reported that identically for fiber content and aspect ratio, the shearing effect decreased as the temperature increased. It implies that this system shows increased pseudoplastic behaviour at higher temperature region.

(b) Effect of fibre length distribution on the steady shear viscosity of semiconcentrated polymer-fibre suspensions has been reported by

Huq and Azaiez[106]. A new model based on averaging approach for milled glass fibre having a wide distribution as well as both uniform and well defined aspect ratios in polyethylene oxide polymer solution has been proposed. It compared the performance of the new method with those of conventional averaging methods, which did not give a good representation of effect of fiber length distribution.

(c) Effect of fibre length orientation and strain rate on mechanical behaviour, and the strength of glass/epoxy composites properties has been reported by Tarfaoui et al.[107]. Seven different orientation of fibres have been studied in thickness direction using split Hopkinson pressure bar (SHPB) for dynamic characterization.

(d) Effect of wool fibres, ammonium polyphosphate and polymer viscosity on flamability and mechanical performance of PP/Wool composites has been reported by Kim et al.[108]. The cone calorimeter and vertical burn tests showed a significant decrease in the heat release rate and direct flame self-extinguishig behavior of the composites.

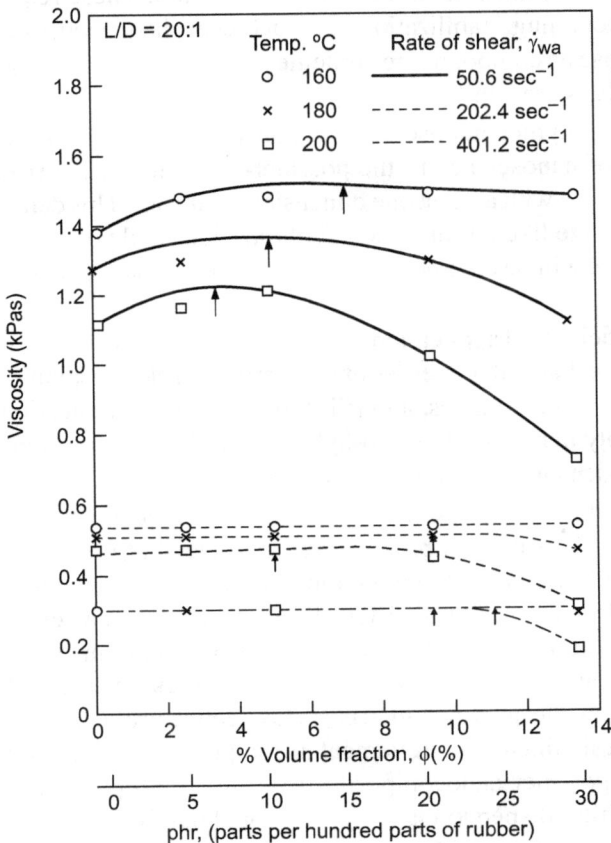

Fig. 4.16 Variation of viscosity with fiber loading for short carbonfiber filled S-I-S system[104]

Other investigators like Roy et al.[109], Murthy et al.[109a], Kitono et al.[109b], Nanguneri et al.[109c] and Gupta et al.[109d] also have reported the similar effects of the use of fibres as fillers in polymers matrix.

4.8.12 Nanoparticle Filled Polymers

The use of nanoparticles has found wide applications as polymer fillers due to the enhancement of mechanical, electrical and other properties of the composites as compared to the micro-fillers at low concentration. The reason for the improvement in properties is due to high surface area of nonoparticles, small particle size and other properties of the nanoscale fillers.

The nanoparticles are used in almost about 95% of the plastics production as reinforcing fillers. Polymer nanocomposites (PNC's) consists of a polymer or copolymers having nanoparticles or nanofillers dispersed in their martrix. E. Manias[110] has pointed out that these composites are better described as "**nanofilled polymer composites**". These fillers may be of different shape but one dimension must be in the range of 1-50 nm. These require controlled mixing/compounding, stabilization of the achieved dispersion, orientation of the dispersed phase as compounding strategies for all multiphase systems (MPS's) including PNC's are similar.

Study of polymer nanoscience is concerned with the investigations of applications of nanoscience to the polymer-nanoparticle matrices. The nano-particles are those with at least one dimension < 100 nm. This definition includes a very wide range like porous media, colloids, gels and polymer. However the discussion here will be confined to polymer-nanocomposites and its rheological behaviour.

Nanoparticles find applications in (i) areas like polymer-filler composites to improve the mechanical properties of the matrix, (ii) electrospinning, production of nanofibres, and nanotubes, and (iii) bioengineering applications to produce bio-hybrid polymer nanotubes, bio-hybrid nanofibres, tissue engineering, drug delivery, bio-sensors applications, and many more.

There are numerous types of nanoparticles, in fact any material, which can be reduced to nanosize can produce nanoparticles but only those which find industrial applications in above mentioned areas are of practical importance, which include some mineral fillers like silica, alumina, ceramics, metals like copper, aluminium, molebdenum, tungston, titanium etc. metal oxides, disulfides, graphene, carbon nanotubes, nano-spheres etc. The addition of these nanoparticles in the polymer matrix at low concentrations (~0.2 weight %) brings about a significant improvement in compressive and flexural mechanical properties of polymer nanocomposites and yield high performance materials provided the filler dispersion is uniform throughout the matrix[111, 112].

It should be noted that the addition of nanoparticles perturb the chain dimensions and expand its dimensions due to increase in the radius of gyration of the molecules with increasing concentration of the fillers as reported by Frischknecht et. al.[112a]. The nanoparticles are assumed to be smaller than the polymer radius of gyration and are attracted to the polymer so that they are miscible. These composites have high potential for application as a novel, mechanically strong, light weight composites as bone implants. Along with uniformity of dispersion, the nanostructure morphology and cross linking density of the polymer matrix as well as the defects therein are the equally important considerations.

4.8.13 Steady-Shear Rheology of Nanofilled Polymers

Rheological investigations of many polymer layered-silicate nanocomposites including intercalated, exfoliated and end-tethered exfoliated (prepared by in situ polymerization from reactive groups tethered to silicate surface) have been carried out using different experimental modes like steady shear and oscillatory using melt state rheometers as reported by Giannelis et al.[112b]. It has been reported that the viscosity of nanocomposites increases considerably with the addition of nanoparticles and keeps on increasing with increasing loading. Pure polymer shows Newtonian behaviour and low loading of layered silicate gives marginally pseudoplastic nature whereas highly filled systems give strong shear thinning nature. However both pure polymer and low loaded samples give highly shear-thinning characteristics beyond $\sim 70\,s^{-1}$.

The variation of viscosity for nanocomposites of polydimethyl siloxane with dimethyl ditallow montmorillonite at 25°C with varying filler loading has been reported. All the systems ie. pure polymer, 6 wt% silicate and 13 wt.% silicate composites show Newtonian behaviour at low shear rates and then pseaudoplastic nature beyond $50\ s^{-1}$ when the viscosity reduces considerably.

Amongst the different nanomaterial the inorganic materials give better reinforcement as compared to carbon based nanoparticles. The use of polymer nanocomposites results in the developments of high quality end products for a wide range of high value added products. These products find applications in the areas of energy conversion and storage, sensing and biomedical tissue engineering etc.[113]. The multi-walled carbon nanotube (MWCNT) nanocomposits have been used to enhance the electrical conductivity[114]. The surface modification of MWCNT using mixed oxidation and CF_4 plasma treatment gave improved mechanical as well as electrical properties of these nano tubes[115]. Method of surface treatment has been reported to have strong influence on electrical properties.

4.8.14 Effect of Plasticizers[75]

Plasticizers or dispersant are the low molecular weight liquid additives, which increase the plasticity and reduce the viscosity of polymer filler composites.

These are used for improving the processability of the polymers and rubbers particularly the filled compounds or the plastics like PVC. The plasticizer most commonly used in PVC applications are phthalate esters i.e. (Bis (2-ethylhexile) phthalate. The chemical structure of phthalate plasticizer is as shown below. The amount of the plasticizer is normally low mostly in the range of 5 to 20 parts per hundred parts of the polymer depending on the requirement for the processability improvement. These liquids tend to dissolve in the polymer or the rubber matrix thereby helping to (i) disentangle and (ii) relax the rubber chains by swelling and therefore increasing the free volume of the matrix and thus reducing the energy required for the deformation of the matrix or effectively reducing the viscosity of the polymers or the filled systems. Kraus and Gruver[75] have reported on the effect of plasticizers on the viscosity of the plasticized rubbers and developed a correlation as,

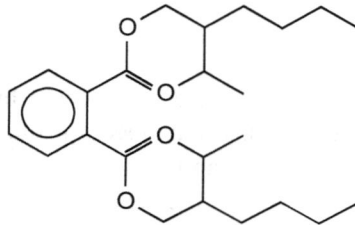

Bis (2-ethylhexile) phthalate

$$\eta_{\theta_p} = \eta_l \, \theta_p^{3.4} \qquad\qquad ...(4.58)$$

where θ_p, η_{θ_p} and η_l are the fraction of polymer in the plasticized system, viscosity of the plasticized polymer and that of the un-plasticized polymer respectively.

4.8.15 Effect of Elastic Parameters[29, 81, 116-118]

High molecular weight polymers and rubbers are the viscoelastic materials that are capable of storing a part of energy of deformation and release it as mechanical work. During the flow through a flow channel this stored energy normally does not interfere with the flow process. However as soon as the material comes out from the channel, the elastic energy is released and causes different effects like die swell, surface distortions and melt fracture etc. These elastic effects are caused due to the development of the normal stresses.

The melt fracture is a phenomenon exhibited by most of the high molecular weight polymers melts at high rates of shear ($MW > 10^5$ at the rates of shear $>10^3$ s^{-1})[81] as shown in Fig. 4.17. At high flow rate of the polymer melt or solutions the extrudate coming out of the flow conduit develops different surface irregularities like beads, sharkskin, wavy type etc. instead of coming out with a smooth and continuous surface. The exact reasons for this phenomenon are not fully understood but it is assumed to be the result of the complex combinations of various forces like normal stresses and adhesion, pressure, elongational and surface forces.

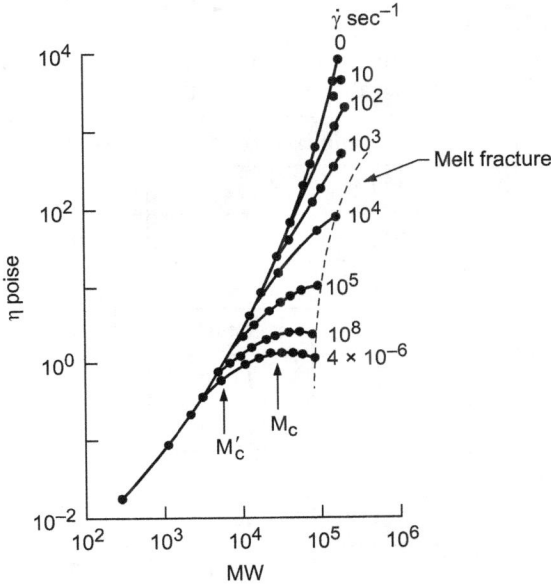

Fig. 4.17 Viscosity versus molecular weight for poly dimethylsiloxane at 35°C showing melt fracture region[117]

PROBLEMS

1. Debye[52] and Bueche[53] have developed a theory for dependence of zero shear viscosity with M.Wt. of the low molecular weight polymers melts and solutions. Refer to these references and derive the said relationship. Discuss the validity of assumptions in this derivation. Explain why the equation fails for high M.Wt. systems?

2. The data of Wagner and Wissbrun[118] is shown in Fig. 4.14. The curve D refers to the data of blend of A and C such that its M.Wt. is almost same as that of B. (A, C and D are pure polymers with similar molecular weight distribution having $\bar{M}_W/\bar{M}_N \approx 2$). Notice that the viscosity of blend D at low shear rate is same as that of pure polymer B and at high shear rate it approaches that of C. Explain the most likely physical reasons for this behaviour.

3. Application of Debye-Bueche theory to both polymer solutions and melts reveals that the slope of ln η_0 versus ln $\bar{M}_W$ plots give slope ≈ 1 for melts but for solutions it is > 1 (PVA-MEK [Onogi et al.50] and PVA-water solutions [Onogi et al., 51]). Explain why should solutions behave differently than the melts?

4. Derive the WLF equation in terms of T_g (Eq. 4.12) from that in terms of T_a (Eq.4.11) by using relationship $T_a \approx T_g + 50$ (°C). Derive the constants of Eq. 4.51 theoretically.

5. Refer to the Einstein's formula (Eq. 4.32) for the viscosity of the colloidal suspensions. This formula was derived on the basis of the considerations of hydrodynamic interaction in the suspension. Study the physical and mathematical arguments given by Einstein and derive the formula yourself.

6. For PVC melt the following shear stress and shear rate data are available at different temperature. Estimate the glass transition temperature using the WLF equation and compare the value reported in the literature.

Rate of shear, $\dot{\gamma}, s^{-1}$	Temperature, °C / Shear stress, $\tau \times 10^{-4}$, Pa			
	170	180	190	200
3.06	95	25.6	10.8	8.0
6.12	124	34.8	14.0	11.0
12.23	162	64.0	50.0	17.0
30.58	237	110.0	56.0	33.0
61.17	325	165.0	95.0	59.0
122.34	438	240.0	150.5	102.0
305.85	650	390.0	263.0	200.0
611.70	825	540.0	392.0	320.0
1223.40	1040	782.0	570.0	490.0
3058.50	1350	1200.0	856.0	780.0

7. Use the WLF equation to predict how the viscosity decreases with an increase in temperature. Obtain the ΔT required for a decade decrease in viscosity for different values of $(T - T_a)$. How does ΔT vary with each decade decrease in viscosity?

8. The zero shear viscosity data for polystyrene has been calculated using the Rheology Software (Appendix II) from Carreau model as

Temperature, °C	210	230	250
Zero shear viscosity, Pa s	8849.0	6160.0	3871.0

Calculate the Glass transition temperature and reference temperature using the WLF equation and compare the values reported in the literature.

9. The zero shear viscosity data for polypropylene has been calculated using the Rheology Software (Appendix II) from Carreau model as

Temperature, °C	190	210	230	250	270
Zero shear viscosity, Pa s	9545	6670	5431	4636	3585

Calculate the Glass transition temperature and reference temperature using the WLF equation and compare the values reported in the literature.

10. The rheological experimental data for LDPE, PP and PS are reported in Appendix III at different temperature. Estimate the temperature shift factor for zero shear viscosity calculated by Ellis model, Carreau model and Meter model. Comment on the variation of shift factor for different models.

11. From the experimental data given in Appendix III establish the shift factor for the viscosity as a function of temperature and rate of shear for polypropylene.

12. H. P. Schreiber, [J. App. Polym. Sci., 9, 2101 (1965)] reported the intrinsic viscosity data for the solutions of linear PE in tetralin at 120°C for a number of samples with different molecular weight. Establish the relationship between the intrinsic viscosity and the molecular weight and calculate the % deviation in the calculated MW with respect to the experimental MW.

13. The experimental values of zero shear viscosities at different temperature and MW are reported for Polybutadienes [J.T. Gruver and G. Kraus, J. Polym. Sci. Part A, 2, 797-810 (1964)] Develop the relationship between zero shear viscosity and M. Wt. at different temperatures.

Zero shear Viscosity, M poise/ Temperature, K

S. No.	$\bar{M}_w$ /1000	235	267	300	339	380	425
1	20.6	—	—	0.001156	0.000280	0.000137	0.0000424
2	61	—	—	0.070	0.0165	0.0055	0.0022
3	145	—	—	1.0	0.235	0.106	0.049
4	238	—	—	5.2	1.24	0.54	0.29
5	278	1210	50	9.4 /10.2	2.5	0.97	0.525
6	298	1120	66	12.6	2.77	1.14	—
7	334	1750	95	18.2	—	1.67	0.92
8	447	4150	264	46	—	4.5	3.0
9	524	7750	386	76	—	—	5.0

14. For the data given in problem No. 13 develop the temperature shift factor for the coefficient of the relations obtained.

15. For the data given in problem No. 13 develop the generalized equation for viscosity as a function of MW and temperature.

16. The experimental data on the variation of molecular weight with the cold mastication of styrene- butadiene rubber on a two roll mill in presence of oxygen has been published by S.K. Bhatnagar and S. Bannerjee, RCT 38(4), 961-96(1965). The data available are for the intrinsic viscosity, Mooney viscosity and (osmotic pressure/concentration) as concentration approaches zero. Calculate the molecular weight from the osmotic pressure data using the relation $(M) = RT/(\pi/c)_{\to 0}$. The osmotic pressure was measured at 35°C.

S.No.	Time of mastication, min.	Intrinsic Viscosity, dl/g	Mooney viscosity	$(\pi/c)_{\to 0} \times 10^4$ lit.atm./g
1	0	2.52	—	—
2	5	2.08	30.0	1.124
3	15	1.73	25.2	1.508
4	30	1.62	22.0	1.805
5	45	1.41	19.9	1.989
6	60	1.30	18.3	2.469
7	75	1.24	17.8	2.418
8	90	1.20	17.5	2.571

Develop the relationship between the time of mastication and the molecular weight.

17. From the data in the problem No.16 plot the intrinsic viscosity versus the Mooney viscosity and develop the relationship between the two viscosities.

18. For the data in problem 16 develop the relationship between the time of mastication and intrinsic viscosity.

19. Develop the relationship between the molecular weight and the intrinsic viscosity for the data in problem 16.

20. Develop the relationship between the molecular weight and the Mooney viscosity for the data in problem 16.

21. Develop the relationship between the mastication time and the Mooney viscosity for the data in problem 16.

REFERENCES

1. B. Hartmann, G. F. Lee, R. F. Cole, Jr., Polym. Eng. Sci., **26**, 554 (1986).
2. B. Hartmann, Lee, G., and W. Wong, Polym. Eng. Sci., **27**, 823 (1987).
    (a) B. Hartmann, and R. F. Cole, Jr., Polym. Eng. Sci., **23**, 13 (1983).
    (b) ibid 25, 65, (1985).
    (c) B. Hartmann, and Lee, G. F. Jr., Polym. Eng. Sci., **31**, 231, (1991)
3. Y. Imai and N. Brown, Polymer, 18, 288 (1977).
4. D.R. Mears, K. D. Pae and J. A. Sauer; J. Appl. Phys., **40**, 4229 (1969).
5. W.A. Spitzig and O. Richmond, Polym. Engg. Sci., **19**, 1129 (1979).
    (a) I. M. Ward, "Mechanical properties of solid polymers," 2nd ed., John Wiley & SonsNew York, 1983
6. W. Whitney and R. D. Andrews, J. Polym. Sci. C, **16**, 2981 (1967).
7. F. C. Chang, J. S. Wu `and L. H. Chu; J. Appl. Polym. Sci. **44**, 491 (1992).
    (a) Macosko, C. W. and Brand, G. J., Polym. Eng. Sci.,**12**, 444 ((1972)
    (b) D. L. Holt, J. Appl. Polym. Sci., **12**, 1653, 1968.

8. C. Bouwens-Crowet, J-C Bouwens and G. Homes, J. Mater. Sci., **7**, 176 (1972)

9. R. W. Truss, P. L. Clarkeand, R. A. Duckett and I. M. Ward, J. Polym. Sci. Phys. Ed. 22, 191 (1984).

10. G. V. Vinogradov and A. Ya.Malkin, Rheology of Polymers, Mir Publishers, Moscow, 1980, pp 105-215.

11. W. I. Kartsovnik and V.V. Pelekh arXiv:0707.0789v1, July, (2007).

    (a) R. M. Neuman, J. Chem. Phys., **66** (2), 870 (1977).

12. A. K. Doolittle, J. Appl. Phys., **22**, 1471 (1951).

13. M. L. Williams, R. F. Landle and J. D. Ferry, J. Am. Chem. Soc., **77**, 3701 (1955).

14. M. L. Williams, J. Phys. Chem., **59**, 95 (1955).

15. T. G. Fox and P. J. Flory, J. Phys. Chem., **55**, 221 (1951).

16. T. G. Fox and P. J. Flory, J. Am. Chem. Soc., **70**, 2384 (1948).

17. T. G. Fox and P. J. Flory, J. Polym. Sci., **14**, 315 (1954).

18. Galimberti, Maurizio; Caprio, Michela; Fino, Luigi (2001-12-21). "Tyre comprising a cycloolefin polymer, tread band and elasomeric composition used therein" (published 2003-03-07). country-code =EU, patent-number =WO03053721

19. Ibeh, Christopher C. (2011). THERMOPLASTIC MATERIALS Properties, Manufacturing Methods, and Applications. CRC Press. pp. 491–497. ISBN 978-1-4200-9383-4.

20. ABS. nrri.umn.edu

21. Nicholson, John W. (2011). The Chemistry of Polymers *(4, Revised ed.)*. *Royal Society* of Chemistry. p. 50. ISBN 9781849733915. Retrieved 10 September 2013.

22. T. G. Fox and V. R. Allen, J. Chem. Phys., **41**, 344 (1964).

23. T. Kataoka and S. Ueda, J. Polym. Sci., **B-4**, 317 (1966).

24. G. V. Vinogradov and A. Ya.Malkin and V. G. Kulichikhin, J. Polym. Sci., **A-2** (8), 333 (1970).

25. R. S. Porter, I. Wang and I. R. Knox, J. Polym. Sci., **B-8** (10), 671 (1970).

26. K. Shirayama, T. Masuda and S. Kita, Macromol. Chem., **147**, 155 (1971).

27. R. L. Combs, D. F. Slonaker and H. W. Coover, J. Appl. Polym. Sci., **13**, 519 (1969).

28. A. B. Bestul and H. V. Belcher, J. Appl. Phys., **24**, 696 (1953).

29. R. S. Lenk, Polymer Rheology, Applied Science Pub. Ltd. London, 1978.

30. V. L. Volt, R. W. Smith and C. E. Wilkes, Rubber Chem. Tech., **44(1)**, 1-11 (1971)

31. F. N. Cogswell, Plastics and Polymers, **39** (Feb. 1973).

32. F. J. Martinez Boza and C. Gallegos, Rheology Vol. 1. High Pressure Rheology. (eolss.net)

33. H. A. Barnes, " A handbook of Elementary Rheology, University of Wales, Institute of Non-Newtonian Fluid Mechanics, Aberystwyth. [An introductory text to rheology.], (2000).

34. P. D. Driscoll and C. D. Bougue, J Appl. Polym. Sci., **39**, 1755, (April 1990).

35. M R Kamal and H. Nyun, Trans. Soc. Rheol., **17(2)** , 271-285 (1973)./Published online in July 2000. http://dx.doi.org/10.1122/1.549315

36. Faezeh Aghazadeh, Ann.Trans. Nordic Rheol. Soc. , **20**, (2012).

37. Chao Li, F, Jang, L. Wu, Xiaoyan and X Li;J. Macromol. Sci., Part B, **54(9)**. 2015.

38. J. M. Smolinski, C. W. Manke and E. Gulary, cmanke@eng.wayne.edu.

39. J. Dealy,and H. E. Park, December 23, 2010. Permalink (PDF Version) also F. Koran and J. M. Dealy, J. Rheol. **43**, 1279-1290, 1999.

40. J. Southern and R. S. Porter, Am. Chem. Soc. Polym. Preprints, **10**, 1028 (1969) and **11**, 266 (1970)., http://doi.org/ACPPAY,CAS

41. T. Niikuni and R. S. Porter, J. Mats. Sci., **9**, 389 (1974)., and W. T. Mead and R. S. Porter, J. Appl. Physics, **47 (10)**, 4276 (1976).

42. W. G. Perkins and R. S. Porter, J. Mats. Sci., **12**, 2355 (1977).

43. R. F. Westover, Polym. Engg. Sci. , **6**, 83 (1966) and **6**, 86 (1966)

44. P. J. Flory and T. G. Fox, J. Am. Chem. Soc., **73**, 276 (1951).
    (a) Z. Jiang, W. Dou, T. Sun, Y. Shen and D. Cao, J. Polym. Research, **22**, 236 (2015).

45. A. Einstein, Ann. Physik, **19**, 289 (1906).

46. A. Einstein, Ann. Physik, **34**, 591 (1911).

47. R. Houwink, J. Prakt. Chem., **157**, 15 (1940).

48. J. Brandrup and E. H. Immergut, Polymer Handbook, 1st Ed., Interscience Pub., New York, 1967.

49. B. Chitrangad, H. R. Osmers and S. Middleman, Polym. Eng. Sci.,17, 806 (1877)

50. S. Onogi, S. Kimura, T. Kato, T. Masuda and N. Miyanaga, J. Polym. Sci., **15** (G), 381(1966).

51. S. Onogi, T. Kobayashi, Y. Kojimaand Y. Taniguchi, J. Appl. Polym. Sci., 7, 847 (1963).

52. P. Debye, J. Chem. Phys., **14**, 636 (1946)

53. F. Bueche, J. Chem. Phys., **20**, 1959 (1952).

54. A. Kumar and S. K. Gupta, Fundamentals of Polymer Science and Engineering, Tata McGraw Hill Pub. Co. Ltd., New Delhi, 1978.

55. T. G. Fox, J. Polym. Sci., **C(9)**, 35 (1965).

56. G. C. Berry and T. G. Fox, Adv. Polym. Sci., **21**, 261 (1968).

57. T. A. Toosi, R. S. Porter and J. F. Jonson, Polym. Prepre., Am. Chem. Soc.Divn. Polym. Chem., **10(1)**, 51 (1969).

58. V. R. Allen and T. G. Fox, J. Chem. Phys, **41**, 337 (1964).

59. J. L. den Otter, Rheol. Acta, **8**, 355 (1969).

60. A. V. Tobolsky, J. J. Aklonis and G. Akovali, J. Chem. Phys. **42**, 545 (1966).

61. E. Bagley and D. West, J. Appl. Phys., **29**, 1511 (1958).

62. G. C. Berry and T. G. Fox, Adv. Polym. Sci., **5**, 261 (1968).

63. D. Gupta and W. C. Forsman, Macromolecules, **2**, 304, (1969).

64. J. D. Ferry, Viscoelastic Properties of Polymers, John Wiley & Sons, New York, 1970.

65. W.W. Graessley, J. Chem. Phys, **43**, 2696 (1965).

66. W.W. Graessley, J. Chem. Phys, **47**, 1942 (1967).

  (a) Richard P. Wool, Micromolecules, **26**, 1564-1569 [1993]

  (b) N.Biswas and A Datta, [arXiv:1403.4721v1[cond-mat.soft]], 2014.

  (c) R. A. Riggleman, G. Toepperwein, G. J. Papakonstantopoulos, J-L., Barret and J. J. de Pablo; J. Chem. Phys.**130**, 244903 (2009).

  (d) A. I. Barzic, R. D. Rusu, Iuliana Stoica and M. N. Damaceanu , J. Mat. Sci., **49(8)**, 3080-3098 (April 2014).

  (e) L. Kong and G. R. Ziegler, J. Visualized Exp., **91**, (2014) :51933. Published online on Sept. 2014 doi:10.3791/51933.

  (f) A. M. Sukhadia, SPE Extrusion Division 1-0-Wiki, March 2015.

  (g) Alison MacArthur (1995) 'The entanglement structure of polymers. Knots and Applications" page 395-426.

67. Y. Ito and S. Shishido, J. Polym. Sci., Polym. Phys. Ed., **11**, 2283 (1973).

68. E. A. Collins W. H. Bauer, Trans. Soc. Rheol., **9(2)**, 1 (1965).

  (a) D. N. Njoroge, Int. J. Eng. Trends and Appl.(IJETA), **4(1)**, 8-12 (2017).

69. A. J. Charlsby, J. Polym. Sci., **17**, 370 (1955).

70. D. P. Wyman, L. G. Elyash and W. J. Frazer, J. Polym. Sci., **A-3**, 681 (1965).

  (a) T. J. Zolper, P. Shiller, M. Jungk, T. J. Marks, Y-W Chung, A. Greco, G. Doll, B. L. Dehkordi, and Q. Wang, J. Tribol, **137(3)**, 031503, (July 01, 2015).

  (b) A. D. Gotsis and B. L. F. Zeevenhoven, J. Rheol., **48(4)**, 895-914 (2004).

  (c) Albena Lederer, Leibnitz Institute of Polymer Rtesearch, Dresden, Germany.

71. G. Akovali, J. Polym. Sci., **A-2** (5), 875 (1967).

72. J. Fujimoto, H. Narukawa and M. Nagasawa, Macromolecules, **8**, 19 (1972)

73. H. P. Schreiber, A. Rudin and e. b. Bagley, J. Appl. Polym. Sci., **7**, 887 (1965).

74. W. W. Graessley, Accounts Chem. Res., **10**, 332 (1977).

75. G. Kraus and J. T. Gruver, J. Polym. Sci., **A-3**, 105 (1965).

76. T. A. Orofino, Polymer, **2**, 305 (1961).

77. G. C. Berry, J. Polym. Sci., **A-2(6)**, 1551(1968).

78. M. Hoffmann and R. Rother, Makromol. Chem., **80**, 95 (1964).

79. G. V. Vinogradov, Pure and Appl. Chem., **26**, 423 (1971); Rheol. Acta, **12**, 357 (1973).

80. IUPAC, Pure Appl. Chem., **81(2)**, 351-353 (2000).

81. Y. Ito and S. Shishido, J. Polym. Sci., Polym. Phys. Ed., **10**, 2239 (1972).

82. I. Ozsoy, A. Demirkol, A. Mimaroglu, H. Unal and Z. Demir, J. Mech. Eng., **61(10)**, 601-609 (2015).

83. H. U. Zaman, P. D. Hun, R. H. Khan and K. B. Yoon, J. Reinforced Plastic and Composites, **31(5)**, 323-329 (2012).

84. Y. Gao, L. liu and Z. Zang, Acta Mechanica Solida Sinica, **22 (6)**, 555-562 (2009).

85. V. A. Agubra, P. S. Owuor and M. V. Hosur, Nanomaterials, **3 (3)**, 550-563 (2013).
86. C. K. Lam, H. Y. Cheung, K. T. Lau, L. M. Zhou, M. W. Ho, and D. Hui, Composites: Part B; Engineering, **36 (3)**, 263-269 (2005).
87. A. Yasmin and I. M. Deniel, Polymer , **45 (24)**, 8211-8219 (2004).
88. M. Sayer, Composites Part B: Engineering, **59**, 12-20 (2014).
89. O. Asi, 28 (23), 2861-2867 (2009).
90. H. S. Yang, H. J. Kim, J. Son, H. J. Park, B. J. Lee and T. S. Hwang
91. M. S. Ibrahim, S. M. Sapuan and A. A. Faieza, J. Mech. Eng. Scis., **2**, 133-147 (2012).
92. M. C. Lee, Y. zhang and U. R. Cho, Materials and Design, **63**, 565-574 (2014).
93. P. E. Imoisili, T. C. Ezenwafor, B. E. AttahDaniel and s. O. O. Olusunle, Am. Chem. Sci. J., **3(4)**, 526-531 (2013).
94. G. B. Jeffery, Proc. Royal Soc. London Series A, **102**, issue **715**, 161-179 (1922)
95. H. L. Goldsmith S. G. Mason, Rheology, F. Eirich Ed., Academic Press, New York, Vol. **4**, 1967.
96. H. Brenner, Intern. J. Multiphase Flow, **1**,195 (1974).
97. T. S. Chong, J. Appl. Polym. Sci., **15**, 2007 (1971).
98. B. R. Gupta, N. K. Purohit and A. N. Roy, Proc., CHEMECA-70, Australia, PP. 25 (1970)
99. R. O. Meshmeyer and R. T. Hill, Trans,. Soc. Rheol., **21**, 183 (1977).
100. G. R. Cotton, Rubber Chem. Technol., **57(1)**, 118 (1984).
101. J. J. Brennen, T.E. Jermyn and B. B. Boonstra, J. Appl. Polym. Sci., **8**, 2687 (1964).
102. D. J. Hine and W. C. Wake, Trans. Faraday Soc., **55**, 1077 (1959).
103. E. M. Dannenberg, Trans. Inst. Rubber Ind., **42**, 726 (1966).
    (a) D. N. Njoroge, Int. J. Eng. Trends and Appl.(IJETA), **4(1)**, 8-12 (2017).
104. N. P. Cheremisinoff, Rubber Mixing Principles in Elastomer Technology Handbook, Ed. N. P. Cheremisinoff, C R C Press Inc., Florida, U S A, Chapter **21**, pp 749-779, 1993.
105. R. Cherizol, M. Sain, J. Tjong; Open J. Polym. Chem.**5**, 1-8 (2015).
106. A. M. A. Huq and J. Azaiez, Polym. Eng. Sci., **45 (10)**, 1357-1368, (Oct. 2005).
107. M. Tarfaoui, S. Choukri and A. Neme, Composite Scince and Technology, **68**, 477-485 (2008).
108. N. K. Kim, R. J. T. Lin and D. Bhattacharya, Polymer degtadation and stability, **119**, 167-177 (2015).
109. D. Roy, A. K. Bhattacharya and B. R. Gupta, Elastomers and Plastics, **25 (1)**, 46-58 (1993).
    (a) V. M. Murthy, B. R. Gupta and S. K. De, Plast. Rubber Proce. Appl., **5(4)**, 307 (1985).
    (b) Kitono, T. Kataoka, T. Nishimura and T. Sakai, Rheol. Acta, **19**, 764 (1980).
    (c) S. R. Nanguneri, N. J. Rao and N. Subramanian, Rheol. Acta, **26**, 301 (1987)

(d) A. K. Gupta, P. Kumar Krishna and B. R. Ratnam, J. App. Polym. Sci., **42**, 2595 (1991).

110. E. Manias, Nature Materials, **6(1)**, 9-11 (2007)

111. G. Lalwani, A. M. Henslee, B. Farshid, Biomacromolecules, **14(3)**, 900-9 (March 2013).

112. ibid Acta Biomaterialia, **9(9)**, 8365-73 (Sept.2013)

(a) A. L. Frischknecht, E. McGarrity and M. Mackay, J. Chem. Phys., **132**, 204901 (2010)

113. G. Teresa, V. Nicola, Mba, Mirium, and M. EnzoEuropian J. Organic Chem., 1071-1090 (2016).

114. B. P. Singh, D.Singh, R. B. Mathur and T. L. Dhami, Nanoscale Research Letters, **3(11)**, 444-453 (2008).

115. Tao Xu, and J. Yang, J. Nanomaterial- special issue on Low-Dimensional Carbon Nanomaterials, 2012, Article No.1, (Jan. 2012).

116. J. P. Tordella, J. Appl. Phys. **27**, 404 (1956).

117. H. P. Schreiber, A. Rudin and E. B. Bagley, J. Appl. Polym. Sci., **9(3)**, 887 (1965).

118. H. L. Wagner and K. F. Wissbrun, Makromol. Chemie, **81**,14 (1965).

❑   ❑   ❑

Rheometry

5

Rheometry refers to the area of the development of experimental techniques for rheological measurements of liquids including polymer melts and solutions. Polymer processing covers a wide range of liquids from low viscosity solutions and latexes to highly concentrated high viscosity suspensions and high MW polymer melts. The energy required for the deformation of low viscosity liquids is much smaller than that for the high viscosity melts. Thus the equipment designed for the low energy input suitable for dilute solutions and latexes cannot be used for high viscosity polymer melts. The polymer melt viscosity is a shear rate dependent property and the viscosity at very low shear rate is much higher than at high shear rate. The processing of polymers in different processing equipment subjects it to different shear rates for example in mixing mills and calenders it may be of the order of a few hundred s^{-1}, in extruders both ram and screw type, a few thousands s^{-1} and in the injection moulding systems of the order of tens of thousand s^{-1}. Thus the design of viscometers or the rheometers vary widely depending on the rate of shear range, the nature of the liquid i.e. solutions, suspensions or the melts and the magnitude of the viscosity. Low viscosity measurements require high degree of accuracy and therefore necessitate the use of precision equipment. Whereas high viscosity and high rate of shear measurements require robust and sturdy equipment to withstand the high stresses. Thus no single apparatus can serve the entire range of liquids and the rate of shear. The term rheometer comes from Greek, meaning device for measuring flow.

The rheometers have been developed based on some basic flow systems[1,2], which impart specific type of deformation to the fluid. These are

(i) *One-dimensional Rectilinear Flow:* The fluid flows through a particular cross section channel (invariably the circular cross section) in one direction only. A number of equipment like Capillary viscometers, Pochettino viscometers, etc. are based on this flow principle.

(ii) *Rotational flow:* The liquids in such a system is subjected to the rotational flow. Some of the equipments in this category are Rotating co-axial cylindrical viscometers, Rotating plate viscometers, Cone and plate viscometers, Double cone viscometers, conicylindrical viscometers etc.

(iii) *Oscillatory Motions:* The fluids are subjected to the oscillatory deformations. The equipments developed based on this flow principle are, Oscillating disc viscometers, Oscillating penetrometers etc.

(iv) *Radial Flow Under Compression:* It includes Parallel plates plastometers, e.g. Williams parallel plate plastometer (plastimeters)

(v) *Sliding Flow Between Plates:* It is relatively a new development in this area.[2a, 2b]

(vi) *Extensional Deformation:* Injection moulding, fiber spinning, extrusion, blow moulding, and coating operations subject the polymers to the extensional deformations. The material in other industries like soap, adhesives, food etc. are also subjected to extensional deformation. Some of the commercially available extensional rheometers are also discussed here.

5.0 VISCOMETERS

The theoretical analysis of some of the important viscometers is discussed in the following sections.

5.1 CAPILLARY VISCOMETERS[3-5]

The capillary viscometers are the most widely used viscometers for the measurements of the viscosity of liquids. These equipments range from a simple glass apparatus for low viscosity Newtonian liquids working under atmospheric pressure like ostwald viscometer to high-pressure systems (operating up to 2000 atm. pressure) used for very high viscosity polymer melts and rubber compounds.

The principle of operation of the capillary viscometers is based on the unidirectional flow of liquids flowing under fully developed laminar, steady, isothermal conditions. The pressure drop, ΔP and volumetric flow rates, Q are measured to calculate the shear stress and rate of shear, which are used for the rheological characterization of the fluids. The liquids may flow under, (i) gravity as in the case of low viscosity fluids in the glass capillary viscometers (Fig. 5.1(a)) or (ii) a gas or hydraulic pressure in a pressurized vessel (Fig. 5.1(b)) or (iii) a mechanically driven plunger as shown in Fig. 5.2(a) (Göttfert Rheograph RG 20). Is dual purpose machine used for both rheological as well as PVT measurements at high pressure. Fig. 5.2(b) shows the details of capillary and chamber (on the right side) etc, which are mentioned in the figure. The chamber on the left is ued for PVT analysis of the system. The flow equations for the capillary flow have been developed in Chapter 3 and only a small discussion about these viscometers is included here.

The flow equations derived for the capillary flow analysis assume a fully developed stream line flow i.e. to say that the pressure gradient, $\Delta P/L$ in the flow channel is constant for a given flow rate. This assumption is quite valid for the major part of the capillary length except at the entrance and the exit where additional energy losses take place due to the change in flow direction and extensional flow as discussed in chapter 3 in the section under Bagley correction[6]. For a given fluid at a particular flow rate the entry and exit losses for the capillaries of same diameter but different lengths will remain unchanged. Hence by carrying out experiments under isothermal conditions using two

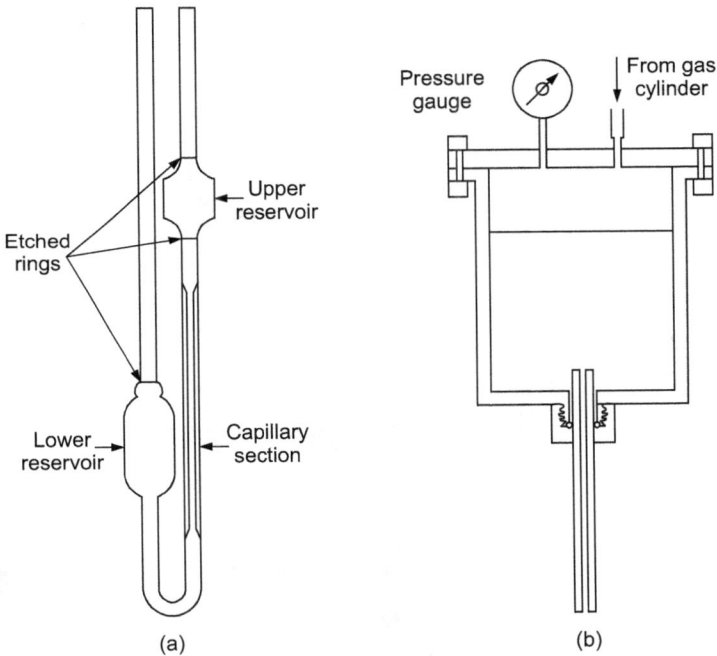

Fig. 5.1   Capillary viscometers (a) a glass capillary, (b) a gas pressurised system[11]

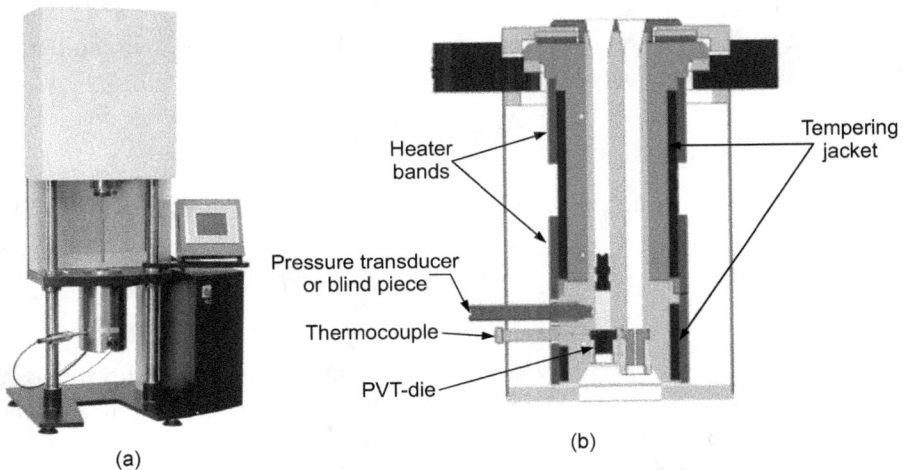

Fig. 5.2   (a) A Götfert Rheograph-RG 20, and (b) a view cut of rheometer barrel with tempering Jacket for isobar measurements [goettfert.com]

different capillaries of the same diameter but different lengths one can estimate the true pressure gradient. The total pressure drop for capillary-1, and capillary-2

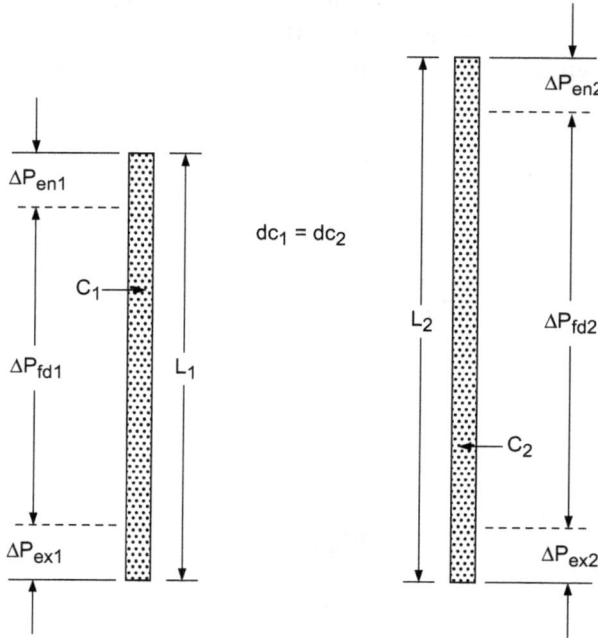

Fig. 5.3 Schematics of capillaries

i.e. ΔP_{T1} and ΔP_{T2} (Fig. 5.3) can be written as

$$\Delta P_{T1} = \Delta P_{en1} + \Delta P_{fd1} + \Delta P_{ex1} \qquad ...(5.1)$$
$$\Delta P_{T2} = \Delta P_{en2} + \Delta P_{fd2} + \Delta P_{ex2} \qquad ...(5.2)$$

where suffixes *en, fd* and *ex* refer to entry, fully developed and exit and 1 and 2 refer to the capillary 1 and 2 respectively. As the diameters of the two capillaries are same the following equalities are satisfied

$$\Delta P_{en1} = \Delta P_{en2} \text{ and } \Delta P_{ex1} = \Delta P_{ex2}$$

Hence $$\Delta P_{T2} - \Delta P_{T1} = \Delta P_{fd2} - \Delta P_{fd1} \qquad ...(5.3)$$

If the lengths of the two capillaries are L_1 and L_2 then the true pressure gradient, $\Delta P/L$ is given by

$$\frac{\Delta P}{L} = \frac{\Delta P_{fd2} - \Delta P_{fd1}}{L_2 - L_1} = \frac{\Delta P_{T2} - \Delta P_{T1}}{L_2 - L_1} \qquad ...(5.4)$$

The corrections for the entry losses are calculated using the Bagley method. These corrections are also known as the Couette corrections. For a capillary of high L/D ratio the corrections become negligible as compared to the fully developed pressure drop[7, 8, 9]. The capillary viscometers are very versatile equipment capable of measuring viscosity in the range of fractions of poise to

10^9 poise in the shear rate range of 10^{-6} to $10^7 \, s^{-1}$[10]. Some of the high-pressure capillary viscometers available are Instron Industrial Products, USA, , Göttfert, and Monsanto viscometers, Kinexus Rheometers and Rosand RH7 and RH10 rheometers from Malvern Instruments Ltd. UK. Also the other manufacturers of high pressure capillary rheometers are Shimatzu Scientific Instruments. Inc, USA, Qualitest Inc. USA, info@qualitest-inc.com, John Morris Scientific, Australia produce Rheo-Tesyer-1000 (info@qualitest-inc.com, John Morris Scientific, Australia, (info@johnmorris.com.au), Brabender GmbH & Co KG, Germany, Dynesco, USA produces Viscosensor Online rheometer and Viscoindicator online rheometer, Rheologica Instruments NJ, USA, Thermo Haake, Karlsruhe, Germany produce Rheoflixer High Pressure Capillary Rheometer (info@thermohaake.com),

5.1.1 Kinetic Energy Correction

When the fluid enters the capillary of smaller cross section from the barrel of a large cross section area the velocity of the fluid increases considerably giving higher kinetic energy of the fluid. This increase in the kinetic energy (KE) is caused due to corresponding reduction in pressure in the system. The KE based on the average velocity and that obtained by the integration over the entire cross section are different and hence a KE correction factor α is applied, which has been reported to vary from 0.5 to 1.55 depending on the flow type i.e. laminar or turbulent flow[11]. This correction is sometimes known as the Hagenbach correction.

5.2 CONE AND PLATE VISCOMETER[12-26]

The studies with cone and plate viscometer has been reported by a number of investigators like Dealy and Wisbrun and other[12-15, 17-22, 24-26]. A circular plate of known diameter is concentrically placed below a cone having a small cone angle in the range of 1 to 5° (Fig. 5.4). The plate is fitted with a constant speed motor and a gear assembly so that it can be rotated with different speeds, which can be accurately recorded with the help of a proximity switch or some other sensor. The cone is supported by a frictionless bearing assembly and is fitted with a calibrated linear torsional spring, a circular scale graduated in degrees and a pointer to measure the deflection.

The fluid is placed in between the gap of cone and plate so that the conical cavity is completely filled. The air gaps must be carefully avoided. The torque generated by the rotation of the plate is transmitted to the cone through the fluid. The rotation of the cone due to this torque is arrested by the counter torque of the same magnitude produced due to the tension in the spring. This torque balance deflects the pointer by a certain degree, and the corresponding magnitude of the torque is obtained using the spring calibration curve. The flow

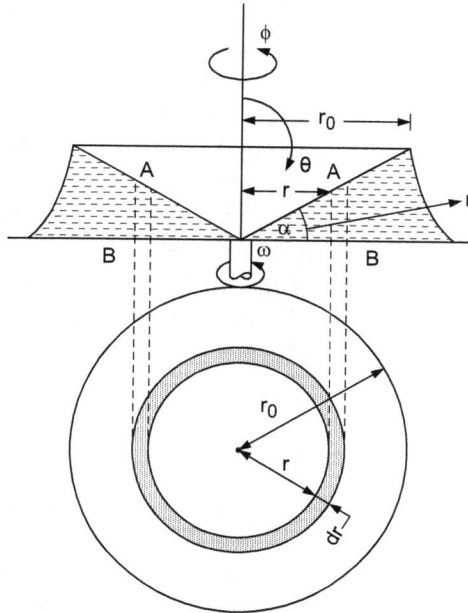

Fig. 5.4 Schematics of cone and plate viscometer[11]

analysis is carried out by considering the forces acting on an elemental circular fluid strip, AB of radius r and thickness δr using the spherical coordinates. The velocity components in r and θ direction i.e. u_r and u_θ are zero and the fluid in contact with the plate at point B rotates only in the ϕ direction with a peripheral velocity, $u_{\phi B} = r\omega$, where, ω is the speed of rotation of the plate. The velocity of the fluid at point A in contact with the cone $u_{\phi A}$ will be zero as the cone is not rotating and also no slip condition is assumed to be applicable. Thus the flow of the liquid is a unidirectional shear flow. As the cone angle α is small (usually $< 5°$) the flow at a particular location i.e. at some value of r, can be approximated to be like torsional flow between the parallel plates.

The difference in the velocity at points A and B subjects the fluid to a rate of shear, $\dot{\gamma}$ which is expressed as

$$\dot{\gamma} = \frac{du_\phi}{dz} = -\frac{\Delta u_\phi}{\Delta z} = -\frac{u_{\phi B} - u_{\phi A}}{z_B - z_A}$$

$$= -\frac{r\omega - 0}{0 - r\tan\alpha} = \frac{\omega}{\tan\alpha} \qquad ...(5.5)$$

where z refers to the vertical position with $z = 0$ at point B. It can be seen from equation 5.5 that the rate of shear is independent of radius giving a constant rate of shear over the total bulk of the fluid. Hence the cone and plate viscometers are known as the constant rate of shear viscometers. The shear stress, τ can be written as

$$\tau = \eta_a \left(\frac{du_\phi}{dz} \right) = \frac{\eta_a \omega}{\tan \alpha} \qquad \qquad ...(5.6)$$

where η_a is the apparent viscosity. The force acting on the cone due to this shear stress is

$$dF = \tau dA = \frac{\eta_a \omega}{\tan \alpha} \left(\frac{2\pi r \, dr}{\cos \alpha} \right) \qquad \qquad ...(5.7)$$

The torque due to this force

$$dT = r \, dF = r \frac{\eta_a \omega}{\tan \alpha} \left(\frac{2\pi r \, dr}{\cos \alpha} \right)$$

$$= \frac{2\pi \eta_a \omega}{\sin \alpha} r^2 \, dr \qquad \qquad ...(5.8)$$

The total torque acting on the fluid is given by integrating equation 5.8

$$T = \int_0^T dT = \frac{2\pi \eta_a \omega}{\sin \alpha} \int_0^R r^2 dr$$

$$= \frac{2\pi \eta_a \omega R^3}{3 \sin \alpha} \qquad \qquad ...(5.9)$$

This is also the torque experienced by the cone and is counter balanced by the tension in the linear spring so as to give deflection, θ to the pointer. If K is the spring constant then the torque is given as

$$T = K\theta = \frac{2\pi \eta_a \omega R^3}{3 \sin \alpha} \qquad \qquad ...(5.10)$$

Therefore the apparent viscosity can be expressed as

$$\eta_a = \frac{3 K\theta \sin \alpha}{2\pi \omega R^3} \qquad \qquad ...(5.11)$$

Thus from the information about the spring constant, K, the deflection, θ, the cone angle, α and the frequency of rotation of the plate, ω the apparent viscosity of the fluid can be calculated using the above equation. The variation in the speed of rotation will give different values of the rate of shear and the dependence of viscosity on the rate of shear or the shear stress and hence the rheological equation of the fluid can be established.

This apparatus can be used in the shear rate range of $10^{-3} \, s^{-1}$ to $10^4 \, s^{-1}$ and gives an excellent reproducibility. At higher shear rates the fluid develops the secondary flows and also has the tendency to be thrown out of the system due to high centrifugal force. However for a relatively low viscosity liquid such a situation may arise even at lower rates of shear. At extremely low rates of shear its performance is not reliable due to the poor response of the system.

It has been reported that the cone and plate viscometer can also be used to estimate the first and the second normal stresses differences by providing the piezometric tubes as shown in Fig. 2.18[23, 24].

Estimation of the Normal Stress Differences: The normal stresses differences can be estimated by using the r–component of equation of motion in terms of stresses in spherical coordinates (ϕ, θ and r), which is

$$\rho\left(\frac{\partial u_r}{\partial t} + u_r\frac{\partial u_r}{\partial r} + \frac{u_\theta}{r}\frac{\partial u_r}{\partial \theta} + \frac{u_\phi}{r\sin\theta}\frac{\partial u_r}{\partial\phi} - \frac{u_\theta^2 + u_\phi^2}{r}\right)$$

$$= -\frac{\partial P}{\partial r} - \left(\frac{1}{r^2}\frac{\partial}{\partial r}(r^2\tau_{rr}) + \frac{1}{r\sin\theta}\frac{\partial}{\partial\theta}(\tau_{r\theta}\sin\theta)\right.$$

$$\left. + \frac{1}{r\sin\theta}\frac{\partial\tau_{r\phi}}{\partial\phi} - \frac{\tau_{\theta\theta}+\tau_{\phi\phi}}{r}\right) + \rho g_r \quad ...(5.12)$$

In this equation $\tau_{\theta\theta}$, $\tau_{\phi\phi}$ and τ_{rr} are the normal stress components in θ, ϕ and r directions respectively, ρ is the density, g the gravity term and P is pressure. Using the conditions that $u_r = u_\theta = 0$, and neglecting the centrifugal forces, the above equation reduces to

$$-\frac{\partial P}{\partial r} - \frac{1}{r^2}\frac{\partial}{\partial r}(r^2\tau_{rr}) + \frac{\tau_{\theta\theta}+\tau_{\phi\phi}}{r} = 0 \qquad ...(5.13)$$

As the system is acted upon by the atmospheric pressure, P, the total normal stresses will be sum of the pressure and the normal stress terms hence introducing the total normal stresses in r, θ and ϕ directions as

$$\pi_{rr} = \tau_{rr} + P$$
$$\pi_{\theta\theta} = \tau_{\theta\theta} + P$$
and
$$\pi_{\phi\phi} = \tau_{\phi\phi} + P \qquad ...(5.14)$$

Substituting for normal stresses in the above equation and simplifying gives

$$-\frac{1}{r^2}\frac{\partial(r^2\pi_{rr})}{\partial r} + \frac{\pi_{\theta\theta}+\pi_{\phi\phi}}{r} = 0 \qquad ...(5.15)$$

Writing the partial differential of the first term and simplifying gives

$$\frac{\pi_{\theta\theta}+\pi_{\phi\phi}-2\pi_{rr}}{r}\,dr = d(\pi_{rr}) \qquad ...(5.16)$$

The term $[-(\pi_{rr} - \pi_{\theta\theta})]$ is the negative of the secondary normal stress difference, which is constant as $\dot\gamma_{\theta\phi}$ is constant[21]. Also the total normal stress, $\pi_{\theta\theta}$ at the plate surface i.e. at $\theta = \pi/2$ is a function of the plate radius. Using these conditions and integrating Eq. 5.16 one gets

$$\frac{[\pi_{\theta\theta}(r) - \pi_{\theta\theta}(R)]_{\theta=\pi/2}}{\ln(r/R)} = [(\tau_{\phi\phi} - \tau_{\theta\theta}) + 2(\tau_{\theta\theta} - \tau_{rr})] \qquad ...(5.17)$$

As the quantity in the left hand side of the equation can be determined experimentally hence the right hand side is determinable.

The normal force on the stationary plate can be expressed as

$$F_N = 2\pi \int_0^R \pi_{\theta\theta} r \, dr - \pi R^2 P_{atm} \qquad ...(5.18)$$

As $P_{atm} = \pi_{rr}(R)$, integrating Eq. 5.18 and substituting from Eq. 5.17 one gets the expression for the first normal stress difference as

$$\tau_{\phi\phi} - \tau_{\theta\theta} = \frac{-2F_N}{\pi R^2} \qquad ...(5.19)$$

The first normal stress difference therefore can be experimentally estimated. Hence the second normal stress difference, $(\tau_{\theta\theta} - \tau_{rr})$ can also be determined using Eq.5.17.

Thus a cone and plate viscometer can be used to experimentally determine the viscosity, and first and second normal stress differences subject to the condition that for melts the rate of shear is below the melt fracture limit and for solution it should be low enough so as not to through the solution out of the viscometer (i.e. low centrifugal force).

Example 5.1. The cone and plate viscometer with a cone radius of 12 cm. and the cone angle of 1° is being used for the measurement of the polymer solution viscosity. For a solution which behaves like a Newtonian liquid and has a viscosity of 0.1 Pa s calculate the torque experienced by the plate if the cone rotates with an angular velocity of 60 rad./mt.

Solution: The angular velocity, $\omega = 60$ rad/mt. $= 1$ rad/s,

Cone angle, $\alpha = 1°$. $\tan \alpha = 0.017455$ and $\sin \alpha = 0.01745$

$$\dot{\gamma} = \frac{\omega}{\tan \alpha} = 57.29 \ s^{-1}$$

Using equation 5.11 the torque,
$$T = 0.021 \text{ Nm}$$

Example 5.2. Using the cone and plate rheometer of example 5.1 calculate the torque for a non-Newtonian polymer solution, which obeys the power law equation $\tau = 47.8 \dot{\gamma}^{0.8}$.

Solution: The rate of shear,
$$\dot{\gamma} = 57.29 \ s^{-1}$$

The Torque, $T = \dfrac{2\pi\eta\omega r^3}{3\sin\alpha}$

$$= \dfrac{2\pi \times 47.8(57.29)^{-0.2} \times 1 \times (12 \times 10^{-2})^3}{3 \times 0.01745}$$

$$= 4.41 \text{ Nm}$$

5.3 CONCENTRIC CYLINDRICAL VISCOMETER[3, 5, 11, 27-34]

The studies using concentric cylindrical viscometer have been reported by a large number of workers[3, 5, 11, 27, 28, 29, 30, 31, 32, 33, 34]. These viscometers deform the fluids due to the drag flow (Couette flow) and are also known as the Couette or the bob and cup viscometers as, the inner and outer cylinders are also referred as bob and cup respectively. If the gap between the two cylinders is very small such that the ratio of the two diameters is close to one, the shear rate becomes nearly uniform in the annular gap and the flow can be approximated to a steady simple shear flow. The measuring system consists of an assembly of two concentric cylinders, a bob of outer radius r_i and a cup of inner radius r_0 with a small annular gap between them (Fig. 5.5). The fluid is filled in the gap such that the bob is completely submerged in it. One of the cylinders is allowed to rotate and the rotation of the other cylinder is arrested with the help of a linear torsional spring and the torque acting on the cylinder is read from the calibration curve of the spring. The flow analysis of such a system is presented in the following sections.

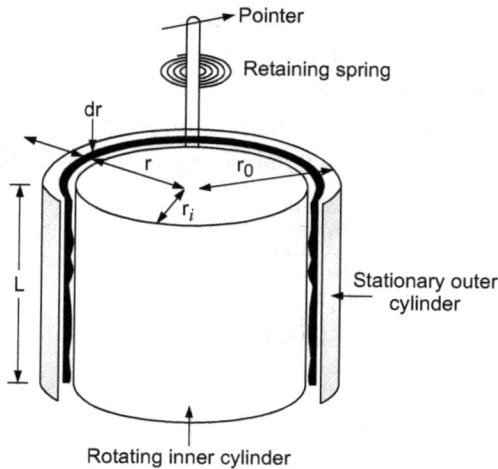

Fig. 5.5   Schematic of concentric cylinder viscometer[11]

Consider the system shown in the Figure 5.5 with inner cylinder or the bob rotating with a speed of rotation say, Ω and the outer cylinder kept stationary with the help of a torsional spring. The fluid in contact with the outer cylinder will have zero rotational velocity as no slip condition is assumed. Hence the fluid in the gap will be subjected to a velocity gradient or the rate of shear, which will be a function of radius. The rate of shear acting on the an elemental fluid cylinder of radius r and thickness δr is given as

$$\dot{\gamma} = -r\frac{d\omega}{dr} \qquad\qquad ...(5.20)$$

where ω is the angular velocity of the fluid elements at radius $r = r$. Integration of this equation leads to

$$\Omega = \int_{\omega_{r_0}}^{\omega_{r_i}} d\omega = -\int_{r_0}^{r_i} \frac{\dot{\gamma}\,dr}{r} \qquad\qquad ...(5.21)$$

The torque due to the rotation of the inner cylinder is transmitted to the outer cylinder through the rotating fluid and hence the torque acting on both the cylinders and on the fluid elements is constant. If the depth of fluid element is say L then writing the balance of moments about the axis of rotation leads to

$$T_i = \tau_i(2\pi r_i L)r_i = \tau_r(2\pi r L)r$$
$$= \tau_0(2\pi r_0 L)r_0 = T_0 \qquad\qquad ...(5.22)$$

Hence we can write,

$$\tau_i\, r_i^2 = \tau_0 r_0^2 = \tau_r r^2 \qquad\qquad ...(5.23)$$

Differentiating the equation 5.22 w.r.t. r and equating it to zero as T is independent of radius one gets

$$dT_i = 0 = 2\pi r^2 L d\tau_r + 4\pi L \tau_r r\, dr \qquad\qquad ...(5.24)$$

or $\qquad\qquad \dfrac{d\tau_r}{2\tau_r} = -\dfrac{dr}{r} \qquad\qquad ...(5.25)$

Substituting equation 5.25 in equation 5.21 and dropping the subscript r we get

$$\Omega = \frac{1}{2}\int_{r_0}^{r_i} \frac{\dot{\gamma}\,d\tau}{\tau} \qquad\qquad ...(5.26)$$

As the stress acting on the inner cylinder is a function of rotational speed, we can differentiate equation 5.26 w.r.t. τ_i to get after using the Leibnitz rule

$$\frac{d\Omega}{d\tau_i} = \frac{\dot{\gamma}}{2\tau_i} - \frac{1}{2}\frac{\dot{\gamma}_0}{\tau_0}\frac{d\tau_0}{d\tau_i} \qquad\qquad ...(5.27)$$

Combining equations 5.27 and 5.23 and eliminating τ_0 one can write

$$2\tau_i\frac{d\Omega}{d\tau_i} = \dot{\gamma}_i - \dot{\gamma}_0 \qquad\qquad ...(5.28)$$

This equation is not easily solvable as the dependence of rate of shear and shear stress on the speed of rotation is not known. It can be solved only numerically however the following special cases with simplified assumption give analytical solutions.

Case 1: When $r_0 \to \infty$, i.e. when the inner cylinder is rotating in the infinite bulk of the fluid. Such a condition is satisfied, if $r_i/r_0 < 0.1$. For this case $\dot{\gamma}_0 \to 0$ and we get

$$2\tau_i \frac{d\Omega}{d\tau_i} = \dot{\gamma}_i = 2T_i \frac{d\Omega}{dT_i} \qquad \qquad ...(5.29)$$

The above equation suggests that a plot of frequency of rotation versus torque will give the slope $d\Omega/dT_i$ and the shear stress at a particular frequency can be expressed as

$$\tau_i = \frac{\dot{\gamma}_i}{4\pi r_i^2 L} \frac{1}{(d\Omega/dT_i)} \qquad \qquad ...(5.30)$$

The apparent viscosity of the fluid can be calculated as

$$\eta_a = \frac{(dT_i/d\Omega)}{4\pi r_i^2 L} \qquad \qquad ...(5.31)$$

Thus the rheological relationship for the fluid can be established. The Brookfield viscometers operate on this principle.

Case 2: When the gap between the cylinders is very small i.e. when $(r_0 - r_i)/r_i \ll 1$, the rate of shear may be assumed to be constant throughout the fluid[28]. It implies that

$$\dot{\gamma}_i = \dot{\gamma}_0 = \dot{\gamma}_{avg} \qquad \qquad ...(5.32)$$

Substituting equation 5.32 in equation 5.26, and integrating and combining with equation 5.23 and simplifying gives

$$\Omega = \left(\frac{1}{2}\right)\dot{\gamma}_{avg} \ln\left(\frac{\tau_i}{\tau_0}\right) = \dot{\gamma}_{avg} \ln\left(\frac{r_0}{r_i}\right) \qquad ...(5.33)$$

As $\ln(r_0/r_i)$ is constant for particular equipment, the rate of shear varies linearly with the speed of rotation. Using equation 5.22 and the spring calibration the shear stress can be calculated for different values of Ω and therefore the rheological relationship for the fluid is completely defined.

These viscometers are quite versatile equipments and can give the viscosity measurements in the rate of shear range of, $10^{-3} < \dot{\gamma} < 10^4 \ s^{-1}$. The higher values of rate of shear may cause the fluid to rise in the cup and flow out of it affecting the submergence of the bob and giving the erroneous results. However at very low rate of shear the response of the equipment is rather poor. These limitations narrow the range of operation of these viscometers. The solution to the general case is rather involved[5, 34] and is not included here.

5.4 DYNAMIC SHEAR RHEOMETER

Dynamic shear rheometer is commonly known as DSR. This type of rheometer is used for research and development as well as for quality control in the manufacture of wide range of materials. These rheometers are in use since 1993 when superpave was used for characterising and understanding high temperature rheological behaviour of asphalt binders in both the molten and solid state and is fundamental in order to formulate the chemistry and predict the end-use performance of these materials.[en.wikipedia.org].

5.5 HIGH PRESSURE SLIDING PLATE RHEOMETER (HPSPR)[35]

A high-pressure sliding-plate rheometer provides an attractive to conventional instruments for measuring impact of pressure and dissolved gases on polymer behaviour. The use of CO_2 and N_2 are preferred as plasticizers and blowing agents over hydrocarbons in plastic processing. It is therefore important to know how the viscosity of molten plastic depends on pressure and dissolved gases. It is well known that the viscosity of molten polymers if often estimated using a high pressure capillary rheometer like Goettfert, Instron or Davenport rheometers. However the data analysis becomes complicated due to the entry and exit effects in these equipments. The high pressure sliding-plate rheometer has proved an ideal tool for such measurements, because pressure, shear rate, shear stress and temperature can be kept uniform throughout the sample. Though the HPSPR is a more complex instrument in comparison to other pressure driven rheometrs but it gives more reliable data and the analysis is much simpler[2b].

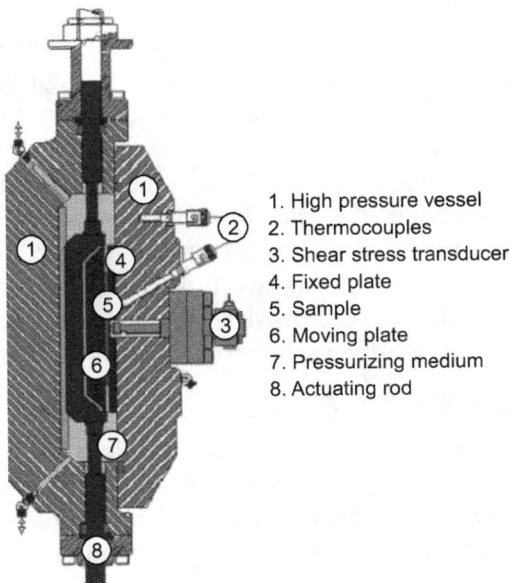

1. High pressure vessel
2. Thermocouples
3. Shear stress transducer
4. Fixed plate
5. Sample
6. Moving plate
7. Pressurizing medium
8. Actuating rod

Fig. 5.6   High pressure sliding plate rheometer (HPSPR)[35]

HPSPR requires components for pressure control and gas insertion. This rheometer can be operated on two modes, (i) for determining the effect of pressure alone, Krytox oil is used to pressurize the rheometer and the bottom inlet is used to introduce the oil and (ii) for determining the effect of dissolved gases the gas is used to pressurize the system, introducing gas through the top inlet. The instrument is capable of measuring viscoelastic properties at temperature up to 240°C and pressures of up to 70 MPa.

A sample is kept between two rectangular parallel plates., where one of the plate moves linearly relative to the other. The fixed plate generates the drag flow and the movement of the moving plate is monitored using a linear variable-displacement transducer. The shear stress is measured at the center of the sample by a shear stress transducer mounted in the fixed plate thus avoiding the end effects.

The HPSPR has been used to determine the slip velocity as it invariably occurs in linear polymers above a critical stress in melt rheometers and processing equipments. The analysis of the flow rate-pressure drop data during stick-slip or spurt in the conventional rheometer was previously thought to be due to change in molecular conformation at a critical shear stress. However it has been established that the spurt is a consequence of system instability resulting out of the combination of different factors involved in capillary flow and not indicative of change in the behaviour of the polymer itself. Another application of HPSPR is that it is possible to analyse the data to estimate the diffusivity of gas in the molten polymer[36].

5.6 HIGH PRESSURE HIGH SHEAR RATE CAPILLARY RHEOMETER[36A]

Although capillary rheometers have been discussed earlier in this chapter, the rheometer being discussed here is a high pressure high shear rate capillary rheometer, which has been specially developed for heavy and extra heavy oil in the viscosity range of 100,000 MPas and rate of shear up to 50,000 1/s and temperature of 200°C and pressure of 1000 bar. The capillary rheometer consists of two high pressure vessels connected by a tube system including a calibrated capillary. Each vessel is fitted with hydraulically operated metal bellow that can be varied in size to make the oil flow from one vessel, to another through a capillary, at a controlled rate of flow. The pressure drop is recorded and from the Q and ΔP data the viscosity can be estimate using standard approach.

5.7 EXTENSION RHEOMETERS

Polymers are subjected to the extensional deformations in different processing operations like fiber spinning, extrusion, injection moulding etc. Therefore it is important to have the knowledge of extensional rheological behaviour of the polymer melts as well as the equipment used to characterise the melts. Polymer melts are high viscosity materials having viscosity >1000 Pas and these materials are best characterized by constant-length devices[36b]. However different polymeric systems have varying viscosity range and rheometers used for different viscosity range are different. Table 5.1 lists different type of commercially available extensional rheometers.

These Rheometers are:

5.7.1 Rheotens Rheometer

Rheotens is a fiber spinning rheometer, suitable for polymeric melts. The material is pumped from an upstream tube, and a set of wheels elongate the strand. A force transducer mounted on one of the wheels measures the resultant extensional force. This system has a disadvantage as the true extensional viscosity is difficult to obtain because of pre-shear induced in the fluid as it is transported through the up-stream tube.

5.7.2 CaBER Rheometer

CaBER is a capillary breakup rheometer. A small quantity of material placed in between the plates and rapidly stretched to a fixed level of strain. The filament thus stretched necks and breaks up under different forces acting on it like surface tension, gravity, and viscoelasticity. Extensional viscosity can be calculated from the data so obtained. Sridhar et al.[37] and Anna et. al.[38] have worked on extensional deformation using this rheometer.

Table 5.1 Different Type of Commercially Available Extensional Rheometers

Instrument Name	Viscosity range [Pa.s]	Flow Type	Status of Manufacturer Rheometer Current/previous	
Rheotens	>100	Fiber spinning	Goettfert	Current
CaBER	0.01–10	Capillary breakup	Thermo Scientific	Current
Sentamanat Extl. Rheom.	>10000	Constant Length	Xpansion Instruments	Current
FiSER	1–1000	Filament Stretching	Cambridge Polymer Group	Current
RFX	0.01–1	Opposed Jet	Rhometric Scintific	Previous
RME	>10000	Constant Length	Rheometric Scientific	Previous
MXR2	>10000	Constant length	Magna Projects	Previous

5.7.3 Fisher Rheometer

Here the fluid filament is driven apart at an exponentially increasing velocity. The force and the diameter of the filament are measured as a function of time and position. The system can monitor the strain dependent extensional viscosity at constant strain rate as reported by McKinley[38] (MIT Web site[38]). Anna et al.[39, 39a] have presented a review about extensional deformations. Jung et al.[40] have reported on the instabilities occurring in extensional deformations. It has been reported that draw resonance and a Hopf bifurcation instability frequently occur in hyperbolic systems.

5.7.4 Sentmanat Extension Rheometer (Fig. 5.7(a) and (b))

Sentmanat extensional rheometer can be installed in fields on shear rheometers. A film of filament is wound on two rotating drums, which apply a constant or variable strain rate extensional deformation on polymer film. The stress is determined from the torque exerted by the drums.

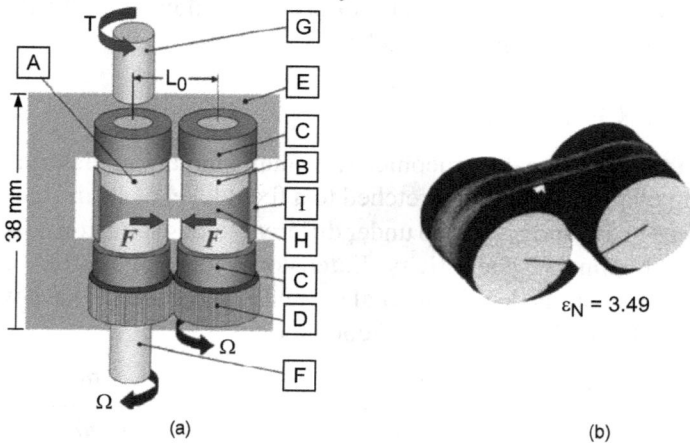

Fig. 5.7 (a) Schematic view of Sentmanat Extensional Rheometer (SER) during operation. A-Master drum, B-slave drum, C-bearings, D-intermeshing gears, E-chesis, F-drive shaft, G-torque shaft, H-sample, I-securing clamps, L_0- unsupported length, Ω-drive shaft rotation rate, T-torque, F-tangential force. (b) Sample under extension

5.8 RHEOMETERS IN RUBBER PROCESSING

In making a rubber based component the rubber compound is mixed with a large number of different additives. Different compounds include, rubber or blend of rubbers like natural rubber or synthetic rubbers like SBR, CR etc. vulcanizing agents like sulfur, reinforcing agent like carbon black and/ or silica fillers, antioxidants like phenolic antioxidant TQ, curing agents or curatives like sulfur, peroxides etc. sulfur donors like dithiodimorpholine (DTDM), or tetramethylthiurium disulfide (TMTD), cure activators like Zink oxide or magnesium oxide, accelerators like benzothiazole (MBT), retarders like cyclohexylthiophthalimide (CTP), peptizers like pentachlorothiophenol derivatives, (PCTP or ZPCTP), softeners like mineral oils or animal and vegetable oils and miscellaneous compounds like pigments, blowing agents etc.

There are a number of parameters like temperature, humidity, time of mixing and storage conditions which influence the performance of rubber products. After mixing these components in two-roll mill and /or Banbury mixer the compound is tested using either by Moony viscometer or other more advanced rheometers like moving die rheometer or oscillating disk rheometer.

5.8.1 Moving Die Rheometer (Gibitre Instruments)[41]

Moving die rheometer is considered the most modern version of rheometers. The sample is placed between the two dies where lower die oscillates with ±0.5 degrees, the upper die is connected to the torque sensor to measure the torque acting on the rubber. The temperature of operation is normally around 170°C to 190°C. For platinum cured silicon rubber it is 100°C to 120°C. To prevent the sample from being contaminated, a plastic film normally polyamide or polyester is placed between the dies and the sample. This equipment has high reproducibility and repeatability with very fast thermal recovery.

5.8.2 Oscillating Disk Rheometer

It is the most traditional rheometer where the sample is placed between the two dies and a rotor placed between the dies oscillates with ±3° or ±1°. A torque sensor measures the torque response of the rubber at deformation. The temperature of operation is in the range between 170°C to 190°C. It takes slightly more time than moving die rheometer to complete the test. To get the repeatable results, the time between tests should be limited. This instrument require very little maintenance and has quite stable calibration.

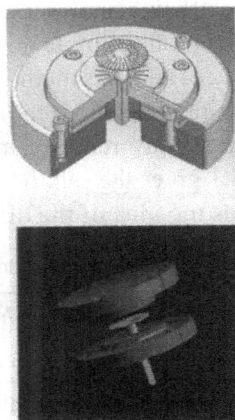

Fig. 5.8   Oscillating Disk Rheometer
[GIBITRE INSTRUMENTS]

5.9 THE IN-LINE/ON-LINE RHEOLOGICAL MEASUREMENTS [mep.net.au]

The rheological controllers can be used in four different modes as the situation demands. They can be used as in-line, on-line, at-line or off-line controllers. These different modes are briefly explained below.

The in-line analysers are simple probes or measuring devices that are placed directly in to the process stream. They are used to measure such variables like pH, temperature, pressure, density and flow. These analysers give direct data without much time loss.

The on-line analysers are fully automated systems used to closely monitor (1-12 results per hour) the concentrations of components that are critical to production process. These analysers are able to control external devices such as pump or valves as part of their analysis sequence and, as the results are exported automatically, the analyser can control the external devices based on their analysis ie turning on/off the dosing pump. The analysers can be controlled remotely via a modem. These analysers may require 5-10 minutes time to analyse the results. These generate results with higher degree of consistency and certainty as the human factor is removed from the analysis procedure, increase the operators safety and also save man-hours by removing routine testing and involvement of operators.

At-line controllers are other types of measuring, controlling and analysing devices in use. These analysers are housed in production area and operated by an operator. These are configured to perform a series of tests on a variety of samples and export the results automatically to the analysing units. These instruments can generate an alarm if any of the result is out of range and automatically turn on/off the pump or valves.These analysers need about half an hour time to analyse the data.

The off-line analysers are designed to be in an environmentally controlled location and used by technically trained personal. These analysers offer most verstality of analysis methods but require most man-hours to perform an analysis and the input results requiring longest result turn around time. The off-line instrument may requirre more than an hours time to complete the analysis. Table 5.1 gives the relative idea of different analysers performance. The figure 5.9 gives the idea of the different process controllers.

Table 5.1 Differences in In-line, Off-line, At-line and On-line analysers (mep.net.au)

System Variables	In-line	Off-line	At-line	On-line
Installation site	In Process	Laboratory	Plant	Process
Analyser Protection	Normal	IP 23 IP54/65	IP66	*H. A. C.
Sampling	Automatic	Manually	Manually	Automatic
Sample points	Continuous	Many	Many	1-10
Analysis frequency	Continuous	Low	Low-Medium	High
Turn around time	Nil	>60 min.	~ 30 min.	5-10 min
Parameters	Many	Many	1-4	1-4
Meas.-Range	Variable	Flexible	Fixed	Fixed
Analysis type	—	—	All Routine	Routine
Sample preparation	Not required	Manually	Manually	Automatic

System Variables	In-line	Off-line	At-line	On-line
Analysis Procedure	Automatic	Manually	Automatic	Automatic
Operation	Automatic	Lab Asst.	Lab Asst.	Proc.Techn.
Maintenance	Plant Techn.	Lab Asst.	Lab Asst.	Proc.Techn.
Control	Plant techn.	No	Simple control	Closed Loop Cont.
Exchange of data	—	LIMS LIMS/ Proc	Sigl. LIMS/	Proc. Sigl.

This table has been modified by adding In-line data.
* HAC-Hazardous area Certificate

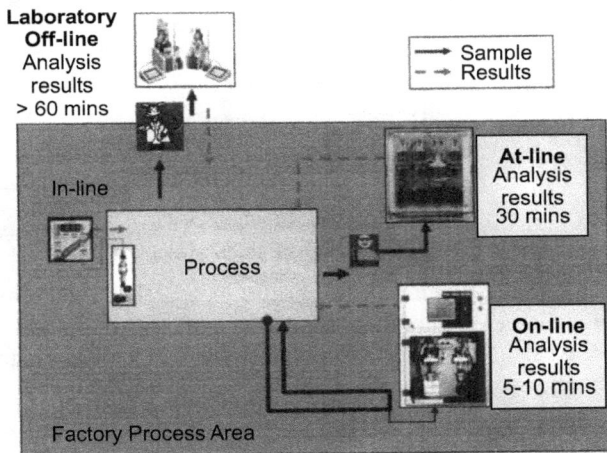

Fig. 5.9 The 4 basic options of instrumentations to perform the data analysis by control equipments [mep.net.au, jw@mep.net.au.]

During the polymer processing like polymerisation, extrusion and melt forming, elastomers compounding, blending of the two or more polymers, dynamic cross linking, reactive extrusion etc. the information about the variation of viscosity with process parameters is very important for quality and process control. The viscosity of high molecular weight polymers melt and solutions is very sensitive to the change in molecular weight, temperature, shear stress and rate of shear and concentration (for the solutions of polymers) and therefore plays an important role in the control of process and the product quality. The on-line and in-line rheological measurements are used to meet these requirements[42].

The on-line rheometers draw the melt or the solution from the process stream using a gear pump through a side stream and the measurements can be carried out at any desired temperature. Such a system suffers from a slight time delay, as the material has to be pumped to the sensor location. In the in-line measurement on the other hand the sensor like stress transducer is placed at a suitable location in the process stream and an instant measurement can be carried out but only at the process conditions of temperature and flow rate.

The capillary viscometer, melt flow indexer and rotational viscometers have been reported for the on-line and in-line measurements applications. Dynisco Kayeness Polymer Test Systems, Morgantown, PA, USA manufacture both capillary rheometers and the melt indexers for on-line control of polymer processing.

PROBLEMS

1. A concentric cylindrical viscometer is used to estimate the rheological behaviour of the LDPE melt at 150°C. The inner diameter of the outer cylinder, $d_0 = 30$ mm and outer diameter of the inner cylinder, $d_i = 28.4$ mm and height of the inner cylinder $h = 50$ mm. The inner cylinder is rotated with different speeds and the torque is measured at the outer cylinder. Determine the rheological nature of the melt and develop the power law equation for the system. The speed of rotation-torque data is

Rotation speed, Ω, rpm	5	15	30	50	100	300	500
Torque, T, Nm	1.22	1.83	2.34	2.81	3.61	5.36	6.44

[**Hint:** As $(r_0 - r_i)/r_i = 0.056 \ll 1$ the case –2 applies]

2. For the LDPE system given in Problem 1 the measurements were also carried out using a large diameter outer cylinder so that the Case-1 applies. Using the data of Problem 1 calculate the apparent viscosity as a function of rate of shear and shear stress and estimate the Ellis model parameters.

3. The LDPE melt in Problem 1 flows through a capillary of $L/D = 20$ and $D = 3.0$ mm. Using the apparent shear stress and rate of shear data calculated in Problem 2 develop the pressure drop- volumetric flow rate plot for the system.

4. A cone and plate viscometer with the diameter of plate equal to 12 cm and a cone angle of 2° is to be calibrated using a standard Newtonian fluid of viscosity 10.0 poises. Develop the speed of rotation versus torque plot.

5. For a cone and plate viscometer derive the expression for the torque for a power law fluid.

6. Use the Carreau model to derive the expression for the torque for a cone and plate viscometer.

7. The cone and plate viscometer given in Problem 4 is used to determine the rheological behaviour of polypropylene melt at 230°C. The frequency of rotation and the torque data as recorded are given. Determine the rheological behaviour of the melt and assuming that the power law model applies determine the model constants.

Frequency of rotation, rad./mt.	6	30	54	90	180	360	600
Torque, Nm	9.0	14.5	17.1	20.0	24.2	30.6	34.4

8. From the geometry of the Mooney viscometer derive an expression for the shear viscosity in terms of the torque and the speed of rotation of the rotor.

9. For a melt flow indexer determine the expression for the melt shear viscosity in terms of MFI value and the dimensions and the load on the melt flow indexer.

10. For a Walace rapid plasticity meter derive an expression for the shear viscosity in terms of the dimensions of the plasticity meter and the plasticity number.

11. The experimental viscosity measurements can be done using a circular parallel plate viscometer. The upper plate of known radius is rotated at an angular velocity of Ω and the lower plate of same radius is kept stationary and the polymer melt or solution fills the gap between the plates. Assuming that the velocity profile is linear at any radial position derive the expression for the torque needed to rotate the upper plate. Explain how will you use this viscometer for generating the rheological data?

12. In the parallel plate viscometer in Pr. 11 what is the essential design feature for the velocity profile to be linear. If the velocity profile is not linear discuss how will you proceed to develop the rheological data?

13. What do you understand by the terms (a) In-line, (b) On-line and (c) Off-line rheological measurements. Discuss different viscometers used for On-line and In-line measurement of viscosity of polymer melts and solutions.

14. Carry out the literature review for On-line and In-line measurement of viscosity of polymer systems containing blowing agents for the production of the polymer foams.

15. Write the balance of moments about the axis of rotation of a cylinder in a concentric cylindrical viscometer and derive equation 5.23.

REFERENCES

1. F. R. Eirich, Rheology, Vol. 3, Academic Press, New York, 1968.
2. J. R. van Wazer, L.W. Lyons, K.Y. Kim and R. F. Colwell, Viscosity and Flow Measurements, 1st. ed., Interscience Publishers Inc., New York, 1963.
   (a) F. Koran and J. M. Dealy; J. Rheol., **43**, 1279-1290 (199).
   (b) H. E. Park, S. T. Lim, H. M. Laun and J. M. Dealy, Rheol. Acta, **47**, 1023-1038 (2008).
3. J. M. Dealy, Rheometers for Molten Plastics, Van Nostrand Reinhold, New York, (1982).
4. R. B. Bird, W. E. Stewart and E. N. Lightfoot, Transport Phenomena, Wiley and Sons, New York, (1960).
5. A. Kumar and S. K. Gupta, Fundamentals of Polymer Science and Engineering, Tata McGraw Hill Pub. Co. Ltd., New Delhi, (1978).
6. E. B. Bagley, J. Appl. Phys., **28**, 624 (1957).
7. F. N. Cogswell, Polymer Melt Rheology, John Wiley & Sons, New York, 1981.
8. H. M. Laun and H. Schuch, J. Rheol., **33**,19 (1989).
9. W. Phillipoff and F. H. Gaskins, Trans. Soc. Rheol., **2**, 263 (1958).
10. Y. Ito and S. Shishido, J. Polymer Sci. Polym. Phys. Ed., **10**, 2239 (1972).

11. J. M. Coulson and J. F. Richardson, Chemical Eng. Vol. 3, The English Language Book Society & Pargaman Press, Oxford, (1971).

12. J. M. Dealy and K. F. Wissbrun, Melt Rheaology and its Role in Plastic Processing, Theory and Applications,Von Nostrand Reinhold, New York, (1990).

13. G. Heuser and E. Krause, Rheol. Acta, **18**, 553, (1979).

14. J. M. H. Fortuin, Chem. Eng. Sci., **40**, 111 (1985).

15. P. T. Gavin and R. W. Whorlow, J. Appl. Polym. Sci., **19**, 567 (1975).

16. D. S. Pearson and W. E. Rochefort, J. Polym. Sci., Polym. Phys., **20**, 83 (1982).

17. L. M. Quinzani and E. M. Vall'es, J. Rheol., **30**, 51 (1986).

18. W. Gleissle, Colloid and Polym. Sci., **252**, 848 (1974).

19. W. Gleissle, Rheol. Acta, **15**, 305 (1976).

20. F. Nazem and M. G. Hansen, J. Appl. Polym. Sci., **20**, 1355 (1976).

21. Z. Tadmor and C. G. Gogos, Principles of Polymer Processing, John Wiley & Sons, New York, pp. 179-181 (1979).

22. R. I. Tanner and M. Keentok, J. Rheol. **27**, 47 (1983).

23. A. S. Lodge, J. Non- Newt. Fl. Mech., **14**, 67 (1984).

24. I. Bruker, Rheol. Acta, **25**, 501 (1986).

25. J. Meissner, J. Appl. Polym. Sci., **16**, 2877 (1972).

26. R. L. Crawley and W. E. Graessley, Trans. Soc. Rheol. **21**, 19 (1977).

27. A. S. Dodge and I. M. Krieger, Rheol. Acta., **8**, 480 (1969).

28. T. M. T. Yang and I. M. Krieger, J. Rheol., **22**, 413 (1978).

29. T. T. Tee and J. M. Dealy, Trans. Soc. Rheol. **19**, 595 (1975).

30. S. Onogi, T. Masuda and T. Matsumoto, Trans. Soc. Rheol., **14**, 275 (1970).

31. T. Matsumoto, Y. Segawa, Y. Waroshina and S. Onogi, Trans. Soc. Rheol., **17**, 47 (1973).

32. S. Onogi, and T. Matsumoto, Polym. Eng. Rev., **1**, 45 (1981).

33. J. M. Dealy and T. K. P. Vu, J. Non-Newt. Fl. Mech., **3**, 127 (1977-78).

34. L. M. Krieger and H. Elrod, J. Appl. Phys. **24**, 134 (1953).

35. J. Dealy and H. E. Park, Permalink, December 23, (2010).

36. H. E. Park and G. M. Dealy, Macromol., **39**, 5438-5442 (2006).

    (a) S. Pollok, S. Hüttemann, S. E. Quiñones-Cisnerose, and E. Weidner, J. Petrom. Sci. Eng. **157**, 581-587 (Aug. 2017).

    (b) C. Tropea (Edr.), A. Yarin (Edr.) and J. F. Foss (Edr.), Springer Handbook of Experimental Fluid Mechanics 2007th edition, Chapter 9.1, (2007)

37. T. Sridhar, J. non-Newtonian Fluid Mech., **40**, 271-280 (1991).

38. G. McKinley, " A decade of filament stretching rheology " web.mit.edu.

39. S. L. Anna, C. Roger and G. H. McKinley, J. Non-Newtonian Fluid Mech.**87**, 307-355 (1999).

    (a) S. L. Anna, G. H. McKinley, D. A. Nguyen, T. Sridhar, S. J. Muller., j. Huang and D. F. James, J. Rheol., **45**, 83-114 (2001).

40. H. W. Jung, J. C. Hyun, Rheology Review , pp131-164 (2006), (www.bsr.org.uk).

41. Gibitre Instruments, "Use of rheometers for rubber componds control".

42. J. F. Dealy and K. F.Wissbrun, 'Melt Rheology and its Role in Processing processing , Theory and Applications", Von Nostrand Reinhold, New York (1990).

❑ ❑ ❑

Viscoelastic Behaviour

6

Polymers and rubbers, particularly amorphous and semicrystalline polymers, biopolymers, ligaments and tendons, metals at very high temperatures and bitumin materials are known to be the viscoelastic materials. This chapter deals with viscoelastic behaviour of polymers and rubbers in general. These materials when subjected to the deformation are capable of storing a part of the deformational energy and dissipate the other part as viscous heat. The stored energy is the elastic component, which is released in the relaxation process. The experimental techniques used for the characterization of either purely elastic materials or purely viscous systems cannot be applied to characterize the viscoelastic materials. As the viscoelastic systems possess both viscous and elastic characteristics, the experimental techniques for their characterization should be capable of giving combined response of both these natures as pointed out by Ferry[1] and Dealy and Wissbrun[2].

When the polymer melt is deformed it tends to acquire a new equilibrium in deformed state due to its elastic nature whereas no such influence is caused due to the viscous energy, which simply gets dissipated as heat. For purely elastic deformation all internal stresses are functions of stress history and instantaneous extent of deformation and for purely viscous deformation these are the functions of the rate of shear. In case of the viscoelastic materials the deformational stresses are function of extent of deformation, rate of deformation and time. This relationship can be expressed as

$$\tau(t) = f(\gamma, \dot{\gamma}, t) \qquad \qquad ...(6.1)$$

As the stress, rate of shear and the viscosity are the complex functions of time, the unsteady state response of the viscoelastic materials is of considerable importance and these materials are characterized mostly based on their transient response. The transient response under static loading gives important bulk properties like modulus, relaxation time etc., and that under the dynamic loading gives the dynamic properties like storage modulus, loss modulus, tan δ etc., which help to understand the molecular interactions. The experimental techniques used for the characterization of these materials are[3],

(i) **Static Stress Relaxation:** Stress relaxation test under static condition involves the application of a step change in the strain and the study of the resulting stress response with time.

(ii) **Static Creep Test:** Static creep behaviour is studied by subjecting the system to a step change in the stress and the study of resulting deformation with time.

(iii) **Dynamic Stress Relaxation:** It involves the stress relaxation study under the dynamic strain loading. The resulting stress also is a dynamic function, which lags the input strain by a certain degree.

(iv) **Dynamic Creep Test:** The creep test is carried out under the dynamic stress application. The resulting deformation is also a dynamic function, which lags the input stress by a certain degree.

6.1 STATIC TESTS

6.1.1 Stress Relaxation

Stress relaxation is the decrease in stress in response to the some amount of strain generated in the system. When the structure is kept in strained condition for some finite time interval it gets subjected to some amount of plastic strain. The stress relaxation is similar to the cold springing process but occurs relatively over a longer period of time. It may be pointed out here that the cold springing is the processes of intentional deformation of piping during assembly to produce a desired initial displacement and stress. This helps the piping system during the erection cycle to permit the system to attain more favorable reactions and stress under the operating conditions.

The stress relaxation describes the process through which the polymers relieve stresses under constant strain. As these are viscoelastic materials, polymers behave in a nonlinear, non-Hookean fashion[3a]. The nonlinearity is described by both, (1) stress relaxation and (2) creep.

Static stress relaxation test is performed when the test piece is subjected to a step input in strain at a particular time and the variation of stress with time is recorded. Figs. 6.1 (a), (b), (c) and (d) show a step input in strain, $\gamma = \gamma_0$ at time, $t = t_0$ and the corresponding response for purely elastic system, for purely viscous liquid and for a viscoelastic material respectively. As can be seen the purely elastic system gets subjected to a finite value of stress, which remains constant over the time of application of the input strain. A purely viscous liquid experiences an instantaneous rise in stress, which falls to zero value simultaneously at the time of the application of strain i.e. at $t = t_0$ and then remains constant at zero value. Whereas the viscoelastic materials show an increase in the stress to some finite value at the time of application of strain and then gradually falls as time passes. For purely elastic solids the governing rheological equation is the Hooke's law i.e. $\tau = G_i\,\gamma$, and therefore the stress acting on the system will be say, $\tau_0 = G_i\gamma_0$ and it will remain constant till the strain continues to be acted upon (Fig. 6.1(b)). As these materials are capable of storing all the deformational energy the stress remains constant over the period.

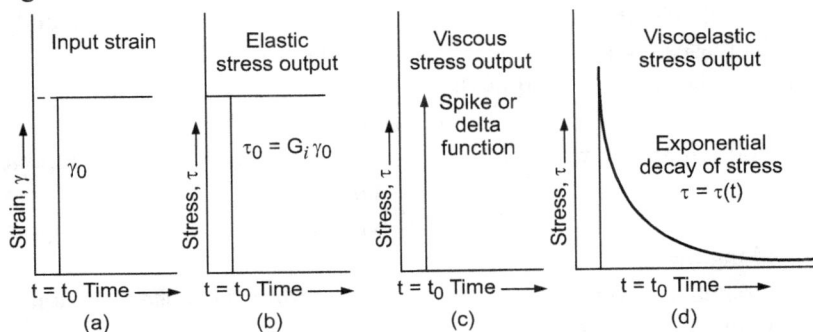

Fig. 6.1 The response of different materials under stress relaxation test

Purely viscous liquids show the stress proportional to rate of shear as in the case of Newton's law for simple fluids. At the time of application of strain the liquid simultaneously experiences a rise to infinity in stress, which falls to zero value instantly thereby giving a delta or spike function in the variation of stress with time (Fig. 6.1(c)). The liquids are incapable of retaining deformation and instantly dissipate energy of deformation as heat hence the stress falls instantly to zero value.

The viscoelastic materials on the other hand have both the characteristics of elastic solids and viscous liquids hence part of the energy is stored giving an instant increase in the stress to some value, which is less than τ_0 as some energy is dissipated.

The polymer molecules store the elastic energy as they are instantly stretched and elongated. As the time passes the molecules slowly disentangle and slip thus giving reduction in the stress. This phenomenon is known as the stress relaxation. As a result of this, the stress continuously decreases to acquire an equilibrium value known as the residual stress (Fig. 6.1(d)). The polymer molecules relax in different modes i.e.

(i) By elementary bond flip (conformational change), as many such bonds flip together it allows the entire molecular chain to return to a random configuration and hence relax stress, leading to equilibrium state or random walk.

(ii) If unentangled, the chain relaxes to their random walk statistics by free Rouse motion[3b].

(iii) If entangled, the chains relax more slowly to their random walk statistics because the Rouse motion is confined to occure inside the tube of surrounding chains.[3b, 3c, 3g]. Entangled ring polymers and, the long linear and branched polymers have a characteristic entanglement plateau and their stress relaxes by chain reptation or branch retraction. In both mechanisms, the presence of chain ends is essential. However the entangled polymers without ends exhibit self-similar dynamics yielding a power-law stress relaxation.[3g]

(iv) Polydispersity greatly increases the longest relaxation time and thus increases the stored elastic energy in the flow.

(v) The relaxation of star-shaped polymer molecules in a melt has been studied by Dubbeldam and Molenaar[3d]. They reported on the mathematical modelling of Rouse equation with time dependent boundary conditions, which arises due to tube dilution. They reported that the tube diameter increases with time depending on the position along the tube.

(vi) Branched polymers relax by *"arm retraction"* because reptation is not allowed by branch points.

(vii) Branching makes for *"more chain ends"* per unit volume. Thus branched chains have more arm retraction relaxations induced by strong flow and exhibit stronger shear thinning than linear polymers.

Stress relaxation dynamics of model branched homopolymer with a range of architectures $A2$-B (T-Shaped) and A-$B2$ (Y-Shaped) asymmetric stars, A-Bn (combs), $B2$-A-$B2$ (H-shaped) and $B3$-A-$B3$ (pom-pom) have been investigated. They have used the tube-based theory to evaluate branch point motion in hierarchically relaxing branched molecule. It has been possible to formulate a self-consistent theory suitable for describing linear viscoelasticity of any branched polymer system.[3e]

The investigations through Brownian dynamics simulations on stresses and configurational relaxation of an initially straight flexible polymer, has been reported by Dimitrakopoulos[3f]. It is reported that after a short-time free diffusion, the configurational relaxation is anisotropic and at intermediate times the chain's width grows while its length is reduced. During the long time the polymer length shows an exponential decay towards the equilibrium coil like shape and after a very long time the chain rotation is found to be significant. In addition to the longitudinal relaxation a quasi-steady equilibrium of the link tension is shown to exist along the chain length.[3g]

Relaxation in oriented polymers like polymer fibers has been reported by Makarov et al.[3h]. It has been found that the stress relaxation is caused by thermal fluctuation, and the thermal conformational transition in stretched chain polymer molecule occurs from the beginning of the stretching of the polymer fiber. For polypropylene fiber the stress relaxation in all initial stages of deformations has been found to be time dependent.

The effect of induced orientation of polymer melts on bidispersed entangled polymer solutions has been reported by Deutsch and Pixley[3i]. The shorter chains, which embed in a majority of longer chains are known to orient by coupling with them. It has been reported by using a computer simulation that the results could be explained in terms of stress fluctuation in polymer melt and chain screening. The chain fluctuations freeze on relaxation time of longer chains, and these would induce strong orientational coupling in shorter chains.

The modulus, which is the ratio of the output stress to the input strain, also follows the same trend as stress. Table 6.1 gives the stress relaxation response of different materials.

Table 6.1   Stress relaxation response of different materials

Time	Materials		
	Elastic	Viscous	Viscoelastic
1. $t < t_0$	$\tau = G_i \gamma;\ G_i = $ Const. $\tau = \gamma = 0$	$\tau = \eta_i\, \dot\gamma = 0;$ $\gamma = \dot\gamma = 0$	$\tau = f(\gamma,\, \dot\gamma) = 0;$ $\gamma = \dot\gamma = 0$
2. $t = t_0$	$\gamma = \gamma_0;\ \tau = \tau_0 = G_i \gamma_0$	$\dot\gamma \to \infty;\ \tau \to \infty;$ $\gamma = \gamma_0$	$\tau = \tau_0 = G_i \gamma_0$ $\gamma = \gamma_0$
3. $t > t_0$	$\gamma = \gamma_0;\ \tau = \tau_0 = G_i \gamma_0$	$\dot\gamma = 0$ and $\tau = 0;$ $\gamma = \gamma_0$	$\tau(t) = [\phi(t) + G_\infty]\gamma_0$ $\gamma = \gamma_0$

6.1.2 Relaxation Modulus

Relaxation modulus, $G(t)$ is defined as the ratio of output stress, $\tau(t)$ to the input strain, γ_0 i.e.

$$G(t) = \frac{\tau(t)}{\gamma_0} = \phi(t) + G_\infty \qquad \text{...(6.2)}$$

where $\phi(t)$, is the relaxation function and G_∞ is the equilibrium or the residual modulus.

The residual modulus represents the stress the system will retain even after infinite time i.e. to say that the system will never acquire a zero stress state even after very long time of relaxation process. Hence as $t \to \infty$ the function $\phi(t) = \phi(\infty) \to 0$

Therefore,
$$G_\infty = \frac{\tau_\infty}{\gamma_0} \qquad \text{...(6.3)}$$

At time $t = 0$ the material will be subjected to the initial stress, τ_0, hence the initial modulus, G_0 will be

$$G_0 = \frac{\tau_0}{\gamma_0} = \phi(0) + G_\infty \qquad \text{...(6.4)}$$

Thus it can be said that the stress experienced by the material under the step change in the strain has two components (i) relaxing component i.e. a time dependent stress and (ii) an equilibrium or the residual component, which the system retains even after the applied strain has been withdrawn.

Example 6.1. A tensile test piece of a rubber vulcanizate having dimensions 2 mm (thickness) × 4 mm (width) × 25 mm (length) is subjected to an elongation of 400% and a time-true stress data is recorded as

Time, t, s	0	10	20	30	40	50	60	70	80	100	130	150
Stress, τ, MPa	55.0	42.2	33.5	27.0	20.8	17.1	13.5	10.5	8.5	5.5	2.6	2.0

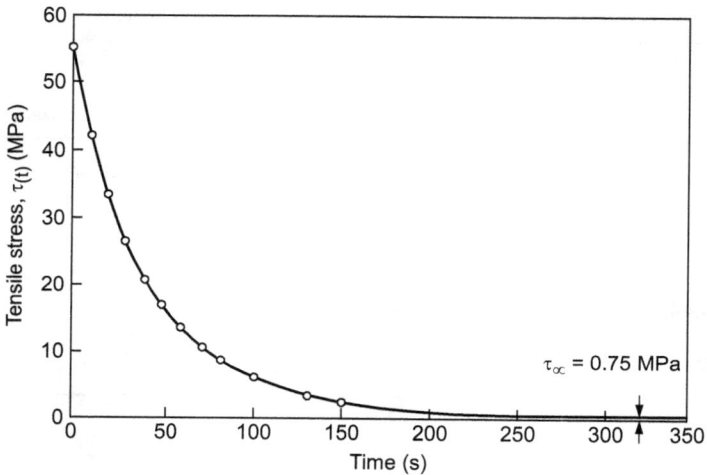

Fig. E-6.1   Stress relaxation plot for example 6.1

From the plot of the stress versus time, calculate (a) Instantaneous stress and modulus and (b) Residual or equilibrium stress and the modulus.

Solution: The strain $\gamma = 400\% = \left(\dfrac{400}{100}\right) = 4.0$

From the plot of true stress versus time the following results can be calculated. The instantaneous stress, $\tau_0 = 55.0$ M Pa, and the residual stress is, $\tau_\infty = 0.75$ M Pa

The instantaneous modulus,

$$G_0 = 13.75 \text{ M Pa and}$$

The residual modulus, $G_\infty = 0.188$ M Pa

6.2 CREEP TEST

Creep is a slow deformation under applied stress. The test piece is subjected to the step change in the stress at a particular time interval and the change in the strain or the extent of deformation is measured over the period of time. The response of different materials under the applied stress considerably differs due to the variation in their rheological behaviour. Figs. 6.2(a), (b), (c) and (d) show the input function, and output functions for elastic, viscous and viscoelastic materials respectively.

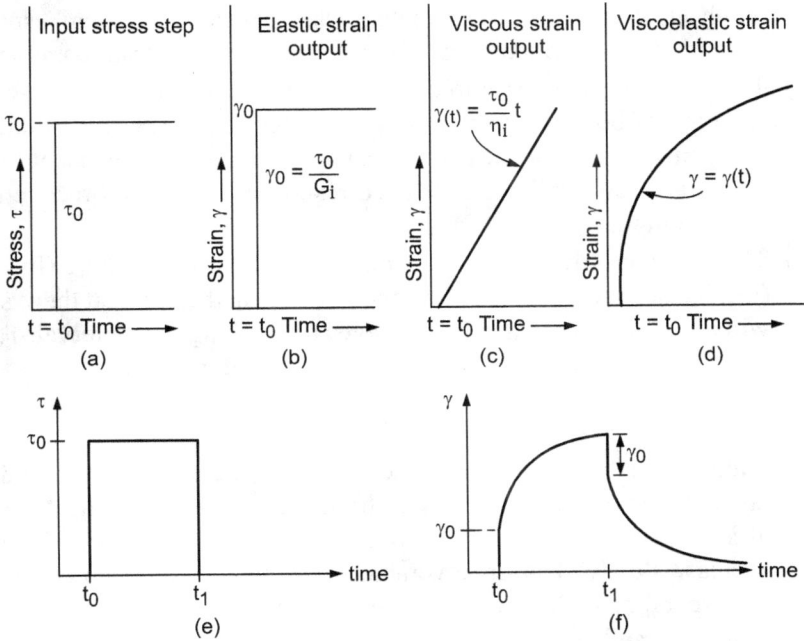

Fig. 6.2   Creep response of different materials (e), and (f) applied stress and induced strain as function of time over a short period for viscoelastic material. [en.wikipedia.org]

Purely elastic materials under the creep deformation get instantly subjected to finite value of the strain, γ_0 at the time, t_0 of application of step change in stress, τ_0. The magnitude of strain depends on the modulus of the system. Thus at time, $t = 0$, $\gamma_0 = \tau_0/G_i$ (Fig. 6.2(b)).

Purely viscous fluids start flowing and the strain continuously increases with time at a constant rate of strain giving linear increase for Newtonian liquids. The output strain therefore is a linear function of time and acquires a final value, which depends on the viscosity of the fluid. Hence at time $t = t_0$ for Newtonian fluids one gets $d\gamma/dt = \dot{\gamma} = \tau_0/\eta_i = $ const. and at some time $t = t_1$ the strain, $\gamma_{t_1} = (\tau_0/\eta_i)t_1$ (Fig. 6.2(c)).

The viscoelastic systems show a non-linear increase in the strain with time and acquire a finite maximum value depending on the extent of energy dissipation due to the viscous drag or the viscosity value. During stain response two situations can be visalised, (i) application of stress (below the critical stress) continuously over a long period of time and (ii) the application of stress over short duration. During the application of stress the materials may fail at some stress i.e. the critical stress and the strain response would not be possible to measure. Under these cases of strain variation the following behaviours are observed.

(a) Under continuous stress application it shows

(i) An initial instantaneous rise at $t = t_0$, which is the consequence of purely elastic response of the material due to which the system responds instantly to the applied stress. This is a recoverable strain component.

(ii) The time dependent rise to a finite value as a result of the combined response of both elastic as well as the viscous natures of the material as it strains due to elastic component and flows due to the viscous part, which is known to have a delayed response. It is also a recoverable strain component.

(iii) The time dependent component due to purely viscous nature (Fig. 6.2 (d)). It is an irrecoverable component implying thereby that the system will retain some permanent strain even after a long time interval. Table 6.2 gives the creep response conditions for different systems.

(b) Under the short duration application of stress:

(i) If the stress is removed after application, the system shows first a sudden fall and then a gradual decrease. The sudden fall is due to the recovery of instantaneous strain and, then the irrecoverable strain continues to decrease slowly. The irrecoverable strain, which equals to $(\tau_0 t/\eta)$, is corresponding to the energy lost due to the viscous nature of the system (Fig. 6.2 e and f)[4].

Polymer molecular weight is known to affect the creep behaviour. The increase in molecular weight of polymer tends to promote secondary bonding between the polymer chains making the system to be more creep resistant. Similarly aromatic polymers are even more creep resistant due to the added stiffness from rings. Both molecular weight and aromatic rings add to the thermal stability increasing creep resistant of a polymer[3a].

MacCrum et al.,[4a] have reported that both polymers and metals can creep. however there are three main differences between metallic and polymeric creep. Metals show different types of creeps ie. Dislocation creep, Nabaro-Herring creep, Coble creep, Solute drag creep, dislocation climb-glide creep, and Harper-Dorn creep. Creep in the metallic structures like chemical plant towers, power towers, ships and tall buildings can have disastrous consequences particularly at higher temperatures.

Table 6.2 Creep response conditions for different systems

Time	Systems		
	Elastic	Viscous	Viscoelastic
1. $t < t_0$	$\tau = G_i\gamma$ $\tau = \gamma = 0$	$\tau = G_i\gamma$ $\tau = \gamma = \dot{\gamma} = 0$	$\tau = f(\gamma, \dot{\gamma})$; $\tau = \gamma = \dot{\gamma} = 0$

2. $t = t_0$	$\gamma = \tau_0/G_i$ $\gamma = \gamma_0$ = const.	$\tau = \tau_0 = \eta\,\dot{\gamma}$ $\tau_0/\eta = \dot{\gamma}$ = const.	$d\gamma/dt$ = finite $\gamma = \gamma_0$ = initial strain
3. $t > t_0$	$\gamma = \gamma_0$	γ = finite	$\gamma = f(t)$; $\dot{\gamma}$ = finite

Polymers experience significant creep at temperatures above ca. −200°C. The polymers show creep basically in two different ways. At typical work loads (5 up to 50%) ultra high molecular weight polyethylene (spectra, Dyneema) will show time-linear creep, whereas polyester or aramids (Twaron , Kevlar) will show time-logarithmic creep. Thus it is important to give due considerations of creep while designing polymer processing equipments, polymer based heavy components and structures. [en.wikipedia.org/wiki/creep_(deformation)]

6.2.1 Creep Compliance

Creep compliance is defined as the ratio of output strain, to the input stress. For a given stress input the resultant strain is a function of time and therefore the compliance also is a function of time. In accordance with the strain the compliance also has three components, (i) instantaneous compliance at time $t = t_0$, (ii) time variant recoverable compliance and (iii) irrecoverable compliance. The strain experienced by the sample is given as

$$\gamma(t) = J_0\tau_0 + \psi(t)\tau_0 + \left(\frac{t}{\eta}\right)\tau_0 \qquad \qquad ...(6.5)$$

The corresponding compliance is given by

$$J(t) = \frac{\gamma(t)}{\tau_0} = J_0 + \psi(t) + \frac{t}{\eta} \qquad \qquad ...(6.6)$$

where, J_0 is the instantaneous recoverable compliance
   $\psi(t)$ is the creep or the memory function
   $J_0\tau_0$ is the instantaneous recoverable deformation
   $\psi(t)\tau_0$ is the time variant recoverable deformation
   $(t/\eta)\tau_0$ is the irrecoverable deformation

The instantaneous compliance determines the deformation at the initial moment i.e. at $t = t_0$. At this instant the time dependent compliance is zero i.e. $\psi(t) = 0$. For the large time interval as, $t \to \infty$ the body will acquire an equilibrium value of the recoverable compliance, which is

$$J_\infty = J_0 + \psi(\infty) \qquad \qquad ...(6.7)$$

The compliance J_∞ corresponds to the maximum recoverable deformation of the body under the specified stress.

6.3 LINEAR VISCOELASTICITY

A linear viscoelastic fluid is a fluid, which has the linear relationship between its strain history and its current value of stress. The material is known to posses the linear viscoelasticity if,

(i) Under the stress relaxation test the out put functions, $\phi(t)$ and G_∞ are independent of the input strain magnitude, γ_0, and

(ii) Under the creep test the creep functions, $\psi(t)$, η and J_0 are independent of the magnitude of the input stress, τ_0.

EXAMPLE 6.2. A tensile test piece of a rubber vulcanizate having dimensions $2 \times 4 \times 25$ mm was subjected to a tensile force of 120 N and gave the following data. The force was removed at $t = 800\,s$ giving an instantaneous recovery in strain.

Time, s	0	10	50	100	200	300	500	600	800
Specimen, length l_e, mm	50.0	75.0	101.0	114.0	130.0	141.0	153.0	158.0	168.0 to 144

Calculate (a) the instantaneous compliance and (b) the compliance at $200\,s$

Solution: (a)
$$\varepsilon_0 = \frac{50 - 25}{25} = 1$$

and
$$\tau_0 = \frac{\text{Force}}{\text{Area}} = \frac{120}{2 \times 4 \times 10^{-6}} = 15.0 \times 10^6 \text{ Pa.}$$

The instantaneous compliance,
$$J_0 = \frac{\text{Strain, } \varepsilon_0}{\text{Stress, } \tau_0} = \frac{1}{15 \times 10^6} = 0.067 \times 10^{-6} \text{ (Pa)}^{-1}$$

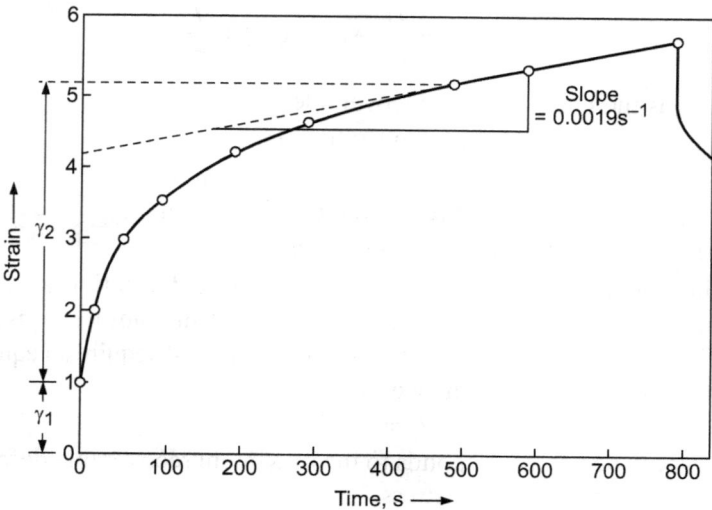

Fig. E-6.2 Creep curve for example 6.2.

(b) Compliance at $t = 200$ s, J_{200}

Length of the specimen = 130 mm

$$\therefore \quad \varepsilon_{200} = \frac{130-25}{25} = \frac{105}{25} = 4.2$$

As the volume of the test piece will remain constant we have

$$A_0 l_0 = A_{200} l_{200}$$

Therefore
$$A_{200} = \frac{A_0 l_0}{l_{200}} = \frac{2 \times 4 \times 25 \times 10^{-9}}{130 \times 10^{-3}}$$

$$= 1.5385 \times 10^{-6}\, m^2$$

Therefore stress, $\quad \tau_{200} = \dfrac{120}{1.5385 \times 10^{-6}} = 78.0 \times 10^6\, Pa$

Therefore the compliance,

$$J_{200} = \frac{4.2}{78.0 \times 10^6} = 0.05385 \times 10^{-6}\,(Pa)^{-1}.$$

6.3.1 Dynamic Tests

Dynamic tests for the viscoelastic materials characterization are carried out under the application of dynamic input function either the sinusoidal stress or the sinusoidal strain. The results obtained from these tests have much wider applicability, as the response parameters can lead to the better understanding of molecular interactions. Experiments can be performed based on the variation of the temperature and frequency and the results reveal the information about the transition temperatures and their frequency dependence corresponding to the long range and short range molecular segmental motions, stretching and bending vibrations of side chemical groups in the backbone of the main chain.

The sinusoidal input functions (Fig. 6.3) are invariably used for the experiments and the mathematical analysis of the response. As the output function is also sinusoidal in nature having the same frequency but lagging the input function by a certain degree, both stress relaxation and creep behaviours of the materials can be predicted from the results of one set of experiments based on either stress or strain input function.

The sinusoidal input function can be represented either as a cosine function or a sine function or a complex function i.e.

The stress function; $\quad \tau(t) = \tau_0 \cos \omega t$ ...(6.8)

The strain function; $\quad \gamma(t) = \gamma_0 \cos \omega t$ ...(6.9)

The complex function; $\quad \tau^*(t) = \tau_0(\omega) \exp j[\omega t + \delta(\omega)]$ ...(6.10)

The out put functions will be as follows

The strain output $\quad \gamma(t) = \gamma_0(\omega) \cos[\omega t - \delta(\omega)]$ ...(6.11)

The Stress out put; $\quad \tau(t) = \tau_0(\omega) \cos[\omega t - \delta(\omega)]$ ...(6.12)

The complex out put; $\quad \gamma^*(t) = \gamma_0(\omega) \exp(j\omega t)$ ...(6.13)

The mathematical analysis using the complex functions gives a better insight to the process of the deformations. A number of important parameters can be calculated, which are discussed in the following sections.

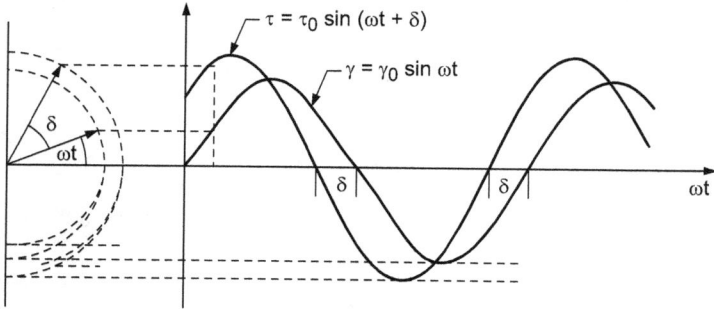

Fig. 6.3   Cyclic input function[4b]

6.3.2 Dynamic Modulus

Under the dynamic loading of a viscoelastic material with a sinusoidal strain the stress experienced by the system always has the two components, (i) in phase with the strain component and (ii) out of phase with the strain component. The in phase component is the residual or equilibrium stress associated with the residual modulus and the out of phase component lags the strain by a certain angle, δ. For a general material the complex stress can be written as

$$\tau^*(t) = G_\infty \gamma^*(t) + \tau'_0 \exp j[\omega t + \delta(\omega)] \qquad \text{...(6.14)}$$

and the complex strain is, $\gamma^*(t) = \gamma_0 \exp(j\omega t)$ ...(6.15)

where τ'_0 is the amplitude value of stress, which lags in phase with respect to the strain by an angle $\delta(\omega)$. The magnitude of the in phase stress component depends on the extent of the elastic behaviour of the viscoelastic material. For liquids for example the value of G_∞ is zero as these materials start flowing under the application of an infinitely small stress and are incapable of retaining residual stresses. Thus for the viscous liquids the first term in the equation 6.14 may be considered as negligible where as for the viscoelastic solids it is quite significant.

For the sinusoidal input function the ratio of the complex stress to complex strain is defined as the complex modulus, G^* i.e.

$$G^* = \frac{\tau^*(t)}{\gamma^*(t)} = G_\infty + \frac{\tau_0}{\gamma_0} \exp j\delta_\omega$$

$$= \left(G_\infty + \frac{\tau'_0}{\gamma_0} \cos \delta_\omega \right) + j \left(\frac{\tau'_0}{\gamma_0} \sin \delta_\omega \right)$$

$$= (G_\infty + G') + jG'' \qquad \text{...(6 .16)}$$

where $G' = \dfrac{\tau_0'}{\gamma_0} \cos \delta(\omega)$ and $G'' = \dfrac{\tau_0'}{\gamma_0} \sin \delta(\omega)$, which are known as the storage

modulus and the loss modulus respectively. The ratio of the loss modulus and the storage modulus is the loss tangent $\tan \delta$, which is also an important property of the viscoelastic materials. Thus

$$\frac{G''}{G'} = \tan \delta(\omega) \qquad\qquad ...(6.17)$$

The storage modulus is the real component of the complex modulus and the loss modulus is the imaginary component. The storage modulus represents the capacity of the viscoelastic materials to store the elastic energy and the loss modulus represents the capacity of the material to loose energy by viscous dissipation. For a particular material both storage modulus and the loss modulus can be estimated from the knowledge of the amplitudes of the input and output functions and the lag in phase angle.

Under the isothermal and isochoric conditions the equation 6.16 and the definitions of the storage modulus and loss modulus imply that if the material does not undergo any physical change in its structure due to the cyclic deformations the response of a particular cycle will be exactly same as that of the previous and of previous to that and so on. It also means that the ratio $\dfrac{\tau_0'}{\gamma_0}$.

and the phase lag angle δ are constant in each cycle and that both the moduli are independent of the amplitude of the input function. Such a stipulation is possible only for the small deformations where the amplitude of the output function changes in the same proportion as that of the input function or in other words the material behaves like a linear viscoelastic system.

6.3.3 Dynamic Compliance

As mentioned earlier under the static test the application of the sinusoidal stress to the system subjects it to the following strain components.

(i) An instantaneous strain component in phase with the input stress, γ_0 just at the instant of the application of the stress, $\tau(t)$ due to the instantaneous response of the elastic nature. It may be expressed as, $\gamma_0 = J_0 \tau(t)$, where J_0 is the instantaneous compliance,

(ii) A sinusoidal strain as a result of combined elastic and viscous effects. This strain is out of phase with the applied stress by an angle, $\delta(\omega)$ and may be written as, $\gamma(t) = \gamma_0' \exp j[\omega t - \delta(\omega)]$ where γ_0' is out of phase strain amplitude, and

(iii) A strain component equivalent to the loss of energy due to the viscous flow, and out of phase with the input stress is

$$\gamma_{vis.}(t) = \int_0^t \frac{\tau(t)}{\eta} dt = \frac{\tau_0}{\eta} \int_0^t \exp(j\omega t) dt$$

$$= \frac{-j\tau(t)}{\omega\eta} \qquad ...(6.17a)$$

Thus the variation of complex strain with time during the sinusoidal loading may be written as the sum of the three strain components as

$$\overset{*}{\gamma}(t) = J_0\tau(t) + \gamma_0' \exp j[\omega t - \delta(\omega)] - \frac{j\tau(t)}{\omega\eta} \quad ...(6.18)$$

The complex compliance, J^* can be written as

$$J^* = \frac{\gamma^*(t)}{\tau^*(t)}$$

$$= J_0 + \frac{\gamma_0'}{\tau(t)} \exp j[\omega t - \delta(\omega)] - \frac{j}{\omega\eta} \qquad ...(6.19)$$

Substituting $\tau(t) = \tau_0 \exp j\omega t$ one gets

$$J^* = \frac{\gamma^*(t)}{\tau^*(t)} = J_0 + \frac{\gamma_0'}{\tau_0} \exp[-j\delta(\omega)] - \frac{j}{\omega\eta} \quad ...(6.20)$$

On simplification it can be expressed as

$$J^* = \left(J_0 + \frac{\gamma_0'}{\tau_0}\cos\delta\right) - j\left(\frac{1}{\omega\eta} + \frac{\gamma_0'}{\tau_0}\sin\delta\right) \quad ...(6.21)$$

or

$$J^* = (J_0 + J') - j\left(\frac{1}{\omega\eta} + J''\right) \qquad ...(6.22)$$

where J' is the storage compliance and J'' is the loss compliance defined respectively as

$$J' = \frac{\gamma_0'}{\tau_0}\cos\delta \quad \text{and} \quad J'' = \frac{\gamma_0'}{\tau_0}\sin\delta \qquad ...(6.22a)$$

The ratio of the loss compliance to the storage compliance is, tan δ i.e.

$$\frac{J''}{J'} = \tan\delta = \frac{G''}{G'} \qquad ...(6.23)$$

The ratio of complex stress to complex strain is complex modulus whereas its inverse is the complex compliance. Hence on can write

$$G^*J^* = 1 \qquad ...(6.24)$$

In equation 6.22 the instantaneous compliance, J_0 is always much smaller than the storage compliance, J' and $1/\omega\eta$ is always smaller than J''. Similarly in equation 6.16 the residual modulus, G_∞ can be considered negligible in comparison to the storage modulus, G'. Hence the complex compliance and the complex modulus can be approximated as

$$J^* = (J') - j(J'') \qquad\qquad ...(6.25)$$

and
$$G^* = G' - jG'' \qquad\qquad ...(6.25a)$$

Combining equations 6.24, 6.25 and 6.25a the following interrelations can be derived between moduli and compliances as

$$G' = \frac{J'}{(J')^2 + (J'')^2}$$

and
$$G'' = \frac{J''}{(J')^2 + (J'')^2}$$

$$J' = \frac{G'}{(G')^2 + (G'')^2}$$

and
$$J'' = \frac{G''}{(G')^2 + (G'')^2} \qquad\qquad ...(6.26)$$

6.4 DYNAMIC VISCOSITY[4]

Vinogradove and Malkin[4] have reported that during the sinusoidal deformation both stress and strain are time dependent. For a sinusoidal stress input the output strain is also sinusoidal. The strain function is

$$\gamma^*(t) = \gamma_0 \exp j(\omega t - \delta) \qquad\qquad ...(6.27)$$

The rate of strain expression can be written as

$$\dot{\gamma}^*(t) = \frac{d\gamma^*(t)}{dt} = \gamma_0 \, j\omega \exp j(\omega t - \delta) \qquad ...(6.28)$$

The ratio of the complex stress to the complex rate of strain can be defined as the complex viscosity as

$$\eta^* = \frac{\tau^*(t)}{\dot{\gamma}^*(t)} = \frac{\tau_0 \exp j(\omega t)}{\gamma_0 \, j\omega \exp j(\omega t - \delta)}$$

$$= \frac{\tau_0 \exp j\delta}{\gamma_0 \, j\omega} \qquad\qquad ...(6.29)$$

Multiplying and dividing by j and simplifying one gets

$$\eta^* = \frac{\tau_0}{\gamma_0 \, \omega}(\sin \delta - j \cos \delta) = (\eta' - j\eta'') \quad ...(6.30)$$

The two viscosity terms on the right hand side of the above equation are the real and imaginary components of the complex viscosity i.e.

Real dynamic viscosity is

$$\eta' = \frac{\tau_0}{\gamma_0 \omega} \sin \delta \qquad\qquad ...(6.31)$$

Imaginary dynamic viscosity is

$$\eta'' = \frac{\tau_0}{\gamma_0 \omega} \cos \delta \qquad\qquad ...(6.32)$$

Using the definition of the storage modulus and the loss modulus as $G' = \frac{\tau_0}{\gamma_0} \cos \delta$ and $G'' = \frac{\tau_0}{\gamma_0} \sin \delta$, we can write the relationship between the components of the complex viscosity and those of the complex modulus as

$$\eta' = \frac{G''}{\omega} \qquad\qquad ...(6.33)$$

$$\eta'' = \frac{G'}{\omega} \qquad\qquad ...(6.34)$$

The quantity η' is known as the dynamic viscosity. In equation 6.33 the loss modulus is proportional to the fraction of the deformational energy, which is lost in the viscous dissipation. Hence the dynamic viscosity must bear a direct proportionality with the shear viscosity. The imaginary component of the complex viscosity given by equation 6.34 represents equivalence with the stored elastic energy.

EXAMPLE 6.3. A polyethylene rod of diameter 7.5 mm and length 50.0 mm at 100°C is subjected to a sinusoidal force of ±100 N at a frequency of 1 Hz. The experimental values of storage modulus, $G' = 0.07$ GPa and tan $\delta = 0.754$. Calculate: (a) The input stress and output strain equations, (b) Maximum deformation in the rod, (c) Input and output complex functions, (d) The values of G'', J' and J''.

Solution: Cross section area of the rod,

$$A = \frac{\pi}{4} (7.5)^2 \times 10^{-6}\, m^2 = 44.18 \times 10^{-6}\, m^2$$

As tan $\delta = 0.754$ therefore, $\delta = 37°$ and sin $\delta = 0.602$ and cos $\delta = 0.799$

Also

$$G' = \left(\frac{\tau_0}{\gamma_0}\right) \cos \delta = 0.07\ \text{GPa} = 0.07 \times 10^9\ \text{Pa}$$

The stress amplitude, $\tau_0 = \left(\frac{\text{Force}}{\text{Area}}\right)$

$$= \frac{100}{44.18 \times 10^{-6}} = 2.26 \times 10^6\ \text{Pa}$$

Therefore, the strain amplitude,

$$\gamma_0 = \frac{\tau_0 \cos \delta}{G'} = \frac{2.26 \times 10^6 \times 0.799}{0.07 \times 10^9}$$

$$= 25.84 \times 10^{-3}$$

(a) The input stress function,

$$\tau_{(t)} = \tau_{(0)} \cos \omega t = 2.26 \times 10^6 \cos \omega t$$

The out put strain function,

$$\gamma_{(t)} = \gamma_0 \cos(\omega t - \delta)$$

$$= 25.84 \times 10^{-3} \cos\left(\frac{t - 37}{360}\right)$$

$$= 25.84 \times 10^{-3} \cos(t - 0.103)$$

(b) The complex input stress function is

$$\tau^*_{(t)} = 2.26 \times 10^6 \exp(jt)$$

and the out put strain function is

$$\gamma^*_{(t)} = 25.84 \times 10^{-3} \exp[j(t - 0.103)]$$

(c) The maximum deformation in the rod

$$= l_0 \gamma_0 = 50.0 \times 25.84 \times 10^{-3} \text{ mm}$$

$$= 1.29 \text{ mm}$$

(d)

$$G'' = G' \times \tan \delta = 0.0528 \times 10^9 \text{ Pa}.$$

$$J' = \cos \delta/G_0 = \cos \delta/(\tau_0/\gamma_0)$$

$$= (0.799 \times 25.84 \times 10^{-3})/2.26 \times 10^6$$

$$= 9.135 \times 10^{-9} \text{ Pa}^{-1}$$

$$J'' = \sin \frac{\delta}{G_0} = \frac{0.602 \times 25.84 \times 10^{-3}}{2.26 \times 10^6}$$

$$= 6.88 \times 10^{-9} \text{ Pa}^{-1}$$

6.5 WORK DONE IN CYCLIC DEFORMATION

When a viscoelastic body is deformed under cyclic loading the energy of deformation is divided in viscous and elastic parts. The work done in one complete cycle of deformation for a unit volume of the material can be expressed as

$$A = \int_0^{2\pi/\omega} \tau(t) \, d\gamma(t) \qquad \qquad ...(6.35)$$

Expressing the stress and strain functions as cosine functions as

$$\tau(t) = \tau_0 \cos \omega t$$

and
$$\gamma(t) = \gamma_0 \cos [\omega t - \delta(\omega)]$$

and substituting in the above equation we get

$$A = \int_0^{2\pi/\omega} \{\tau_0 \cos(\omega t)\} \gamma_0 \omega \{-\sin[\omega t - \delta(\omega)]\} dt$$

...(6.36)

Simplification of the above equation results in

$$A = \frac{\tau_0 \gamma_0}{4} \cos \delta \left[\cos^2 \omega t - \sin^2 \omega t \right]_0^{2\pi/\omega}$$

$$+ \tau_0 \gamma_0 \sin \delta \left[\frac{1}{4} \sin 2\omega t + \frac{\omega t}{2} \right]_0^{2\pi/\omega} \quad ...(6.37)$$

On further simplification it can be seen that the first term is equal to zero and equation 6.37 reduces to

$$A = \pi \tau_0 \gamma_0 \sin \delta \qquad ...(6.38)$$

At the end of each cycle the system is expected to go back to its original state provided the deformation conforms to the conditions of linear viscoelasticity. It means that the system will be free of any strains or stresses imposed on it due to the deformation. Therefore the work done, A per cycle on the system during the cyclic deformation is irreversibly lost or dissipated during every cycle of deformation. Thus the rate of work done per unit time per unit volume or the energy dissipated per unit volume, which is also known as the dissipative function can be written as

$$D = A(\text{No. of cycles per unit time})$$

$$= \frac{A\omega}{2\pi} \qquad ...(6.39)$$

Using equation 6.38 in the above equation and simplifying one gets

$$D = \frac{\tau_0 \gamma_0 \omega \sin \delta}{2} \qquad ...(6.40)$$

Using the expressions of loss modulus and loss compliance in the above equation we get

$$D = \frac{\gamma_0^2 \omega}{2} G'' = \frac{\tau_0^2 \omega}{2} J'' \qquad ...(6.41)$$

It is important to note here that during the cyclic deformation no elastic energy must remain stored at the end of each cycle otherwise with progressive increase in the number of cycles it will increase to a very large value. A closer look at the two energy terms of equation 6.37 at the end of each one fourth of the cycle (as shown in Table 6.3 and Figure 6.4) will make this point clear. It can be seen that during the first quarter of the deformation strain increases continuously to become maximum (equal to the amplitude of the strain) at the end of the $1/4^{th}$ of the cycle. The work done on the system during this period is stored and is consumed when the strain reduces to zero in the next $1/4^{th}$ cycle.

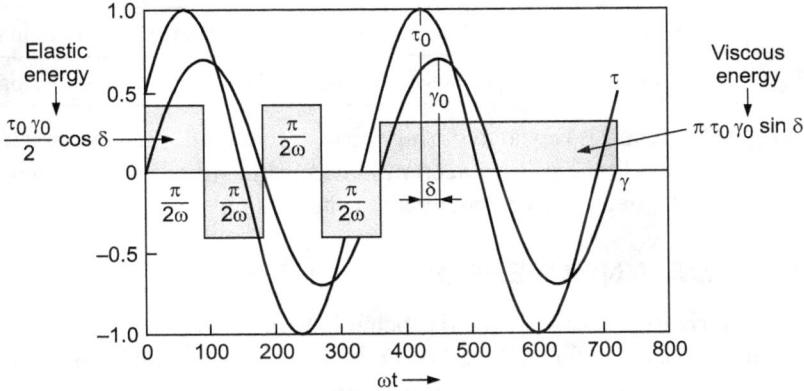

Fig. 6.4 Deformational energy during one cycle of deformation

EXAMPLE 6.4. For the system given in the Example 6.3 calculate the power required to sustain the oscillations.

Solution: The work done per unit volume per cycle of the oscillation

$$A = \pi \gamma_0 \tau_0 \sin \delta$$
$$= 3.14 \times 25.84 \times 10^{-3} \times 2.26 \times 10^6 \times 0.602$$
$$= 110.45 \times 10^3 \, \text{Nm/m}^3$$

The volume of the rod, $V = $ Area $\times$ length

$$= 44.18 \times 10^{-6} \times 50 \times 10^3$$
$$= 2.209 \times 10^{-6} \, \text{m}^3$$

The work done on the system per cycle,

$$w = 110.45 \times 10^3 \times 2.209 \times 10^{-6}$$
$$= 0.244 \, \text{Nm}$$

The work done per second or the power,

$$\dot{w} = w \times \text{frequency} = 0.244 \, \text{Nm/s}$$
$$= \frac{0.244}{0.10197} = 2.39 \, \text{Watts}$$

Table 6.3   Energy terms of equation 6.37

Fraction of cycle	Cycle time or upper limit of integrations	Value of energy in / Status	
		First term (elastic)	Second term (viscous)
$1/4^{th}$	$\pi/2\omega$	$-(\tau_0\gamma_0/2)\cos\delta$ stored	$(1/4)\pi\tau_0\gamma_0\sin\delta$ dissipated
$1/2$	π/ω	$(\tau_0\gamma_0/2)\cos\delta$ consumed	$(1/4)\pi\tau_0\gamma_0\sin\delta$ dissipated
$3/4^{th}$	$3\pi/2\omega$	$-(\tau_0\gamma_0/2)\cos\delta$ stored	$(1/4)\pi\tau_0\gamma_0\sin\delta$ dissipated
1	$2\pi/\omega$	$(\tau_0\gamma_0/2)\cos\delta$ consumed	$(1/4)\pi\tau_0\gamma_0\sin\delta$ dissipated

This phenomenon is repeated during the negative half of the cycle. Thus at the end of each cycle there is no accumulated elastic stress in the system and the system is assumed to have come back to its original state.

6.6 BOLTZMANN SUPERPOSITION THEORY[2, 3, 4C, 5]

The characterization methods for the behaviour of the viscoelastic materials have been discussed so far. All methods either static or dynamic represent the relationship between the stress and the strain under the small and reasonably slow deformations so as to conform to the linear viscoelastic behaviour. Each method leads to some fundamental material properties based on the response of the material for example stress relaxation test gives relaxation function and relaxation time, creep test gives the creep function and the retardation time and the dynamic analysis gives loss and storage moduli and compliances and loss tangent. As all these parameters relate to the stress or strain it may be of interest to know if they are independent of each other or there exists an inter relationship between them.

Boltzmann used the superposition principle to formulate theory of linear viscoelasticity, which states that, "All the effects of the mechanical history on the body may be considered independent and additive and that the response of the body to these effects is linear". He used this theory to explain the different behaviours in polymer viscoelasticity like relationship between shear modulus and Young's modulus, shear deformation and shear flow etc.

The superposition principle is the principle derived from physics and system theory[6, 7]. This principle is also known as superposition property and states that "For all linear systems the net response at the given place and time caused by two or more stimuli is the sum of responses that would have been caused by each stimulus individually". So that if input A produces response X and input B produces response Y then input $(A + B)$ will produce response $(X + Y)$.

The homogeneity and additivity properties of linear functions together are called the superposition principle. A linear function is the one that satisfies the properties of superposition. It is defined as

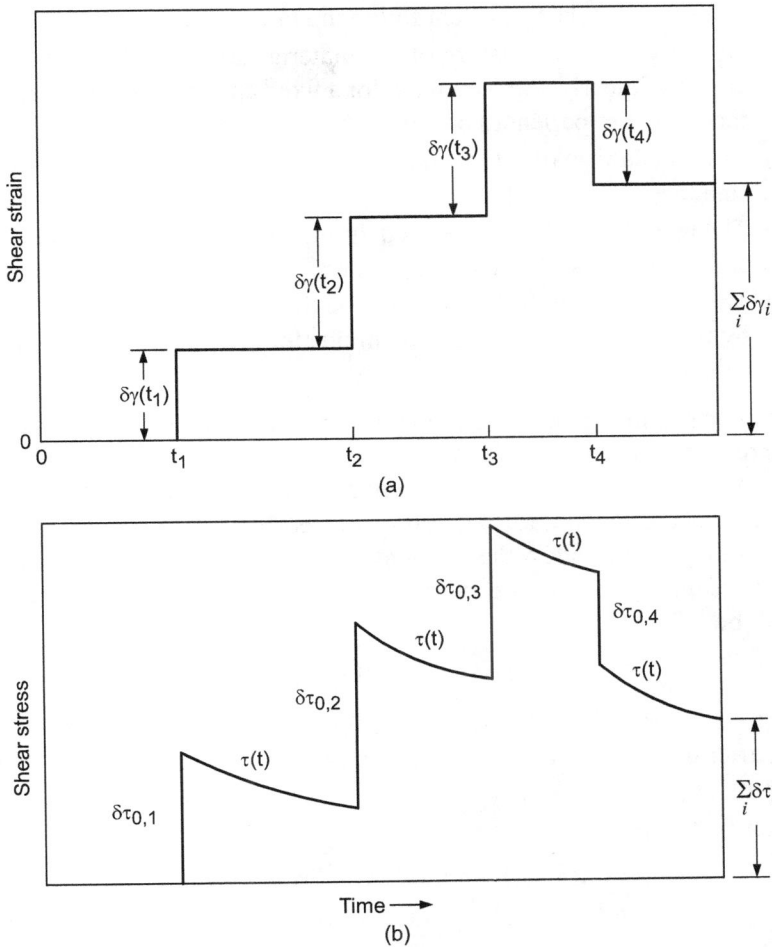

Fig. 6.5(a) and (b) Step strain input and stress relaxation at different time intervals [2]

$$F(x_1 + x_2) = F(x_1) + F(x_2) \qquad \rightarrow \text{Additivity}$$
$$F(ax) = aF(x) \text{ for scalar a } \rightarrow \text{Homogeneity}$$

The superposition principle applies to any linear systems, however the physical systems are only approximately linear (including the so called linear polymeric systems) and the application of superposition principle is only an approximation of true physical behaviour.

To understand it, let us consider a sequence of small step changes in the shear strain, $\delta\gamma$ (Fig. 6.5(a)) at different time intervals say t_1, t_2, t_3 and son on, given to a viscoelastic material. The response of the system will be (i) to experience stress increment, $\delta\tau$ every time the strain is input and (ii) to relax it by a certain amount, $\delta\tau'$ during the interval of successive strain applications (Fig. 6.5(b)). The magnitude of the relaxed stress depends on the material characteristics.

For a linear viscoelastic system following must apply

(i)   As the viscoelastic nature of the material does not change, the stress changes linearly with strain i.e. for a fixed change in strain at any time, the system experiences constant change in stress,

(ii) The mechanism of relaxation at any level of extension and stressing is same,

(iii) The magnitude of stress relaxed during the fixed time interval after the strain application is always the same,

(iv) The total stress in the system will be the sum of the stresses generated by the successive strain inputs minus the sum of the stresses relaxed (Fig. 6.5 (b)).

As indicated in Fig. 6.5 the step inputs in the strain at time intervals t_1, t_2, t_3, etc. are respectively $\delta\gamma_1, \delta\gamma_2, \delta\gamma_3$ etc. and the corresponding stresses developed are $\delta\tau_1, \delta\tau_2, \delta\tau_3$ and so on and say the relaxed stresses are $\delta'\tau_1, \delta'\tau_2, \delta'\tau_3$ etc. Assuming that the magnitudes of the strains are equal then the developed stresses also will be equal and follow the relationship $\delta\tau = G\delta\gamma$. The extent of relaxation of the stress during the interval of time, say Δt after application of each strain step will be

$$\delta'\tau_i = \delta\tau_i - \delta\gamma_i[\phi(t - t_i)] \qquad \qquad ...(6.42)$$

where $(t - t_i) = \Delta t$ and ϕ is the relaxation function. Therefore the stress in the system after the lapse of time t after the application of a step change in strain is given by

After first step,

$$\tau(t - t_1) = \delta\tau_1 - [\delta\tau_1 - \delta\gamma_1\{\phi(t - t_1)\}]$$
$$= \delta\gamma_1\{\phi(t - t_1)\} \qquad \qquad ...(6.43)$$

After second step,

$$\tau(t - t_2) = \delta\gamma_1\{\phi(t - t_1)\} + \delta\gamma_2\{\phi(t - t_2)\} \qquad \qquad ...(6.44)$$

After third step,

$$\tau(t - t_3) = \delta\gamma_1\{\phi(t - t_1)\} + \delta\gamma_2\{\phi(t - t_2)\}$$
$$+ \delta\gamma_3\{\phi(t - t_3)\} \qquad ...(6.45)$$

or it can be written as,

$$\tau(t - t_3) = \Sigma\text{Stress buildup} - \Sigma\text{Stress relaxed} \qquad ...(6.46)$$

i.e. after three step changes

$$\tau(t - t_3) = \sum_1^3 \delta\gamma_i\{\phi(t - t_i)\} \qquad \qquad ...(6.47)$$

and so on. Hence after i^{th} step the stress will be

$$\tau(t-t_i) = \sum_i \delta\gamma_i\{\phi(t-t_i)\} \qquad ...(6.48)$$

For smooth strain history each finite step strain inputs must tend to zero, hence we have $\delta\gamma_1 = \delta\gamma_2 = \delta\gamma_3 \to 0$, and the summation is replaced by the integration such that

$$\tau(t-t_i) = \int_{-\infty}^{t} \{\phi(t-t_i)\} d\gamma_i \qquad ...(6.49)$$

It can also be expressed in terms of rate of strain, $\dot{\gamma}_i = \dfrac{d\gamma_i}{dt}$ as

$$\tau(t-t_i) = \int_{-\infty}^{t} \{\phi(t-t_i)\}\dot{\gamma}_i \, dt_i \qquad ...(6.50)$$

The systems, which have a finite residual modulus, G_∞, equation 6.50 changes to

$$\tau(t-t_i) = \int_{0}^{t} \{\phi(t-t_i)\}\dot{\gamma}_i \, dt_i + G_\infty \qquad ...(6.51)$$

The application of the lower limit of minus infinity takes in to account the strain history of the material in the most general case of system being far away in the past. Such a stipulation however does not apply when one considers the present state of the material to be a stress free state. Under such a case the lower limit of the integral becomes zero as the starting time of the experiment will be, $t = 0$, and the stress, $\tau(0) = 0$, and the equation 6.50 is modified as

$$\tau(t-t_i) = \int_{0}^{t} \{\phi(t-t_i)\}\dot{\gamma}_i \, dt_i \qquad ...(6.52)$$

The linear viscoelastic characteristics of the system implies that its response is independent of the kinematics of the deformation i.e. the response will be linear for all types of strains and stresses be it elongational, shear or any other kind. Thus the equations 6.49 and 6.50 respectively can be written for the most general case of deformation by using the stress and strain tensors as

$$\tau_{ij}(t-t_i) = \int_{-\infty}^{t} \{\phi(t-t_i)\} d\gamma_{ij} \qquad ...(6.53)$$

$$\tau_{ij}(t-t_i) = \int_{-\infty}^{t} \{\phi(t-t_i)\dot{\gamma}_{ij} \, dt_i \qquad ...(6.54)$$

The tensorial form of the Boltzmann's superposition principle leads to develop the interrelations between the response parameters obtained from different types of deformations. For example extensional flow properties can be predicted from the shear data and vice versa. Similar relationships for the creep test and the dynamic tests also have been developed and are available in the literature[4, 5] but are not included here.

EXAMPLE 6.5. A carbon black filled rubber vulcanizate has elongation at break of 120%. The stress strain curve is given by $\tau_{(\gamma)} = 26.7\,\gamma^{0.85}$ (stress is in MPa) and its stress relaxation is given by $\tau(t) = \tau_0 e^{-0.01t}$. The system does not show any hysteresis during repeated elongation. The sample is subjected to the following step changes in strain at different time interval.

Time, s	$t < 0$	$0 < t \leq 50$	$50 < t \leq 150$	$150 < t \leq 250$	$250 < t \leq 300$	$300 \leq t$
Strain, γ	0	0.40	0.55	0.65	0.75	0.70
Step in γ	0	+ 0.4	+ 0.15	+ 0.10	+ 0.10	–0.05

Using the Boltzman Superposition Principle, calculate the stress at, (a) $200\,s$ and (b) $350\,s$.

Solution: The relaxation modulus is given as

$$G_{(t)} = \frac{\tau_{(t)}}{\gamma_0} = \frac{\tau_0}{\gamma_0} e^{-0.01t}$$

For the given system $\gamma_0 = 0.4$ therefore $\tau_0 = 26.7\,(0.4)^{0.85} = 12.254$ M Pa

$$\therefore \qquad G_{(t)} = \left(\frac{12.254}{0.4}\right) e^{-0.01\,t} = 30.635\,e^{-0.01t};$$

the residual modulus is zero for the present case. The BSP equation is given as

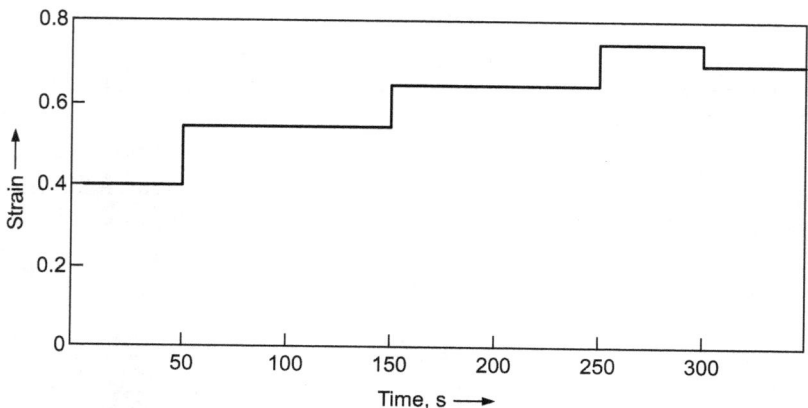

Fig. E-6.3 Step changes in strain in example 6.5.

$$\tau_{(t)} = \sum_{i}^{n} \Delta\gamma G(t - t_i) \qquad\qquad t_n < t$$

(a) When $t = 200\,s$, $n = 3$, $\Delta\gamma_1 = +0.4$, $t_1 = 0$; $\Delta\gamma_2 = +0.15$, $t_2 = 50$; $\Delta\gamma_3 = +0.1$, $t_3 = 150$

$\therefore$ $\quad \tau_{(200)} = \Delta\gamma_1 \times G(200 - 0) + \Delta\gamma_2 \times G(200 - 50)$

$$+ \Delta\gamma_3 \times G(200\text{-}150)$$

$$= 0.4 \times 30.635\, e^{-0.01(200 - 0)} + 0.15 \times 30.635\, e^{-0.01(200-50)}$$

$$+ 0.41 \times 30.635\, e^{-0.01(200\text{-}150)}$$

$$= 30.635(0.4/e^2 + 0.15/e^{1.5} + 0.1/e^{0.5})$$

$$= 4.54 \text{ M Pa}$$

(b) At $t = 350\,s$, $n = 5$ and $\Delta\gamma_1 = +0.4$, $t_1 = 0$; $\Delta\gamma_2 = +0.15$, $t_2 = 50$; $\Delta\gamma_3 = +0.1$, $t_3 = 150$; $\Delta\gamma_4 = +0.1$, $t_4 = 250$; $\Delta\gamma_5 = -0.05$, $t_5 = 300$

Substituting the values and solving gives $\tau_{(350)} = 1.86$ M Pa

6.6.1 Applications of Superposition Principles

This principle has many applications in physics and engineering:

(a) In different physical Systems

(i) *Wave motion:* This principle is only approximately applicable in super position of waves from a distant source and waves from wake of ducks. [8, 9, 10]

(ii) *Rolling motion:* Rolling motion as super position of two motions— The rolling motion of the wheel can be described as combination of two different motion : translation without rotation and rotation without translation.

(iii) *The deflection of a beam:* A beam can be modelled as a linear system where input stimulus is the load on the beam and the output response is the deflection of the beam.

(iv) In electrical engineering, in a linear circuit.

(v) In physics Maxwell's equations.

(vi) Loading of Beams[11]

(b) In mathematical systems

Linear systems are easier to analyse mathematically for example in frequency domain linear transform, Fourier, Laplace transforms and linear operator theory, etc.

(c) Algebraic systems

Linear algebraic systems like algebraic equations, linear differential equations and other systems of equations of those forms.

(d) Other general systems

The stimuli and responses could be numbers, functions, vectors, vector fields, time varying systems or any other systems that satisfies certain axioms. It should be noted that when vector and vector fields are involved a superposition is interpreted as a vector sum.

Although these examples are not directly related to the polymer rheology and processing, the above examples are given to understand the wide areas of applications of the superposition principle.

(e) In polymeric systems

1. **Relationship between shear modulus and Young's modulus:** For a simple elongational deformation of an incompressible material undergoing a step change in the strain, it can be proved from the consideration of stress and strain tensors that (i) the net stretching stress is equal to the normal stress difference, $(\tau_{11} - \tau_{22})$ and (ii) the Young's modulus, $E(t)$ is three times the shear modulus, $G(t)$ i.e.

$$\sigma_E = \tau_{11} - \tau_{22} = 3\gamma_0 G(t) \qquad ...(6.55)$$
$$E(t) = 3G(t) \qquad ...(6.56)$$

2. **Shear flow:** For a shear deformation of a fluid under steady simple shear with given strain rate components, the relationship between shear stress and modulus and that between zero shear viscosity and the modulus can be established. For simple shear deformations, only two shear stress components τ_{12} and τ_{21} are nonzero whereas the normal stresses are zero. These can be expressed from equation 6.51 as (with G_∞ being negligible),

$$\tau_{12} = \tau_{21} = \tau(t) = \dot{\gamma}_i \int_{-\infty}^{t} [\phi(t - t_i)]dt_i \qquad ...(6.57)$$

The zero shear viscosity, η_0 is given by

$$\eta_0 = \int_{-\infty}^{t} [\phi(t - t_1)]dt_i \qquad ...(6.58)$$

The viscoelastic materials under the shear flow very quickly reach the steady shear condition, normally within the range of a few seconds or fraction of minutes, hence the limits of the integral could be changed to lie in the true experimental time period range, of zero to infinity. The upper limit of infinity is to suggest that the characteristics of the shear deformation are valid over an infinite period of time provided the deformation is slow and small to conform to the linear nature. Hence

$$\eta_0 = \int_0^\infty [\phi(t - t_i)]dt_i \qquad \qquad ...(6.59)$$

The time-temperature superposition (TTS also frequency-temperature or the method of reduced variables has been reported to be applicable to polymeric systems by a number of workers,[1, 12-15]. It is an important approach applied to either to determine the temperature dependence of rheological behaviour of polymeric liquids or to expand the time or frequency regime at a given temperature at which the material behaviour is studied. The effect of temperature will be included under the effect of various parameters on the viscoelastic functions (Chapter 7).

6.7 THE MECHANICAL MODELS OF VISCOELASTIC SYSTEMS

The viscoelastic characteristic of the material encompasses both the limiting behaviours i.e. the purely elastic and purely viscous natures. The purely elastic behaviour is characterized by linearity between the stress and strain whereas the purely viscous response relates stress with the rate of shear linearly. Mechanically the elastic nature is represented by a linear spring, which shows the stress to be directly proportional to the strain and the viscous behaviour by a dashpot where stress varies linearly with the rate of shear. The viscoelastic nature is obtained when these two basic elements are either combined in series or in parallel. Maxwell first described the series combination, which is popularly known as the Maxwell model. The parallel combination was given by Voigt and Kelvin and is known as the Voigt-Kelvin (VK) model. The deformations of the Maxwell and the Voigt-Kelvin models under static and dynamic conditions are discussed next.

6.7.1 The Maxwell Model

6.7.1.1 Static Stress Relaxation

The Maxwell model is a combination of purely viscous damper and a purely elastic spring in series as shown in Fig. 6.6. it should be noted that this model does not predict creep behaviour accurately as under constant-stress condition this model postulates that the strain will increase linearly with time, where as polymers for most part show the strain rate to be decreasing with time[16].

The stress relaxation behaviour of the model is carried out by subjecting it to a constant strain at a particular time interval and then by studying the variation of stress with time. The response of the model can be mathematically derived. Fig. 6.6 shows the schematics of the Maxwell model. A step change in strain equal to γ_0 is applied to the model at a time interval, $t = t_0 = 0$. As a result of the strain both spring and the dashpot experience the stress, $\tau(t)$, which reduces with time. The strain, which remains constant at the initial value, gets distributed in both the elements depending on their characteristics i.e. the modulus of the spring and the viscosity of the fluid in the dashpot such that the stress in both the elements is same. At any instant of time the total strain of the model will be the sum of the strain in the elements and will be equal to the input strain. Similarly the rate of strain of the model will be the sum of the rates of strain of individual elements. The following equations apply for strain and rate of shear in the model,

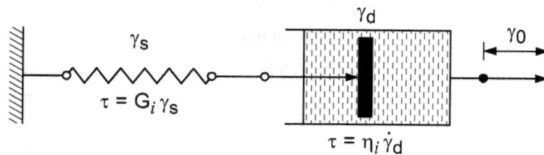

Fig. 6.6   Maxwell model element under stress relaxation[4c]

$$\gamma_m = \gamma_s + \gamma_d = \gamma_0 \qquad \qquad ...(6.60)$$

$$\dot{\gamma}_m = \dot{\gamma}_s + \dot{\gamma}_d = 0 \qquad \qquad ...(6.61)$$

as γ_0 is constant for time, $t \geq 0$. The suffixes m, s and d refer to model, spring and dashpot respectively. The governing equation for the spring is, $\tau(t) = G_i \gamma_s$ and that for the dashpot is, $\tau(t) = \eta_i \dot{\gamma}_d$. Hence the rate of strain experienced by the spring,

$$\dot{\gamma}_s = \frac{d\tau(t)/dt}{G_i} = \frac{\dot{\tau}(t)}{G_i} \qquad \qquad ...(6.62)$$

and by the dashpot, $$\dot{\gamma}_d = \frac{\tau(t)}{\eta_i} \qquad \qquad ...(6.63)$$

Substituting the equations 6.62 and 6.63 in 6.61 one gets the governing equation of the Maxwell's model as

$$\dot{\gamma}_m = \frac{\tau(t)}{\eta_i} + \frac{\dot{\tau}(t)}{G_i} = 0 \qquad \qquad ...(6.64)$$

Separating the variables in the above equation and integrating gives the expression for the stress as a function of time as

$$\tau(t) = C \exp\left(\frac{-G_i}{\eta_i} t\right) \qquad \qquad ...(6.65)$$

where C is the constant of integration, which must be evaluated from the experimental boundary conditions. As the response of the purely viscous fluids to the step input in the strain is a spike or a delta function in stress these materials do not undergo any displacement at the instant of the application of strain i.e. the dashpot gets frozen and only shows the delayed movement. Hence the total strain will be taken by the spring. Thus at the time of the application of strain, stress in the system is given as

$$\text{at } t = 0, \ \tau(t) = \tau_0 = G_i\gamma_0 = C \qquad \qquad ...(6.66)$$

Substituting it in the equation 6.65 we get

$$\tau(t) = \tau_0 \exp\left(\frac{-G_i}{\eta_i}t\right) = G_i\gamma_0 \exp\left(\frac{-G_i}{\eta_i}t\right) \qquad ...(6.67)$$

The ratio (η_i/G_i) has the unit of time and is the characteristic of the material known as the relaxation time, θ. In terms of θ Equation 6.67 becomes

$$\tau(t) = \tau_0 \exp\left(-\frac{t}{\theta}\right) = G_i\,\gamma_0 \exp\left(-\frac{t}{\theta}\right) \qquad ...(6.68)$$

By expressing the time in terms of θ i.e. $t = \theta, 2\theta, 3\theta$ and so on one can estimate the extent of relaxation of the stress. Table 6.4 gives the fraction of stress retained and relaxed by the system after the laps of such time interval and Fig. 6.7 gives the plot corresponding to Equation 6.68.

EXAMPLE 6.6. For the data in Example 6.1 apply the Maxwell model and calculate the model constants

Solution: The data given is:

$$\text{Test piece size} = 2 \times 4 \times 25 \text{ mm and}$$
the strain applied $= 400\%$ and the stress relaxation data is

Time, t, s	0	10	20	30	40	50	60	70	80	100	130	150
Stress, τ_r MPa	55.0	42.2	33.5	27.0	20.8	17.1	13.5	10.5	8.5	5.5	2.6	2.0

The Maxwell model is

$$\tau(t) = \tau_0 \exp\left(-\frac{t}{\theta}\right)$$

Applying the exponential regression the values of the Maxwell model constant are

$$\tau_0 = 52.49 \text{ M Pa}$$

and Relaxation time, $\theta = 44.47 \ s$;

the correlation coefficient,

$$r = 0.9991$$

At the time of application of strain the stress,

$$\tau_{0,\text{ exptl}} = 55 \text{ M Pa};$$

therefore the instantaneous modulus,

$$G_0 = \frac{55.0}{4} = 13.75 \text{ MPa} = \text{spring constant}$$

The relaxation time, $\theta = \dfrac{\text{Viscosity of the dashpot liquid}}{\text{Spring constant}}$

$$= \frac{\eta}{G_0} = 44.47\,s$$

Therefore $\eta = 44.45 \times 13.75 = 611.2$ MPa s

Thus the Maxwell model for the system is completely defined.

From Table 6.4 it can be seen that the stress decays exponentially with time. The relaxation time represents the time required to relax the stress by a constant factor of **e** between the successive relaxation time intervals. It characterizes the rate of approach to the equilibrium when the stresses disappear. The rate of relaxation of stress depends on the magnitude of the relaxation time. For small value of θ the rate of relaxation is high and for high values it is low. It is important to note that the fraction of stress relaxed is very high in the beginning and then reduces progressively to become insignificant. As a result of this mechanism of relaxation the stress reduces to zero after a very long time. Some materials retain some stress, which is known as the residual or the equilibrium stress.

Table 6.4   Stress relaxation for Maxwell model

Time, t	t/θ	Stress retained, τ(θ)	Stress relaxed [τ₀ − τ(θ)]
θ	1	$\tau_0/e = 0.3676\,\tau_0$	$0.6322\tau_0$
2θ	2	$\tau_0/e^2 = 0.1353\,\tau_0$	$0.8650\,\tau_0$
3θ	3	$\tau_0/e^3 = 0.0498\,\tau_0$	$0.9500\tau_0$
4θ	4	$\tau_0/e^4 = 0.0183\,\tau_0$	$0.9820\tau_0$
5θ	5	$\tau_0/e^5 = 0.0067\,\tau_0$	$0.9930\tau_0$
6θ	6	$\tau_0/e^6 = 0.0025\,\tau_0$	$0.9970\tau_0$

The relaxation modulus, $G(t)$ is given by

$$\frac{\tau(t)}{\gamma_0} = G(t) = G_i \exp\left(-\frac{t}{\theta}\right)$$

$$= G_0 \exp\left(-\frac{t}{\theta}\right) \qquad\qquad ...(6.69)$$

At time, $t = 0$ the modulus, $G(t) = G_0 = G_i$, is the instantaneous modulus. The modulus also falls exponentially as the stress. It can be seen that the modulus does not depend on the initial strain magnitude hence the Maxwell's model represents the linear viscoelastic liquids. The relaxation process is a characteristic of the purely viscous materials like the fluid in the dashpot due to which it is capable of slowly absorbing the strain energy and dissipating it as heat. Various polymer melts and solutions show this behaviour to a varying degree depending on their viscosity.

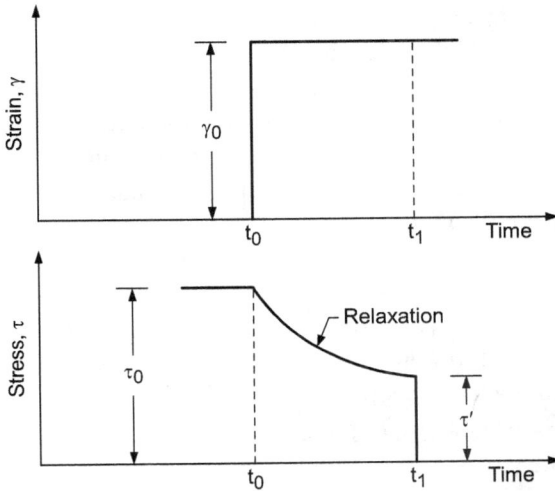

Fig. 6.7 Stress relaxation curve for Maxwell model[17]

During the process of relaxation if the strain is withdrawn at any instant of time t_1 the response of the system will be to go back to its original state of no stress. The removal of the strain is equivalent to subjecting the system having a stress history to a negative strain of the magnitude equal to $(-\gamma_0)$ and therefore the dashpot will freeze again and the spring will instantly go back to its original length and the dashpot will slowly reverse its direction of the movement and will acquire its original state after the lapse of some time. Thus the strain in the system will take some time to equilibrate. Accordingly the magnitude of the stress also will fall instantly by an amount equivalent to the strain in the spring at time, $t = t_1$ and then continue to decrease further till it approaches zero.

EXAMPLE 6.7. From the Maxwell model developed in Example 6.6 calculate the stress at 30 s, 80 s, and 130 s and estimate the % deviation w r t the experimental value.

Solution: The Maxwell model developed is

$$\tau_{(t)} = 52.49\, e^{-(t/44.45)}$$

The calculated values of the stress are given below

S.No.	Time	$\tau_{(t),\,cal}$, MPa	$\tau_{(t),\,exptl}$, MPa	%δ
1	30	26.72	27.0	−1.03
2	80	8.68	8.5	+2.12
3	130	2.82	2.6	+8.46

6.7.1.2 Static Creep Behaviour[4, 17]

The viscoelastic materials when subjected to a constant stress experience a time-dependent increase in strain. This phenomenon is known as viscoelastic creep. It is also referred as cold flow. It is defined as the tendency of the solid material to move slowly or deform permanently under the influence of mechanical stresses. It can occur as a result of long term exposure to high stresses, which are below the yield strength of the material. Creep is more severe in materials that are subjected to thermal stresses for long periods, and generally increases as the material nears its melting point.

The rate of deformation is a function of material properties, exposer time, exposer temperature and applied stress or load. Depending on the magnitude of applied stress the strain may become very large so much so that the component becomes unable to perform satisfactorily in its practical applications particularly at high temperature and high stresses.

The static creep behaviour is analysed by subjecting the test system to a step change in the stress say τ_0, which is maintained constant over the period of time, and studying its response with respect to the variation in strain, $\gamma(t)$ with time. The Maxwell model's elements will be subjected to a time variant strain, the magnitude of which depends on the modulus of the spring and the viscosity of the fluid in the dashpot. Figs. 6.8(a) and (b) show the schematics of the creep test and its response respectively. With respect to the stress and strain acting on the system following conditions apply,

At $\quad t < 0 \quad$ strain, $\gamma = 0 \quad$ and $\quad$ stress, $\tau = 0$

At $\quad t \geq 0 \quad$ strain, $\gamma = \gamma(t) \quad$ and $\quad$ stress, $\tau = \tau_0$.

As in the case of stress relaxation in this case also the stress acting on both elements of the system will be same and equal to τ_0 and the total strain and the rate of strain on the system will be the sum of those acting on the individual elements. Hence we modify equation 6.64 (by using, $\dot{\tau}(t) = 0$ as stress is constant and rate of strain to be a function of time and not zero as it changes with time) as,

$$\dot{\gamma}(t) = \frac{d\gamma(t)}{dt} = \frac{\tau_0}{\eta_i} \qquad ...(6.70)$$

Separating the variables and integrating results

$$\gamma(t) = \frac{\tau_0}{\eta_i}t + C_1 \qquad ...(6.71)$$

where C_1 is the constant of integration which must be evaluated from the boundary conditions as at $t = 0$. At this time the dashpot will be frozen and the total initial strain will be experienced by the spring alone. Hence at the instant of the application of stress the strain in the system is

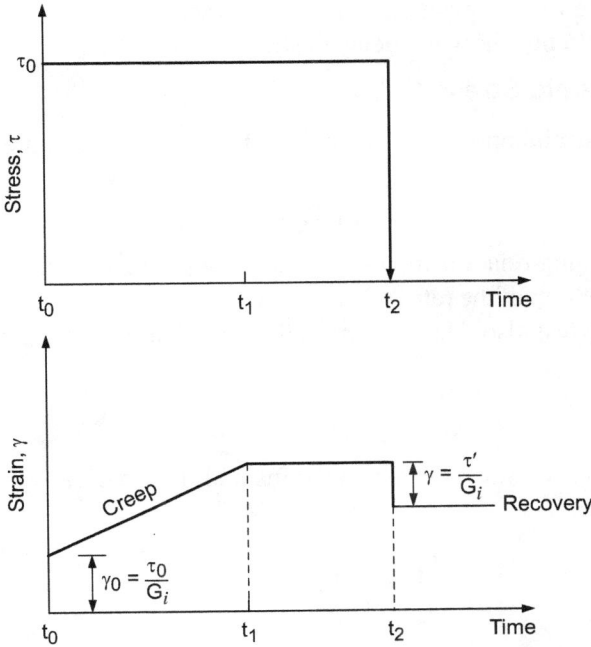

Fig. 6.8 Schematics of creep test and its response curve[17]

$$\gamma(0) = \frac{\tau_0}{G_i} \qquad \qquad ...(6.72)$$

Substituting it in equation 6.71 for $t = 0$ one gets $C_1 = \dfrac{\tau_0}{G_i}$, hence we get

$$\gamma(t) = \frac{\tau_0}{G_i} + \frac{\tau_0}{\eta_i}t \qquad \qquad ...(6.73)$$

Equation 6.73 is an equation to a straight line giving a linear variation of strain with time. At the time of application of stress the system gets subjected to an instantaneous rise in the strain due to the spring and then keeps on increasing as the time passes due to the movement of the dashpot. The creep compliance is given by the ratio of the out put strain to the input stress i.e.

$$\frac{\gamma(t)}{\tau_0} = J(t) = \frac{1}{G_i} + \frac{t}{\eta_i} = \frac{1}{G_i}\left[1 + \frac{t}{\psi_i}\right] \qquad ...(6.74)$$

where $\psi_i = [\eta_i/G_i]$ is the retardation time.

It can be seen from the above equation that the Maxwell fluids have only two components of the compliance i.e. (i) the recoverable instantaneous compliance $1/G_i$ corresponding to the elastic behaviour and (ii) the irrecoverable compliance, t/η_i corresponding to the viscous nature of the material. These fluids do not show the recoverable time dependent viscoelastic component of the creep function and therefore do not show the delayed strains.

6.7.1.3 Dynamic Stress Relaxation

The dynamic strain input function applied is $\gamma^*(t) = \gamma_0 e^{j\omega t}$ and the out put stress will be

$$\tau^*(t) = G^*(\omega)\,\gamma^*(t) = G^*(\omega)\,\gamma_0 e^{j\omega t} \qquad ...(6.75)$$

The governing relationship of the static test giving the rate of strain of the model to be the sum of the rates of strain of the individual elements is applicable to the dynamic test also. Hence writing the rate of strain equation as

$$\frac{d\gamma^*(t)}{dt} = \frac{1}{G_i}\left(\frac{d\tau^*}{dt}\right) + \frac{\tau^*}{\eta_i} \qquad ...(6.76)$$

Substituting for stress, strain functions and their derivatives in the above equation one can get

$$\frac{d(\gamma_0 e^{j\omega t})}{dt} = \frac{1}{G_i}\left(\frac{d\{G^*(\omega)\gamma_0 e^{j\omega t}\}}{dt}\right) + \frac{G^*(\omega)\gamma_0 e^{j\omega t}}{\eta_i}$$

$$...(6.77)$$

Simplification of this equation results in

$$G^* = \frac{G_i\,j\omega}{j\omega + 1/\theta_i} = \frac{G_i\,j\omega\theta_i}{j\omega\theta_i + 1}$$

Multiplication and division by $(j\omega\theta_i - 1)$ gives.

$$G^* = \frac{G_i\,j\omega\theta_i}{j\omega\theta_i + 1}\left(\frac{j\omega\theta_i - 1}{j\omega\theta_i - 1}\right) \qquad ...(6.77(a))$$

$$= \frac{G_i\,j\omega\theta_i(j\omega\theta_i - 1)}{-\omega^2\theta_i^2 - 1} \qquad ...(6.78)$$

where θ_i is the relaxation time. Further simplification gives

$$G^* = \frac{G_i\omega^2\theta_i^2}{\omega^2\theta_i^2 + 1} + j\frac{G_i\,\omega\theta_i}{\omega^2\theta_i^2 + 1}$$

$$= G' + jG'' \qquad ...(6.79)$$

where the storage modulus,

$$G' = \frac{G_i\,\omega^2\theta_i^2}{\omega^2\theta_i^2 + 1} \qquad ...(6.80)$$

and the loss modulus,

$$G'' = \frac{G_i\,\omega\theta i}{\omega^2\theta_i^2 + 1} \qquad ...(6.81)$$

The loss tangent is $\tan \delta = \dfrac{G''}{G'} = \dfrac{1}{\omega \theta_i}$ ...(6.82)

For a given Maxwell model having known spring constant and the viscosity of the dashpot fluid, the two moduli are the functions of the frequency of oscillations only. For low frequency tests the samples have sufficient time to relax and the viscous nature dominates whereas at high frequency the experimental time scale becomes too short as compared to the relaxation time of the model hence the dashpot gets completely frozen in its movement and behaves like a rigid body. Thus at high frequency the system behaves like elastic solid[17] giving a constant storage modulus and zero loss modulus (Fig. 6.9). The loss tangent i.e. the tan δ curves do not pass through a maximum unlike the actual polymeric materials. Thus the Maxwell's model does not represent the actual behaviour of the systems. The low frequency region is known as the terminal region or the terminal zone and the high frequency region is known as the transition region.

EXAMPLE 6.8. A certain linear polymer obeys the Maxwell model having element constants as $G_i = 60.0$ MPa and $\eta_i = 6.0$ MPa s. If it is subjected to a sinusoidal loading calculate the storage modulus, loss modulus and tan δ for the system at frequencies of 0.1, 1.0, 10.0, 25.0, 50.0 and 100.0 Hz and comment on their variation with frequency

Solution: The storage modulus, loss modulus and tan δ for the Maxwell model are given by Equations 6.80, 6.81 and 6.82 respectively. Using these equations the calculated values of the parameters are tabulated below

S. No.	Frequency, ω, Hz	Storage modulus, G', M Pa	Loss modulus, G'', M Pa	tan δ
1	0.1	0.006	0.60	100.0
2	1	0.594	5.94	10.0
3	10	30.0	30.0	1.0
4	25	51.72	20.7	0.4
5	50	57.69	11.54	0.2
6	100	59.41	5.94	0.1

The storage modulus increases with a much faster rate at low frequency and relatively slowly at higher frequency. The loss modulus initially increases goes through a maximum at 10 Hz and then decreases whereas the tan δ continuously decreases with increase in frequency. Thus the Maxwell model is incapable of giving peaks in tan δ with frequency.

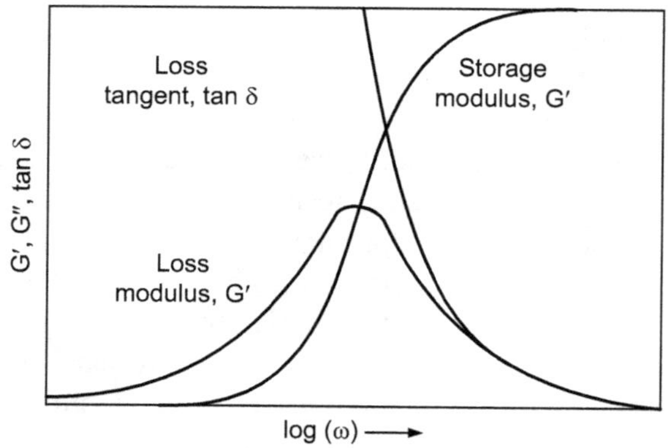

Fig. 6.9   Variation of storage and loss moduli with frequency[17]

6.7.1.4 Dynamic Creep Behaviour

Input dynamic stress function

$$\tau^*(t) = \tau_0 e^{j\omega t}$$

Output dynamic strain function

$$\gamma^*(t) = \gamma_0 e^{j(\omega t - \delta(\omega))}$$
$$= J^* \tau^* = J^* \tau_0 e^{j\omega t}$$

Applying the basic equation of the Maxwell model $\dot{\gamma}_m = \dot{\gamma}_d + \dot{\gamma}_s$ and substituting for the required parameters and simplifying, it results

$$J^* = \frac{1}{\eta_i j\omega} + \frac{1}{G_i} = \frac{1}{G_i}\left(1 + \frac{G_i}{\eta_i j\omega}\right)$$

$$= \frac{1}{G_i}\left(1 + \frac{1}{(\eta_i/G_i) j\omega}\right)$$

$$= \frac{1}{G_i}\left(1 + \frac{1}{\psi_i j\omega}\right)$$

$$= \left(\frac{1}{G_i} - j\frac{1}{G_i \psi_i \omega}\right)$$

$$= J' - jJ'' \qquad \qquad ...(6.83)$$

where the retardation time, ψ_i is

$$\psi_i = \frac{\eta_i}{G_i} \qquad \qquad ...(6.84)$$

the storage compliance

$$J' = \frac{1}{G_i} \qquad \qquad ...(6.85)$$

and the loss compliance

$$J'' = \frac{1}{G_i \omega \psi_i} \qquad ...(6.86)$$

The loss tangent

$$\tan \delta = \left(\frac{J''}{J}\right) = \frac{1}{\omega \psi_i} \qquad ...(6.87)$$

The storage compliance is independent of the frequency of the oscillation and depends only on the modulus of the spring, whereas the loss compliance is a function of viscosity of the fluid in the dashpot and the frequency. The retardation time is the characteristic of the material and represents its delayed response in straining.

6.8 VOIGT- KELVIN MODEL

The parallel combination of a spring and a dashpot results in the Voigt-Kelvin (V-K) model (Fig. 6.10). It represents the viscoelastic solids with spring giving purly elastic response and dashpot giving purely viscous response. Like Maxwell model this model also has limitations. It is very good in modelling creep in materials but with respect to relaxation it is much less accurate[17a]. Following analysis gives its response to the static and dynamic stress relaxation and creep tests. The stress experienced by the model is shared by both spring and the dashpot. Hence

$$\tau_{v,k} = \tau_s + \tau_d \qquad ...(6.88)$$

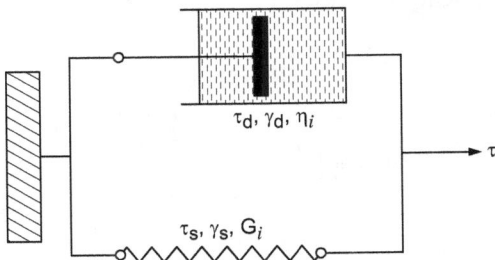

Fig. 6.10 Schematics of Voigt-Kelvin Model[4]

Both the elements will be strained equally, hence

$$\gamma_{v,k} = \gamma_s = \gamma_d$$

and

$$\dot{\gamma}_{v,k} = \dot{\gamma}_s = \dot{\gamma}_d \qquad ...(6.89)$$

$$\therefore \quad \tau_{v,k} = G_i \gamma_s + \eta_i \dot{\gamma}_d \qquad ...(6.90)$$

or

$$\tau_{v,k} = G_i \gamma_{v,k} + \eta_i \dot{\gamma}_{v,k} \qquad ...(6.91)$$

The suffix v,k refers to the Voigt-Kelvin model. Equation 6.91 is the governing equation of this model.

6.8.1 Static Stress Relaxation

As the strain is held constant $\dot{\gamma}$ is zero and the equation 6.91 becomes

$$\tau_{v,k} = G_i \gamma_{v,k} \qquad \qquad ...(6.92)$$

The total stress is taken by the spring and, the dashpot remains unstressed. The response of the model is same as that of the spring hence there is no relaxation.

6.8.2 Static Creep Test

In this test a constant magnitude of stress say, τ_0 is applied. Applying equations 6.88, 6.89 and 6.91 and dropping the suffix v, k the governing equation becomes

$$\dot{\gamma}(t) + \frac{G_i}{\eta_i}\gamma(t) - \frac{\tau_0}{\eta_i} = 0 \qquad \qquad ...(6.93)$$

The solution to this first order linear differential equation is

$$\gamma(t) = \frac{\tau_0}{G_i} + \left(C_2 e^{-\frac{G_i}{\eta_i}t} \right) \qquad \qquad ...(6.94)$$

At the instant of application of stress the dashpot will not be able to move as it has the delayed response i.e. it will be frozen and therefore the spring also will not move. Thus the system will have zero strain at the time $t = 0$. Using this boundary condition C_2 can be calculated i.e. at $t = 0$, $\gamma(t) = 0$

$$\therefore \qquad \qquad C_2 = -\frac{\tau_0}{G_i}$$

$$\text{Hence} \qquad \qquad \gamma(t) = \left(\frac{\tau_0}{G_i} - \frac{\tau_0}{G_i} e^{-\frac{t}{\psi_i}} \right)$$

$$= \frac{\tau_0}{G_i}\left(1 - e^{-\frac{t}{\psi_i}} \right) \qquad \qquad ...(6.95)$$

where $\psi_i = \eta_i/G_i$ is the retardation time. Using the retardation time as a unit of time and expressing $t = \psi_i, 2\psi_i, 3\psi_i$ and so on, the magnitude of the strain developed in the Voigt-Kelvin model can be calculated as shown in Fig. 6.11. After a long time interval the system will acquire a limiting strain i.e. as $t \to \infty$; $\gamma(t) = \gamma(\infty) \to (\tau_0/G_i)$. Thus the system will reach an asymptotic strain value equal to (τ_0/G_i) if not constrained by the dashpot reaching its end position. Table 6.5 gives the increase in strain with retardation time.

Table 6.5 shows that the rate of increase of strain progressively decreases to give the asymptotic variation of the strain to reach the final value of (τ_0/G_i).

Table 6.5   Increase in strain as a function of the retardation time

Time, t	t/ψ_i	Strain acquired, $\gamma(\psi_i)$	Strain increase, $\gamma(\psi_{i+1}) - \gamma(\psi_i)$
ψ	1	$0.631\,(\tau_0/G_i)$	$0.631\,(\tau_0/G_i)$
2ψ	2	$0.865\,(\tau_0/G_i)$	$0.224\,(\tau_0/G_i)$
3ψ	3	$0.950\,(\tau_0/G_i)$	$0.085\,(\tau_0/G_i)$
4ψ	4	$0.982\,(\tau_0/G_i)$	$0.032\,(\tau_0/G_i)$
5ψ	5	$0.993\,(\tau_0/G_i)$	$0.011\,(\tau_0/G_i)$
6ψ	6	$0.997\,(\tau_0/G_i)$	$0.004\,(\tau_0/G_i)$

EXAMPLE 6.9: For the creep data given in example 6.2 discuss the possibility of application of the Voigt- Kelvin model to the system. Give physical reasons for its suitability or otherwise.

Solution: The given data gives a finite value of strain at the time of application of the force giving an instantaneous strain. The Voigt-Kelvin model does not show any response at the time $t = 0$ due to the delayed response of the dashpot fluid, hence this model will not be suitable for the given system. More suitable models for such systems will be those where a spring element is available to respond independently to the applied force. Such models may be, (i) a Maxwell and a V-K model in series or (ii) Zener model.

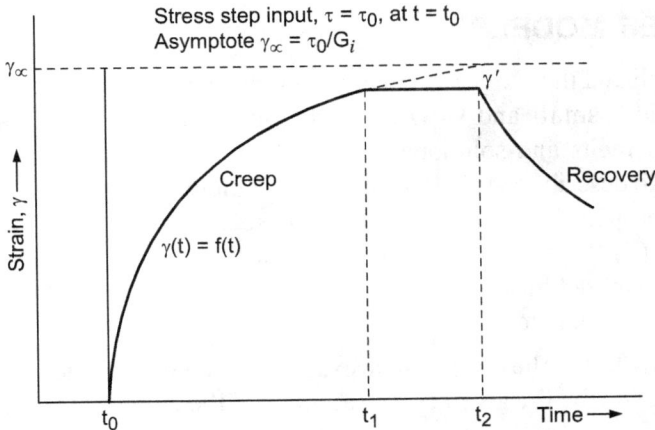

Fig. 6.11   Creep response of Voigt-Kelvin Model[17]

6.8.3 Dynamic Stress Relaxation

The V-K model does not give the stress relaxation and the stress in the model varies sinusoidally along with the strain.

6.8.4 Dynamic Creep Test

The input dynamic stress is $\tau^*_{(t)} = \tau_0 e^{j\omega t}$ and out put strain function is $\gamma^*(t) = J^*(\omega)\tau_0 e^{j\omega t}$. Applying the governing equation as $\tau^*(t) = \tau^*_s + \tau^*_d$ and substituting for different quantities in the governing equation and simplifying one gets

$$G_i J^*(\omega) + j\eta_i \omega J^*(\omega) = 1 \qquad \qquad ...(6.96)$$

$$\therefore \qquad J^*(\omega) = \frac{1}{G_i + j\eta_i \omega} = \frac{1}{G_i}\left(\frac{1}{1 + j\omega\psi_i}\right)$$

$$= \frac{1}{G_i}\left(\frac{1 - j\omega\psi_i}{1 + \omega^2\psi_i^2}\right)$$

$$= J'(\omega) - jJ''(\omega) \qquad \qquad ...(6.97)$$

where $\psi_i = \eta_i/G_i$ is the retardation time.
the storage compliance

$$J'(\omega) = \frac{1}{G_i}\left(\frac{1}{1 + \omega^2\psi_i^2}\right) \qquad \qquad ...(6.98)$$

the loss compliance

$$J''(\omega) = \frac{1}{G_i}\left(\frac{\omega\psi_i}{1 + \omega^2\psi_i^2}\right) \qquad \qquad ...(6.99)$$

and the loss tangent $\tan\delta = \dfrac{J''(\omega)}{J'(\omega)} = \omega\psi_i \qquad \qquad ...(6.100)$

6.9 OTHER MODELS

The Maxwell and the Voigt-Kelvin models represent simple linear viscoelastic systems under small and slow deformations. However in actual practice the polymer melts and solutions undergo large deformations under varying frequency. To describe such behaviours other combinations of both these models have been proposed (Fig. 6.12). Some of these are, (i) two Maxwell's models in parallel, (ii) Zener model or the standard linear solid, (iii) Maxwell and Voigt-Kelvin model in series, (iv) generalized Maxwell model, (v) generalized Voigt-Kelvin model etc.

EXAMPLE 6.10. For the creep data in example 6.2 determine the suitable model which will describe the given creep response of the material.

Solution: The data given and the calculated strain values are

Time, s	0	10	50	100	200	300	500	600	800
Specimen, length l_t, mm	50.0	75.0	101.0	114.0	130.0	141.0	153.0	158.0	168.0 to 144
Strain, γ	1.0	2.0	3.04	3.56	4.2	4.64	5.12	5.32	5.72 to 4.78

The dimensions of the sample are $2 \times 4 \times 25$ $(mm)^3$ and the force applied $= 120$ N

The plot of strain versus time shows (Fig. E-6.2) three distinct regions (i) instantaneous rise in strain at $t = 0$, (ii) exponential rise up to about 500 s and (iii) a linear rise beyond 500 s. From the creep response of Maxwell and V-K models it can be said that the 1st and 3rd region are due to the Maxwell model and the 2nd region is due to the V-K model. Hence it can be argued that a series combination of these models may be a suitable model for the system in question. Such a combination is shown in Fig. 6.12 (iii). The parameters of the Maxwell model are G_1, and η_1 and those for the V-K model are G_2 and η_2. These constants are estimated from the corresponding responses of these models from the Fig. E-6.2.

The creep response of the combined model can be calculated to be

$$\gamma_{(t)} = \frac{\tau_0}{G_1} + \frac{\tau_0 t}{\eta_1} + \frac{\tau_0}{G_2}\left(1 - e^{-\frac{G_2 t}{\eta_2}}\right) \qquad ...(E-1)$$

The applied stress, $\quad \tau_0 = \dfrac{F}{\text{Area}} = \left[\dfrac{120(N)}{2 \times 4 \times 10^6}\right] = 15 \times 10^6$ Pa.

From the value of instantaneous strain $= 1$, the modulus

$$G_1 = \frac{\tau_0}{1} = 15 \times 10^6 \text{ Pa.}$$

From the linear region the strain rate,
$$\dot{\gamma} = 0.0019\,s^{-1}$$

Hence, $\quad\quad \eta_1 = \dfrac{\tau_0}{\dot{\gamma}} = \dfrac{15 \times 10^6}{0.0019}$

$$= 7894.7 \times 10^6 \text{ Pa s}$$

The V-K model parameters are obtained from the region 2 of the graph. It can be assumed that the 2nd region ends at about 500 s. Therefore the maximum strain that would have been instantly experienced by the spring of the V-K model if it was not retarded by the dashpot $= \gamma_{(500)} - \gamma_{(0)} = 5.15 - 1 = 4.15$

Hence $\quad\quad G_2 = \left[\dfrac{\tau_0}{\gamma_{(500)} - \gamma_{(0)}}\right]$

$$= \frac{15 \times 10^6}{4.15} = 3.61 \times 10^6 \text{ Pa}$$

The value of η_2 is calculated by using Equation E-1 in the 2nd region. Taking the strain value at a time where it appears to be most dominating, say at $s = 100$, $\gamma_{(100)} = 3.55$

Substituting different values in Equation E-1 and solving for η_2 gives

$$3.55 = \frac{15 \times 10^6}{15 \times 10^6} + \frac{15 \times 10^6 \times 100}{7894.7 \times 10^6} + \frac{15 \times 10^6}{3.61 \times 10^6}\left[1 - e^{-\left(\frac{3.61 \times 10^6 \times 100}{\eta_2}\right)}\right]$$

Solving for η_2 gives

$$\eta_2 = 430 \times 10^6 \text{ Pas}$$

6.9.1 Two Maxwell Models in Parallel

(i) Static analysis

This combination gives two relaxation times, $\theta_1 = \eta_1/G_1$ and $\theta_2 = \eta_2/G_2$. The governing equations of the individual models are

$$\dot{\gamma} = \frac{\dot{\tau}_1}{G_1} + \frac{\tau_1}{G_1}$$

$$= \frac{\dot{\tau}_2}{G_2} + \frac{\tau_2}{G_2} \qquad ...(6.101)$$

and

$$\tau = \tau_1 + \tau_2 \qquad ...(6.102)$$

Elimination of τ_1 and τ_2 from the above equations results in the governing equation of the combined model, which is a second order linear differential equation between stress and strain. It is

$$\left(\frac{G_1 + G_2}{G_1 G_2}\eta_1\eta_2\right)\frac{\delta^2\gamma}{\delta t^2} + (\eta_1 + \eta_2)\frac{\delta\gamma}{\delta t}$$

$$= \left(\frac{\eta_1\eta_2}{G_1 G_2}\right)\frac{\delta^2\tau}{\delta t^2} + \left(\frac{G_1\eta_2 + G_2\eta_1}{G_1 G_2}\right)\frac{\delta\tau}{\delta t} + \tau \qquad ...(6.103)$$

The solution of equation 6.103 will result in the following relationships between stress and strain in two Maxwell elements

$$\tau_1(t) = G_1\gamma_0 e^{(-t/\theta_1)}$$

and

$$\tau_2(t) = G_2\gamma_0 e^{(-t/\theta_2)} \qquad ...(6.104)$$

The modulus expression for the model can be written as

$$G(t) = \frac{[\tau_1(t) + \tau_2(t)]}{\gamma_0}$$

$$= \left(G_1 e^{(-t/\theta_1)} + G_2 e^{(-t/\theta_2)}\right) \qquad ...(6.105)$$

The retardation time for the constant stress input can be calculated from

$$\psi_i = \left(\frac{G_1 + G_2}{\eta_1 + \eta_2}\right)\left(\frac{\eta_1\eta_2}{G_1 G_2}\right) \qquad ...(6.106)$$

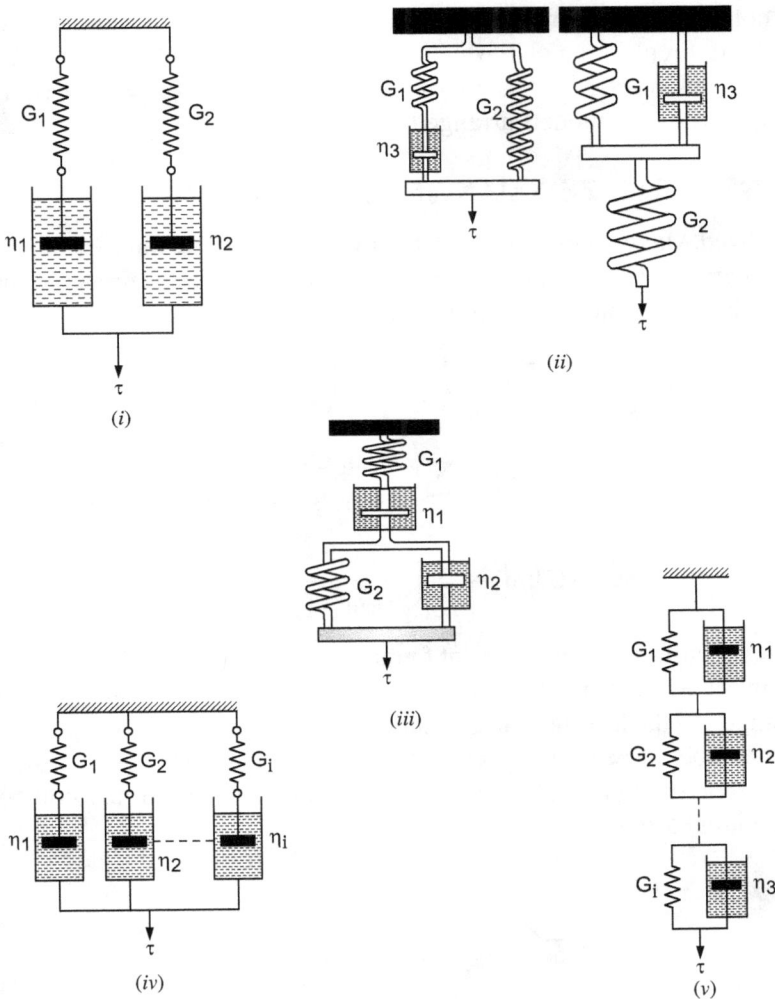

Fig. 6.12 Other combinations of Maxwell and Voigt-Kelvin models.

(ii) Dynamic Analysis

The dynamic response of this system can be analysed following the procedure given earlier. The expressions for the storage modulus and the loss modulus are

$$G'(\omega) = \left(\frac{G_1 \omega^2 \theta_1^2}{1 + \omega^2 \theta_1^2} \right) + \left(\frac{G_2 \omega^2 \theta_2^2}{1 + \omega^2 \theta_2^2} \right) \qquad ...(6.107)$$

and

$$G''(\omega) = \left(\frac{G_1 \omega \theta_1}{1 + \omega^2 \theta_1^2} \right) + \left(\frac{G_2 \omega \theta_2}{1 + \omega^2 \theta_2^2} \right) \qquad ...(6.108)$$

Another combination with two relaxation times and one retardation time is the Burgers-Frenkel model in which one Maxwell and one Voigt-Kelvin model are connected in series[4c,p-217]. This model is mathematically quite equivalent to the two Maxwell models arranged in parallel.

6.10 GENERALIZED MAXWELL MODEL

Generalized Maxwell model is obtained when an infinite number of Maxwell elements are connected in parallel (Fig.6.12(iv)). The static and dynamic analyses give the following modulus expressions

$$G(t) = \sum_{i=1}^{\infty} G_i \exp\left(\frac{t}{\theta_i}\right) \qquad ...(6.109)$$

$$G'(\omega) = \sum_{i=1}^{\infty} \left(\frac{G_i \omega^2 \theta_i^2}{1+\omega^2 \theta_i^2}\right)$$

and
$$G''(\omega) = \sum_{i=1}^{\infty} \left(\frac{G_i \omega \theta_i}{1+\omega^2 \theta_i^2}\right) \qquad ...(6.110)$$

The moduli are the functions of frequency, modulus and relaxation times of different Maxwell elements in the system. Fig. 6.13 gives the variations of both the storage and the loss moduli as functions of frequency. The loss modulus plot gives a number of peaks at different frequencies of $\omega = 1/\theta_1, 1/\theta_2, 1/\theta_3, ...$ etc. and the strength of the peaks depends on the moduli of the springs constituting the individual Maxwell elements.

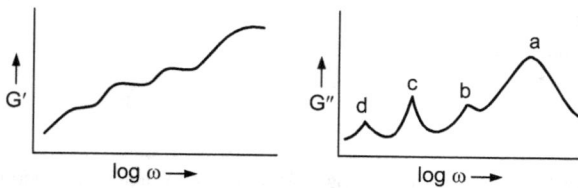

Fig. 6.13   Response of the generalised Maxwell model[3]

The small peaks refer to the shorter relaxation times (peak-b) and vice versa (peaks-c and d). The broader peaks (peak-a) represent a number of Maxwell elements with similar relaxation time, the peaks being reinforced together. The tan δ plots also show a number of peaks corresponding to each Maxwell elements. These peaks occur at the values of the frequencies slightly displaced from those in the loss modulus peaks.

6.11 ZENER MODEL OR THE STANDARD LINEAR SOLID

This model is represented either (i) by a Maxwell model in parallel with a spring, or (ii) a Voigt-Kelvin model in series with a spring (Fig. 6.12 (ii)). The governing equation for the model may be derived from the following equilibrium relations.

$$\tau_M = \tau_1 + \tau_2 \text{ and } \tau_1 = \tau_3$$
$$\gamma_M = \gamma_2 = \gamma_1 + \gamma_3$$

and
$$\dot\gamma_M = \dot\gamma_2 = \dot\gamma_1 + \dot\gamma_3$$

where the suffix M stands for the model. The required governing relationship is a first order differential equation in stress and strain

$$\eta_3 \dot\tau_M + G_1 \tau_M = \eta_3 (G_1 + G_2)\dot\gamma_M + (G_1 G_2)\gamma_M$$

or
$$a_1 \dot\tau_M + a_0 \tau_M = b_1 \dot\gamma_M + b_0 \gamma_M \qquad\qquad ...(6.111)$$

where a_1, a_0, b_1 and b_0 are constants obtained from different characteristic parameters of the elements of model.

6.11.1 Stress Relaxation

Input strain, $\gamma = \gamma_0 = \gamma_M = $ const. At the instant of application of strain the dashpot will be frozen and the strain will be taken by the springs 1 and 2. Therefore the following boundary conditions (B.C.) will apply

At, t = 0,

$$\gamma_0 = \gamma_1(0) = \gamma_2(0)$$

and
$$\tau_0 = \tau_M(0) = \tau_1(0) + \tau_2(0)$$

Substituting these B.C. in the governing equation of the model, integrating and simplifying gives the stress relaxation equation as

$$\tau(t) = \left(\frac{\tau_0}{G_1 + G_2}\right)\left(G_2 + G_1 e^{-t/\theta_i}\right) \qquad\qquad ...(6.112)$$

where $\theta_i = \eta_3/G_1$ is the relaxation time for the Zener model which is same as that of the Maxwell model in the given combination. The stress relaxation for this model is schematically presented in the Fig. 6.14.

6.11.2 Creep Test

The input stress at $t = 0$ is, say $\tau = \tau_0 = $ constant, therefore $\dot\tau = 0$. The B.C. is, at $t = 0$, $\gamma(0) = \tau_0/(G_1 + G_2)$. Substituting these in the governing equation of the model, integrating and simplifying we get the required expression for the time dependent strain as

$$\gamma(t) = \frac{\tau_0}{G_2} - \left(\frac{\tau_0 G_1}{(G_1 + G_2)G_2}\right) \qquad\qquad ...(6.113)$$

$$\left(e^{-\frac{G_1 G_2 t}{\eta_3(G_1 + G_2)}}\right)$$

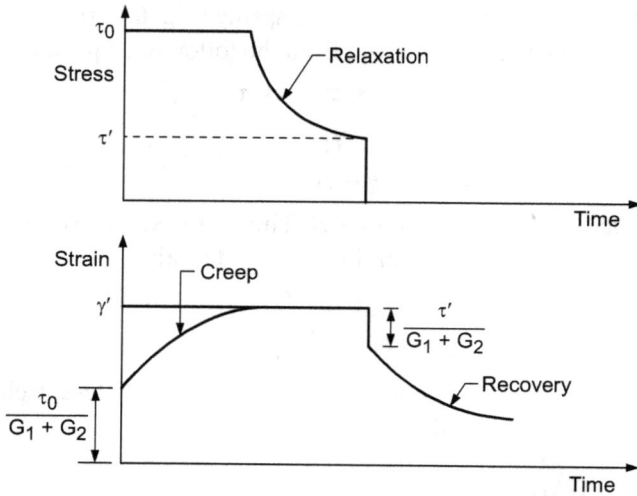

Fig. 6.14   Stress relaxation and creep behaviour of Zener Model[17]

This equation predicts that at the instant of application of stress i.e. at $t = 0$ the system will experience an instantaneous rise in the strain of magnitude $\gamma(0) = \tau_0/(G_1 + G_2)$. The creep behaviour of this model is shown in Fig. 6.14. If the stress is withdrawn after some time the system will recover the instantaneous strain instantly and then the strain will decay exponentially.

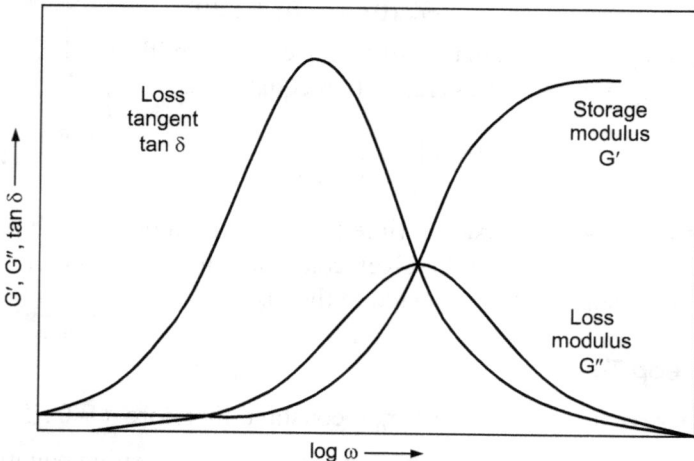

Fig. 6.15   The dynamic response for a standard linear solid or Zener model[17]

6.11.3 The Dynamic Response

The dynamic response of the Zener model can be obtained by substituting the complex stress and strain functions in the equation 6.111 and subsequent simplification.

The storage modulus, the loss modulus and the loss tangent expressions can be derived as the storage modulus $$G' = \frac{G_1^2 G_2 + \eta^2 (G_1 + G_2)\omega^2}{G_1^2 + \eta^2 \omega^2} \qquad \text{...(6.114)}$$

the loss modulus

$$G'' = \frac{G_1^2 \eta \omega}{G_1^2 + \eta^2 \omega^2} \qquad \text{...(6.115)}$$

and the loss tangent,

$$\tan \delta = \frac{G_1^2 \eta \omega}{G_1^2 G_2 + \eta^2 \omega^2 (G_1 + G_2)} \qquad \text{...(6.116)}$$

The variation of these parameters is shown in Fig. 6.15. Using relations in equations 6.26 the expressions for the storage and loss compliances can be easily derived.

6.12 POLYMER VISCOELASTICITY

The viscoelastic characteristics of the polymer melts and solutions are the direct consequence of wreathing and wriggling motions of coiled and entangled long chain molecules due to the Brownian motions caused by the van der Waals forces acting on them. The presence of the heavy groups or the branches or the grafted structures in the backbone of the molecules hinder these movements resulting in the requirement of higher energy to accomplish them. The momentum of the movement of the side groups is absorbed by the main chain bonds in the vicinity of the group to produce the localized restorative rate of change of momentum. Some of the typical polymer molecules are shown below.

Polyethylene, HDPE:	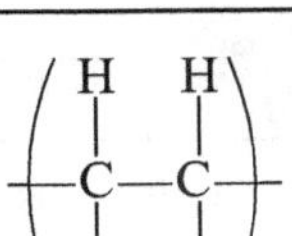 $\left(\begin{array}{cc} H & H \\ \mid & \mid \\ -C-C- \\ \mid & \mid \\ H & H \end{array} \right)_n$
Polyethylene, LDPE:	$-(CH_2-CH-CH_2-CH-CH_2-CH_2)_n-$ with side chains CH_2, CH_2 and CH_2, CH_2, CH_2, CH_2
Polystyrene, PS	$-(CH-CH_2)_n-$ with phenyl group

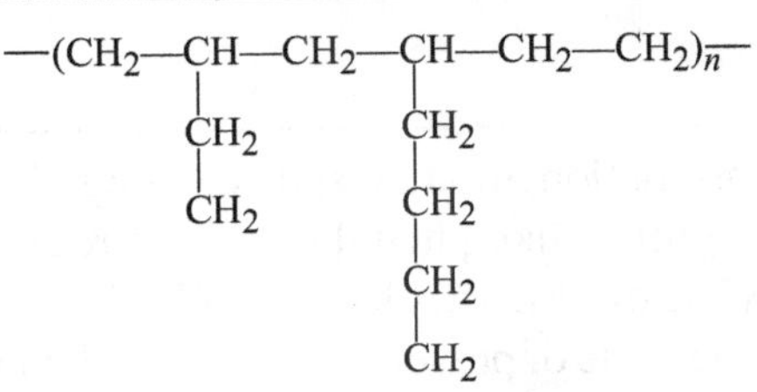

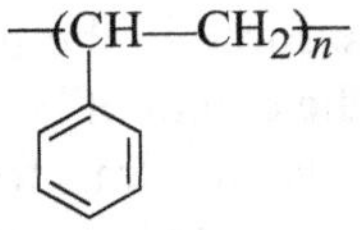

Polymethylemethacrylate, PMMA:	
Polyvinylchloride, PVC:	
Polyetheleneterephthalate: PET	
Polyisoprene or Natural Rubber: NR	
Polydecylacrylate:	
Vinylchloridedecylacrylate Copolymer:	

The oscillation, rotation and swinging, flapping and twisting movements of the bulky groups like phenyl of polystyrene, branching of LDPE, methyl and methacrylate of PMMA, chlorine of PVC, terephthalate of PET, methyl of NR, and decylacrylate of polydecylacrylate and vinyl-decylacrylate copolymer cause the bending of the C-C bond, which joins these groups with the main chain, and give rise to the restorative force in the bond. The magnitude of this force increases with the increase in the extent of bending of the bond hence the nature of this force is elastic. Similarly the drag force experienced by these groups during different types of motions is of viscous type. Thus the polymer molecules manifest both elastic and viscous characteristics during the deformations. As the effect of the movement of the groups is only felt locally, each position of the groups in the main chain acts either as a Maxwell or a Voigt-Kelvin element having a specific modulus, G_i and viscosity, η_i and hence a typical relaxation and retardation times. Therefore from the rheological point of view the polymeric system as a whole can be considered to be consisting of an infinite number of Maxwell elements in parallel or the infinite number of the Voigt-Kelvin elements in series giving the spectrum of relaxation and retardation times.

Apart from the above mentioned localized motions the polymer molecules are known to possess the coordinated long-range motions of relatively large blocks of the main chain. The onset of the long-range motion represents the transition from the glassy state to the rubbery state and the temperature at which it occurs is known as the glass transition temperature. The energy required for these motions is far in excess of that required for different motions of the groups or the short-range motions. The long-range motions subject the molecules to large changes in the spatial configurations characterised by large drag force and weak elastic force. Whereas the short-range motions are characterised by relatively higher elastic restorative force as compared to the viscous force. It can be argued that the relaxation time for the long-range motion must be much higher than the short-range motions as $\theta_i = \eta_i/G_i$.

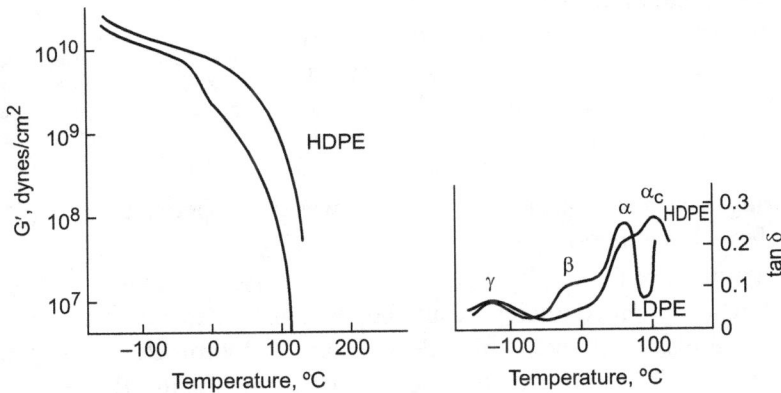

Fig. 6.16 Isochronal data for polyethylenes at 1 Hz.[4b]

The dynamic mechanical measurements over a wide range of temperature offer an excellent means to study these behaviours. The steep drop in storage and loss moduli and a large peak in tan δ around the glass transition temperature reflect the effect of the long-range motions whereas the variations in tan δ below Tg represent the influence of the short-range motions. The tan δ versus temperature plot at a particular frequency below glass transition temperature gives a number of small peaks to represent the short-range motions of the side groups present in the main chain. Figure 6.16 gives a plot of storage modulus and tan δ at different temperatures for HDPE and LDPE[4b,18].

6.13 THEORETICAL DEVELOPMENTS

Long-Range Motions- The Rouse Theory

Theoretical developments so far for the prediction of the viscoelastic behaviour of the polymer melts and solutions, are highly inadequate and are incapable of quantitative prediction of the rheological properties of polydisperse high molecular weight polymers. Nevertheless these are helpful in understanding the trends in behaviour and in estimating the longest relaxation time, zero shear viscosity and steady state compliance. The Rouse model describes conformational dynamics of an ideal chain (Fig. 6.17). In this model the single chain diffusion is represented by Brownian motion of beads connected by harmonic springs also referred as the Rouse-dumb-bell model. There are no excluded volume interactions between the beads and each bead is subjected to a random thermal force as in Longevin dynamics. This model was proposed by Prince E. Rouse in 1953.[19]

Rouse[19] first proposed the long-range motion theory especially for the dilute solutions where the polymer molecules are completely relaxed and are not influenced by the presence of other molecules. He considered the polymer molecule to be consisting of *n* statistical segments or "submolecules," which are capable of taking part in long-range coordinated motion. These segments are considered to behave as Hookean springs and are connected by "beads" in which the mass of the molecules is proposed to be concentrated (Fig. 6.17). The model of beads

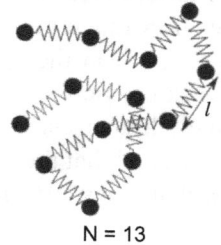

Fig. 6.17 Schematic view of the Rouse model with beads and springs connecting them for $N = 13$ beads and l = average distance between them

and springs was not proposed by Rouse but was developed latter to explain his theory [en.wikipedia.org]

The motion of the beads in the solvent gives rise to the viscous resistance defined by a friction coefficient, ξ and bending and extension of the segments produce the elastic effects, which depend on the coordinates of a particular segment as well as those of two neighbouring segments. Because of this interdependence on the coordinates, the equation of motion of each segments cannot be solved independently. Rouse defined a new set of position coordinates known as the normal coordinates and the long-range motion with respect to these coordinates as the normal mode of motion (Fig. 6.18). The first normal mode of motion corresponds to the rotation of entire molecule, including all the *n* segments, about its centre involving only one node at the middle or at $n/2$ position. It implies that if we consider that the entire molecule consisting of *n* segments takes some short of spherical shape then the rotation is around its center as the nodal point then it refers to first normal mode of motion. The second normal mode involves two nodes at $n/4$ and $3n/4$ positions and the

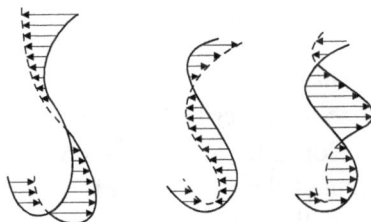

Fig. 6.18 First three normal modes of motion[20]

longest length of the chain that moves in the coordinated fashion is a block of $n/2$ segments. The third normal mode of motion in the same way will involve three nodes, which will occur at $n/6$, $n/2$ and $5n/6$ positions and the longest length that will move in the coordinated fashion is a block of n/3 segments. Thus it can be argued that the ith normal mode will involve the block of n/i segments in the coordinated motion. These polymer chain motions have been modelled and their mathematical analysis was worked out. Although Rouse developed his theory for dilute polymer solutions he did not apply it for concentrated solutions and melts. This was done by Zim[20] which is discussed in brief below.

6.13.1 Polymer Dynamics

In dynamic analysis three basic assumptions are made

(a) *Gaussian chains:* Where the force on a subchain segment *n* is the net entropic force from its neighbors. In continuum representation this has been adopted as equivalent to thermodynamic force at each point of the chain i.e.

$$\frac{\partial}{\partial n}\left(\frac{k\,\partial R}{\partial n}\right) = -\frac{K\partial^2 R}{\partial n^2} \text{ with } K = \frac{3k_BT}{b^2}$$

Here K is the force constant with b representing the root-mean-square size of an elementary polymer segment while R is the position vector of segment n at time t, T is the absolute temperature and k_B is the Boltzmann constant.[29]

(b) *Local drag:* The drag force on a subchain segment comes from the frictional drag against background without long range mediated back force (this works in melts where all long range mediated back flows are screened). This force is $\xi\,\partial R/\partial t$ where ξ is a drag coefficient per chain segment.

(c) *Brownian motion:* A random force $f(n, t)$ acts on each subchain with correlation time much faster than any polymer dynamics to be modelled by the theory.

The application of the equations of motion to the deformation of the Rouse chain model results in the expressions for modulus, relaxation time and zero shear viscosity but as this model does not take care of the hydrodynamic interaction of one chain with the other, the results obtained are not realistic. Zimm[20] modified it to account for the hydrodynamic interaction and the resultant expressions gave good representation of the experimental results for the deformational behaviour of the dilute solutions. The behaviour of polymer melts and concentrated solutions however could not be described by such a proposition.

The Rouse theory developed for the solutions can be applied to the polymer melts by taking in to account the interaction between the different polymer molecules essentially below the critical molecular weight as this theory is not applicable for the entangled molecules. The expressions for relaxation times, modulus, and storage and loss moduli derived from the Rouse model are

$$\theta_i = \frac{n\xi\langle L^2\rangle}{6\pi^2 kTi^2} \; ;$$

$$G_i = NkT = \frac{\rho RT}{M}$$

$$G'(\omega) = \frac{\rho RT}{M}\sum_{i=1}^{n}\left(\frac{\omega^2\theta_i^2}{1+\omega^2\theta_i^2}\right)$$

and
$$G''(\omega) = \frac{\rho RT}{M}\sum_{i=1}^{n}\frac{\rho RT}{M}\left(\frac{\omega\theta_i}{1+\omega^2\theta_i^2}\right) \qquad ...(6.117)$$

where, M is the molecular weight of the segment and the other terms have their usual meanings.

The storage and loss moduli are the functions of frequency, relaxation time, segmental molecular weight, and temperature. For a given material under isothermal conditions the plots of storage and loss moduli versus frequency (Fig. 6.19) results in two distinct zones, These are (i) the terminal zone corresponding to the low frequency and (ii) the transition zone at high frequency. Figure 6.19 shows the dynamic data of polystyrene-chlorinated diphenyl solutions[1] at 25°C. The molecular weight of polystyrene and its concentration in the solution were below the critical values corresponding to the onset of entanglement. The experimental data and the theoretical line are shown in the figure. The agreement

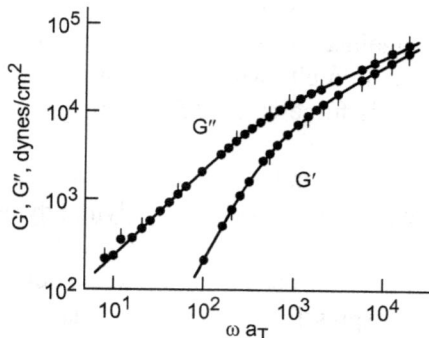

Fig. 6.19 Dynamic data of polystyrene –chlorinated diphenyl solution
(M = 267,000, c = 0.124 g/cc at 25°C, Solid lines correspond to the Rouse theory)[1]

between the experimental data and the Rouse theory is excellent. Zim[20] has reported similar results using polystyrene of much higher MW [$MW = 860000$] in decalin and dioctyl phthalate (DOP) solutions, which show much lower values of G' and G'' as compared to those in chlorinated diphenyl solutions.

The theoretical lines were calculated using the equations 6.117. The following observations are important.

(i) Both G' and G'' plots show two regions with different slopes.

(ii) In the high frequency region both storage and loss moduli are approximately equal and have the same slope of 0.5.

(iii) The slope of 1/2 is predictable from the Rouse theory as the experimental data matches well with the theoretical prediction.

(iv) In this region the value of both the moduli go down from around 10^6 kPa to 10 kPa i.e. from the one, characteristic of the glassy state to that of rubbery state as the frequency is reduced. Hence this region is known as the transition region

(v) In the low frequency region (the terminal zone) both storage and the loss moduli follow different lines. The storage modulus increases with much higher rate with frequency than the loss modulus. The magnitude of the loss modulus is higher than that of the storage modulus.

(vi) The slope of the G'' line is unity and that of the G' line is 2.

(vii) Both these slopes are predicted by the Rouse theory.

In the high frequency region or in the transition zone the force responsible for the deformation changes the sign frequently and the time available between the successive changes of sign of the force is rather small. For the long-range coordinated motion of the large blocks of the main chain the cycle time is much larger than that of the high frequency motion. Hence under the high frequency operation before the long-range sequences can move very far the direction of the force reverses. Thus the motions of the first few normal modes or the motions with high relaxation time are completely eliminated. It is equivalent to saying that for the Maxwell model representing the material the movement of the dashpot is completely frozen as it represents the high relaxation time systems. All other motions involving the movements of the smaller groups in the backbone do occur, which form the essential characteristics of the transition zone.

Low frequency operation gives the large time between the reversal of the direction of the driving force hence the coordinated motion of the large blocks has sufficient time to complete the cycle and therefore this zone is characterized by the first few normal modes of the motions. The viscous drag or the viscous energy loss is much higher as compared to the stored elastic energy during these deformations therefore the magnitude of the loss modulus is higher than that of the storage modulus. As the frequency is increased further the elastic nature increases with faster rate due to progressive freezing of the first few normal modes of motions. Thus the slope of the G' curve (≈ 2) is almost twice as high as that of the G'' curves (≈ 1).

At much higher frequencies beyond the transition zone the system behaves essentially as a stiff spring and the variation of the frequency does not give any further information and the Rouse theory fails to predict the behaviour of the systems. For the very low relaxation time motions the temperature serves as an important variable and the isochronal tan δ- temperature curves are more useful.

After Rouse proposed the theory of long-range motions of polymer molecules in solutions it has drawn a lot researchers to propose its extensions to other systems particularly, dynamics of concentrated solutions and melts of entangled molecular systems. The results have shown that this theory is capable of explaining behaviour of other systems satisfactorily with few exceptions.

Malevanets and Yeomans[21] have reported numerical and analytical simulation describing the effect hydrodynamic interactions on dynamics of short polymer chains in solution. The analysis has given explicit expression for velocity autocorrelation function of center of mass of polymer, which agrees well with numerical results if Brownian dynamics, hydrodynamic correlation and sound wave scattering effects are included.

Berciea and Rusu[22] have reported that the physical properties and processing behavours of polymers are related to the broad range of length and time scale over which dynamical processes occure. It also gives a brief review of the background of polymer dynamics of solution as well as recent developments in this area. It is suggested that the understanding of polymer dynamics will help predict their flow behaviour and also guide one to better design the composition of ingredients to produce high-performance polymers, with tailord properties. Other authors like Burkhard[23], Lemak and Balabaev[24], Fujiwara and Sato[25], also have reported on the simulation of polymer molecules in solution.

Molecular chain relaxation studies have been reported by Kalathi et al.[26] for homopolymer melts by applying molecular dynamics simulations of the Kremer Grest (KG) bead-spring model of polymer chains of length between 10 and 500. The relaxation rates of the chains show two limiting effective monomeric functions, with the local modes experiencing much lower effective friction than the longer modes. The monomeric relaxation rate of longer modes vary with chain length due to kinetic confinement effects.

Phillies[27] has shown that treatments of the dynamics of a polymer chain contain a fundamental mathematical error, which causes them to lose, the zero-relaxation-rate, whole-body rotation motions dominant in viscosity and dielectric relaxation. As a result, the oft-cited mode solutions for these models are qualitatively incorrect. Comparison with the Wilson-Decius-Cross[26a] treatment of vibrational modes of polyatomic molecules reveals qualitatively the correct form for the solutions to the Rouse model.

6.13.2 High Molecular Weight Entangled Melts

Increase in the molecular weight of the polymer beyond the critical value causes considerable change in the rheological response of the material. These are attributed to the formation of physical entanglements between the molecules, which hinder their free movement. Following effects have been reported

(i) Shear viscosity varies as 3.5 power of MW as against to unity for low MW systems (Fig. 4.7).

(ii) The relaxation modulus-frequency plots (Fig.6.20(a)) passes through a plateau in the transition region at intermediate frequencies and then goes to the terminal region at higher frequencies. It is in contrast to the low MW system, which does not show any plateau effects[2].

(iii) Creep compliance curves also show the plateau effects for the entangled melts as compared to the low MW melts (Fig. 6.20(b))[2].

(iv) Storage modulus-frequency curves in the terminal region shift to lower frequency values as the MW increases and then pass through a plateau before entering the transition region (Fig. 6.20(c)). The width of the frequency range in the plateau region increases with increase in the MW[3].

(v) The loss modulus-frequency curves also shift to the lower frequency values in the terminal region as in the case of storage modulus (Fig. 6.20(d)). These curves show a shallow valley before entering the transition region[3].

(vi) Storage compliance and the loss compliances show plateau and a broad peak in compliance-frequency plots (Fig. 6.20(e))[3].

The deviations in the rheological response of the high molecular weight polymer as against that of the low MW systems has been explained on the basis of the entanglement formation. The coiled long chain molecules tend to physically entangle with each other due to the Brownian motions caused due to the van der Waals forces. It is postulated that the entanglements suppress the

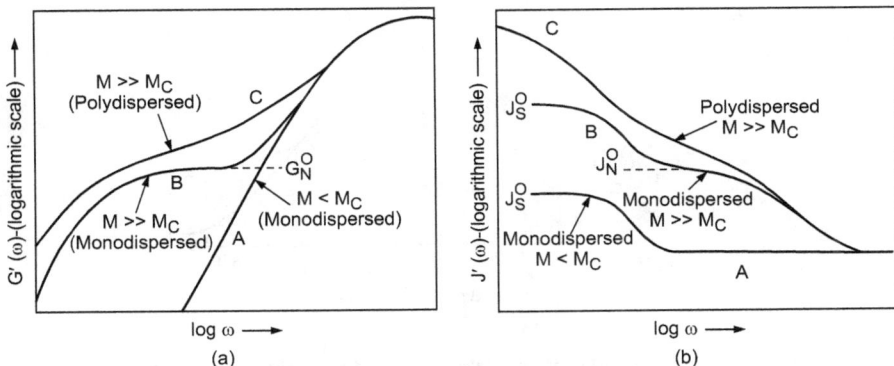

Fig. 6.20

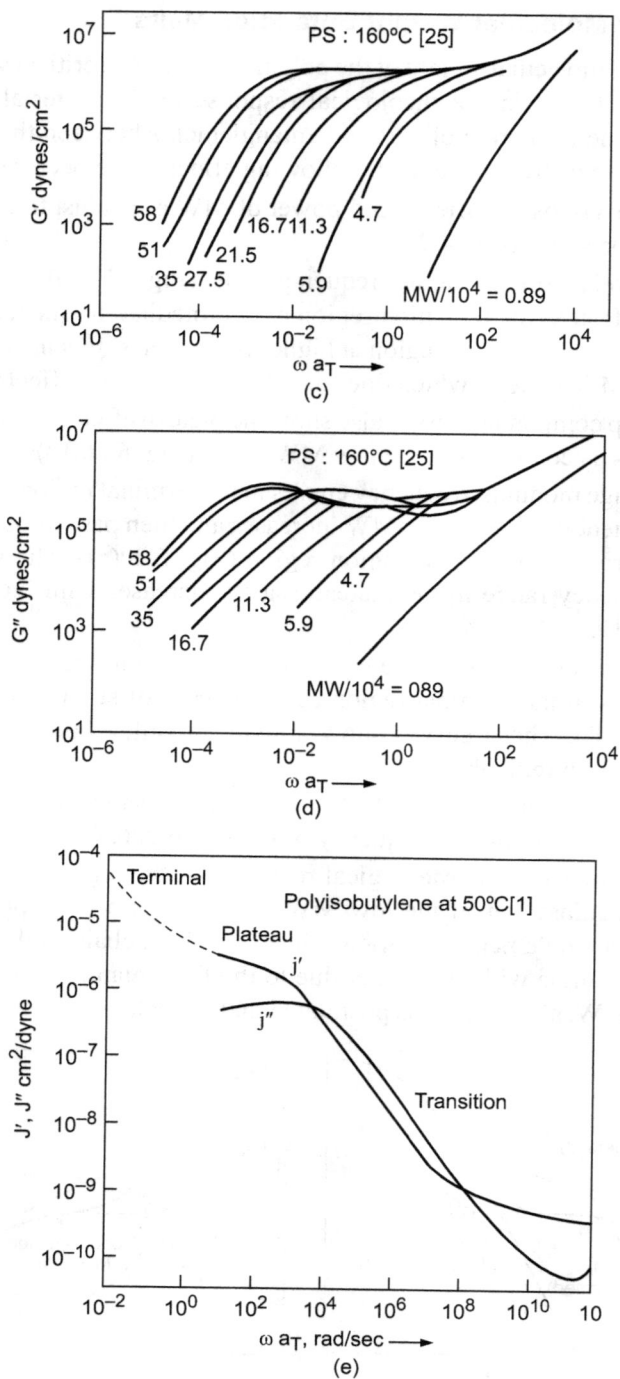

Fig. 6.20 Effect of high MW or entanglement on different rheological parameters.

first few normal modes of motions and make the relaxation phenomena slow and increase the relaxation time. As a consequence of this the terminal zone is shifted towards the lower frequency but the other zone remains unaffected as the relaxation time of the higher normal motions are not altered. This argument essentially implies the following

(i)   The freedom of movement of the chains is hindered due to the presence of other similar molecules and the long-range motions are suppressed.

(ii) The free space available for the long-range motion is reduced.

(iii) Only those motions are possible, which can take place in the limited space.

6.13.3 Doi-Edwards Theory

Based on these observations and arguments Doi and Edwards[28, 29] proposed a theory, which is based on the concept of "Reptation". The theory of reptation was first proposed by Piere Gill de Gennes in 1971[30, 31] and latter extended to the tube model by Doi and Edwards. The model describes the thermal motion of long polymer chains in concentrated solutions and melts. Doi and Edwards postulated that the motion of an entangled poymer molecule resembles the motion of a polymer in a tube that is, they assumed that the long range motion of polymer molecules is only allowed essentially along its own pass. Since the diffusion path resembles a wiggling snake, de Gennes termed the movement reptation, which is the root for reptile (reptation is the latin word for creeping).

Under the applications of the stresses causing deformations it is postulated that the system responds in two ways (i) a quick response and (ii) a slow response. The quick response corresponds to shorter relaxation time or the changes take place at high rate. Under these conditions the only reaction that can possibily occure is the redistribution of the extensions among the segments between the locations, where the topological constraints or the entanglements are present. The theory predicts that this relaxation process has a characteristic time θ_e called the "equilibrium time". After this process is over the remaining stresses are relaxed slowly having higher characteristic relaxation time θ_d. This process corresponds to the process of diffusion and as the diffusion is a slow process it take long time giving higher relaxation time. For long chains $\theta_d > \theta_e$. It is also possible that the molecule may get retracted within the tube over a long period of time but this process does not contribute significantly to the change in the properties.

The prediction of rheological properties for time θ_e corresponds to those obtained by Rouse theory. The application of the mathematical treatment to slow relaxation process leads to the expression for zero shear viscosity to increase with MW raised to power $3^{[32]}$. Thus, the tube model is based on two assumptions:

(a) the movement of each molecule is independent of those of neighbouring molecule, implying that no cooperate motion of polymer molecules takes place and (b) the lateral motion of molecules can be neglected, meaning that the molecules stay entirely within the virtual tube formed by the surrounding chains [Fig. 6.21].

Fig. 6.21   Schematic of a typical molecule hindered by other segments.[2]

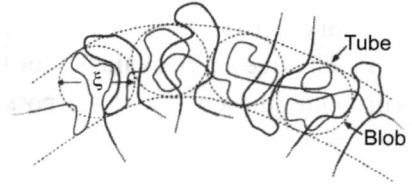

Fig. 6.22   Sketch showing the virtual tube formed by surrounding molecules. [polymerdatabase.com, 29]

6.13.4 The Mathematical Analysis

The shortest path between the end groups of the polymer chains is known as the primitive path, which coincides with the average position of the monomers along the tube. It can be seen from the Fig. 6.22 that the path constitutes the trajectory of a hypothetical chain of thickness ξ. Concept of Kuhn chain has been used to replace the real chain for simplification. Kuhn chain is assumed (i) to consist of N statistical segments of length a, i.e. the contour length, and (ii) these chains forms "blobs" of diameter ξ. Treating both the actual Kuhn chain of contour length, Na, and the primitive length L as random coils in the melt, it can be said that contour length of the primitive path equals the number of blobs multiplied by its average diameter ξ. It results in equation (6.118) where N_e is assumed to be equal to the number of segments between two successive

$$L \approx \frac{N\xi}{N_e} \approx \frac{Na}{(N_e)^{1/2}} \qquad ...(6.118)$$

entanglements and ξ is equal to the average end-to-end distance of the subchain,

$$\xi \approx a(N_e)^{1/2} \qquad ...(6.119)$$

The time required for disentanglement or the relaxation time, which is the time needed for the chain to diffuse out of the initial tube can be shown to be proportional to N^3. The mean squared displacement of molecules, assuming random motion, i.e. the Brownian motion, is given by

$$L^2 \approx D_t \tau \qquad ...(6.120)$$

where D_t is the diffusion coefficient, which can be calculated using Einstein relation:

$$D_t = \frac{kT}{\mu_t} \qquad \qquad ...(6.121)$$

where μ_t is the coefficient of friction of the creeping chain. This coefficient is N times higher than that of an individual link, $\mu_t = N\mu$. The time necessary for Kuhn chain to displace the length of the tube will be

$$\tau \approx \frac{L^2}{D_t} = \frac{L^2 \mu_t}{kT} = \frac{L^2 N\mu}{kT} \qquad \qquad ...(6.122)$$

where μ is coefficient of friction for each link and k_B is the Boltzmann's constant

As $\qquad \qquad L \approx \dfrac{Na}{(N_e)^{1/2}}$ it gives

$$\tau \approx \frac{a^2 N^3 \mu}{N_e T - N^3 \mu} \qquad \qquad ...(6.123)$$

Since the relaxation time determines the viscosity (η) the reptation model predicts

$$\eta \approx N^3 \approx M^3 \qquad \qquad ...(6.124)$$

where M is the molecular weight of the polymer chain and N is the number of statistical segments in the Kuhn chain. This finding is in good agreement with the experimental results

$$\tau \approx \eta \approx M^{3.4} \qquad \qquad ...(6.125)$$

Since the reptation model describes the motion of chains through entanglement, it is only valid for polymer long chains with $M \gg M_e$. For short chains which normally do not entangle the viscosity is proportional to MW:

$$\eta \approx N \approx M \qquad \qquad ...(6.126)$$

It implies that according to the reptation model the transition from non-entangled to entangled polymer leads to the change in exponent of MW from 1 to 3 which is very close to experimental exponent of 3.4.

This analysis has a number of limitations, which have been discussed by a number of workers ie Doi, Roovers, Doi and Graessley, Doi and Kuzuu, Pearson and Helfandand Mcleish. These limitations have been taken care by Bird and Curtiss[34] in its further development.

The viscoelastic behaviour of entangled linear polymers in terms of dynamics of single macromolecule has been explained by Pyshnograi et al.[34a]. These authors have developed the constitutive equations, which describe the mobility of the very long macromolecules during the longest observation times. They have shown that the additional introduction of local anisotropy of mobility and relaxation times of macromolecular coil and, beyond it, to estimate the transition time between weakly (the length of macromolecule $M < 10$ Me, no reptation) and strongly (length of macromolecules $M > 10$ Me, reptation) entangled polymer systems give the results of the conventional reptation-tube model. M is the MW of the polymer and Me is the molecular weight of segment of the polymer molecule between two entanglement points.

6.13.5 The Curtiss-Bird Model

Curtiss and Bird[33, 34] developed a theory that incorporates the concept of anisotropic hydrodynamic drag in addition to the anisotropic Brownian motions (reptation) accounted by the Doi-Edward theory. They introduced two empirical Parameters to account for this drag. These are (i) "Link tension coefficient", ε and (ii) "Chain constant exponent", β. Both these parameters are estimated by fitting the experimental data and it has been reported that they have the values in the range of 0.3 to 0.5. For $\varepsilon = 0$ Doi-Edwards results are reported. For monodispersed systems, the theory predicts that the zero shear viscosity is proportional to the longest relaxation time and both are proportional to the $(3 + \beta)$ power of the MW. Thus if β is taken as equal to 0.4 the experimental results of viscosity variation with the MW will be established.

6.14 NONLINEAR VISCOELASTICITY

The nonlinear viscoelastic behaviour is also the consequence of the combined elastic and viscous response of the polymeric materials but under large deformations at high deformation rates. The discussion so far is applicable to small and slow deformations where the basic material characteristics do not depend on the magnitude of the input functions. Almost all processing steps except the storage involve large and rapid deformations and thus both elastic and viscous characteristics of the viscoelastic materials show nonlinear behaviour and as a consequence of which the total response becomes nonlinear viscoelastic in nature. For such deformations the linear theory is no longer valid and response to the imposed deformation depends on

    (i) The extent of deformation,

    (ii) The rate of deformation and

    (iii) The kinematics of deformation, and

    (iv) Stress history

    This implies the following

    (i) It is not possible to use the results of the response of one type of deformation to predict the response of the same type unless both rate and magnitude of deformation are the same in both cases although the kinematics of deformation are same.

    (ii) It is not possible to use the result of one type of deformation, say extensional type to predict the response of other type say the shear deformation as the kinematics are different.

Many rheological phenomena, which are of considerable practical and scientific importance, cannot be foreseen from the linear viscoelastic theory. These are

(i) Nonlinear variation of shear viscosity with rate of shear (linear approach predicts constant viscosity).

(ii) Appearance of a nonzero first normal stress difference or the Weissenberg effect.

(iii) The appearance of the "Stress overshoot", where on the application of the rate of deformation the stress increases sharply to a maximum value then reduces to a constant value. This behaviour is in contrast to the response of Newtonian fluids or polymer melts at low deformation rates where the stress increases monotonically and asymptotically to a steady state value.

(iv) Non-linear dependence of extensional viscosity on the rate of strain.

(v) Nonlinear dependence of compliance on the magnitude of input stress and that of the modulus on the magnitude of input strain.

(vi) Non-sinusoidal response for the sinusoidal input function. For example, in the case of sinusoidal strain input the output stress may be non-sinusoidal (in the case of linear response the out put stress is sinusoidal but lags by a certain degree). Thus the basic properties like storage and loss moduli no longer remain useful to describe the material characteristics.

For simple static stress relaxation test the difference between linear and nonlinear behaviours is demonstrated using the Fig. 6.23.

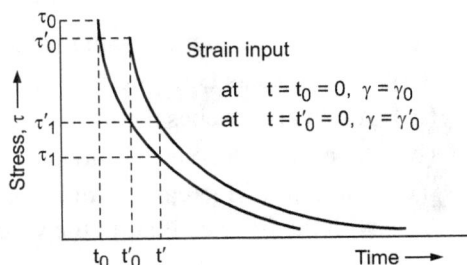

Fig. 6.23   Stress relaxation response

Consider the application of the input strains γ_0 and γ'_0 to a particular system at time $t = 0$ and estimation of modulus at a time interval $t = t'$. If the stresses for the two input functions are τ_1 and τ'_1 respectively then the two moduli are

$$G_1 = \frac{\tau_1}{\gamma_0} \quad \text{and} \quad G'_1 = \frac{\tau'_1}{\gamma'_0}$$

Then for the linear case $G_1 = G'_1$

And for the nonlinear case $G_1 \neq G'_1$

Payne[35, 36], and Freakley and Payne[37] published a number of experimental results showing the nonlinear behaviour of filled rubber vulcanizates for NR, IIR and other rubbers under the dynamic deformation. The storage shear modulus has been reported to be a function of double strain amplitude (DSA) and the filler loading at a particular temperature and frequency of oscillation. The DSA is given as, DSA = $2\delta l/l$, where δl is the amplitude of input strain function and l is the length of the specimen. Fig. 6.24 shows the variation of the dynamic storage modulus as a function of DSA at different filler loadings.

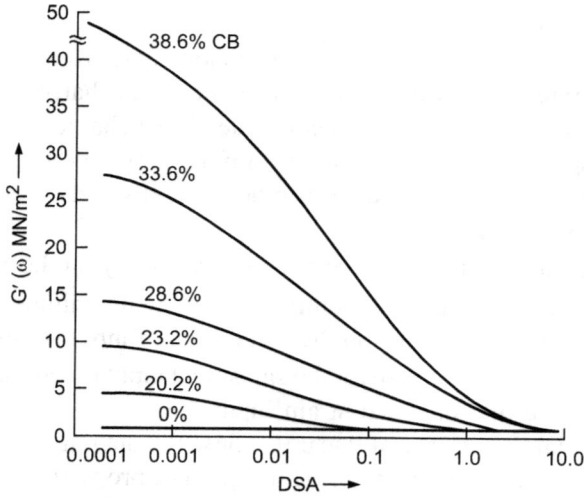

Fig. 6.24   Variation of the storage shear modulus for IIR vulcanizates[35]

It has been reported that the storage modulus, $G'(\omega)$

(i) Remains constant at some value say G'_0 at small DSA,

(ii) Decreases as DSA increases at intermediate values of DSA,

(iii) Again approaches a constant value, G'_∞ at high DSA,

(iv) At low filler loading the modulus is independent of the magnitude of DSA,

(v) The rate of increase of the modulus with the filler loading increases with the filler content in the vulcanizate, and

(vi) The onset of decrease in the modulus lies in the range 0.001 < DSA < 0.01 for different filler loading.

The magnitude of the storage modulus depends on the DSA as well as the filler loading. As the filler loading increases G'_0 increases with a much faster rate as compared to both $G'(\omega)$ and G'_∞.

At low strain amplitude the carbon black aggregate structure and the rubber matrix deform reversibly together to the same extent to give much higher modulus in this range of DSA. The structure of carbon black breaks down reversibly to give equilibrium in the rates of breakdown and rebuilding resulting in a constant value of the modulus.

At the intermediate strain amplitudes the rate of carbon black structure breakdown is higher than the rate of its build-up and therefore as the DSA increases further the carbon black structure breakdown also increases giving a continuous decrease in the storage modulus.

As the strain amplitude is further increased to some high value the carbon black aggregates are completely broken down and the modulus attains a constant value, which is just slightly higher than that for the gum vulcanizates.

All the modulus curves are similar in nature and therefore can be superimposed on each other and a generalized curve valid for all the filler loadings can be generated. Define the following parameters

$(G'_0 - G'_\infty) \rightarrow$ Represents the total capacity of the structure (both carbon black and rubber matrix) to store strain energy,

$(G'_\omega - G'_\infty) \rightarrow$ Represents the remaining capacity of the structure to store strain energy after certain amount of straining, and

$(G'_\omega - G'_\infty)/(G'_0 - G'_\infty) \rightarrow$ Represents the fractional capacity remaining to store the strain energy.

A plot of $(G'_\omega - G'_\infty)/(G'_0 - G'_\infty)$ versus strain energy $(\tau\gamma)$ gives a unique single curve for all vulcanizates (Fig. 6.25). The curve is represented by a Gaussian function as given below (Equation 6.127)

$$\frac{(G'_\omega - G'_\infty)}{(G'_0 - G'_\infty)} = \frac{1}{\alpha\sqrt{2\pi}} \int \exp\left[-\frac{(y - \beta)^2}{2\alpha^2}\right] dy \qquad ...(6.127)$$

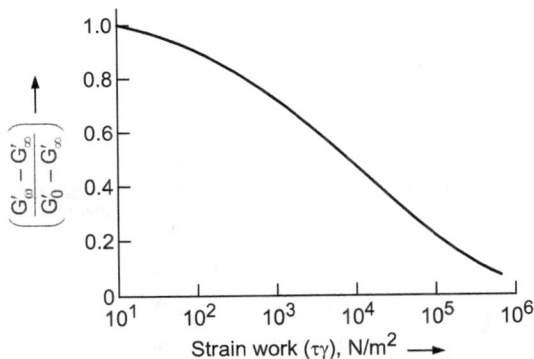

Fig. 6.25   A plot of $(G'_\omega - G'_\infty)/(G'_0 - G'_\infty)$ versus strain work $(\tau\gamma)$ for IIR vulcanizates[35]

where y is the log (strain amplitude) and α and β are constants. α represents the slope of the curve and β is the value of the energy to break down half of the black aggregate structure. Some important conclusions have been drawn from the above results. These are

(i) It has been reported that the pure rubber vulcanizates have more or less a constant value of $\dfrac{(G'_\omega - G'_\infty)}{(G'_0 - G'_\infty)}$, hence it can be said that the increase in the strain energy with increase in the filler loading is essentially due to the carbon black structure.

(ii) Different carbon blacks give similar curves indicating that these blacks do not differ in basic energy storing mechanism.

(iii) Different carbon blacks give different values of $(G'_0 - G'_\infty)$ indicating that the capacity of storing energy varies from black to black.

(iv) It is postulated that the carbon black structure must consist of continuous chain of particles with van der Waals bonds of attractions.

(v) The variation of loss modulus G'' shows maxima at some intermediate value of strain energy. This is due to the higher rate of change of structure breakdown as compared to the structure build-up, which reduces progressively with amplitude. At high amplitudes where the rate of breakdown of carbon black is much higher, the material flow occurs hence the increase in the loss modulus.

(vi) The magnitude of the viscoelastic function increases with increase in the surface area of the carbon black implying that it increases with increase in the structure (Fig. 6.26).

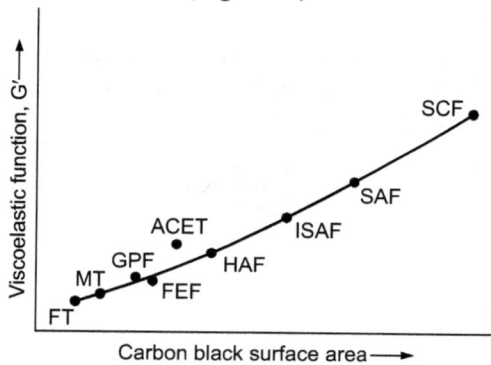

Fig. 6.26 Variation of viscoelastic function G' with carbon black surface area[35]
Surface area m^2/gm (Nitrogen absorption) of different carbon blacks:
FT-17, MT-9, GPF-36, FEF-40, ACET-75, HAF-82, ISAF-122, SAF-145, SCF-200.

PROBLEMS

1. An acrylic test piece of dimensions $4 \times 8 \times 25$ mm^3 is subjected to a tensile force of 448 N. The creep response of the system gives three zones (i) at $t = 0$ an instantaneous rise in strain, (ii) a retarded strain up to 200 hrs. and (iii) a linear increase in the strain beyond 200 hrs. The following data applies (i) at $t = 0$ the strain= 0.5% (ii) The maximum retarded strain = 0.7% (ii) the linear strain rate beyond 200 hrs = 1.167×10^{-5} hr^{-1} and (iv) the strain in 2nd region at 50 hr. = 0.65%. The system is to be described by a 4- parameter model consisting of a M-model and a V-K model in series calculate the model constants

2. A certain plastic test specimen of size $4 \times 8 \times 25$ mm^3 is subjected to an elongational force of 320 N for 200 s and then removed. Apply the Maxwell and V-K models to calculate the strain at 100 and 230 s. The model elements have the constants of 1 GN/m^2 and 50 GN s/m^2.

3. When a plastic test specimen of size $4 \times 8 \times 25$ mm^3 was subjected to a force of 100 N the following creep data was obtained. Develop the creep response equation of the V-K model and calculate the strain after 3000 s if the force applied was doubled. Calculate the retardation time and the model constants.

Time, s	Strain	Time, s	Strain
0.0	0.0	5000.0	9.75/(1000)
500.0	3.1/(1000)	7000.0	9.94/(1000)
1000.0	5.2/(1000)	100000.0	9.99/(1000)
3000.0	8.9/(1000)	150000.0	9.99/(1000)

4. A Maxwell model having its element constants of 200 MN/m^2 and 100 MN s/m^2 is subjected to a constant strain of 50%. Calculate and plot the stress relaxation curve. How will the relaxation curve change if the model constants are changed to (a) 100 MN/m^2 and 300 MNs/m^2 and (b) 500 MN/m^2 and 50 MN s/m^2. Comment on the variation of the relaxation curves.

5. A rubberised plastic material is subjected to creep response by the application of a force. The V-K model and the Maxwell (M) model are used to model the creep data. The V-K model constants are 0.2 GN/m^2 and 10 GN s/m^2 One of the element constant for the M model is 20 GN s/m^2. If both the models are to predict the same creep strain value after 50 s, estimate the other element constant for the M model.

6. A Maxwell model with constants 100 MN/m^2 and 1000 MN s/m^2, a Zener model with constants 100 MN/m^2 and 1000 MN s/m^2 and 500 MN/m^2 and a four parameter model of V-K and M models in series of constants 100 MN/m^2 and 1000 MN s/m^2 and 100 MN/m^2 and 1000 MN s/m^2 are subjected to strain of 50% at a certain time interval. Calculate the time required to relax 50% stress. Comment on your result.

7. A rubberised plastic test piece of dimensions $4 \times 8 \times 30$ mm^3 is subjected to a force, which increases with time. The force time relationship is given by $F = 3.2\,t$ for 100 s and then the force is reduced according to $F = -6.4\,t$. A Maxwell model with the constants of 2.5 GN/m^2 and 120 GN s/m^2 describes the creep curve. Calculate the strain in the material after 60 s, 120 s and 160 s.

8. Derive the governing equations of series combination of a V-K and M models. Develop the response equations for the stress relaxation and creep behaviours.

9. If the model in problem 8 is subjected to the sinusoidal stress then derive relations for the storage and loss moduli and compliances and the tan δ.

10. For the system in the problem 9 calculate and plot tan δ, storage and loss modulus as a function of the frequency if the M model constants are 100 MN/m^2 and 1000 MN s/m^2 and those of V-K model are 50 MN/m^2 and 500 MN s/m^2.

11. For the system given in problem 10 calculate and plot storage and the loss compliance as a function of the frequency.

12. Derive the dynamic response equations for the standard linear solid for complex stress and strain, storage and loss moduli and compliances and the tan δ.

13. Calculate and plot the tan δ. versus frequency graphs for the Zener model having the constants of 10 MN/m² and 100 MN s/m² and a spring constants of 50 MN/m².

14. If the Zener model of problem 13 is subjected to a sinusoidal stress of frequency 10 Hz calculate the power required to sustain the oscillations.

15. A Maxwell model having the element constants of 10 MN/m² and 100 MN s/m² is subjected to the dynamic response of the frequency 1, 10, 50 and 100 Hz. Calculate and plot the power required to sustain the oscillations as a function of the frequency.

16. Repeat the problem 15 for a V-K model of the same element constants.

17. Discuss the difference in the power required to sustain the oscillations in the problems 15 and 16 for the two models. Give physical reasons for the difference in the power consumption.

18. A monodispersed polymer behaves like a linear viscoelastic material. A Maxwell model having constants 10 MN/m² and 100 MN s/m² represents the stress relaxation response of this polymer. It is subjected to the following step changes in strain.

$t < 0\,s$	$\gamma = 0$
$100\,s \geq t > 0\,s$	$\gamma = 0.5$
$300\,s \geq t > 100\,s$	$\gamma = 1.5$
$500\,s \geq t > 300\,s$	$\gamma = 3.5$
$t > 500\,s$	$\gamma = 3.0$

Calculate the stress in the sample using B S P at t = 350s and 550s.

19. The polymer in example 18 is subjected to the following step change in the stress

$t < 0\,s$	$\tau = 0$ MPa
$100\,s \geq t > 0\,s$	$\tau = 3.0$ MPa
$300\,s \geq t > 100\,s$	$\tau = 5.0$ MPa
$500\,s \geq t > 300\,s$	$\tau = 5.5$ MPa
$t > 500\,s$	$\tau = 5.0$ MPa

Calculate the strain using BSP in the sample at $t = 150\,s$ and $550\,s$ assuming that the M model is applicable for the creep response of the material.

20. Use the V-K model for the creep response of the material and solve the problem 19 using BSP.

21. Derive the relationship for the creep response of a power law fluid.

22. If in a Maxwell model the dashpot contains a liquid, which follows a power law model and the spring is linear spring derive the stress relaxation relationship for this model. How will the stress relaxation time be expressed?

23. In the model in Prob. 22 what will be the effect of variation of the n value on the relaxation time?

24. Calculate and plot the relaxation time as a function of rate of shear at a particular value of n.

25. Calculate and plot the relaxation time as a function of n at a particular value of rate of shear.

26. How a pseudoplastic fluid in a dashpot of a V-K model will affect its creep response? Derive the corresponding equation for strain as a function of time.
27. Derive expression for the retardation time and discuss the effects of variation of n value of the dashpot fluid.
28. Discuss the stress relaxation behaviour of the V-K model in Prob. 26 and compare its response with a similar model having a Newtonian liquid in the dashpot.
29. Derive the expressions for storage modulus, loss modulus and tanδ for Maxwell model in Prob. 22 and discuss the effect of variation of n value on these parameters.
30. Derive the expressions for storage compliance, loss compliance and tan δ for a V-K model in Prob. 26 and discuss the effect of variation of n value on these parameters.

REFERENCES

1. J. D. Ferry, Viscoelastic Properties of Polymers, Third Ed., John Wiley & Sons, New York, 1980.
2. J. M. Dealy and K. P. Wissbrun, Melt Rheology and its Role in Plastic Processing, van Nostrand Reinhold, New York, 1990.
3. A. Kumar and S. K. Gupta, Fundamentals of Polymer Science and Engineering, Tata McGraw Hill Pub. Co. Ltd., New Delhi, 1978.
   (a) M. A. Mayers and K. K. Chawla, "Mechanical Behaviour of Materials" Cambridge University Press, P-573, (2001). ISBN 978-0-521-86675-0.
   (b) A. P. Chatterjee, P. L. Geissler and R. F. Loring, J. Chem. Phys., **104** (13), 5284 (1996)
   (c) zues.pimsc.psu/~manias/MatSE447/16_Molecular structure Effects.pdf.
   (d) Johan L.A. Dubbeldam and J Molenaar, Macromolecules, **42**, 6784-6790 (2009).
   (e) J. H. Lee, L. J. Fetters and L. A. Archer, Macromolecules, **38 (26)**, 10763-71 (2005).
   (f) P. Dimitrakopoulos, j. Fluid Mech., **523**, 265-286 (2004).
   (g) M. Kapnistos, M. Lang, D. Vlassopoulos, W. Pyckhout-Hintzen, D. Richter, D. Cho, T. Chang, and M. Rubinstein, Nature Materials, **7**, 997-1002 (2008).
   (h) A. G. Makarov, G. Ya. Slutsker, I. V. Gofman and V. V. Vasil'eva, Physics of Solid State, **58 (4)**, 840-846 (2016).
   (i) J. M. Deutsch and J. H. Pixley, Physical Review **E 80**, 011803 (2009).
4. D. V. Rosato, Plastic Design Handbook, Publishers : Springer , 2001. ISBN 10: 0792379802 / ISBN13: 9780792379805
   (a) N. G. McCrum, C. P. Buckley and C. B. Bucknall, Principles of Polymer Engineering, Oxford Science Publications, (2003). ISBN 0-19-856526-7
   (b) N. G. McCrum, B. E. Read and G. Williams, "Anelastic and dielectric effects in polymer solids1st Ed. , John Wiley and SonsInc. , New York, 1967.
   (c) G. V. Vinogradov and A. Ya. Malkin, Rheology of Polymers, Mir Publishers, Moscow, 1980.

5. N. W. Tschoegl, The Phenomenological Theory of Linear Viscoelasticity: An Introduction, Springer- Verlag, Berlin 1980.

6. en.wikipedia.org.

7. The Penguin dictionary of Physics, ed. Valerie Illingworth, 1991, Penguin Books, London.

8. R. Feynman, Lectures in Physics, Vol.1, pp30, Addition Wesley Publishing Company, Reading, Masss. (1963)

9. N. K. Varma, Physics for Engineers, PHI Learning Pvt. Ltd., pp 361, (2013).

10. T. Freegard, Introduction to the Physics of Waves, Cambridge University Press,

11. J. E. Shigley, C. R. Mischke and R.G. Budynas, Mechanical Engineering Design, Published, McGraw-Hill Professional, page,192, (2004).

12. R. M. Christensen, "Theory of Viscoelasticity: An Introduction , New York: Academic Press (1971), pp-92

13. P. C. Hiemenz, C. Paul, P.T. Lodge; Polymer Chemistry, 2nd ed., Florida:Taylor & Francis Group. (2007), pp-486-491.

14. R. Li, Materials Sci. and Eng., **A- 278 (1-2),** 36-45, (2000).

15. Van G. Marniz and J. Palmen, Rheology Bulletin, **67(1),** 5-8 (1998).

16. N. G. McCrum, C. B. Bucknall and C. P. Buckley "Solutions Manual to Accompny Principles of Polymer Engineering" Oxford University Press , USA.] (2003).

17. R. J. Crawford, Plastics Engineering, Butterworth / Heinemenn, Oxford, 1998, p 114.

    (a) I. R. Tanner , Engineering Rheology, Oxford University Press.P. 27 (1988). ISBN 0-19-856197-0

18. H. A. Flocke, Kolloid Z., **180,** 118 (1962).

19. P. E. Rouse, Jr., J. Chem. Phys., **21,** 1272 (1953).20. B. H. Zimm, J. Chem. Phys., **24,** 269 (1956).

20. B. H. Zimm, J. Chem. Phys., **24,** 269 (1956).

21. A. Malevanets and J. M. Yeomans, EPL (Europhysics Letters), **52,** number 2 (2000).

22. M. Berciea and D. Rusu, in book: New Trends in Nonionic (Co) Polymers and Hybrid, Chapter: The dynamics of Polymer Chains in Solution, Publisher: Nova Science Publishers, Editor: E. S Dragan. Pp 79-199, January 2006.

23. Burkhard Dünweg, AIP The J. Chem. Phys., **99,** 6983 (1993)

24. A. S. Lemak and N. K. Balabaev, J. Computnl. Chem., **17** (15), 1685-1695 (1996).

25. S. Fujiwara and T. Sato, researchgate.net, August 2002

26. J. K. T. Kalathi, S. K. Kumar, M Rubinstein and G. S. Grest, Macromolecules , **47 (19),** 69025-69031 (2014).

    (a) E.B. Wilson, J.C. Decius and P.C. cross, Molecular vibrations, New York, Dover-ISBN 048663941X, (1995).

27. G. D. J. Phillies, researchgate.net, 23 July 2009. [arxiv:0907.4172v1].

28. M. Doi and S. F. Edward, J. Chem. Soc., Farade Trans. II, **74,** 789, 1802, 1808 (1978) and **75,** 38 (1979).

29. M. Doi and S. F. Edward, The Theory of Polymer Dynamics, Oxford Univ. Press. 1986. [Polymer Properties Database, polymerdatabase.com].

30. P. G. De Gennes, J. Chem. Phys. **55**, 572 (1972).

31. P. G. De Gennes, Scalling concepts in polymer physics, Cornell University Press, New York, 1979.

32. D. S. Pearson and E. Helfand, Macromolecules, **17**, 888 (1984).

33. C. F. Curtiss and R. B. Bird , J. Chem. Phys., **74**, 2016 (1982).

34. R. B. Bird, C. F. Curtiss, R. C. Armstrong and O. Hassager, Dynamics of Polymer Liquids Vol. 2, Second Ed., John Wiley & Sons, New York, 1987.

    (a) G. Pyshnograi, H.NA. Al Joda, and I. Pyshnograi, World J. mechanics, **2**, 19-27, 2012.

35. A. R. Payne, J. Appl. Polym. Sci., **6**, 57 (1962).

36. P. K. Freakley and A. R. Payne, Theory and Practice of Engineering with Rubber, Applied Science Pub., Ltd. London, 1978, pp. 74-84.

37. A. R. Payne, J. Appl. Polym. Sci., **7**, 873 (1963).

❏   ❏   ❏

Viscoelastic Functions: Effect of Various Parameters

7

A number of physical and processing parameters influence the viscoelastic behaviour of the polymer materials. The important variables are temperature, concentration, cross-link density, crystallinity, copolymerization, branching etc. This chapter includes the discussion on the effect of different parameters on the viscoelastic properties of the polymers.

7.1 EFFECT OF TEMPERATURE

Studies to investigate the effect of temperature on the rheological properties of viscoelastic materials are of continuous interest to the researchers as temperature has varied effects under different processing conditions. Different polymers and rubbers have typical response to the temperature.

Drodzov and group[1-4] in their number of studies have reported on the temperature effects on polymer behaviour:

(i) Rubbery polymers at finite strain[1] The long chains of the polymers, which are connected to temporary junctions tend to free from the junctions which are treated as transition from their active state to the dangling state. These chains capture a new junction in the vicinity of its free end at random and returns to its active state. The model developed by simulation of testing uniaxial tensile strength of carbon black filled rubber at different temperatures gave good agreement with experimental data.

(ii) Isothermal torsional oscillation test is reported for (isotatic PP) iPP and LDPE in the temperature range of 120 and 190°C[2]. The constitutive equations are developed with reference to the concept of transient networks for the response of polymer melts at three dimensional network deformations with small strains. The three adjustable parameters like plateau modulus, the average activation energy for rearrangements of strands and standard deviation of activation energies have been determined. The two polymers give different constants, which are because of difference in their molecular structure.

(iii) Isothermal torsional relaxation test for PP, LDPE and LLDPE in the temperature range of room temperature to 120°C has been published[3].

Constitutive equations have been derived for viscoelastic behaviour of semicrystaline polymers at small strain. It has been assumed that the polymers are equivalent to the network of strands bridged by temporary junctions i.e. entanglement, physical crosslinks on the surface of the crystallites and lamellar blocks. The network is thought of as an ensemble of mesoregions with various activation energies for detachment of active strands from their junctions. The time-dependent response of the ensemble reflects thermally induced rearrangement of strands (separation of active strands from temporary junctions and merging of dangling strands with the network). This study focuses on the influence of temperature and crystalline morphology of polyolefins on the material parameters in the constitutive relations.

(iv) For isotactic polypropylene the tensile test data with various cross-head speeds and relaxation tests in a wide interval of temperatures ranging from room temperature to 120°C are reported[4]. A constitutive model is derived for the responses of a semicrystalline polymer at arbitrary deformations with small strains. The stress–strain relations involve 6 adjustable parameters that are found by fitting the experimental data. The presence of a critical temperature is demonstrated at which some parameters of the model reach their maxima. This temperature is associated with the α-relaxation temperature of polypropylene.

Hiremath et al.[5] have published on the effect of post curing temperature on viscoelastic and flexural properties of epoxy/alumina polymer nanocomposites (PNCs) using dynamic mechanical analysis (DMA) and three point bend flexural tests over the temperature range of 25 to 200°C to determine the viscoelastic properties. The results indicated that the post curing at a temperature below the glass transition temperature of epoxy enhanced the viscoelastic and flexural properties of PNCs and post curing temperature above the glass transition temperature of epoxy had a detrimental effect on the properties for both neat epoxy as well as nanocomposites. As the post curing temperature was increased the glass transition temperature increased for both the neat epoxy and nanocomposites, however the increment in glass transition temperature of PNC was more in comparison to that of neat epoxy. Flexural modulus and strength of epoxy and PNC decreased with the increase in post cure temperature.

Pandini and Pegoretti[6] have reported on the variation of Poisson's ratio of polymeric materials, although generally assumed as a constant, it is known to display a viscoelastic dependence on time, temperature, and strain. This article investigates the phenomenology of this dependence on two cross-linked epoxy systems with different glass transition temperatures. Poisson's ratio measurements are performed by contact extensometers simultaneously measuring the axial and transverse deformations under two different tensile testing conditions: (i) constant deformation rate, in which the effects of strain,

strain rate, and temperature are highlighted; (ii) stress relaxation (or constant deformation), where the dependence of Poisson's ratio on time is studied at various strain levels. The viscoelastic Poisson's ratio increases as strain, temperature, and time increases, with trends markedly depending on the materials glass transition.

Papadogiannis et al.[7] have studied the effect of temperature on viscoelastic properties of nano-hybrid composites polymer resins. The objective was to to examine the viscoelastic properties of nanofilled dental composites under both static and dynamic testing and to determine the influence of temperature, medium of storage and storage time in the temperature range of 21–50°C after being stored for 24 h and 1 month under both static and dynamic conditions. All materials tested showed a significant decrease in their moduli with the increase of temperature, while the effect of water storage was different among the composites. Most of the materials tested did not have elastic moduli near to that of dentin, making them less satisfactory in posterior restorations. The materials possessing nano-size filler particles had different elastic properties amongst the materials tested, which implies that filler size is not the only factor that affects the elastic behavior of dental composites.

Kao et al.[8] have reported on the effect of temperature on the viscoelastic properties of model and industrial dispersions. The systems investigated are polystyrene gelatin and industrial photographic coupler dispersions. Time temperature superposition (TTS) principle has been successfully applied in dispersions in the sol-state as well as gel-like state. It allowed the estimation of the rheological properties of the dispersions over the frequency range, which is otherwise inaccessible to the range of experimental measurements. Therefore, the linear viscoelastic properties of these model and photographic coupler dispersions at very low frequency (which is useful in predicting the stability of the product), as well as properties at very high frequency (or large deformation, e.g., during coating or pumping processes), could be estimated.

Driscoll et al.[9] have studied the effect of temperature on optical and viscoelastic properties of newly synthesized homogeneous thermotropic liquid crystalline polymers and reported about its dynamic properties. The unique behaviour of this material is highlighted by the appearance of an intermediate temperature in dynamic viscosity vs temperature curve. This is indicative of the liquid crystall nature of the material and may be related to the structural differences. Viscoelastic results were therefore compared with optical micrographs taken in the static state over the range of temperature. Polarized light optical microscopy identifies the so-called isotropitization temperature, which signals the beginning of the isotropic rheological behaviour, but optical micrographs show little apparent corresponding structure change below this temperature, in the range where the intermediate temperature rheological transition is located.

Dos Reis[10] published the results of an experimental program to investigate the effect of temperature on the performance of epoxy and unsaturated polyester polymer mortars (PM). PM is a composite material in which polymeric materials are used to bind the aggregates in a fashion similar to that used in the preparation of Portland cement concrete. For this purpose, prismatic and cylindrical specimens were prepared for flexural and compressive tests, respectively, at different temperatures. Measurements of the temperature-dependent elastic modulus and the compressive and flexural strength were conducted using a thermostatic chamber attached to a universal test machine for a range of temperatures varying from room temperature to 90°C. The flexural and compressive strengths decrease as temperature increases, especially after matrix heat distortion temperature (HDT). Epoxy polymer mortars are more sensitive to temperature variation than unsaturated polyester ones.

Caputo et al[11] have studied the effect of temperature on the viscoelastic behaviour of entangled solution of multisticker associating polyacrylamides (PAM). Hydrofobically modified PAM with small blocks (~ 3 monomer units) of N, N-dihexilacrylamide (2% mole) has been investigated in the entangled regime as a function of temperature by steady flow and oscillatory experiments. The effect of temperature was also studied in presence of surfactants. In both cases the results were compared to those obtained for unmodified analogue. The experimental data were analysed by using a simple Maxwell model and the time-temperature superposition principle. The results indicated that the main effect of temperature variation and/or addition of surfactant is to modify the lifetime of physical crosslinks without changing their density, as indicated from quasi-constant value of plateau modulus. The modified system shows much higher value of plateau modulus as compared to the unmodified system.

The review of literature shows that change in the temperature influences all rheological parameters to a considerable extent. The response of a polymer may be assumed to be equivalent to the response of a combination of a number of Maxwell models exhibiting both elastic and viscous effects. Both these characteristics change with temperature giving the variation in the relaxation time, retardation time, modulus, compliance and in the dynamic parameters[12]. The expressions for the storage and loss moduli and the relaxation time can be used to understand the influence of the temperature on these quantities.

Storage modulus is given as

$$G' = \sum_{i=1}^{\infty} \frac{G_i \omega^2 \theta_i^2}{1+\omega^2 \theta_i^2}$$

...(7.1)

Loss modulus is given as

$$G'' = \sum_{i=1}^{\infty} \frac{G_i \omega \theta_i}{1+\omega^2 \theta_i^2}$$

...(7.2)

The expression for the modulus G_i is

$$G_i = NkT = \frac{cRT}{M} \qquad \text{...(7.3)}$$

where N is the number of moles per unit volume, c is the concentration, M is the molecular weight, k is the Boltzman constant, and T is the absolute temperature. The modulus for a given material can be assumed to be constant. The relaxation time in terms of the viscosity and the activation energy, E can be written as

$$\theta_i = \frac{6\eta M}{\pi^2 i^2 cRT} = \frac{6\eta}{\pi^2 i^2 NkT} = \frac{6\,Ae^{E/RT}}{\pi^2 i^2 NkT} \qquad \text{...(7.4)}$$

Substitution of equations 7.3 and 7.4 in equations 7.1 and 7.2 one gets the expressions for storage and loss moduli as functions of temperature. The storage modulus thus can be written as

$$G' = \sum_{i=1}^{\infty} \frac{G_i\, \omega^2 \theta_i^2}{1 + \omega^2 \theta_i^2}$$

$$= NkT \sum_{i=1}^{\infty} \left[\frac{\omega^2 \left(\dfrac{6\,Ae^{E/RT}}{\pi^2 i^2 NkT} \right)^2}{1 + \omega^2 \left(\dfrac{6\,Ae^{E/RT}}{\pi^2 i^2 NkT} \right)^2} \right] \qquad \text{...(7.5)}$$

A similar expression for loss modulus also can be written. It can be seen that the moduli are a complex function of the temperature. As the temperature of the system increases the storage modulus, (i) increases due to the term (NkT) and (ii) decreases due to term under the summation sign. The effect of the exponential term or the viscosity in the summation term will tend to reduce the storage modulus to a greater extent as compared to the increase due to the N k T term at some suitable temperature where the viscosity is most sensitive. One such temperature is glass transition temperature at which the glassy state changes to the rubbery state and long range molecular motions dominate giving a sharp reduction in the storage modulus as shown in Fig. 7.1[13]. At temperatures below T_g the effect of the two terms seems to more or less balance out and the modulus either remains constant or reduces only slightly with increase in the temperature. At higher temperature where the polymers exhibit the fluid like behaviour with large relaxation time the moduli keep on decreasing.

The loss modulus also follows the similar pattern except that it gives a peak just before the glass transition temperature indicating the onset of the long-range motions and increased loss of energy due to the viscous drag. At temperature beyond T_g it falls regularly with the increase in the temperature. Below T_g the loss modulus keeps falling and there appear a number of small broad peaks in the curve corresponding to the short-range motions of different side groups, branches and grafted segments.

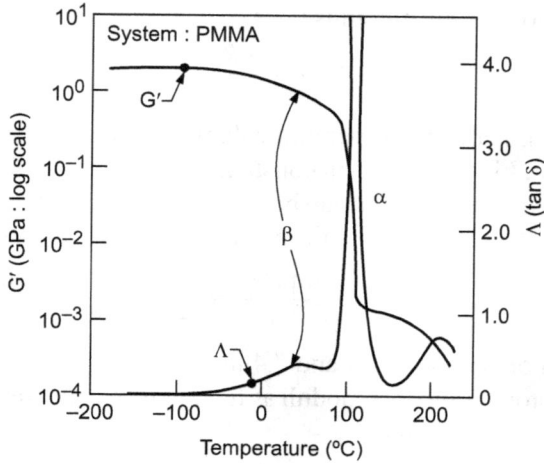

Fig. 7.1   Effect of temperature on storage modulus and tan δ[13]

The loss tangent traces as a function of the temperature also show a considerable decrease in the loss tangent value and show a number of small peaks corresponding to the short range motions. These peaks are much sharper than those in the case of loss modulus. Therefore the tan δ curves are much more useful in understanding the short range molecular motions.

Estimation of the magnitude of energy associated with these different peaks leads to the assignment of the peaks to short range motion of a particular group. The peaks occurring progressively at lower temperature would correspond to the similar level of reduction in the energy loss due to the motions. It therefore implies that the corresponding peaks are associated with the motions of smaller and smaller groups. Polystyrene gives four peaks at 0.5 Hz as follows (Fig. 7.2)[14, 15, 16].

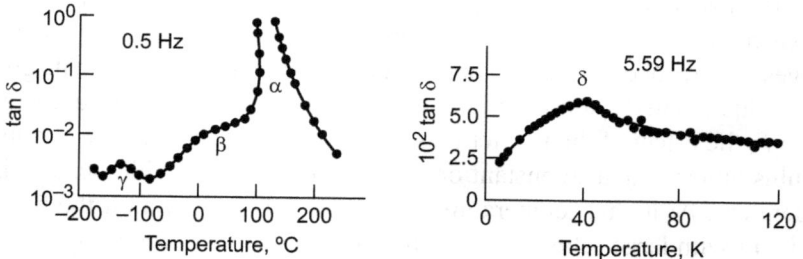

Fig. 7.2   Tan δ versus temperature plots for polystyrene[17]

  (i) α-Peak at around 110°C known as α transition corresponding to the onset of long range coordinated motions of molecular segments. The storage modulus registers a steep fall around this temperature.

 (ii) β-Peak at around 50°C. It has been interpreted to be due to the hindered rotation of the phenyl group

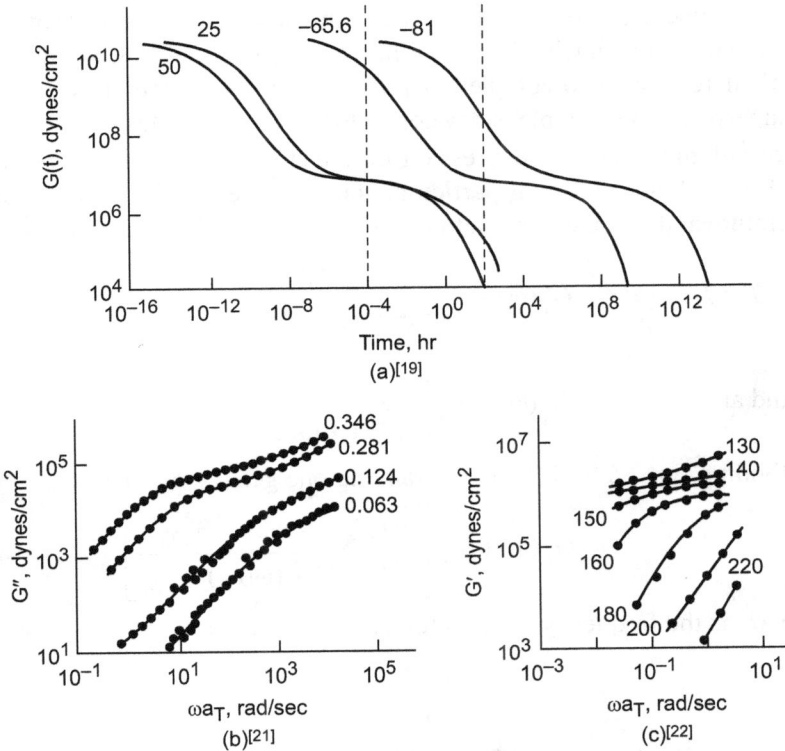

Fig. 7.3 Different viscoelastic functions. (a) Stress relaxation data for Polyisobutylene at different temperatures, (b) Loss modulus for polystyrene (M = 267000) solution in partially chlorinated diphenyl at different concentrations of polystyrene (g/ml) (c) storage modulus for polystyrene (M = 167000) at different temperatures.

(iii) γ-Peak at around −120°C has been thought to be associated with two types of motions, (a) local torsional modes of small sequences of main chain carbon atoms about the backbone bonds and (b) rotation of the phenyl groups. And

(iv) δ-Peak at around −235°C (38°K) has been assigned to the coupled oscillatory and wagging movement of the phenyl groups.

7.1.1 Data Superposition

A number of linear viscoelastic functions for example modulus versus time (Fig. 7.3(a)),[18, 19, 20] loss modulus versus frequency (Fig. 7.3(b))[21] and storage modulus versus frequency (Fig. 7.3(c))[22] etc. show similar nature of dependence at different temperatures. Therefore it is possible to calculate a shift factor, which will be a function of temperature (Figs. 7.4 & 7.5), to get a generalized curve to predict the value of the function at different temperature from its value at some other temperature. This is known as the data superposition principle. Use of this approach enables to predict the data in the time or frequency scales, which are not possible to achieve by normal experimental techniques for example the relaxation experiment (Fig. 7.3(a)) for polyisobutylene at $-81°C$ would require more than 10^{12} hours to complete. Application of the temperature-frequency data superposition principle is developed for storage modulus.

For polymer melts the expression for the storage modulus can be written from Eqs. 7.1 and 7.3 at a particular temperature. If T_0 is any reference temperature and T is the temperature at which the data is to be predicted then

$$\text{At } T = T \qquad G'_T(\omega) = \frac{RT}{M} \sum_{i=1}^{\infty} \frac{\omega^2 \theta_i^2}{1 + \omega^2 \theta_i^2} \qquad \qquad ...(7.6)$$

$$\text{And at } T = T_0 \qquad G'_{T_0}(\omega) = \frac{RT_0}{M} \sum_{i=1}^{\infty} \frac{\omega^2 \theta_i^2}{1 + \omega^2 \theta_i^2} \qquad ...(7.7)$$

Dividing Eq. 7.6 by 7.7 and rearranging one gets

$$G'_T(\omega) = \frac{T}{T_0} \left[\frac{RT_0}{M} \sum_{i=1}^{\infty} \frac{(\omega \alpha_T)^2 \theta_{i,T_0}^2}{1 + (\omega \alpha_T)^2 \theta_{i,T_0}^2} \right] \qquad ...(7.8)$$

where α_T is the frequency shift factor, which is given as

$$\alpha_T = \frac{T_0 \, \eta_T}{T \, \eta_{T_0}} \qquad \qquad ...(7.9)$$

The term inside the parentheses of Eq.7.8 represents the expression for the storage modulus at frequency of $(\omega \alpha_T)$ hence

$$G'_T(\omega) = \frac{T}{T_0} G'_{T_0}(\omega \alpha_T) = \beta_T G'_{T_0}(\omega \alpha_T) \qquad ...(7.10)$$

where β_T is the temperature shift factor given as $\beta_T = T/T_0$. Fig. 7.4 shows the concept of data superposition for storage modulus-frequency data at two different temperatures[12]. With reference to this data α_T is the horizontal shift factor and β_T is the vertical shift factor.

Figure 7.3(c) gives the storage modulus versus frequency data for polystyrene at a number of temperatures. This data has been shifted with reference to the temperature of 160°C and the shift factor as a function of temperature is shown in the Fig.7.5[22]. This figure also gives the shift factors for two other molecular weight of PS [(i) MW = 14800 and (ii) MW = 8900]. The shift factor is independent of the MW at higher temperatures and marginally deviates at lower temperatures indicating its viscosity sensitive characteristics at low temperatures only.

Thus the application of the superposition principle can be extended to different MW system so long as the mechanisms associated with the response vary in the same manner with the temperature. It is possible if the polymer phase does not undergo any change with the temperature. In case of the crystalline polymers however this condition will keep on varying as the extent of crystallization varies with the temperature affecting the mechanism of the response. Therefore it is not expected to be valid for the crystalline polymers.

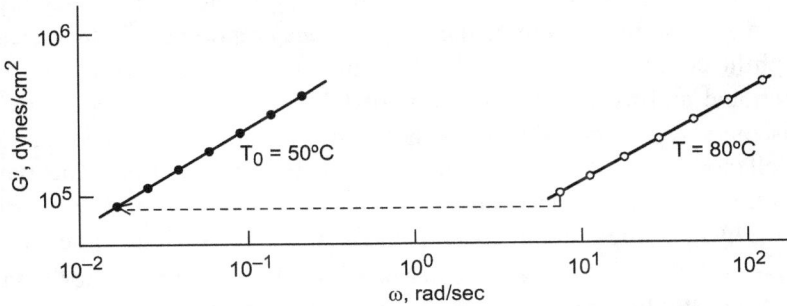

Fig. 7.4 Shifting of experimental data of commercial polypropylene ($T = 80°C$ to $50°C$)[12]

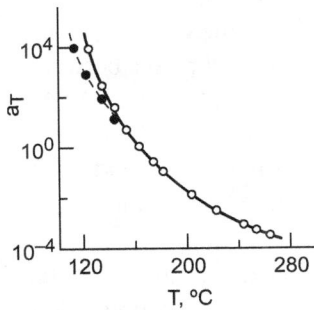

Fig. 7.5 Plot of shift factor, a_T versus temperature for polystyrene. (Ref. Temp. = 160°C, ...o... for M = 148000....•...for M = 8900----)[22]

7.2 EFFECT OF MOLECULAR WEIGHT AND OTHER PARAMETERS

The investigations on the effects of concentration, molecular weight and other parameters have been presented by many researchers.

Colby et al.[23] have presented results of studies on the solutions of a high molecular weight polybutadiene covering the entire range of concentration from pure polymer to pure solvent at 25°C. The solvents used are (a) dioctyl phthalate, which is nearly a θ solvent and (b) phenyloctane, which is a good solvent. In dilute solutions the viscosities are well represented by Huggins equation although the viscosities and Huggins coefficients in two solvents are quite different. In entangled semidilute and concentrated solutions the free volume adjusted viscosity is identical in two solvents and obeys power laws of $c^{4.7}$ in semidilute solution and $c^{3.6}$ in concentrated solution. The elastic response is also identical in two solvents with dynamic modulus varying as $c^{2.3}$ for c > 0.02. The obsevations in these studies suggest that the concentration dependence of characteristic molecular weight are somewhat different.

Barmer et al.[24] investigated the response of series of hydrophobically modified ethoxylated urethanes (HEURs) systems in aqueous solutions. Both steady shear and oscillatory shear was determined as a function of hydrophilic chain length and polymer concentration. A dramatic increase in viscosity was observed with decreasing molecular weight of the prepolymer (with decrease of hydrophilic components' size and at the same time an increased ratio between hydrophobic and hydrophilic sections of HEURs). Also a steep increase in the viscosity with increasing thickener concentration is obtained. A simple Maxwell model describes the rheological properties of the system. The dynamic measurements verified the results obtained from steady state measurements about the hydrophilic section size and its effect on the association phenomenon.

Jing Tao[25] in his MSc. degree in Material Science submitted to the faculty of the Worcester Polytechnic Institute, investigated on the effects of molecular weight (MW) and concentration, c on electrospinning of PVA. The results indicate that the minimum MW and c corresponding to capillary number, $Ca = [\eta^* \, V/\sigma) - 0.5]$ (where η^* is the dynamic viscosity, V is the characteristic volume and σ is the interfacial tension between the two liquid phases) is necessary for forming a fibrous structure. As MW or c increases the fiber diameter becomes larger and a broader distribution of fibers may be obtained. Round fibers may be obtained at low MW and c where as flat fibers are observed at high MW and c. The addition of NaCl lowers the average fiber diameter in PVA samples with high MW.

The dynamic viscoelasitc behavior of lentinan, one triple helical β-(1 $\rightarrow$ 3)-d-glucan from the fruiting body of *Lentinus edodes (an edible mushroom)*, in water was investigated as a function of concentration and molecular weight at 25°C by using dynamic rheology[26]. It was revealed that the shear storage moduli (G'), viscous loss moduli (G''), and the dynamic complex viscosity (η^*) exhibited strong dependence on concentration and molecular weight. At low concentrations, the lentinan/water systems displayed liquid-like behavior with G' lower than G'' at low frequencies and crossing-over at high frequencies. With increasing concentration, the elastic response of the lentinan/water system was stronger than the viscous response, leading to the conclusion that the lentinan/water systems displayed a predominantly solid-like behavior. The gel point (c_{gel}) was determined from Winter–Chambon method (frequency-independence of tan δ). Winter and Chambon developed a gel equation to describe linear viscoelastic behaviour of critical gel [W.B. Yoon and S. Gunasekaran, Food Sci. Biotechnol., 16(6), pp. 1069-1071 (2007)]. The most important point is that the c_{gel} was much lower than some synthesized polymers and other flexible polysaccharides, which may be attributable to the high stiffness of triple helical lentinan and strong intra- and intermolecular interactions among polysaccharide chains. Furthermore, a decrease in molecular weight leads to a sharp increase of c_{gel}. The dynamic strain sweep measurements proved that the gelation of lentinan in water is induced by the extremely entangled and stiff triple helices forming continuous network, and the lentinan gel is structurally more like a solution that is unable to flow within a timescale of usual observation.

Dynamic viscoelastic behavior was investigated by Osaki et.al. for solutions of polystyrene in tricresyl phosphate, a good solvent, at concentrations, c, less than the coil-overlapping concentration, c^{*}[27]. At the infinite dilution limit, the behavior was in accord with the theory of Doi and Edwards involving the excluded volume potential and hydrodynamic interaction (HDI). Thus, the viscoelastic functions were completely derived from the intrinsic viscosity–molecular weight relation. At finite c, the complex modulus was represented by the sum of two terms. One was a Rouse–Zimm (RZ) term conveniently represented by the Zimm theory with an arbitrarily chosen value of the HDI parameter. The other was a term with a single relaxation time, longer than the longest RZ relaxation time, and with a high-frequency modulus proportional to the square of c [the long-time (LT) term]. The behavior of the RZ term indicated the stronger screening of HDI with increasing c. Using the experimental c dependence of the longest RZ relaxation time to get the relevant parameter, the authors compared the RZ viscoelastic function with the Muthukumar–Freed theory. The agreement was good at low concentrations, $c < c^{*}$. The contribution of the LT term, which was not included in the theory, was quite significant at low frequencies; about 60% of the Huggins coefficient was attributable to this term.

Clasen et al.[28] have reported on the effects of polymer concentration and molecular weight on the dynamics of visco-elasto-capillary break up behavior of a series of polystyrene solutions with narrowly-distributed molecular weight. These solutions were progressively diluted in solvents with different viscosities. It has been demonstrated that there exists a critical dilution concentration below which the dominant stress resisting breakup was carried solely by the solvent. Increase in the molecular weight results in a larger relaxation time and (at any fixed concentration) a slightly higher shear viscosity. For all fluids, the relaxation time was found to be a power-law function of the concentration, even significantly below critical overlap concentration c*. The physical origin of this phenomenon is not fully understood but must result from hydrodynamic interactions between the extended and overlapping macromolecules.

Verani[29] has submitted a thesis on the effect of polymer concentration and molecular weight on the dynamics of visco-elastic-capillary breakup using a Capillary Breakup Extensional Rheometer (CABER). The test fluids used are series of diluted polystyrene Boger fluids ('PS025' and its diluents and new polystyrene Boger fluid (MV1) comprising of lower molecular weight solute. The persistent dependence of the measured relaxation time on concentration, even in the dilute regime predicted from theory, is demonstrated both experimentally and numerically. Below a critical dilution, solely the solvent carries the stress in the necking thread with no appreciable contribution from the polymer chains, and the dynamics of the necking process change appreciably.

Gupta et al.[30] have reported a detailed investigation of the steady shear and extensional properties of mono-disperse polystyrene solutions for a range of molecular weights from 1.95 to 20 million and a range of concentrations from 69 ppm to 777 ppm. The steady shear and dynamic properties are reasonably well described by Zimm model. The relaxation time and polymer viscosity evaluated using the Zimm model exhibit the expected scaling with concentration and molecular weight for theta solutions. In an extensional flow, these solutions show strain hardening and need about 5.5 to 6 strain units to reach steady state. The stress growth depends both on strain rate and strain. However, at moderate values of strain and when the Weissenberg number exceeds 6, the extensional stress growth depends only on total strain. The steady state extensional viscosity for each fluid was observed to depend on the magnitude of Weissenberg number (Wi). For Wi < 6 the steady state extensional viscosity was observed to be an increasing function of the strain rate. At a Wi ~ 10, the steady state extensional viscosity exhibits a maximum and surprisingly, for Wi > 10 the steady state extensional viscosity is a decreasing function of the strain rate. The results indicate that even dilute solutions, with concentrations as low as 0.21 times the critical concentration, do show extension thinning under certain conditions. The transient and steady state extensional viscosities are found to be proportional to molecular weight (MW) and concentration (c). This result is rather unexpected as the Zimm model would predict a scaling with c and $M_w^{0.5}$ for the transient viscosity and scaling with c and $M_w^{1.5}$ for steady state viscosity. On the other hand, the Rouse model predicts the scaling with c and M_w (for transient extensional viscosity) and with c and M_w^2 (for steady state extensional viscosity). Hence the data appear to follow a Rouse-like behavior for the transient viscosity. This implies that the advent of the stretching reduces the hydrodynamic interaction and a free draining behavior is obtained. As a result, predictions using the Zimm model parameters, estimated from shear data, are unable to predict the transient extensional viscosity. The data are analyzed using a constitutive equation that incorporates an anisotropic drag coefficient along with a Rouse spectrum of relaxation time. Such a model captures the extensional behavior of the solutions.

Graessley[31] has reported on the effects of concentration, molecular weight and solvent power on polymer chain dimensions and viscoelastic properties. Based upon the screening principle of Edwards and de Gennes, it appears that chain dimensions should approach their unperturbed values for many systems at polymer volume fractions in the range of 0.05–0.20, independent of molecular weight. From these results a method for correlating viscoelastic properties in the semi-dilute region, which takes into account the contraction of dimensions with concentration has been proposed. Also the use of a concentration-molecular weight diagram to distinguish the several regions of viscoelastic behaviour has been suggested.

The viscoelastic properties like modulus, relaxation time etc. of polymer solutions vary considerably with the variation of both concentration of polymer and its molecular weight. The relationships for the storage modulus for the polymer solutions can be considered for discussion.

The storage modulus is

$$G'(\omega) = \frac{cRT}{M} \sum_{i=1}^{\infty} \frac{\omega^2 \theta_i^2}{1 + \omega^2 \theta_i^2} \qquad ...(7.11)$$

And the relaxation time is

$$\theta_i = \frac{6 \eta M}{\pi^2 i^2 cRT} \qquad ...(\text{Ref. eq. } 7.4)$$

The zero shear viscosity of the polymer solution depends on the concentration and molecular weight as

$$\eta_0 = K M^a c^b \qquad ...(7.12)$$

The constants a and b depend on the magnitude of both M and c i.e. whether these are below or above the critical values. For a constant M polymer the value of b lies between 1 and 2.5 for $M < M_c$ and $c < c_c$, whereas for $M > M_c$ it has been reported to be about 6, and increases with increase in the value of M.[32–35] Thus it can be said that the viscosity is much more sensitive to the concentration than the molecular weight. The storage modulus thus can be visualized to vary in a complex fashion with the concentration together with the molecular weight. However the following qualitative observations can be made.

(i) For dilute solutions of a polymer having a particular molecular weight storage modulus varies linearly with concentration and

(i) For concentrated solutions it varies with much higher power to the concentration

Application of the superposition principle to the variations of the storage modulus with concentration at a particular temperature and molecular weight will result in the horizontal shift factor as $\alpha_c = (c_0/c)(\eta_c/\eta_{c0})$ and a vertical shift factor $\beta_c = (c_0/c)$. Similarly other viscoelastic functions also can be treated. The dependence of the loss modulus as a function of frequency and concentration for polystyrene-Arochlor 1232 (partially chlorinated diphenyl) is shown in Fig. 7.3(b).

The combined effect of temperature and concentration can be treated in the similar manner and the expressions for the horizontal and vertical shift factors can be derived. These are

Horizontal shift factor

$$\alpha_{T,c} = \frac{c_0 T_0}{cT}\left(\frac{\eta_{T,c}}{\eta_{T_0,c_0}}\right) \qquad\qquad ...(7.13)$$

And the vertical shift factor

$$\beta_{T,c} = \left(\frac{c_0 T_0}{cT}\right) \qquad\qquad ...(7.14)$$

7.2.1 Effect of Cross-Linking[36-44]

The polymers like rubbers and thermosets are cross-linked to impart strength either chemically under higher temperature and pressure or by irradiation. The viscoelastic response of the cross-linked polymers is quite different than that of the uncross-linked ones. The following observations and explanations may be helpful in understanding their behaviour

 (i) The increase in the cross-link density increases the modulus and tensile strength and reduces the elongation at break.
 (ii) The viscosity of the cross-linked system increases.
 (iii) The MW of the system increases but the MW of the segments progressively reduces as the segmental length reduces with the increase in the extent of cross-linking (Fig. 7.6).

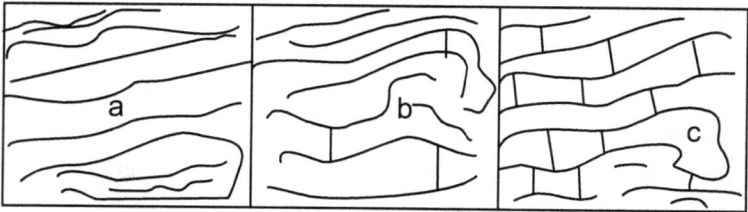

a–uncross-linked, b-light cross-linked, c-highly cross-linked

Fig. 7.6   Schematic of cross-linking of rubbers

 (iv) The terminal region in the moduli-frequency and compliance-frequency curves progressively reduces with the extent of cross-linking.
 (v) These systems show an extended plateau region.
 (vi) The transition and the glassy regions remain relatively unaffected but tend to shift to the lower frequency region.
 (vii) The numerical value of the storage modulus or the relaxation modulus in the plateau region has been reported to increase from about 10^5 Pa for uncross-linked or slightly cross-linked systems to about 10^8 Pa for the highly cross-linked ones. The modulus values for thermosets and glassy systems are in the range of $10^8 - 10^9$ Pa.
 (viii) The compliance values accordingly go down to 10^{-8} to 10^{-10} cm^2/dyne at low frequencies (Fig. 7.7).

As the long chain polymer molecules progressively get cross-linked in the three- dimensional structure the length of the chain segments between the cross-links keeps on reducing. This results in the reduced capacity of the segment,

   (i)   To have large segmental mobility,

   (ii) To acquire large number of spatial configurations, and

   (iii) To have first few normal modes of long-ranged coordinated motions.

For highly cross-linked systems, where the segmental MW may be as low as 200 to 400, the first few normal modes of coordinated motions are more or less absent and only the short-range motions of the side groups are present. As a consequence of this, the terminal zone completely disappears and the plateau region broadens considerably. The magnitude of the storage modulus increases and that of the loss modulus decreases resulting in the considerable reduction in the loss tangent peak value. The compliances are also affected accordingly.

Indei et al.[39] have reported on the effect of fluctuations of cross-linking points on the dynamic mechanical and viscoelastic properties of associating polymer networks formed by multisticker associating polymers on the basis of a single-chain approach. The effects of either neglecting or including the direct contribution of virtual springs are reported. If neglected, the fluctuation of junction decreases (or softens) the dynamic modulus over a wide range of frequencies. On the other hand if the direct contribution of the virtual springs is included in the stress, thermodynamics is satisfied but the plateau modulus does not change. The softening occures only at low-frequency regime. Thus both approaches that is either including the virtual spring contributions or neglecting, have their merits and demerits, hence the treatment of the junction fluctuations in the framework of single chain approach should be applied with care.

The effect of cross-link density on creep and recovery behaviour of epoxy-based shape memory polymers (SMEPs) are gaining importance in the area of aerospace structures due to their high strength and stiffness, which is a primary requirement for an SMEP in structural applications[40]. The understanding of viscoelastic behavior of SMEPs is very essential to assess their shape memory effect. Three types of SMEPs with varying cross-linking densities were developed by curing an aromatic epoxy resin with aliphatic amines. Glass transition temperature (T_g) was measured for these SMEPs using advanced rheometric expansion system, and from the T_g measurements, a range of temperatures from glassy to rubbery regimes were chosen. At selected temperatures, creep-recovery tests were performed in order to evaluate the viscoelastic behavior of SMEPs and also to investigate the effect of temperature on creep-recovery. Further, a three-parameter viscoelastic model (Zener) was used to fit the data obtained from experiments. Model parameters like moduli of the springs and viscosity of the dashpot were evaluated by curve fitting. Results revealed that Zener model was well suited to describe the viscoelastic behavior of SMEPs as a function of test temperatures.

Krumova et al.[41] have reported on how the crosslinking influences the mechanical and thermal properties of poly(vinyl alcohol). They investigated using a broad range of degree of crosslinking from 1.7 to 74 mol % of reacted hydroxyl groups. The effect of crosslink density showed an initial decrease in properties and then an abrupt rise at about 20% and then leveling off independently on the increase of crosslinking. The initial decrease is attributed to the diminution of the crystallinity of the system, caused by the crosslinking. The observed behaviour is explained on the basis competitive action of atleast three factors during the crosslinking: (i) weakening of existing physical network due to hydrogen bonding; (ii) formation of a chemical network; and (iii) introduction of flexible moieties. The last factor is closely linked with the specific chemical structure of chemical crosslinker itself.

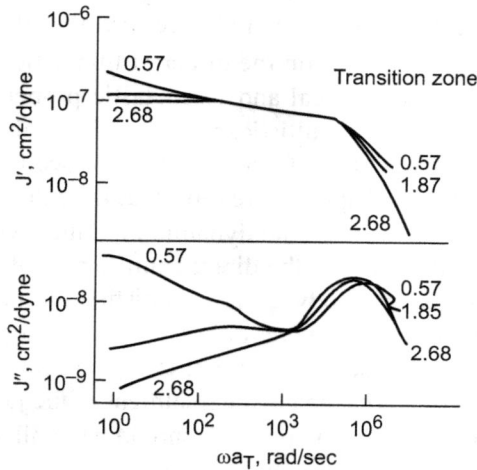

Fig. 7.7 Dynamic compliance for polybutadiene cross-linked with DCP[36]. (Cross-link density, $v \times 10^4$ is indicated on the plot: $T = 25°C$.)

Plazek[42] has studied the torsional creep behaviour of natural rubber vulcanizates in the temperature range of −50 to 90°C. The results of creep measurements are in agreement with stress relaxation results of other workers carried out on similar samples. Apart from response through reduction of temperature during studies, a superposed curve was obtained for long time response using apparent molecular weight between crosslinks, M_{CL}, as a reduction variable. The variation in viscoelastic response with crosslink density is interpreted as a restrictive action of chemical crosslinks in transient entanglement network.

Hagen et al.[43] have investigated viscoelastic properties of cross-linked NR using different methods of crosslinking like sulfur cross-linked or peroxide crosslinks. The glass transition temperature T_g, was found to depend both on crosslink density and crosslink type. T_g for sulfur crosslinked NR was higher than that for peroxide cross-linked rubber for the same crosslink density. The greater effect of sulfur content on T_g may be attributed to polysulfidic crosslinks and cyclic sulfide structure favoured at high sulfur contents. Sufur vulcanized natural rubber with monosulfidic crosslinks favoured at relatively high acceleratortor/sulfur ratio, have properties more similar to the peroxide cured NR with simple (C—C) crosslinks covalent bonds, resulting in only small shifts in T_g. Storage modulus in rubbery plateau region increased with increasing crosslinks density. However the crosslinks types did not influence the moduli as much as it influenced T_g values. Nielsen[44] of Monsnto Corp. has presented a report on the crosslinking effect on physical properties of polymers including both rubbers and rigid polymers.

7.2.2 Effect of Crystallinity [13, 32, 45, 46, 47-53]

Polymers are available in (i) crystalline, (ii) amorphous and (iii) mixed crystalline and amorphous states. However 100% crystalline state has never been encountered. Thus they can be considered to have a semi-crystalline state where the degree of crystallinity may vary from being very low (less than 2%) for rubbers like natural rubber and polystyrene to as high as 99% for high density polyethylene (HDPE).

There are a number of other polymers like polyethylenes, polyamides, polyesters etc. with varying degree of crystallinity. The crystalline and the amorphous states coexist within the same matrix in different zones. As the polymer molecules have very long chains, same set of molecules is known to pass through several of these zones (Fig. 7.8). The crystallites are the well-oriented cluster of a number of parallel chains placed so close to each other that they form a single unit as if all the chains have been made to pass through a narrow tube thus, the density of the crystalline region is much higher than the amorphous region.

Fig. 7.8   Morphological structure of the semi-crystalline polymer[13]

The viscoelastic response of the semi-crystalline polymers is the combined response of both the phases. The molecules in the amorphous phase will be easily subjected to the long range coordinated motions where those in the crystalline phase will resist them. However as the crystallites contain the segments of the same group of molecules, which lie on its either sides, these crystallites act as the dampers to the long-range motions corresponding to the low frequency.

Thus first few normal modes of motions will tend to get suppressed or shift progressively to the lower frequency region i.e. to say that the terminal region will shift to the left in the property-frequency curve.

Increase in the extent of crystallization will increase the number density of the crystallites per unit volume of the polymer and will introduce the effect of interaction between them. This increases the relaxation modulus and consequently decreases the capacity of the system to creep. The variation of the relaxation modulus is shown in the Fig. 7.9.

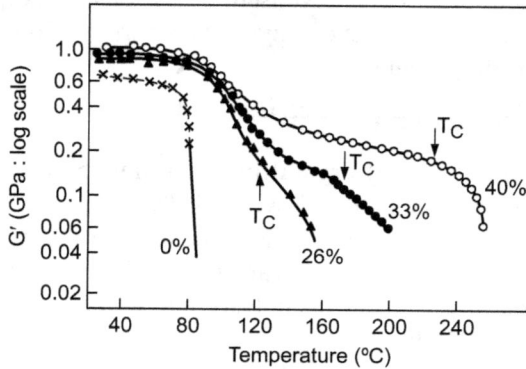

Fig. 7.9   Relaxion modulus (tensile) of PE with different degree of crystallization[13].
(T = 40°C. Degree of crystallinity is indicated in the figure)

The effect of the crystallization is most important in the range between glass transition and melting temperatures i.e. between T_g and T_m. Below T_g all the long-range motions freeze and only the short-range motions remain effective and above T_m all crystals melt away and the system's response corresponds to that of the amorphous material. The extent of crystallization and the size of the crystals depend on the rate of cooling during crystallization that means the thermal history has a considerable effect on the structure of the matrix and therefore on its viscoelastic response. For polymers cooled slowly from above the melting temperatures, the equilibrium degree of crystallinity increases very drastically a few degrees below the T_m and then remains almost constant. The viscoelastic functions show a most significant change around T_m and the behaviour much below this temperature depends mainly on the structural features of the polymer like branching, steric hindrance to the rotation etc. However for the polymers, which are quenched, the extent of crystallization and the crystal size are below that of the equilibrium value. Such systems behave differently and the behaviour is closer to that of the amorphous polymers.

The increase in the extent of crystallization due to stretching of semi-crystalline polymers is a common occurrence and increases the strength and modulus of the system. This phenomenon is exploited in the manufacture of the high strength fibers. Many invastigators have reported on the effects of crystallinity on viscoelastic behaviour of polymers.[47–62]

Olasz and Gudumundson[47] have reported on the effect of crystallinity and temperature on the viscoelastic properties of crosslinked polyethylene, (XLPE). They developed a mathematical model to describe its mechanical behaviour. The measurement, using a commonly used high voltage cable insulation, was performed by an extensive set of isothermal uniaxial tensile relaxation tests. These experiments were complemented to P-V-T experiments as well as density and crystallinity measurements. Based on the results of experiments a four parameter viscoelastic power law model has been formulated. It was found that a master curve can be developed by both horizontal and vertical shifting of relaxation curves. The model represented the behaviour quite well.

Sakai et al.[48,48a] have discussed the effect of crystallinity and fiber volume fraction on creep behaviour of glass fiber reinforced polyamide. Fiber reinforced thermoplastic polyamide (PA) or nylon is one of the engineering plastics. The knowledge of its viscoelastic properties is very important from this point of view. Also as PA is crystalline material it is of interest to know how the crystallinity will affect its mechanical properties. To know the effect of crystallization on the creep response suitable materials with adjusted crystallanity on each fiber volume fraction was chosen. Also it conformed to the Arrhenius type of time-temperature superposition principle. Using superposition approach master curves for crystallinity and fiber volume fraction were generated. It is clear from the curves that creep behaviour under different conditions have the same characteristics and also it was possible to estimate creep behaviour with effect of arbitrary time, temperature, crystallinity and fiber contents.

The effect of crystallinity on bending creep behaviour was examined by Yamada and Somiya[49] using the thermoplastic polyoxymethylene (POM) and glass fiber reinforced POM (GFRPOM). The following conclusions were obtained.

1. Crystallinity of the POM and GFRPOM was adjusted by changing the heat treatment temperature from 140°C to 170°C and it has been understood to be able to adjust the crystallinity of POM and GFRPOM within the range of 5% compared with the as-received materials.
2. According to the result of three points bending test, it became clear that the bending strength and bending modulus increased as the fiber content and the crystallinity increased. However, the effect of increasing fiber volume fraction on mechanical properties was larger than the effect of increasing crystallinity.
3. Time-temperature dependency was confirmed in creep behavior of POM and GFRPOM on various degree of crystallinity. Moreover, master curve was able to be drawn from creep compliance curve of each temperature and conformed to the time-temperature reciprocation law of Arrhenius type.

Tanaka et al.[50] have carried out investigations to study the effect of crystallinity on nonlinear viscoelastic and optical properties of undrawn and drawn poly (ethylene terephthalate) films. Two different types of nonlinear properties were found in undrawn films, these are: (i) slipage of amorphous chains in low crystalinity region and (ii) slipage of crystallites in the lammellae in a high crystallinity region. In the intermediate crystallinity region, on the other hand, linear behaviour was found. In the case of drawn films two different nonlinearities were clearly observed. One found in low crystalinity region was also associated with slipage of amorphous chains, which is almost same as the one found in undrawn film. Another was associated with structural changes such as opening of cracks and cavitation in amorphous region between spherulites or between crystallites and was found in intermediate and high crystallinity region.

Feng et al.[51] have reported about the effect of crystallinity on compressive properties of Al-PTFE. Al-PTFE (Al-polytetra fluoroethene) is an important kind of reactive material (RM), however only limited importance was placed to the effect of crystallinity of PTFE. Experiments in the strain rates range from 10^{-2} to 3×10^3 s^{-1} were carried out to study the proposed effect. Two kinds of samples were prepared by different sintering procedures to acquire different crystallinity. Low crystalline samples have consistently higher strength and toughness than the high crystalline samples. An "elastic-plastic network" model combined with the effect of chain entanglement density explained the phenomenon. Finally, a close connection between fracture initiation of Al-PTFE was confirmed in quasi-static tests, split hopkinson pressure bar (SHPB) tests, and drop weight tests. It was hypothesized that the high temperature at the crack tips of PTFE is an important promoting factor of crack initiation.

Sung et al.[52] have studied the effect of crystallinity and crosslinking on thermal and rheological properties of EVA copolymer. It has been found that the storage modulus decreases with increase in VA in the temperature range of -70 to $50°C$. It suggests that the crystallinity of EVA is affected due to the week crosslinking of EVA by DCP. The complex viscosity of EVA in the molten state, has been affected by crosslinking with DCP.

Armsgan[53] has reported on the effects of crystallization on mechanical properties of carbon fiber reinforced poly (ether-ether ketone) (C-PEEK) composites. PEEK is a new generation engineering polymer having good high temperature properties. It is semi-crystalline thermoplastic polymer. For this polymer it is possible with heat treatment to get trans-crystalline interphase, which is very important for material performance. The size of the spherulites are controlled by controlling the heat treatment during the manufacturing of composites[54, 55, 56], and by the microstructure of the fibre, which determines the ability to develop a transcrystalline layer[57]. There are several indications that slow-cooled or annealed materials, which have a fully developed crystalline structure, exhibit better mechanical properties[58]. Likewise the presence of transcrystallinity also has been shown to influence the mechanical performance through its effect on fibre/matrix bonding and on the stress transfer mechanism[59, 60].

Khalil et al.[60a.] have worked on effect of coupling agent and crystallinity on viscoelastic properties of composites of rice hull ash-filled semi-amorphous polypropylene (PP). Rice hull ash (RHA) is a silica rich product from agricultural waste, which has been used as filler in PP. Addition RHA increased storage modulus and onset Temp. of crystallinity, T_{co} and peak of crystallinity temperatures, T_c and decreased the degree of crytallanity of the system. Two coupling agents (i) maliated Polypropylene, MAPP, and an aminofunctional silane were used to improve the interfacial adhesion of PP and RHA. Addition of MAPP increased G', T_c and T_{co} and the crystalline phase of the system. Addition of silane had mixed effects. It increased the rheological properties at higher concentrations, increased T_c and T_{co} at all the concentrations and increased crystallinity at lower concentrations and decreased it at higher concentrations.

Dusunceli and Colak[61] reported on the mechanical behaviour of semicrystalline polymer. They used viscoplastic theory based on overstress (VBO), which is one of the unified state variable theories to account for crystallinity ratio (ϕ). VBO theory has been extended to the semicrystalline polymer system assuming it as a composite of amorphous and crystalline phases. Amorphous and crystalline phase resistances are arranged in two different analog models: amorphous stiffness and flow are in parallel and series with crystalline phase. It is shown that when amorphous and crystalline phase resistances acting in parallel are considered in the model, creep, relaxation and uniaxial loading and unloading behaviours can be simulated well using the modified VBO. A good match with experimental data has been reported for many systems like ultra-high molecular weight PE (UHMWPE), and PTFE with different crystallinity ratios.

Toft[62] investigated the effect of crystalline morphology on the glass transition and ageing characteristics of semi-crystalline PEEK. It is shown that an increasing degree of crystallinity acts to raise the glass transition temperature of the polymer and reduce the overall degree of enthalpic relaxation. For an equal degree of crystallinity the glass transition temperature is also shown to be sensitive to isothermal crystallisation temperature. By compensating for shifts in T_g and the influence of crystalline content, samples of varying morphology were produced and physically aged at undercoolings tailored to the T_g of the system. A greater degree of enthalpic relaxation was observed in cold crystallised samples where the degree of constraint of the amorphous fraction at the crystal/amorphous interface is thought to be greater.

Bartczak and Galeski[62a] have studied the plasticity of semicrystalline polymers. Semi-crystalline polymers can be deformed up to a very high strain. The deformation process involves frequently a complete molecular rearrangement of the chain-folded lamellar morphology into a more or less chain-unfolded fibrillar microstructure. This transformation is likely to occur through an intermediate state of high molecular disorder at a local scale. It led to the formulation of a concept of strain-induced melting-recrystallization process as a main mechanism of the structure transformation. In contrast, several structural features occurring at moderate plastic strains are relevant to strictly crystallographic processes. The plastic deformation process of semicrystalline polymers and the micromechanisms involved are discussed. A critical discussion of experimental findings is made to point out the strength or the deficiency of the various argumentations. It is demonstrated that the crystallographic slip mechanisms, including slips: transverse and along the chains are the basic deformation mechanisms in the deformation sequence, active at all strain levels. Direct microscopic evidence of chain slip activity even at well advanced stages of the deformation process is presented. In contrary, the melting-recrystallization seems to be restricted to the high-strain stage accompanied by chain unfolding and perhaps limited to only a small fraction of the crystalline phase. This can be effectively accomplished with only crystallographic mechanisms employed. A very important role in the deformation sequence is played by the partially reversible shear deformation of amorphous interlamellar layers, producing not only high orientation of amorphous component but also influencing deeply the deformation of crystalline phase, since both phases are strongly connected and must deform simultaneously and consistently.

7.2.3 Effect of Branching[63, 64, 65, 66, 67]

Branching in the polymers increases the inter-chain distance and therefore increases the free volume and also lowers the degree of crystallization and thus is expected to reduce the viscoelastic modulus. However the different lengths of branches are expected to give different effects as the small and medium branch lengths will show the above effects but the large branch lengths tend to entangle and thus expected to give the higher moduli. The variation of the viscoelastic parameters with the degree of branching as well as with the branch lengths has been reported to be rather erratic due to the complex interaction between the entanglements and the branching. It is expected that the viscoelastic functions of high concentration solutions and melts will be the functions of structural parameter, g as well as exact topology of the branches. The parameter g is the ratio of the mean square radii of the branched and linear polymer.

For very dilute solutions however the effect of interaction between branches and the entanglements can be ruled out as the molecules are far apart and these are expected to give similar variations as the un-branched systems except that the magnitude of the moduli may be smaller for the branched system due to its lower density.

Wood-Adams and Dealy[68] have reported on effects of weight average MW, and short and long chain branching on the linear viscoelastic behaviour of polyethylenes (and ethylene-α-olefin copolymers). Short chain branching gave no effect up to a comonomer (butene) content of 21.2 wt %. The zero shear viscosity of the linear polyethylenes varied in the expected manner with MW. Using a high molecular weight linear polyethylene with narrow molecular weight distribution (MWD), an estimate of the plateau modulus and molecular weight between entanglements (M_e) was obtained. The extent of long chain branching (LCB) well below $1LCB/10^4C$ in polyethylene has been quantified. The long chain branching increased the zero shear viscosity as compared to that of a linear material of the same molecular weight. LCB also broadened the relaxation spectrum by adding a long time relaxation mode that was not present for the linear polyethylene with the same MWD. The ^{13}C NMR has been successfully used for measuring LCB levels.

It has been reported[68a, 69] that the brittle cracking of many ductile polymers, polyethylenes being no exception, is stabilized by the ability of the material to undergo microstructural changes (inelastic deformation) at the crack tip. This inelastic deformation is in the form of crazes, which usually initiate at a stress concentration site. Thermodynamic analyses[70, 71, 72, 73, 74] conclude that, lower the rate of energy dissipation in craze formation and growth, the slower the rate of crack propagation. Under identical loading conditions and specimen configuration, the rate of energy dissipation is determined by viscoelastic processes contributing to the craze formation and rupture. These processes are expected to be controlled by the morphological hierarchy, which in turn affected by the chain microstructure. In a short chain branched polyethylene, the type of copolymer determines the type and size of short chain branches. Thus, an ethylene-hexene copolymer has butyl branches, while an ethylene-butene copolymer has ethyl branches. It has been shown that an ethylene-hexene copolymer outperforms an ethylene-butene copolymer in terms of fatigue lifetime,[71] while the same ethylene-hexene copolymer behaves differently when quenched and slow cooled as reported by Strebel[75] and Brown and Lu[76]. In a related study by Brown and Lu[77], it is noted that a four-fold rise in the branch density increased the time to failure by four orders of magnitude. These and many connected studies suggest that a specific chain microstructure imparts a particular viscoelastic character that controls the rate of slow crack growth.

Vega et al.[78] have reported on the effect of short chain branching in molecular dimensions and Newtonian viscosity of ethylene/1-hexane copolymers, with reference to the molecular dimensions in dilute solutions. The experiments show a molecular contraction as branching level increases. They observed that there was a clear dependence of Newtonian viscosity with short chain branching (at $T = 463°C$). This dependences can be related to the changes observed in the macromolecular conformational features, in the equilibration entanglement time and the molecular weight between entanglements, as the number of short chain branches increased.

Askadkii and Kovriga[79] have reported the effects of the length and character of the branching distribution on a number of physical characteristics of polymers. Different parameters analysed were glass transition temperature, the coefficient of bulk thermal expansion, the refractive index, the Hildebrand solubility parameter, and the surface energy. It is concluded that short chain branches have the greatest effect on the physical characteristics, since they lead to a significant change in the chemical structure of the polymer system. The effect of long chain branching depends on the differences in their chemical structure from that of the main chain and to a lesser extent on the degree of long chain branching. The physical characteristics of the polymer show the sharpest changes when tree-like branching occurs.

Former et al.[80] worked on the effect of molecular branching on the poly(butyl acrylate) using samples of different MW and varying branch lengths. The effect of branching showed that highly branched samples give lower storage and loss moduli however at low frequency, effects of both MW and total degree of branching could be inferred. Although the studies could not give any specific conclusive trends but the low frequency trends could be semiquantitatively fitted with reptation and retraction theory if it was assumed that an increased degree of short chain branching led to an increased tube size.

Doelder et al.[81] have developed computational schemes based on tube theories that enable calculations of rheological properties for polymers of arbitrary topology. A scheme is used to systematically explore key rheological features of model long-chain branched systems. Empirical relations between molecular structure and rheology, typically use overall molar mass and branching averages as structural variables.

7.2.4 Effect of Copolymerization[18, 68a, 68b, 82]

Simultaneous polymerization of two or more monomers to produce a polymer containing all the monomers is known as the copolymerization. The monomers may occur randomly in the chain giving random copolymer or in blocks as in the block copolymers or alternatively as in the alternating copolymers. The copolymerization is carried out to produce the polymers having mixed properties for example if it is desired to have rubbery nature of polybutadiene and glassy nature of polystyrene then styrene and butadiene monomers are co-polymerized to get styrene-butadiene copolymer. The viscoelastic response of the copolymers depends on the type of copolymers and level of compatibility of the monomers.

Different random copolymers show different responses depending on the nature of the response of the pure components. It means that if the shapes of the pure component response curves are similar and give peak at different temperature as in the case of polystyrene and polybutadiene then the random copolymer curve will give a single peak[68a]. The temperature corresponding to the peak value of the property say loss tangent or the loss modulus will shift horizontally (Fig. 7.10) from pure component (I) to the pure component (II) as the composition of (II) increases in the copolymer of (I) and (II). On the other hand if the pure component response curves are different as in the case of polyvinyl alcohol (crystalline) and polyvinyl formate (amorphous) then the response curves of the copolymer are different giving the drastic changes in its shape at some critical composition[68b].

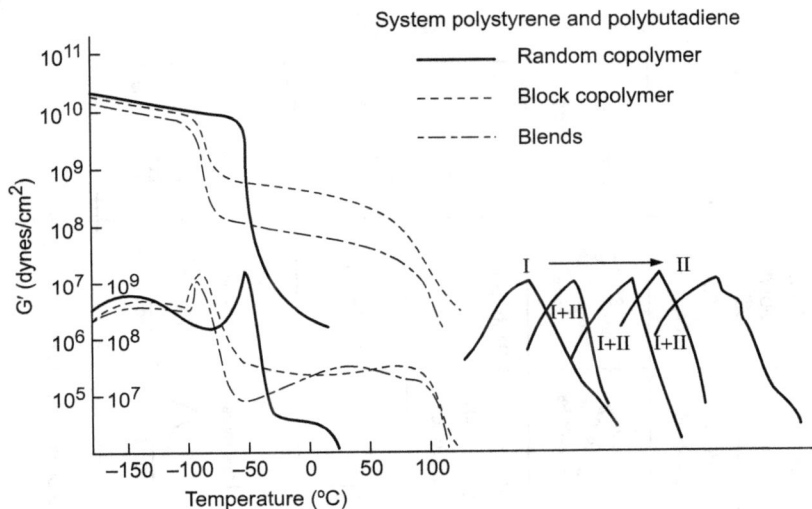

Fig. 7.10   Viscoelastic functions for copolymers[82]

The viscoelastic functions of copolymers are similar to those of the physical blends of the two homo-polymers if these are mutually soluble or are compatible with each other. However if the homo-polymers are incompatible then the isochronal data of the copolymers will give a single peak but the physical blends will give two peaks.

The block copolymers of the type $(AAA\text{—}AAA)_n (BB\text{----}BB)_m$ are formed in general by the block copolymerization of incompatible pure components and thus consist of two phases, one phase is continuous and other is dispersed. The pure component with lower viscosity forms continuous matrix in which the other component with high viscosity is dispersed. The dispersed phase acts like a thermally sensitive deformable filler and its effect is most prominent in the region of two glass transition temperatures of the pure components. Some of these copolymers act like thermoplastic elastomers, which at lower temperatures behave like rubber vulcanizates and at higher temperatures act like thermoplastics. Thus these systems will give two peaks corresponding to two glass transition temperatures of pure components as shown in Fig. 7.11 for S-B-S copolymer[2,82,83].

The graft copolymers of (B) grafted on (A) are obtained by grafting the segments of (B) on the backbone of (A) (Fig. 7.11a). The response curves of these are similar to those of the block copolymers. The grafting is easily achieved on the natural polymers whereas the block co-polymerization is important for the synthetic polymers.

Monomer(s)	A	A and B			B
Class of polymer	Homo-polymer	Random copolymer	Block copolymer	Graft copolymer	Homo-polymer
Chemical name	Poly A	Poly (A-co-B)	Poly (A-b-B)	Poly (A-g-B)	Poly B
Schematic chemical structure					
Example	Polybuta-diene	Poly (buta-diene-co-styrene)	Poly (buta-tadiene-b-styrene)	Poly (buta-diene g-styrene)	Poly-styrene
Variation of shear modulus G' and log dec Λ with temperature	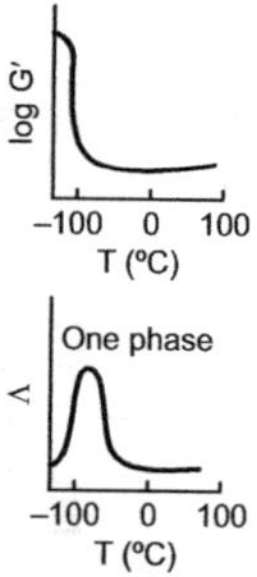	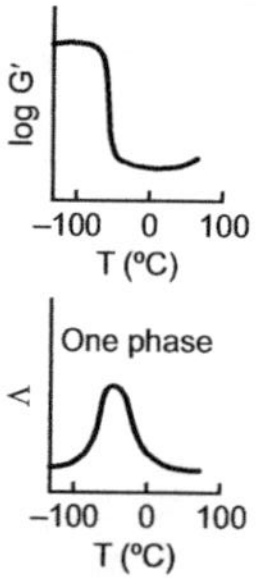	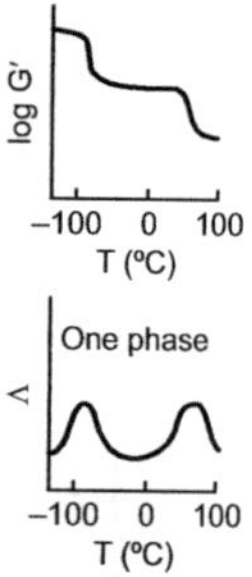	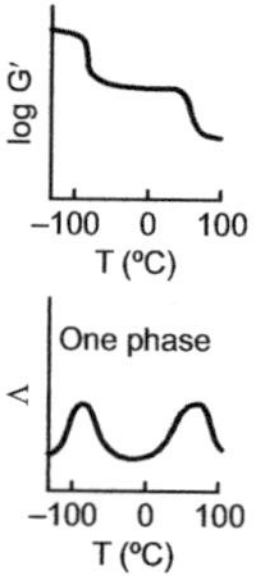	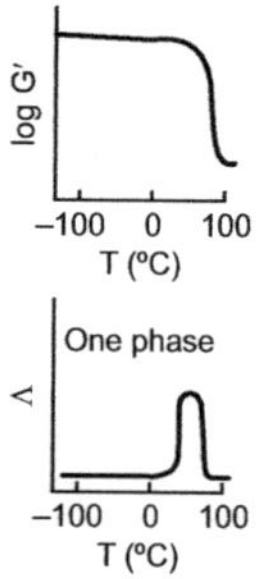

Fig. 7.11   The viscoelastic functions for copolymers at 1 Hz[13].
$(\Lambda = \pi \tan \delta$ for $\Lambda \ll 1)$

Ghose et al.[84] have reported on the shear stability and thickening properties of homo and copolymers of methyl methacrylate. The degradation stability towards mechanical shearing (shear stability) of poly(methyl methacrylate) (PMMA) and its copolymer with styrene at different level of concentrations has been investigated, with a view to understand the relationship between the thickening performance and shear stability of the polymer. The thickening abilities of the polymer were also determined and compared. Pour point depressant properties of the copolymers in comparison to the homopolymer were also tested in different base stocks.

Koh et al.[85] have reported on the effect of copolymerization with methacrylic acid on poly(butyl acrylate) film properties. The influence of methacrylic acid (MAA) on physical properties of polymer made by emulsion copolymerization of butyl acrylate (BA) and MMA were examined. The presence of methacrylic acid had only a small effect on gel fraction and on MW, but profound effect on the film properties. Changing the synthesis temperature was found to slightly alter the properties of the copolymer films. Latex containing MAA formed much stronger films (from creep test), and significantly increased tack and peel adhesion. This was attributed to inter molecular dipole-dipole interaction of acid groups. Conductometric titration measurements revealed that the acid groups were predominently located inside the latex particle, with only a small proportion in the aqueous phase and on the particle surface. Temperature had only little effect on partitioning of acid groups in latex.

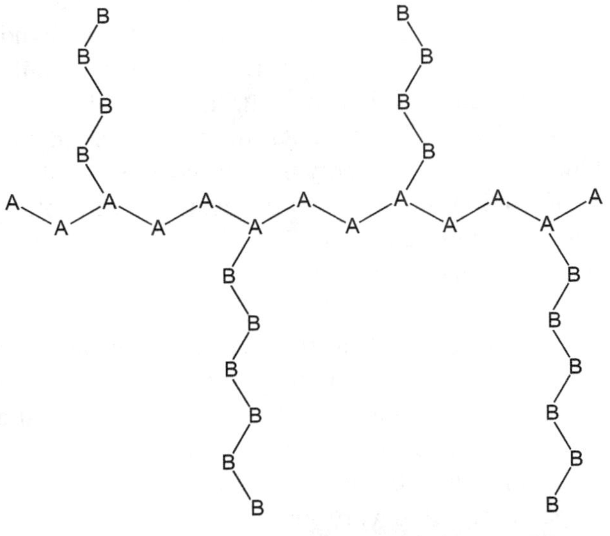

Fig. 7.11(a) Graft copolymer havig main chain backbone of (A) and covalently bonded to one or more side chains (B) [en.wickipedia.org].

Riahinezhad et al.[86] reported on the possibility of copolymerization kinetics to help tailor the properties of poly(acrylamide (AAm)/acrylic acid (AAc) for enhanced oil recovery (EOR). This product is a water-soluble polymer and is most widely used in this area. However it has a disadvantage of poor stability in viscous and viscoelastic properties when subjected to the combined effects of temperature, shear and extentional stresses. The copolymerization kinetic studies were used to manipulate the chain properties such as MW and copolymer composition, which in turn helps to control the microstructure bulk properties of the copolymer.

Diaz-Selvestre et al.[87] have described effect of concentration of N-isopropylacrylamide (NIPAM) on viscoelastic properties of hydrosoluble thermo-thickening copolymeres. These copolymers were produced through free radical polymerization in aqueous media. It has been reported that turbidity point rises according to the quantity of poly(acrylamide), PAM incorporated in to copolymers. The MW of each copolymer decreased when the amount of NIPAM was increased. The growth of viscosity of copolymers compared well with the increase in temperature from 25 to 70°C. However in the case of copolymer with highest NIPAM concentration the viscosity decreased in the temperature range of 60 to 70°C.

Jahani et al.[88] have reported on the effect of copolymerization condition on rheology, morphology and thermal behaviour of polypropylene heterophasic copolymers. It was found that the ethylene propylene rubber (EPR) content increased up to 32 wt% by increasing the copolymerization time to 20 min. At a fixed copolymerization time of 10 min, the addition of 50 ppm hydrogen (H_2), increased the EPR content from 9.7 to 12.8 wt%. By doubling H_2 concentration, no considerable change in EPR wt% was observed. It was found that the zero shear viscosity of the alloys varied significantly under the influence of EPR wt%, but not the molecular weight of matrix. The molecular weight of PP matrices determined by rheological data, mildly decreased from 463000 to 458000 g/mol by increasing the copolymerization time from 10 min to 15 min. At high copolymerization time/high H_2 concentration, a melting peak in the differential scanning calorimetry test around 165°C for isotactic PP and also an endothermic peak around 127°C for the block copolymer with long ethylene segments, was observed.

Liao et al.[89] have reported on the dynamic mechanical behaviour of terpolymers based on butadiene, isoprene and styrene. The results showed that addition of structure modifier, tetrahydrofuran (THF) had a great effect on the polymerization rate of three monomers, especially styrene, and could tune copolymer composition by using different addition time of THF. The copolymer composition could vary gradually along the chain from block copolymers (isoprene-butadiene) to gradient copolymers (isoprene-butadiene-styrene-isoprene-butadiene). It resulted in broadening of glass transition region and damping range. The monomer ratio and molecular weight that affected the mechanical and dynamic mechanical properties of copolymers were also investigated. At the same time, the microstructure, morphology and mechanical properties were compared with random terpolymer and solution styrene-butadiene rubber/butadiene rubber/natural rubber blend.

Roos and Creton[90], worked on the effect of the presence of diblock copolymer on nonlinear elastic and viscoelastic properties of elastomeric triblock copolymers. The mechanical properties of blends of triblock and diblock copolymers of polystyrene (PS) and polyisoprene (PI) containing 15 wt% PS have been investigated with tensile tests, relaxation tests, dynamic mechanical tests, and hysteresis tests. The properties of binary triblock/diblock blends have been investigated in parallel with those of the same blends diluted with 60 wt% of a low molecular weight but high-T_g solvent, miscible with the PI component but not with the PS domains. Both binary and ternary blends show a nonlinear elastic behavior at large strains, which can be analyzed with rubber elasticity models, and a pronounced nonlinear viscoelastic behavior at intermediate strains. The analysis of the viscoelastic behavior beyond the linear regime shows that the proportion of PI blocks effectively bridging two PS domains controls the deformation of the blends at large strains, while the small and intermediate strain behavior is controlled by the density of entanglements in the rubbery component and by the relaxation of the PI block of the PS–PI diblock. Both of these nonlinear features depend on the diblock/triblock ratio in the blend and will have direct consequences on more complex properties such as fracture or adhesive debonding.

Styrene/butyl acrylate copolymers with layer morphology were synthesized using an emulsion copolymerization process[91]. The effects of kinetic parameters like monomers composition, initiator and transfer agent content, feeding time, and monomer addition sequence were investigated. The final product consists of a homopolymer nucleus surrounded by concentric shells of copolymers with different composition. The initial composition is quite rich in the monomer that forms the nucleus while the process ends with an enriched layer of the second homopolymer. The composition effect on the viscoelastic properties shows, on one hand, that an increase in the butyl acrylate content lowers the elastic response of the final product, and on the other hand, the elasticity increases with copolymer content. As the initiator content in the reaction media increases, the viscosity of the coreshell products decreases because of the existence of a media flooded with free radicals. If the butyl acrylate is first added, a graft polymerization is favored because of the polar nature of this homopolymer and, therefore, the molecular weight increases.

7.2.5 Effect of the Blending

Blending of two or more polymers is an important approach to get the product of the desired viscoelastic properties. Different variable like the degree of crystallinity, branching, cross-linking etc., tend to alter the molecular behaviour of the polymers because theses are purely chemical in nature. Blending on the other hand is purely physical and the pure constituents do not change their molecular structures. The properties of the final product depend on whether the polymers being blended are compatible or incompatible in nature. The pure constituents form a blend of single homogeneous phase if they are mutually soluble in each other and the viscoelastic properties of the blend correspond to a new pure system giving a single point transitions. The plots of G' and G'' with frequency give the curves with the slopes of 2 and 1 in the terminal region and 1/2 in the transition region with plateau and valley as in the case of pure components. Fig. 7.12 gives a plot of loss modulus versus frequency for the blends of polystyrene (PS) and polyphenylene oxide (PPO)[92]. Similarly at a particular frequency the tan δ of such systems for example polyvinyl acetate-polymethylacrylate blends give a single sharp peak at the glass transition temperature.

The blends of polyvinylchloride and a copolymer of butadiene and acrylonitrile (I) or styrene with styrene-butadiene copolymer (II) or polystyrene and polybutadiene (III) are not single phase systems as the constituents forming these are not completely soluble in each other but may only have a limited solubility or may be completely incompatible. Such systems give either a broad maxima at the glass transition as in (I) or separate peaks corresponding to individual constituents for (II) and (III)[2, 7]. Such cases are shown in Fig. 7.11.

Kim and Chung[93] have extended the concept of reptation of de Gennes and tube model theory of Doi and Edwards[22, 35] to explain terminal viscoelastic properties of binary blends in the highly entangled state of two linear monodisperse polymers with different molecular weights M_1 and M_2. A modified tube model has been proposed, which considers the significance of the constraint release by local tube renewal in accounting for the relaxation process of the higher molecular weight chain. Its relaxation by both reptation and the constraint release is remodeled as the disengagement by pure reptation of an equivalent primitive chain. It is important to have information about the relaxation times of both constituents i.e. which component has longer and which one has shorter relaxation time.

From knowledge of the longest relaxation times of the blend components, the stress equation is formulated, from which blending laws of viscoelastic properties for the binary blends are derived. To force better agreement between theory and experiment at the pure monodisperse limits of the blends, a crude treatment to include the effect of contour-length fluctuation in the equivalent-chain model is also discussed. Theoretical predictions of the zero-shear viscosity and steady-state shear compliance are shown to be in good agreement with literature data on undiluted polystyrenes and polybutadienes over a wide range of the blend composition and M_2/M_1 ratio. The laws for blends with $M_2/M_1 \to 1$ and 0 are comparable to those from the relaxation spectrum proposed by others earlier on the basis of the tube model.

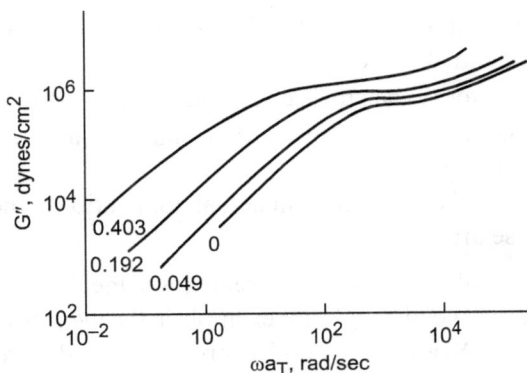

Fig. 7.12 Loss modulus-frequency plots for PS and PPO blends at 220°C.[92]. (Numbers indicate the weight fraction of PPO: Mppo = 69000, Mps = 97000).

Read et al.[94] have developed a full-chain tube based constitutive model in the lines of Graham et al.[94a], for nonlinear rheology of bidisperse blends of long and short linear polymers. The physical picture of a fat tube was assumed to represent long-lived entanglements with long chains and a thin tube model was used for all other chains. The model includes the processes of reptation, contour length fluctuations (CLF), constraint release and stretch relaxation thus defining a new relaxation process: CLF along the fat tube contour, achieved via a combination of chain motion along the thin tube, and local constraint release of thin tube as it explores the width of the fat tube. This process is assumed to be very fast to relax a significant portion of the long chains before raptation. It provides an explanation of the decrease in terminal time of long chains upon dilution with short chains in a framework where motion along the thin tube is the dominant reptation mechanism. Once the linear rheology is matched, nonlinear rheology is predicted with no further adjustments to the model. This model represents the different experimental data quite well.

Auhl et al.[95] have reported on the effective stretch relaxation time for elongational flow of blends of long and short polymers. They studied the onset of chain stretch and emergent extension hardening in the nonlinear rheological response of molten binary blends. It has been predicted that, upon dilution with short chains, the effective stretch relaxation time of long chains initially increases in proprtion to the volume fraction of the long chains and dilution exponent of entanglements. Such a behaviour has been confirmed experimentally as well.

7.2.5.1 Time-Temperature Superposition (TTS) Principle applied to Polymer Blends

Von Gurp and Palmen[96] have analysed the applicability of TTS principle (also known as frequency-temperature superposition or method of reduced variables) to the polymer blends. According to Ferry[97], TTS holds when:

(i) Exact matching of shapes of adjacent (time or frequency dependent) curve is obtained,

(ii) The shift factor α_T has the same value for all viscoelastic functions,

(iii) The temperature dependence of α_T has a reasonable form (WLF, Arrhenius type).

However there exist conditions in actual applications when this principle does not hold, these are

(i) Occurrence of more than one relaxation mechanism with different temperature dependence[98] for example in the vicinity of glass transition temperature. As around Tg both energetically and entropically induced relaxations play the role,

(ii) If the systems or materials change their physical or chemical nature during the rheological measurements,

Polymer blends are mainly of three types

(a) Homogeneous blends with a distribution of molecular chain length having a single-phase structure and a single glass transition temperature.

(b) Immiscible blends or heterogeneous polymer blends of two polymers will show two different glass transition temperatures.

(c) Miscible or compatible polymer blends are blends that exhibit macroscopically uniform physical properties. The macroscopically uniform properties are caused due to sufficiently strong interactions between the components[99]. In case of inhomogeneous polymeric systems the different components in general display a different temperature dependent rheology, so the TTS should not hold for these blends. However many workers have reported that the principle holds for a number of such blands[100] . The systems for which either TTS holds or fails are given below:

• Miscible blends for which TTS holds are, PS/PVME, SAN/SMA, SAN/PCA, PMMA/PVDF, PPO/PS, PS/PCHMA etc. and

• Immiscible blends, where it holds are SB/PB, PS/PC, HDPE/LDPE, PA6/EVA, PS/PB, PE/PP, PP/EVA etc.

• The Miscible blends for which the TTS fails are, PEO/PMMA, 1,2PB/ PIP

• Immiscible blends for which the principle does not hold are PS/ PMMA, PS/LDPE, PE/EVA, PS/PEA, SAN/PMMA.

Homogeneous Blends

The polymers having distribution in chain length do not normally influence the relaxation mechanism and also there is no evidence to suggest that it has any effect on activation energy. So, for such blends the TTS principle will hold. The branched polymers, which are considered as blends of polymers with various amount of long chain branching, are likely to show different behaviour. A very high MW polymer having very long branches of chains will have very inhomogeneous branch distribution, which may cause the failure of TTS principle (LDPE and EPDM may be considered as examples for such a case).

Miscible Blends

Miscible or compatible polymers form blends due to their intimate mixing and as Tsenoglou[101] reported that it was considered till late 1980's that components in such blends would have same temperature dependence and TTS should hold. However EO/PMMA system, which is considered to be miscible does not follow the TTS principle[102]. Arendt[103] explained that the constituents retain their temperature sensitivity in blends at very small scale, as the system possibly behaves as heterogeneous mixture at that level. As in these blends a strong intercomponent interaction plays an important role and as such a strong coupling between contituents may result in one single temperature dependence[104]. As the blend behaves as heterogeneous system at a small scale the failure of TTS is considered to be very sensitive to the difference in Tg of the constituents[105]. However SAN/SMA system seems to obey TTS quite well even though the difference in T_g of these polymers is about 47°C (T_g for SMA is in the range of 130-160°C and that for SAN is around 100°C).

Immiscible Blends

Immiscible or heterogeneous polymer blends show different temperature dependencies of individual components. However a number of such blends like SB/PB[106], PS/PC[107, 108], LLDPE/LDPE[96], PE/PP[96], PP/EVA[109] etc. hold TTS principle whereas others like PS/PMMA[110], PS/LDPE[96], PE/EVA[111], SAN/PMMA etc. do not obey this princple. The likly reason for such behaviour of polymer blends may be: (i) The activation energies or WLF parameters of components are not very different, (ii) Experimental errors and (iii) Change of blend morphology during experimental measurements particularly those, which take very long time, (iv) When the contribution of measured stress signal of one of the two components is very small, which may be caused due either to small volume fraction or low viscosity of one of the constituents of the blend or if the relaxation time of one of the component falls outside the experimental time regime.

It is important to note that for compatible systems the physical blending is the best options to produce the blends of the desired viscoelastic properties whereas the copolymerization route results in controlled properties for the incompatible constituents.

7.3 IONIC POLYMERS AND THEIR BLENDS[110A, 110B, 110C, 110D]

Introduction of a small amount of ionic groups into the hydrocarbon polymers yields ionomers. The viscoelastic properties of ionomers are different from those of the parent polymers as the ionic groups form the ionic aggregates in the matrix under suitable condition of the temperature and chemical environment and behave like the reinforcing fillers. It has been reported that the viscosity of the ion aggregate containing polymers is higher than that of the original polymer because of the strong intermolecular ionic interaction.

The viscoelastic behaviour of the ionic polymers depends on the number of parameters like type of polymer backbone, types of counter-ions, degree of neutralization, presence of plasticizers, concentration of the ionic groups and the type of the cations, sulfonation of the ionic polymers, ionic poly-blending and the addition of the reinforcing fillers. Presence of the Zinc stearate has been reported to have profound effect on the viscoelastic properties of the ionomers. Zinc stearate acts as mild reinforcing filler reducing the magnitude of tan δ peak indicating the increase in the storage modulus and corresponding decrease in the loss modulus implying thereby an increase in the matrix stiffness. Fig. 7.13 gives the effect of Zinc stearate on the dynamic properties of ionic thermoplastic elastomers (TPE) based on the Zinc salts of the maleated EPDM (Zn-s-m-EPDM).

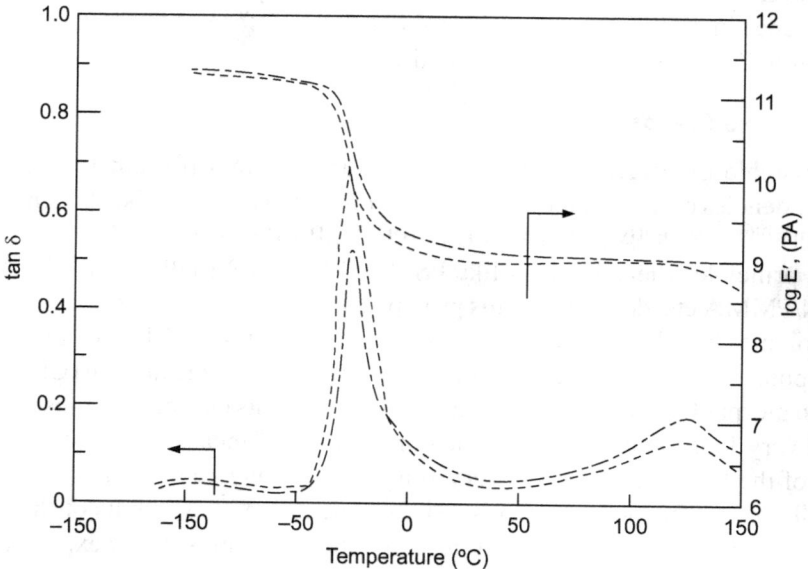

Fig 7.13 Plots of tan δ and storage modulus versus temperature at 10 Hz.
(--- for Zn-s-m—EPDM, ------- for Zn-s-m-EPDM+30 phr of Zn stearate)[110b].

Poly-blends generate the ionic clusters having varying number of the ions as shown in Fig. 7.14 and affect the viscoelastic functions corresponding to the increased chain stiffness (Fig. 7.15). The tan δ peak is maximum for Zn-XNBR and reduces progressively as the chain stiffness increases due to blending and addition of Zinc Oxide.

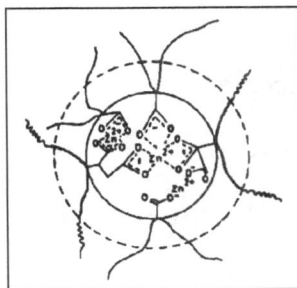

Fig. 7.14 Schematics of ionic cluster formation in ionic poly-blends.
(— for Zn- XNBR and ... for Zn-m-HDPE)[110c]

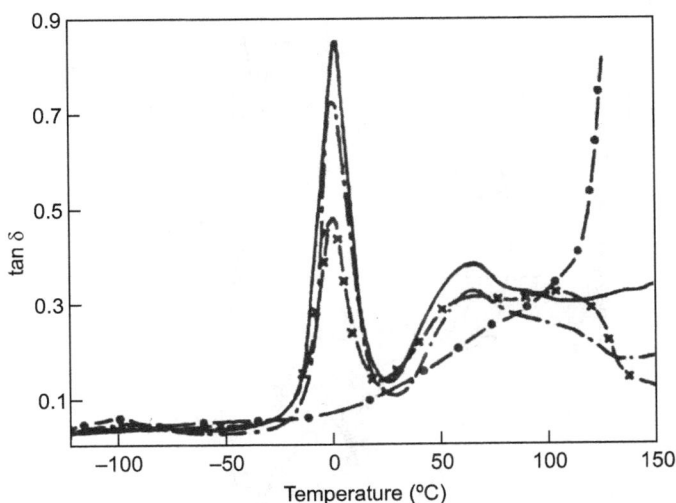

Fig. 7.15 Plots of loss tangent versus temperature for ionic polyblends.
(—for Zn-XNBR; —•—for Zn-m HDPE + Zn-XNBR/Zn-mHDPE blends (B); ––– for (B)
+ 80/20 XNBR/mHDPE(I) and ...×... for 40/40 (I) with 12 phr ZnO)[110c]

The addition of the reinforcing fillers like HAF, Silica and ISAF increases the chain stiffness further. The ionomers interact with the free radicals present on the filler surface to bind the polymers to it as shown in Fig. 7.16 for the ionic TPE based on the Zinc salt of sulfonated maleated EPDM rubber. The effect of fillers on the viscoelastic properties of the filled ionomers is shown in Fig. 7.17 for the above mentioned TPE.

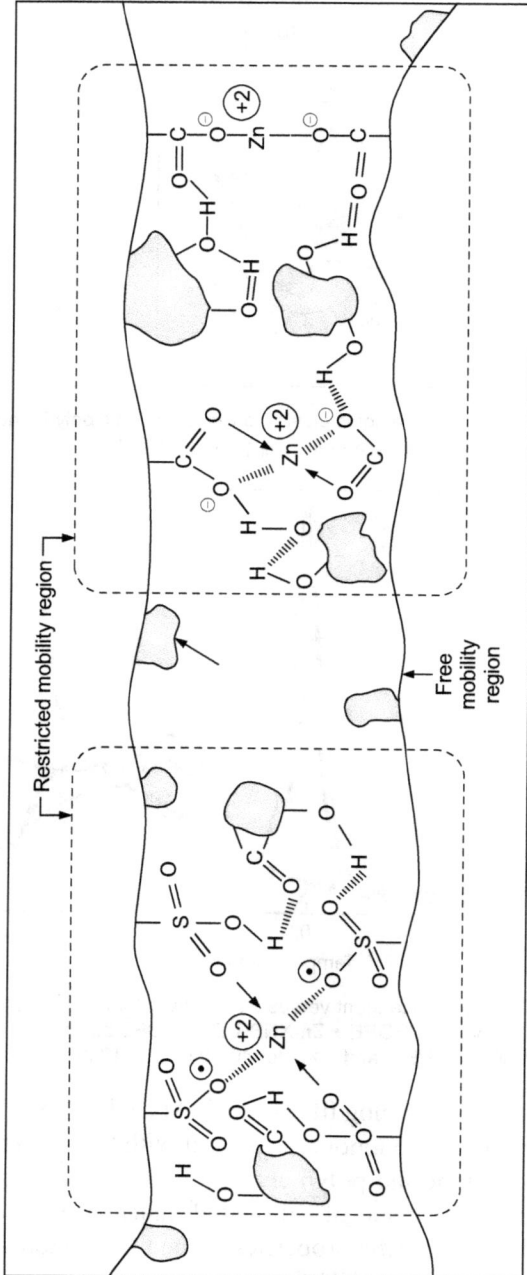

Fig. 7.16   Interaction between filler surface and ionomers[110d]

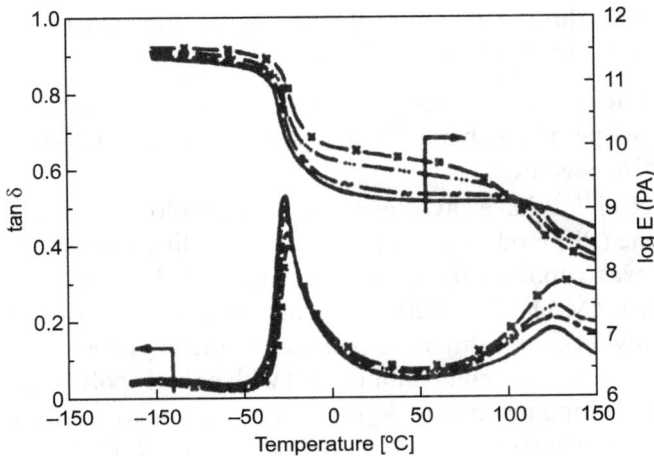

Fig. 7.17 Viscoelastic functions as functions of fillers for ionic polymer. (—for Zn-s-m-EPDM +30phr ZnSt (I); —~—~— for I + 20 phr SRF Black; —···— for I + 20 phr silica; -x- for (I) + 20 phr ISAF black).

7.4 EFFECT OF GRAFTING[12, 35, 112-121, 121A]

Grafting is a method of modifying the viscoelastic properties of the polymers particularly the natural polymeric products like silk, cotton etc. The grafted products show the similar response as block copolymers. The response varies with extent of grafting and the length of the grafted chains. Low degree of grafting of small chains may not show appreciable change in the properties but grafting of long chains makes the system more flexible however very long chain grafting is likely to present the problems of entanglements.

Grafted polymers are segments copolymers with a linear backbone of one composite and randomly distributed branches of another composites. Fig. 7.11a shows how grafted chains of species B are covalently bonded to polymer chain A. One of the well known example of graft polymer is high impact polystyrene, which consists of polystyrene backbone and poly butadiene grafted chains. Some of the characteristics of graft copolymers are:

1. Graft copolymers containing a larger quantity of side chains are capable of wormlike conformation, compact molecular dimension, and notable chain end effects due to their confined and tight fit structures[112]. They are used as[113]:
2. Impact resistant polymers,
3. Thermoplastic elastomers,
4. Compatibilizers,
5. Emulsifiers for preparation of stable blends and alloys.
6. Generally, grafting method for copolymer synthesis results in products, which are more thermostable than their homopolymer counterparts[114].

7. There are three methods of synthesis (i) grafting onto, (ii) grafting from and (iii) grafting through[115]

Detailed description of these methods and approaches to synthesizing are beyond the scope of this book. Rheological studies carried out by different people are discussed next.

Jahani et al.[116] have investigated on the rheology and morphology of polypropylene (PP) modified by grafting and blending with vinyl monomers. PP powder was copolymerized by impregnating it by a mixture of methyl methacralate (MMA)/n-butylacrylate in presence of azobis-isobutyro-nitrile (n-BA) in a twin-screw extruder in presence of dicumyl peroxide resulting in its grafting, blending and compatibilization with polar copolymers. The results indicate that grafting reaction took place with chain scission of PP molecules and also cross-linking reactions. Also the degree of grafting on to PP increased with increasing acrylate monomers. The dynamic viscosity and modulus depend on the degree of grafting and angular frequency. The modulus increased with frequency, and dynamic viscosity shows usual trend with frequency and level of grafting. The complex viscosity was correlated with angular frequency and relaxation time.

Lin et al.[117] have reported on linear rheology of comb-like copolymer melts with high grafting density: PVSt-g-(PS-b-PE) comblike block copolymers. The following conclusions were drawn:

1. Comb-like block polymers PVSt-g(PS-b-PE) with high density were synthasized.
2. Microphase structure of PVSt-g-(PS-b-PE) at the melt state depends on the volume fraction of PS.
3. PVSt-g (PS-b-PE) shows a typical rheological behaviour of network-like structure.
4. Low polystyrene (11%) vol. content gave spherical structure and high PS (55%) has lamellar structure.
5. Storage modulus, G' > G" at both low and high PS content.
6. At relativly lower PS content both G' and G" are closer to each other in magnitude
7. At high PS loading G'> > G"
8. The reason for such a behaviour is the formation of different microphase-separated structure types in the PVSt-g-(PS-b-PE).

Wong and Baker[118] studied the low frequency rheological behaviour of a family of glicidyl methacrylate (GMA) and styrene graft modified polypropylene in presence of two initiators. It was found that significantly different copolymers in terms of the degree of grafting, molecular weight, and rheological properties were produced. Rheological evidence shows that β-scission occurred during the grafting for the PP-g-GMA copolymers. An increase in the elastic response was observed most likely due to crosslinking via epoxy ring opening. The elastic response of the styrene grafted copolymers was enhanced significantly, and the copolymers behaved like typical crosslinked materials at higher levels of grafted styrene. The low frequency storage modulus (G') increased with the addition of styrene as a comonomer for the grafting of GMA. At higher levels of grafting, both the PP-g-S and PP-g-(GMA-S) copolymers were sufficiently crosslinked showing G' far greater than G'', over the entire frequency range examined. However, these samples contained no gel material.

Lu et al.[119] studied the influence of an *in situ* reactive interphase composed of graft copolymer on the melt shear and extensional rheology with a model bilayer of polyamide-6 (PA6)/maleic anhydride (MAH) grafted poly(vinylidene fluoride) (PVDF-g-MAH). An *in situ* experiment to probe using a small oscillatory shear and studies on shear and extensional rheology was performed. Interestingly, shear stress relaxation and particularly extensional rheology were pretty sensitive to the presence of reactive interphase but start-up shear gave negligible sensitivity. The stress relaxation of healed bilayers was retarted and extensional rheology indicated a more direct and quantitative view of the contribution of reactive interphase in both linear and nonlinear regimes. The transient extensional viscosity and strain hardening increased very much due to the interfacial reaction in bilayer and the interfacial stress increased with reaction time. These changes were attributed to the *in situ* formed interphase with graft copolymers and increased entanglements between neighboring layers.

Vargas et al.[120] heve reported on the viscoelastic behaviour of asphalt, modified with grafted malic anhydride triblock copolymers. These blends were compared with those made with copolymers without modification. The use of grafted polymers improves the conventional properties of blends made with the asphalt and precursor polymers, such as penetration, softening point, storage stability at high temperature and the rheological properties. The blends made with the chemically modified polymers produced a biphasic morphology with improved dispersion and interconnected phases.

Sailer et al.[121] have reported on the studies of melt rheology of multigrafted copolymers with a styrene-acrylonitrile malic anhydride (SANMA) terpolymer backbone and randomly grafted polyamide 6 (PA6) chains. During the melt blending the maleic anhydride groups of SANMA and the amino end groups of PA 6 reacted to form multigrafted chains in the interphase. Because of the phase separation of SANMA and PA 6, the grafted SANMA backbones formed circular domains (of diameter 20 to 40 nm) which were embedded in the PA 6 melt. The linear viscoelastic behaviour of PA 6/SANMA blends at a sufficiently large SANMA concentration displayed a critical gel state under which, the PA 6/SANMA blend showed a non-linear strain hardening behaviour at a very small Hencky strain at the critical gel state. The elasticity of the PA 6/SANMA blends was clearly apparent as against PA6. Increasing the SANMA concentration led to a larger melt strength and a reduced drawability. The occurrence of the critical gel state can be due to the cooperative motion of molecules, which develops between the grafted PA 6 chains of neighbouring micelle-like SANMA domains.

Pino-Ramos et al.[121a] have reported on the radiation grafting for the functionalization and development of smart polymeric materials using gamma radiation. It produces active sites (free radicals) on polymeric backbone due to high energy radiation exposure to initiate the reaction between monomer and/or polymer leading to propagation to form side chain graft. The radiation grafting, which requires no catalyst or any additive usually produces no change of mechanical properties over original polymer. They also reported on the synthesis of smart polymers, coating with gamma radiation and their application in drug delivery and other biomedical areas.

7.5 EFFECT OF FILLERS[122-127]

Fillers for polymers and rubbers can be broadly classified as, (a) active fillers and (b) inactive fillers. Active fillers, in general increase the stiffness due to two factors (i) reduction in the polymer volume per unit total volume thereby stretching the polymer chains and (ii) by forming the bound rubber (in case of carbon black-rubbers and silica-rubber systems) on filler particles due to the chemical binding between rubber molecules and the free radicals on the fillers. In case of inactive fillers only the first of the above reason applies. Figure 7.18 gives loss compliance for glass beads filled polyisobutylene, showing progressive decrease in compliance with filler loading.

Fillers in general are classified as (a) microfillers, where the size of the filler particle is expressed in microns, μ. Very fine particles having size of the order of nanometer are nanoparticles, which when mixed with polymers give considerable increase in the mechanical and flexural properties at very low filler concentration as compared to the micro fillers essentially due to very high surface area of the nanofillers. This is the reason why nanofillers find very wide applications as fillers in rubbers and plastics now a days.

Further, apart from particle size, fillers have been subdivided based on different criterion for example, (a) physical nature ie rigid or flexible fillers; (b) physical form or shape ie (i) three dimensional (ii) two dimensional (iii) one dimensional fillers; (c) based on chemical form i.e. organic or inorganic fillers; (iv) hybrid fillers. Best source for information on filler are (i) *Handbook of fillers*[125] and also *Rheology of filled polymer systems by A. V. Shenoy*[126] who has presented a reasonable discussion about fillers used in polymers and rubbers. All the above fillers are micro size particle. Nanoparticle or nanofillers are relatively of recent origin and are a separate class.

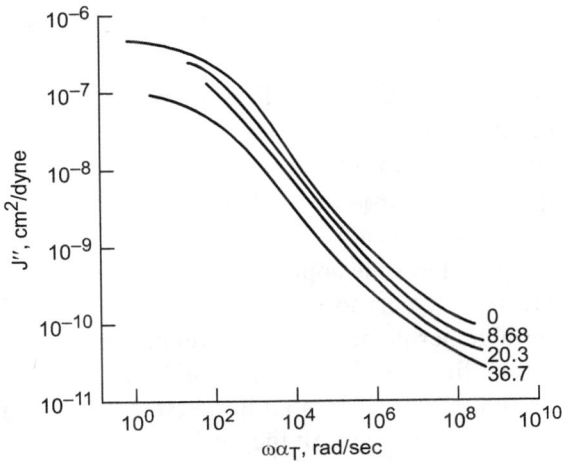

Fig. 7.18 Loss compliance for glass beads filled polyisobutylene at 12.5°C[122].
(M = 1.56 × 10⁶, Avg. dia. of glass beads = 40μ, % vol. fraction is indicated in the figure)

7.5.1 Fillers Based on Physical Nature

Rigid fillers: Some of the examples of the rigid fillers are Aluminium oxide (Al_2O_3), Barium carbonate ($BaCO_3$), Calcium carbonate ($CaCO_3$), Calciumhydroxide $Ca(OH)_2$, Calcium silicate (Ca_2SiO_4), Clay, glass fiber, Magnesium hydroxide $Mg(OH)_2$, metal fiber, mica, talc, wollastonite ($CaSiO_3$). *Flexible fillers:*Asbestos fibers, cotton linters, jute fibers, kevlar fibers, nylon fibers, polyester fibers, sisal etc. Fillers based on physical form:

(i)   Three dimentional fillers -spherical, ellipsoidal, glass beads and wood flour;

(ii) Two dimensional fillers - flakes, platelets, mica, and clay;

(iii) One dimensional fillers - fibers, whiskers, glass fibers and wollastonite (Calcium silicate- $CaSiO_3$, an important industrial mineral and an acicular (needlelike) polymer filler; and

(iv) Organic fillers- Cellulosic, lignins proteins,

(iv) Inorganic fillers- Carbonates, hydroxides, oxides, silicates, sulfates, metals and miscellaneous.

Important functions of fillers - Different fillers are used to fulfill different requirements like:

(i)   *Cost reduction:* Wood flour, saw dust and cotton flocks.

(ii) *Reinforcement:* (a) plastic reinforcement-Glass fibers, cellulosic fibers, synthetic fibers, asbestos fibers and (b) Rubber reinforcement- Carbon black and silica.

(iii) *To impart hardness:* Metalic powders, mineral powders, silica and graphite

(iv) *Chemical resistance:* Glass fibers, synthetic fibers, metalic oxides and graphites.

In addition to these functions the fillers are required to increase stiffness, polymer composite strength and dimensional stability of parts, enhance toughness and impact strength, increase heat deflection temperature, improve mechanical damping, to reduce permeability to gases and liquids and to modify electrical properties of composites.

In filled polymer systems the applied load is mostly born by the fillers and the polymer phase merely acts to transmit the load to the filler matrix. The filler and polymer together must respond simultaneously to the load so as to fulfill this requirement for proper functioning of the system[127]. The interphase between the filler and polymer plays an important role during application of the components to efficiently transmit the load to the filler network so that the system gives maximum benefit of strength of the matrix. The polymer matrix and the filler network combination must provide a uniform distribution of stress in the entire structure and there should not develop any stress concentration in the matrix which may result in its failure. Thus a uniform polymer-filler interaction is vital, which can be effectively achieved if the filler surface is completely wetted by the polymer.

Polymer-filler interaction can be visualised as (i) the filler is mearly present without interaction, (ii) the filler is totally wetted by the polymer either due to its affinity to the polymer or due to its surface treatment, (iii) the filler has established a chemical bond with polymer giving high strength to the matrix for example carbon black-rubber or a glass fiber-plastic system. The factor that affect filler polymer interaction are (a) filler geometry, (ii) filler volume fraction, (iii) filler surface characteristics and its affinity to the polymer, (iv) extent of filler suface wettability with the polymer. To achieve a good mechanical strength and satisfactory performance of the components it may be required to subject the filler to a surface treatment in the form of surface cleaning, or the use of surface modifiers. Some typical surface modifiers are: Silanes, Titanates, Zirconates, Zircoaluminates and hydrophobic wetting agents[126].

7.5.2 Nano-Fillers[127a-127v]

Fillers play a very important role in polymer composites. Addition of fillers improves tensile strength and improves the beneficial chemical interactions between host material and the fillers. This has led to the development of surface treatd fillers particularly nano-fillers. Different types of fillers used in polymers are (i) Extenders essentially used to cheapen end products. These are mostly calcium carbonate – ground calcium carbonate (GCC)[127a] mostly used by plastic industry and (ii) Precipitated calcium carbonate (PCC) mostly consumed by paper industry. Wood flour and saw dust are used in thermosetting plastics. Considering the relatively recent origin of nano-fillers and their effectiveness to give high mechanical and flxural properties the characteristics of these fillers are discussed in some details.

7.5.2.1 Nanoparticles

Nanoparticles and other particles are classified as

1. Ultrafine particles and nanoparticles have same meaning having size between 1 and 100 nm with a surrounding interfacial layer. The interfacial layer is an integral part of nanoscale matter, fundamentally affecting all of its properties. The interfacial layer generally consists of ions, organic or inorganic molecules. Organic molcules coating inorganic nanoparticles are known as stabilizers, capping and surface ligands, or passivating agents[127b]. In nanotechnology, a particle is defined as a small object that behaves as a whole unit with respect to its transport and other properties. Particles are further classified according to diameter[127c].

2. IUPAC definition of nanoparticle- Particle of any shape with dimensions in the range of 1×10^{-9} and 1×10^{-7} m.[127d, 127e]. Figure 7.19 and 7.19*a* show respectively 15 nm silica particle and TEM images of prepared mesoporous silica nanoparticles with mean outer diameter: (a) 20 nm, (b) 45 nm, (c) 80 nm. (d) SEM image corresponding to (b) . The insets are high magnification of mesoporous silica particles.

3. The term "nanoparticle" is not usually applied to individual molecules; it usually refers to inorganic materials.

4. Nanoparticles can exhibit size related properties significantly different from either fine particles or bulk materials.[127f, 127g]

5. *Nanoclusters:* These have at least one dimension between 1 and 10 nanometers and narrow size distribution.

6. *Nanopowders:* These are agglomerates of ultrafine particles, nanoparticles or nanoclusters.[127h]

7. *Nanometer* sized single crystals, or single-domain ultrafine particles, are often referred to as *nanocrystals*.

8. Fine particles are sized between 100 and 2500 nm

9. Coarse particles cover the size range between 2500 and 10,000 nm.

Some of nanoparticles are shown in figures 7.19 and 7.19(a).

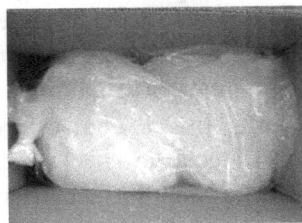

Fig. 7.19   Nano SiO$_2$/Silicon Dioxide nanoparticle (15 nm) [EPRUI Nanoparticles & Microspheres Co. Ltd.][nanoparticles-microspheres.com]

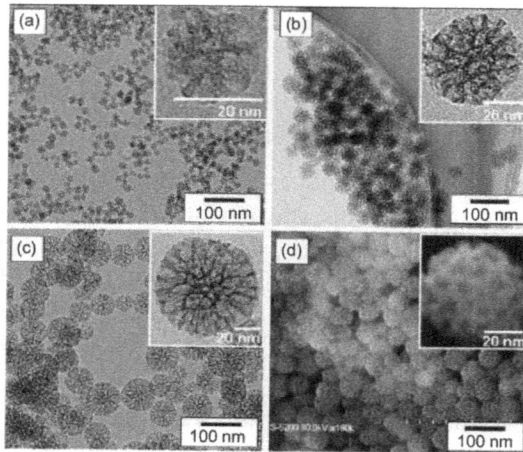

Fig. 7.19(a)   TEM (a,b,c) images of prepared mesoporous silica nanoparticles with mean outer diameter: (a) 20 nm, (b) 45 nm and (c) 80 nm. SEM (d) image corresponding to (b). The insets are high magnification of mesoporous silica particle. [en.wikipedia.org- Sept. 2017]

7.5.2.2 Properties of Nanoparticles

Nanoparticles act as a bridge between bulk materials and atomic and molecular structures. Nanoparticles are some time referred as ultrafine particles. The Bulk materials normally have uniform properties through the bulk but at nano-scale, size-dependent properties are often observed. It is essentially due to the very high surface area in nanoparticles as theses systems have very high surface to the volume ratio in its bulk, for example 1 kg of 1 mm^3 particle have same surface area as 1 mg of particle of 1 nm^3.

Polymer nanocomposites offer new technological and economic benefits. The addition of nanoparticles as fillers in polymers may dramatically increase the mechanical properties, reduce the permeability and improve the flame retardency[127h1].

Nanoparticles often show unexpected properties for example yellow gold nanoparticles appear deep-red to black in solution and have much reduced melting point of 300°C for 2.5 nm size as against 1064°C for gold slabs[127i]

Absorption of solar radiation by nano-composites is much higher as compared to thin films of continuous sheets of material. In solar photovoltaic systems as well as in solar thermal applications, by controlling the size and shape and material of the particles, it is possible to control absorption.[127j, 127k, 127l, 127m]

Recently it has been reported by Yu, et al.[127n] that metal-core dielectric-shell nanoparticle has demonstrated a zero backward scattering with enhanced forward scattering on Si substrate when surface plasmon is located in front of a solar cell. The core-shell nanoparticles can support simultaneously both electric and magnetic resonances, demonstrating entirely new properties when compared with bare metallic nanoparticles if the resonances are properly engineered.

Other size- dependent property changes include (a) quantum confinement in semiconductor particles, (b) surface plasmon resonance[127-o] in some metal particles, and super-magnetism in magnetic materials. However ferromagnetic materials smaller than 10 nm can switch their magnetisation direction using room temperature thermal energy thus making them unsuitable for memory storage[127p].

Clay nanoparticles when incorporated into polymer matrices increase reinforcement, leading to stronger plastics, verifiable by a higher glass transition temperature and other mechanical property tests. These nanoparticles are hard, and impart their properties to the polymer (plastic). Nanoparticles have also been attached to textile fibers in order to create smart and functional clothing[127q].

Nanoparticles with one half hydrophilic and the other half hydrophobic are termed Janus particles and are particularly effective for stabilizing emulsions. They can self-assemble at water/oil interfaces and act as solid surfactants.

Metal and other dielectric, and semiconductor nanoparticles have been formed, as well as hybrid structures (e.g., core–shell nanoparticles)[127r]. Nanoparticles made of semiconducting material may also be labeled quantum dots if they are small enough (typically sub 10 nm) such that quantization of electronic energy levels occurs. Such nanoscale particles are used in biomedical applications as drug carriers or imaging agent.

Hydrogel nanoparticles made of N-isopropylacrylamide hydrogel core shell can be dyed with affinity baits internally. These affinity baits allow the nanoparticles to isolate and remove undesirable proteins while enhancing the target analytes[127s].

Ceramic silicon carbide nanoparticle dispersed in magnesium produce a strong light weight material.[127t]

Silicate nanoparticles can be used to provide a barrier to gases (for example oxygen), or moisture in a plastic film used for packaging. This could slow down the process of spoiling or drying of packed food material.

Zinc oxide nanoparticles can be dispersed in industrial coatings to protect wood, plastic, and textiles from exposure to UV rays.

Silver nanoparticles in fabrics are used to kill bacteria, making clothing odor-resistant

Silicon dioxide crystalline nanoparticles can be used to fill gaps between carbon fibers, thereby strengthening tennis rackets.

Huang etal.[127Y] have reported on the dispersion rheology of carbon nanotubes in a polymer matrix. As the dispersion of nanotubes in polymer matrix presents considerable problem, these studies are quite significant. They have used untreated nanotubes to understand the mechanizm of dispersion and re-aggregation. The viscosity of the mixture has been found to depend directly on the spatial and orientational distribution. Such a distribution has been proposed to act as an indication of quality of dispersion provided the rheological signal is properly calibrated. The time required for dispersion of nanotubes has been reported to be very long and mixing shear is so high that the quality of dispersion reported by all previous studies could be questioned. Thus to avoid re-agglomeration, the mixing was carried out at very high speed of rotation of mixer (1000 rpm) and at relatively lower temperature (about $30 \pm 0.5°C$).

7.5.3 Properties Usefull in Specific Areas

Nanotechnology find applications in drug delivery using nanoparticles. Specific applications are in cancer therapy, to reduce toxicity and side effects of drugs. A number of substances are currently under investigations for preparation of nanoparticles for drug delivery, varying from biological substances like albumin, gelatin and phospholipids for liposomes, and nanoparticles containing chemical substances like various polymers and solid metals and polymeric micelle nanoparticles to deliver drugs to tumors[127u].

The use of polymer coated iron oxide nanoparticle to breakup clusters of bacteria, possibly allowing more effective treatment of chronic bacterial infections.

The surface charge of protein filled nanoparticles has been shown to affect the ability of the nanoparticle to stimulate immune responses. Researchers are thinking that these nano-particles may be used in inhalable vaccines.

Researchers at Rice University have demonstrated that cerium oxide nanoparticles act as an antioxidant to remove oxygen free radicals that are present in a patient's bloodstream following a traumatic injury. The nanoparticles absorb the oxygen free radicals and then release the oxygen in a less dangerous state, freeing up the nanoparticle to absorb more free radicals.

Researchers are developing ways to use carbon nanoparticles called nanodiamonds in medical applications. For example nanodiamonds with protein molecules attached can be used to increase bone growth around dental or joint implants.

Researchers are also testing the use of chemotherapy drugs attached to nanodiamonds to treat brain tumors. Other researchers are testing the use of chemotherapy drugs attached to nanodiamonds to treat leukemia.

There are many more applications in different area for example in manufacturing and materials, environment protection, energy and electronics.

7.5.4 Production of Nanoparticles

The nano particles are manufactured using different methods for example gas condensation, attritions, chemical precipitation, ion implantation, pyrolysis and thermal synthesis, sol-gel or chemical solution deposition method. The detailed discussion of these methods is beyond the scope of this book. Interested reader may refer to Nanoparticles in Wikipedia at [en.wikipedia.org] , November 2017.

7.6 RHEOLOGY OF NANOPARTICLES FILLED POLYMERIC SYSTEMS[127w-z; 127A-Z; 127Z1-5]

A number of workers have carried out research on the rheological response of nanoparticle filled polymer composites. Krishnamurthy and Giannelis[127w] reported the nonterminal low-frequency rheological behaviour of nanocomposites containing layered silicates. Zang et al.[127x] reported a transition to solid like response at low oscillation frequencies for poly (ethyl oxide)/silica nanocomposites. Lozano et al.[127y] and Potschke et al.[127z] also reported the similar solidlike nonterminal rheological behaviour in polymer nanocomposites containing carbon fibers or multiwall carbon nanotubes (MWCNT). Such a nonterminal solidlike behaviour in polymer nanocomposites has been attributed to a filler network formed in nano composites and melt state rheology studies provide insight about such a behaviour.

Giannelis etal.[127A] have reported on the linear dynamic oscillatory shear measurements of various polymer layered silicate nanocomposites (prepared by in-situ polymerization from reactive groups/tethered to silicate surface). The storage and loss moduli (G' and G'' respectively) exhibit a plateau at low-frequencies. Further as observed with steady shear measurements a monotonic increase in both G' and G'' at all frequencies is observed with increasing silicate content. Such a behaviour is attributed to the alteration of relaxation dynamics due to the intimate contact between polymer and inorganic sheets, which leads to the appearance of low-frequency plateau in moduli and low-shear rate non-Newtonian behaviour.

As against this behaviour a series of polydimethysiloxane based delaminated hybrids (dimethyl ditallow montmorillonite), shows an increase in shear viscosity with respect to that of the pure polymer at low shear rates but still obeys Newtonian type behaviour even up to the loading of 19% silicate. The increase in viscosity with silicate loading has been found to be linear. Similar effects have also been observed in linear dynamic oscillatory shear measurements, where storage and loss moduli for delaminated hybrids show similar frequency dependence as pure polymer, with a monotonic increase in the magnitude of moduli with increasing silicate loading. It seems therefore that the process of ralaxation of polymer chains does not change in these delaminated hybrids (at least within the sensitivity of measurements) due to the presence of silicate layers. This relaxation response is similar to the relaxation response of a typical filled polymers systems, wherein the relaxation dynamics of polymer chains does not change considerably due to the presence of non-interacting fillers[127B].

Gianellis et al. also reported the results of two end-tethered hybrid systems prepared by in-situ polymerization- (a) Poly (ε-caprolactone)-montmorillonite (PCLC) and (b) nylon-6-montmorillonite (NCH)-wherein the polymer chains are end-tethered to silicate surface via cationic surfactants[127w] as shown in Fig. 7.20. As a result of the scheme used to prepare the nanocomposites the average molecular weight of polymer chains reduced sharply at low silicate loading up to 2 wt.% but remained fairly constant for higher loadings. It has been reported that there was a change in the rheological response indicating the possibility of silicate layers being aligned because of the application of large amplitude oscillatory shear (needed to get clear signal to be recorded by the transducer[127A]).

The storage modulus for nanocomposites show a monotonic increase with increase in silicate loading at all frequencies. However at highest frequency investigated (reduced frequency range from 10^{-3} to 10^2 rad/s) 2% silicate sample showed slightly lower value than 1% sample.

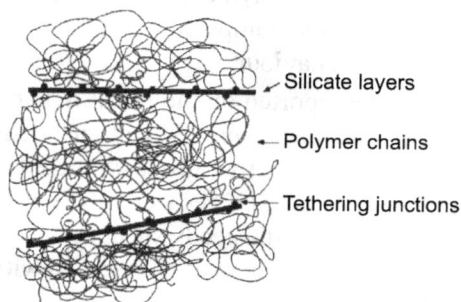

Silicate layers

Polymer chains

Tethering junctions

Fig. 7.20 Schematic diagram describing the end-tethered nanocomposites. The layered silicate is highly anisotropic with a thickness of 1 nm and lateral dimensions (length and width) ranging from 100 nm to few microns. The polymer chains tethered to the surface via ionic interactions between the silicate layer and the polymer-end[127A, 127w]

The loss modulus however gave some what ambiguous results with 1% silicate loading showing higher value than 2% and 3% loading but samples from 2% to 10% silicate loading gave regularly increasing loss moduli with increasing silicate loading. The reason for such a variation is due to the reduction in the molecular weight of the chains, which occurs mainly around 1 to 2 % silicate loading.

For Nylon-6 silicate hybrids no such ambiguity in the variation of moduli has been observed and these have been found to increase with increase in silicate loading. However the loss modulus varies over a wide range than storage modulus particularly at lower frequency.

It has been postulated that when the molecular relaxation process is retarded then the non-terminal low-frequency behaviour is observed in these systems i e. PCL and Nylon-6 based nanocomposites. Such an effect is produced by tethering of one-end of these molecules to the silicate surface.

7.6.1 Effect of Alignment of Nanocomposites

Giannelis et al.[127A] have reported the effect of alignment of nanocomposites on the rheological response of PCLC samples. Under the application of large strain-amplitude oscillatory shear the sample gets aligned in the direction of shear. It was observed that during shear, moduli show a decrease with continuous shearing and finally reach a plateau value for 3, 5 and 10 wt.% silicate samples. The effect of alignment becomes very clear after first few cycles particularly on 3 and 5% samples. For these samples the moduli decrease monotonicaly and the stress signal does not undergo any distortion but remains sinusoidal. In the first few cycles of shear for 3 and 5 wt% the moduli show a maximum before decreasing continuously possibly due to the transient state of process of alignment.

The small-strain moduli after shear alignment for PCLC with10 wt% silicate (PCLC10) (experimental conditions $-T = 70°C$, $\omega = 1$ rad/s, $\gamma_0 = 120\%$ and time 3 h) show a considerable reduction in the magnitude of both storage and loss moduli as compared to those for unaligned values. The reasons for such a reduction in the magnitudes of moduli is that the aligned samples store and discipate relatively smaller amount of energy as compared to the unaligned samples. Another influence of the alignment is that both the moduli appear to be more sensitive to the frequency and rise with much faster rate as compared to the unaligned samples[127A].

Chow and Mohd Ishak[127C] have reported on the mechanical, morphological and rheological properties polyamide 6/organo-montmorillonite nanocomposites. The following conclusions were drawn:

(a) Modulus and strength of Polyamide 6 (PA6) was improved significantly in presence of organo-montmorillonite (OMMT) attributed to the stiffness, rigidity, reinforcing effect and exfoliation of OMMT.

(b) X-ray diffraction (XRD) and atomic force microscopy (AFM) techniques were used to study the morphology of PA6 /OMMT nanocomposites. The OMMT was seen to be exfoliated in the PA6 matrix. This is attributed to the strong interaction of PA6 and OMMT, and intercalation capability of PA6 in silicate layers of OMMT.

(c) Based on the XRD and AFM results of OMMT dispersion in PA6 nanocomposites, the following rheological parameters, (a) the storage modulus G' and its slope and (b) complex viscosity η^* and its slope were found to be good indicator for better exfoliation of clay. The table 7.1 gives the improvement in tensile and flaxural propertis of PA6 on the addition of 4 wt.% OMMT,

Table 7.1 Improvement in Tensile and Flexural Properties of PA6 on addition of 4% OMMT[127C]

Properties	Units	PA6	PA6/OMMT
Tensile Modulus	GPa	2.3±0.02	3.0±0.02
Tensile Strength	MPa	40.1± 0.05	68.7±0.10
Elongation at break	%	58.4 ±0.05	3.5±0.02
Flexural modulus	GPa	2.1±0.02	2.9±0.01
Flexural strength	MPa	94.2±0.10	105.3±0.10

It is important to note here that the flexural and tensile strengths and moduli of PA/OMMT nanocomposite increased although the increase in strengths is far more pronounced. The significant improvement in both tensile and flexural strength has been attributed to the delaminated clay formation. XRD and AFM results revealed the exfoliation of OMMT in the composites. As the OMMT has a plate like structure irrespective of its degree of exfoliation, the silicate layers when exfoliated get aligned in mold flow direction (MFD) especially in the skin layers. The injection-molding direction, being parallel to the long axis of layered silicate, produces the forced orientation of layers resulting in production of skin-core morphology, which is the main cause of increased flexural strength.

Chong et al.[127D] have reported on the mechanical and rheological properties of maleated polypropylene-layered silicate nanocomposites (PLSNs) with different morphology. Three types of maleated polypropylene-layered silicate nanocomposites with different dispersion states of layered silicate i.e. (a) deintercalated, (b) intercalated and (c) exfoliated states were used to investigate the effect of final morphology of nanocomposites on the rheological and mechanical properties. In theses studies two types of polypropylenes (PP) with different molecular weight (MW), organically modified layered silicate (C18M) and pristine montmorillonite (MMT) were used. High MW PP intercalates slowly whereas the one with low MW exfoliates fast into the organophilic layered silicates. The state of dispersion of layered silicate affects the oscillatory storage modulus, nonterminal behaviour, and relative viscosity. The exfoliated nanocomposites gave the largest increase in these parameters whereas deintercalated nanocomposites showed almost no change in relative shear and complex viscosities with clay content. The exfoliated noancomposites undergo the shear alignment of clay layers giving considerable reduction in complex viscosity. The dynamic storage moduli gave similar behaviour as the relative shear and complex viscosities.

The final morphology influences the mechanical and rheological properties of PLSNs. The investigations on PLSNs have revealed a significant variations in viscoelastic properties depending on their microstructure and interfacial characteristics between polymer and layered silicates.[127 (E-P)] It has been reported by Hoffmann et al.[127M] and Lim and Park[127O], that the linear viscoelastic behaviour in nanocomposites was considerably affected by the final dispersion state of clay in polymer matrix.

Lim et al.[127R] have reported on ultrasonic effect on dispersion quality and rheological properties of poly(ethylene oxide) (PEO), (PEO)/organophillic montmorillonite (OMMT) nanocomposites. It has been observed that the ultrasonic technique plays an important role in preparation of fine dispersion of inorganic layered silicate and enhancement of swelling to intercalate polymer chains. A systematically controlled sonicating procedure was applied to differentiate the level of dispersion quality for different systems used. It was observed that the level of dispersion plays an important role in rheological properties of composites. It was reported that (a) shear-thinning nature increased with degree of dispersion due to the alignment of clay layers under shear[127I, 127S], i.e. the suspension microstruture changes from a random to an ordered orientation[127T], and (b) the poor dispersion and enhanced swelling results in agglomeration of clay particles. It is possible for layered silicate to align more effectively in optimized dispersion medium for example the dispersion of platelets can effectively vary the flow characteristics under shear. The shear viscosity variation with rate of shear for different samples prepared under different conditions are given in Fig. 7.20(a).

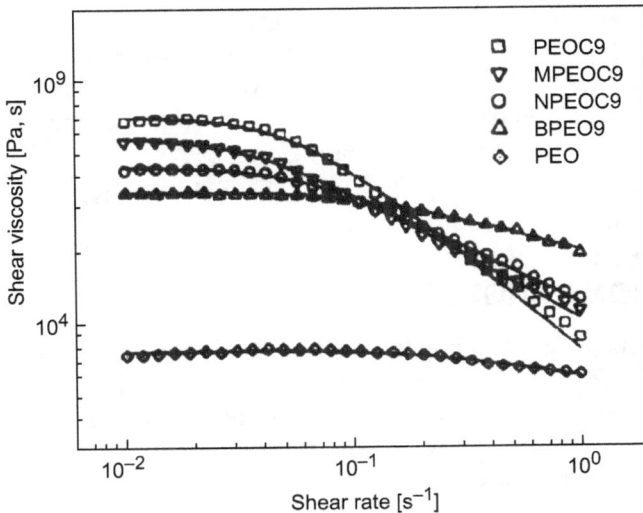

Fig. 7.20(a) Shear viscosity vs. rate of shear for various PEO/OMMT nanocomposites obtained from different preparation procedures. Symbols represent the experimental data and the solid lines indicate the best fits obtained from Carreau model. [127R]. (PEOC9- sonicated clay dispersion system); NPEOC9 (non-sonicated clay dispersion); MPEOC9-PEO solution and non-sonicated clay suspension, sonicated after mixing; BPEO9- an immiscible PEO-clay blend). [Adopted from Ref. 127R]

Mishra et al.[127U] have reported on the mechanical and rheological properties of a thermoplastic polyolefine (TPO)/organoclay nanocomposites with reference to the effect of maleic anhydride modified polypropelene as a compatibilizer. Thermoplastic elastomer compositions based on uncured ethylene propylene dien terpolymer (EPDM) and polypropylene (PP) is referred as TPO. The compatibilizer not only enhances the intercalation of polymer chains inside the clay gallery but also changes the composition of the TPO/clay nanocomposites. The tensile modulus as well as storage modulus of the nanocomposite was substantially higher over their pristine counterpart. The increase in storage modulus in the melted state and decrease in terminal slope of nanocomposite confirms the strong interaction between clay platelet and polymer molecule in the melted state. The nanocomposites and their pritine counterpart show strong shear thining behaviour at the melted state.

Incarnato et al.[127V] have reported on rheology, structure and properties of silicate/polyamide nanocomposites. Melt compounded nanocomposite systems based on nylon6 (Ny6) and a copolyamide at 5% w/w of the comonomer 1,1', 3 trimethylcyclohexane-3-methylamine-5-isophthalamide(ADS) were obtained using a counter rotating intermeshing extruder. Rheological results, X-ray and DSC analysis have evidenced that a larger extent of exfoliation of silicate platelets can be obtained by increasing the clay percentage and extrusion rate. The non-Newtonian behaviour for blends at 6% and 9% of cloisite show low frequency η^* comparable or even higher with respect to Ny6-based nanocomposites at the same percentage of silicate, even though ADS has lower viscosity than Ny6. This phenomenon is due to different viscosity and chemical nature of two matrixes. Also the ADS based blends at 6% and 9% of clay extruded at 100 rpm exhibit an increase in elastic modulus of 20% and 30% respectively, compared to the matrix.

Many other researchers[127V1-127V13] also have also reported different aspects of polymer nanocomposites.

7.7 CARBON NANOMATERIALS

Different carbon based nanoparticles are (a) Graphene (b) Nanodiamonds (c) Nanofibers (d) Nanocones (e) Fullerenes (f) Nanotubes as shown in Fig. 7.21. In addition to it nanoparticles can be in the form of double wall carbon nanotubes (DWCNTs), Single wall carbon nanotubes (SWCNTs), Functionalized nanotubes, platelet, fibers, rods, disks, nanobuds and nanohorns.[127X, 127W] Most single wall nanotubes have diameter close to 1 nanometer and can be many millions of times longer. The structure of SWCNT can be conceptualized by wrapping a one-atom-thick layer of graphene in to a seemless cylinder. Multi-walled carbon nanotubes (MWCNTs) consists of multiple rolled layers (concentric tube) of graphene. The interlayer distance in multi-walled nanotubes is close to the distance between graphene layers in graphite, approximately 3.4 Å (angstrom), (1 angstrom = 0.1 nm).

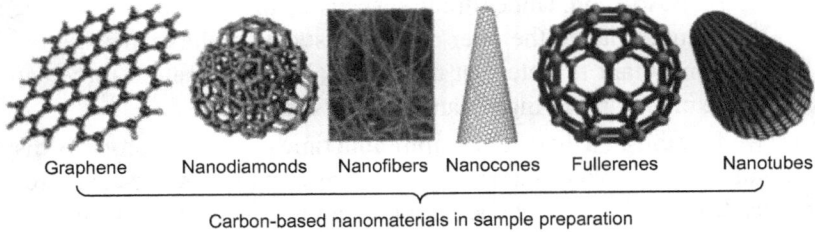

| Graphene | Nanodiamonds | Nanofibers | Nanocones | Fullerenes | Nanotubes |

Carbon-based nanomaterials in sample preparation

Fig. 7.21 Carbon based nanomaterials of different types. [127W] [sciencedirect.com]

The rheological properties of nanoparticle filled systems are extremely important from the point of view of processing and applications. Fangming Du et al.[127v] have reported on the rheological properties of carbon nanotube (CNT) polymer composites. The composite properties are related to the materials' microstructure, the state of dispersion of nanoparticles (CNT), the aspect ratio and orientation of nanotubes and interaction between polymer chains and nanotubes.

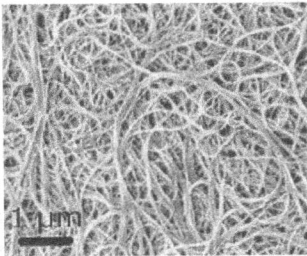

Scanning Electron Microscopy image
of Carbon nanotube bundle

A stable carbon
nanobud strucutre

Fig. 7.22 Carbon nanotubebundle and carbon nanobud structure [en.wilipedia.org]

Huang et al.[127Y] have reported on the dispersion rheology of carbon nanotubes (prepared by catalytic vapour deposition – CVD method) in polydimethylsiloxane (PDMS) polymer matrix. An uniformly dispersed nanotubes in the polymer matrix is the essential requirement for correct, efficient and economically viable processing of nanotube-polymer composites. Proper dispersion of the nanotube is also a primary condition for the accurate measurements of mechanical and rheological properties of the matrix. The problems associated with the nanotubes are:

(a) These are already aggregated when procurred for the application,

(b) In the polymer matrix there is a severe problem of reaggregation after dispersion, as the particles are likely to reaggregate particularly at higher nanotube loading. Under dilute suspension the chances of reaggregation are reduced due to the inter particle distance in the matrix being large. It is important to note that the external hydrodynamic force can also cause clustering in highly anisotropic suspensions.[127 Y1]

It is well established that there is minimum time required to properly disperse the nanotubes in a polymer matrix. The time after which the viscosity remains practically constant (i.e. the dispersion plateau viscosity) is considered as the critical time (t^*) of dispersion, which depends on the type of matrix, nanotube concentration, shear stress in the mixing unit and external hydrodynamic forces. The critical time has to be determined experimentally for the system. It has been reported by Huang et al. that the critical time varies linearly with wt% of nanotube loading.

The rheological response for time $t_{mix} < t^*$ is nonconsistent and the viscosity measurements give erroneous data. The viscosity measured in poorly dispersed samples is sometimes several order of magnitude higher than normally expected value. It is due to the cluster of nanotubes in the system that arrests the functioning of the rheometer giving highly erratic viscosity values. It is therefore essential to ensure complete dispersion of the nanotubes in the medium. Fig. 7.22(a), 7.22(b), 7.22(c), 7.22(d), 7.22(e) show respectively the SEM image of cluster of nanotubes, dispersion plateau viscosity as a function of nanotube loading, viscosity as function of frequency for poorly dispersed nanotubes $(t_{mix} < t^*)$, viscosity vs frequency for well dispersed $(t_{mix} > t^*)$ nanotubes and storage modulus versus frequency plots for well dispersed nanotube suspensions having different concentration of nanotubes.

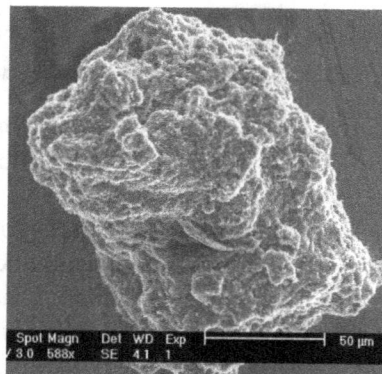

Fig. 7.22(a)   SEM image of a typical compact nanotube cluster found in a sample mixed only for a few minutes [adopted from Ref. 127Y]

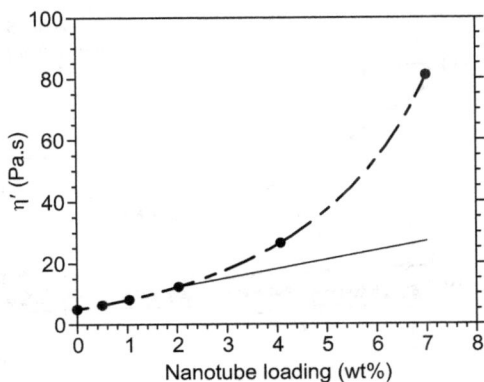

Fig. 7.22(b) Plot of the dispersion plateau viscosity (at 50 Hz.) against the weight fraction of nanotubes in the composite. The solid line shows the initial linear (non-interacting) regime; the dashed line is a guide to an eye in the nonlinear (entangled) regime[127Y].

A properly dispersed nanotube system ($t_{mix} > t^*$) will give reproducible rheological measurements for example viscosity value will be same from all different measurements. However it should be kept in mind that very small nanotube clusters which do not get dispersed are always present and it is assumed that these do not interfere with measurements and the system is assumed to be well dispersed for all practical purposes.

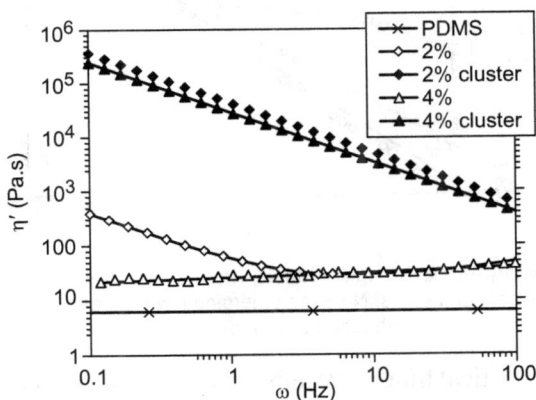

Fig. 7.22(c) Frequency sweep of viscosity for 2 and 4 wt% samples mixed for $t_{mix} < t^*$. The bold symbols show the result when the sample aliquot in rheometer was found to contain visible nanotube clusters: note the extremely high viscosity values of (arrested) η', the same for different samples.[127Y]

At low concentration, nanotubes are homogeneously dispersed in polymer matrix and are quite stable against reaggregation. Here the rheological measurements correspond to that of viscous fluid and the viscosity is a linear function of concentration of the tubes. However due to the profound anisotropy of the nanotubes the slope of this function is normally much higher than the conventional Einstein equation slope of 2.5.

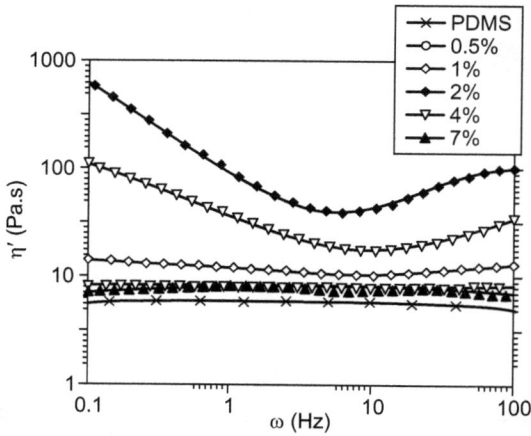

Fig. 7.22(d) Summary of dynamic viscosity, $\eta'(\omega)$ against frequency for well dispersed samples of different concentrations[127Y].

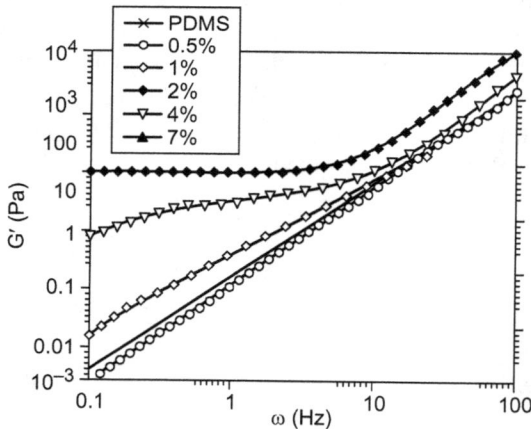

Fig. 7.22(e) Summary for the storage modulus, $G'(\omega)$ against frequency for well disperesed samples of different concentrations. Note the emerging low frequency rubber plateau in G' at high tube concentrations[127Y].

Similar to the critical time there appears a threshold concentration (2 to 3% in this case) above which there emerges an elastic gel of entangled nanotubes in their homogeneously dispersed state. The rheological characteristics of these composites show a rubber modulus in the limit of zero frequency. These systems also show a characteristic superposition between mixing time and frequency of rheological testing, similar to the time/temperature in the classical glass-forming polymers.

Single wall carbon nanotubes are in the form of bundle (Fig. 7.22), which has an isotropic orientation. These nanotubes when suspended in polymer give at low concentration the rheological and electrical properties of the composite, which are comparable to those of pure polymer, and as the concentration increases the onset of solid like viscoelastic behaviour occurs when the size of polymer chain is somewhat large to the separation between nanotube bundle. At still higher concentration of the nanotubes, the nanotubes are close enough to form a percolating conducting path along the nanotubes. Thus it can be seen that the behaviour of the system changes progressively from normal fluid like to the formation of rheological percolation threshold and then at still higher loading it reaches to the electrical percolation threshold. These stages are shown in the Fig. 7.23.

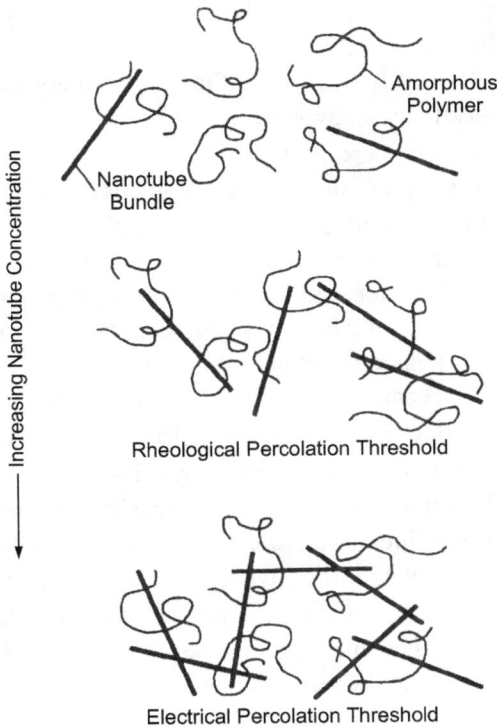

Fig. 7.23 Schematic of SWNT/polymer nanocomposites in which the nanotube bundles have isotropic orientation. (top) At low nanotube concentrations, the rheological and electrical properties of composite are comparable to those of host polymer . (middle) The onset of solidlike viscoelastic behaviour occurs when size of the polymer chain is somewhat large to the separation between the nanotubes. (bottom) The onset of electrical conductivity is observed when nanotube bundles are sufficiently close to one another to form a percolating conductive path along the nanotubes.[127v]

Kota et al.[127Z] have reported on electrical and rheological percolation in polystyrene multiwall carbon nanotube (MWCNT) nanocomposites. Electrical conductivity gave a power law fit indicating that the microstructure starts to percolate at 1 volume %. The electrical conductivity of composite increased considerably (by 20 order of magnitude) as compared to that of pure PS. The rheological parameters like storage modulus exhibited a low frequency plateau at concentration above 2 vol. %, which suggests the formation of a pseudo-solid-like network of percolated MWCNTs.

It is reported by Shaffer and Sandler[127 Z1] that the dynamic viscosity, η^* and loss modulus, G'' are less sensitive to the percolation as compared to G', G'/G'' and experimentally determined electrical conductivity, σ_c of the system. Microstructurally, this indicates that the elastic load transfer and electrical conductivity are far more sensitive to the onset of percolated MWCNTs than discipation mechanism that affects the viscous response. Amongst the two solvents used in the study the dimethylformamide (DFM) imparts higher electrical percolation at lower concentration of MWCNT as compared to tetrahydrofuran (THF). DMF influences the electrical conductivity more than the elastic behaviour of nanocomposites.

Lawrence Livermore National Laboratory (LLNL) in US has developed carbon nanotubes over 50,000 times thinner than human hair that can "Separate Salt from Sea Water", an advance that may help solve the global water crisis. Increasing demand for fresh water pose a global threat to sustainable development, resulting in water scarcity for six billion people. Ramya Tunuguntla a postdoctoral researcher at a laboratory reporeted that carbon nanotubes with diameter smaller than a nanometer bear the key structural feature that enable enhanced transport. [Deccan Heral, Bangaluru; www.decanherald.com].

Huang and Terentjev[127 Z2] have studied the dispersion and rheological behaviour of carbon nanotubes in polymer matrix. They selected two regimes one in a low viscosity solvent and other in a high viscosity polymer using nanotubes without any surface-modification. An estimate of the van der Waals energies involved has been made in nanotube aggregate using ultrasonication and shear mixing. Also the tube scission during cavitation created by ultrasound has been analysed and estimated. The characteristic nanotube length L_{lim}, below which no further scission of nanotubes occurs has been predicted. For shear mixing it has been reported that it was possible to disperse the non-parallel bundled nanotube aggregate in high viscosity polymers, once the critical mixing time t^* was reached. The characteristic features of nanotube-polymer composite rheology and its ageing/stability against reaggregation also were examined. It has also been reported that at the nanotube loading above overlap concentration, the tubes form an elastic network, where the physical junctions are quite strong and stable to provide a rubber-like elastic response with very low relaxation.

Chaterjee and Krishnamoorti[127 Z3] have reported on rheology of polymer carbon nanocomposites. The most important parameters influencing the rheology of nanotubes-polymer composites are (a) concentration of nanotubes and their state of dispersion. Huang and Terentjev[127Z2] also reported the similar observations earlier. The percolation threshold and concentration for isotropic to oriented transition are strong functions of inverse of effective aspect ratio of dispersed nanotubes and therefore restrict the range of concentration over which such nanocomposites can be deployed.

Wu et al.[127 Z4] have reported on the rheology of carbon nanotubes-filled poly(vinylidene fluoride) composites. The carbon nanotubes (CNTs)–filled poly(vinylidene fluoride) (PVDF) composites (PCT*s*) were prepared by melt compounding for rheological study. The steady and oscillatory flow behaviours were then explored. The results show that the presence of CNTs enhances the pseudoplastic flow accompanied by the increased flow activation energy. However, the linear flow region is not sensitive to the temperature whether driven by shear rate or by strain. During oscillatory shear flow, the solid-like response is attributed to the percolation of CNTs, but the formation of a percolated CNT network is temperature-dependent, and the percolation threshold values reduce with an increase in temperature. The two-phase viscoelastic model was then used to further describe the linear responses of composites, aiming at relating the hierarchical structures of the CNTs to flow behaviours of the composites.

Zhang et al.[127 Z5] have used the dynamic rheological behaviour to estimate the dispersion of carbon nanotubes in carbon nanotube/polymer composites. Well-dispersed multiwalled carbon nanotube (MWNT)/polystyrene composites have been prepared. Transmission and scanning electron microscopy were employed to observe the distribution of the MWNTs in the composites in a microscopic scale, indicating a nanotube network formed in the matrix. The dispersion of the nanotubes in the polymer was monitored by oscillatory rheology. It was found that the addition of MWNTs in the polymer had a drastic influence on the rheological behavior of the composites. As the MWNT loading increased, Newtonian behavior disappeared at low frequency, suggesting a transition from liquid-like to solid-like viscoelastic behaviour. A more homogeneous dispersion or a greater loading of the nanotubes in the matrix produced stronger solid-like and nonterminal behaviour, and the composites exhibited reduced temperature dependence at elevated temperature, compared to the matrix melt.

Rochfort and Avouris[127 Z6] have reported on the twisting strength of Multiwall carbon nanotubes and electrical and mechanical properties of the twisted tubes. Kulshrestha et al.[127 Z7] have worked on the internal structure of the MW nanotube-Polypropylene composites produced using continuous chaotic blender. The percolation was obtained at a minimum composition of 1%. The structure in the nanocomposite film has been reported to be related to process condition and electrical resistivity.

7.8 FILLER CHARACTERIZATION

Fillers are used in polymers for a variety of reasons: cost reduction, improved processing, density control, optical effects, thermal conductivity, control of thermal expansion, electrical properties, magnetic properties, flame retardancy and improved mechanical properties, such as hardness and tear resistance etc.

For example, in cable applications, fillers such as metakaolinite are used to provide better electrical stability while others, such as alumina trihydrate, are used as fire retardants. Each filler type has different properties and these in turn are influenced by the particle size, shape and surface chemistry. Filler characteristics are important from cost consideration to particle morphology. Particle specific surface area, filler loading and packing are equally important aspects. It is therefore essential that the fillers must be properly graded and listed. For example, the use of an average particle size on data sheets can be misleading as it may not accurately reflect particle size distribution.[128]

The important characteristics of fillers, particularly for rubber applications, are reported by Hofmann[129] which include:

(a) Filler activity by estimating bound rubber,
(b) Particle size and particle size distribution using electron microscopy under the magnification of 50000-75000.
(c) Filler surface area; using diameter measured as mentioned above and using BET method[130] through nitrogen absorption as m^2/gm.
(d) Filler structure measurements by Oil or dibutyl phthalate adsorption method, which expresses the structure through estimation of aggregate structures.
(e) Black-tinting-intensity and
(f) pH-value

It is important to note that these tests are only guides and in no way allow the final judgement as to the behaviour of fillers in rubber/polymer mixture and the practical experience is most important guide to arrive at the final composition.

7.9 FILLERS FOR PLASTICS

Calcium carbonate (limestone), hydrous magnesium silicate (talc), mica, calcium metasilicate (wollastonite), calcium sulfate (gypsum), calcium sulfate fibers (also known as Franklin fiber), glass beads and aluminium trihydrate etc. are the most common fillers used in plastics. Calcium carbonate is used maximum in different industries.[131]

Mineral fillers like mica, clay or calcium carbonate are used to tailer their conductivity, shrinkage and mold times.

Alumina trihydrate for flame retrdancy

Clay fillers improve electrical properties and processability of resins.

Mica improves mechanical properties (used in the form of small brown flakes).

Glass beads, silica, calcium carbonate and talc usually weaken plastics.

Mineral fillers with length to width ratio 10:1 (low aspect ratio) contribute significantly to isotropic or uniform shrinkage and reduce warping in composites.

Those fillers with high aspect ratio help to maintain or increase the tensile strength and other mechanical properties.[132]

7.10 SELECTION OF POLYMERS[133–136]

Dependence of the polymer properties on different operating parameters and their response to the applied stresses vary widely. The selection of the polymers for a specific application is guided by application environment like temperature, pressure, humidity, pH, extent and nature of the stresses acting, frequency and magnitude of deformation and contact with oils and greases. If the material is to be used outdoors for a long period, an ultraviolet resistant (UV) material must be specified. Thus before selecting a polymer for a particular application it is of utmost importance that the essential properties of the component critical to its performance must be defined and the polymers response to these properties should be analysed. Some times if the application requires subjecting the component to say highly corrosive environment like a chemical plant handling concentrated acids at higher temperature or a very wide fluctuations in the temperature or high frequency loading then the choice of the material is limited to only one or two polymers and the cost of the material may not play a deciding role in the selection. However for moderate to low risk environment conditions there may exist a number of polymers, which can be suitable for a particular requirement. Under such conditions the cost of the material may become the deciding factor for the selection of the material. Thus for proper material selection, a careful planning, a thorough understanding of plastic material's viscoelastic characteristics (having both solid like and liquid like behaviours under different operating conditions) and reasonable prototype testing are essential requirements.

During the process of selection of plastic material, a design engineer will review the material data sheets provided by material suppliers. It is important to understand that the data sheets are useful only for the comparison of property values of different plastic materials for example tensile strength of nylon verses polycarbonate or impact strength of polystyrene versus ABS. These data sheet values do not give any indications about how the material will behave under long term exposer to external environment or in the severe acidic or alkaline conditions or exposer to relatively high or very low temperature. Thus the data sheets should never be used for engineering design and final selection of polymers.

Thus it can be said that if the design engineer selects the material solely on the basis of published data in data sheets and does not give due considerations to the actual conditions of application then the recommendations by the design engineer will invariably result in a disaster.

Vishu Shah, John Wiley & Sons[136] has discussed the following approach to the selection of materials. These are reproduced below.

The key considerations for the polymeric material selection are:

7.10.1 Mechanical Properties

1. Tensile strength and Modulus
2. Flexural strength and Modulus
3. Impact Strength
4. Compressive strength
5. Fatigue endurance
6. Creep
7. Stress relaxation

Both short term and long term data should be evaluated. Short term data is useful for quick comparison and screening and long term data is helpful for final selection of the material. Creep and stress relaxation data, which represents deformation under load over a long period needs to be scrutinized over the usable range of temperature. Isochronous stress–strain curves should be evaluated for comparing on the equal time basis. Multipoint impact data gives meaningful information about the energy at a certain strain and total energy at break must be considered. Plastic parts suddenly loose the impact strength in very cold environment and fail. It is needed to give proper consideration and multi-point impact data should be obtained from the material suppliers.

7.10.2 Thermal Properties

Heat distortion temperature and Vicot softening point are the short term tests. These tests are conducted without consideration of time and the values derived are instantaneous and the data reported on the data sheets are also derived from the single point measurements. Such data should be used only for initial screening. The continuous use of temperature and coefficient of thermal expansion are more helpful for final material selection.

7.10.3 Expansion/Contraction

Plastic materials tend to expand and contract anywhere from seven to ten times more than conventional materials like metals, wood and ceramics. Designers must be well aware of this and special consideration must be given if dissimilar materials are to be assembled. The thermal expansion differences can develop internal stresses from push-pull effect along with internal stresses can cause the parts to fail prematurely. Restraining the tendency of a piping system to expand/contract can result in significant stress reactions in pipe and fittings, or between the piping and its supporting structure. The allowing of a moderate change in length of an installed piping system as a consequence of a temperature change is generally beneficial, regardless of the piping material, in that it tends to reduce and redistribute the stresses that are generated should the tendency for a dimensional change be fully restrained. Thus, allowing controlled expansion/ contraction to take place in one part of a piping system is an accepted means to prevent added stresses to rise to levels in other parts of the system that could compromise with the performance of, or cause damage to the structural integrity of a piping component, or to the structure, which supports the piping.

7.10.4 Exposer to Chemicals

Many plastics materials are susceptible to chemical attack and therefore behaviour of plastics material in chemical environment is one of the most important considerations in selecting material. No single property defines material's ability to perform in a given chemical environment and factors such as external or molded-in stresses, length of exposure, temperature, chemical concentration etc. should be carefully scrutinized. One of the most important considerations in selecting the right material is it's resistance to various chemicals. The resistance of plastics to various chemicals is dependent on time of contact with chemicals, temperature, molded-in or external stress, and concentration of the chemical. Part design and processing practices play a major role in material's ability to withstand chemical attack. For example, stress concentration factor increases significantly for the parts designed with radius to wall thickness ratio of less than 0.4. As a rule, crystalline polymers are more resistant to chemicals when compared to amorphous polymers (and therefore if the application requires the parts to be constantly exposed to chemicals, crystalline materials should be given serious consideration. Chemical exposure to plastic parts may result in physical degradation such as stress cracking, softening, swelling, discoloration, and chemical attack in terms of reaction of chemicals with polymers and loss of properties.

7.10.5 Environmental Considerations

Plastic materials are sensitive to environmental conditions. Environmental considerations include exposure to UV, IR, X-ray, high humidity, weather extremes, pollution from industrial chemicals, micro-organisms, bacteria, fungus, and mould. The combined effect of various factors may be much more severe than any single factor and the degradation process is accelerated many times. It is very important to understand that the published test results do not include synergistic effects of various environmental factors, which almost always exist in real life situations. Designers should consider exposing fabricated parts to environmental extremes much similar to the ones encountered during the actual use of the products.

7.10.6 Regulatory Approval Requirements

Material selection may be driven by the regulatory requirements put forth by agencies such as Underwriters Laboratories (UL), National Sanitation Foundation (NSF), Food and Drug Administration (FDA) in terms of flammability, pressure ratings, and toxicological considerations.

7.10.7 Economics

Material selection should not be based completely on the cost considerations. A systematic approach to the material selection is essential in order to select the best material for any application. The most logical approach calls for choosing 3 to 4 most suitable materials based on the requirements and then apply the cost consideration to select the lowest cost material.

7.10.8 Other Considerations

Material selection approach should give due consideration to the type of fabrication process, secondary operations and component assembly.

   For a design engineer it is important to use all the information about the behaviour of different polymers under (i) the application of various stresses under static and dynamic conditions, (ii) various operating conditions like temperature and pressure, (iii) varying electrical field and (iv) different manufacturing processes for the same component. In addition to these factors it is also important to consider the solubility, compatibility and swelling characteristics in different industrial solvents and different mechanical properties as pointed out above (like tensile strength, tear strength, abrasion, heat build up, impact and bursting strength, fatigue behaviour and stress-strain characteristics). Information about the dimensional stability, expected service life, air and gas permeability, inter-phase migration of additives etc., toxicity and hazardous nature and cost are equally important in arriving at an appropriate choice of the polymer for a particular application.

7.11 POLYMERS IN AUTOMOBILE INDUSTRY

Body panels-materials for the body panel must have sufficiently high softening temperature to withstand the conditions in the paint oven. A number of polymeric materials are being tried such as glass fiber reinforced polyester resins, reaction injection moulded polyurethane etc.

   (a) *Rear light lenses:* The material for these must have high transparency therefore PMMA is the choice for these.
   (b) *Head lamps*: Material for these must have high softening temperature and high fracture resistance, therefore PMMA is unsuitable, and polycarbonate may be a possible choice.

(c) *Radiator header tanks:* Resistance to the high temperature is the main consideration therefore the glass fiber filled nylon is selected for this purpose.

(d) *Cooling fans:* Both resistance to temperature and strength under dynamic conditions are important and nylons and polyethylenes have been used.

(e) *Engine mounts:* The vibration isolation and flexibility are the important considerations over a wide range of temperature therefore natural rubber with excellent damping behaviour is the choice.

(f) *Tire tread:* For the tire tread the important considerations are high wear and abrasion resistance, low heat build-up, high fatigue to failure resistance, high cut growth resistance, high chipping and chunking resistance, high wet grip and low skid character, low compression set and high resilience. No single rubber is capable of meeting all these requirements. Therefore the blend of different rubbers like polybutadiene (BR) rubber having high wear resistance, styrene-butadiene rubber (SBR) with high wet gripping and natural rubber (NR) with high resilience are used in different proportions.

(g) *Tubes:* Tubes used for the inflation of tires use the compressed air for inflation. These are enclosed within the tire and do not come in direct contact with the road therefore have no problem with respect to wear and tear, chipping and chunking etc. The most important consideration for the tubes is the air permeability and flexibility. Therefore air impermeable rubbers like butyl rubbers are the obvious choice.

(h) *Storage containers*: Industrial containers are required for the storage of the raw material and finished products of varying characteristics. The materials may be neutral non-interacting type or may have corrosive nature. The containers are expected to last minimum 5 to 7 years and therefore must not deform or fail under the weight of the material contained. Apart from the good tensile strength the most important consideration to be given in selecting the polymeric material for these applications is the tensile creep compliance. Within this time frame the materials like polydimethylsiloxane and polyisobutylene, which show terminal region are likely to show high creep compliance and therefore are likely to deform and are unsuitable for the containers. On the other hand uncross-linked natural rubber though shows a plateau region, its thermal sensitivity and oxidative degradations are highly negative parameters, which make it undesirable for the containers. The cross-linking will improve these behaviours but it will be soft like a ball and therefore will be unsuitable. However the materials like polystyrene, polymethylmethacrylate and polyethylene show the glassy region and therefore are structurally highly stable giving practically no or insignificant deformation and therefore are most suitable for the storage vessels.

(i) *Non-stick coating:* Teflon is used as it can withstand high temperature.

(j) *Raincoats:* Plasticized PVC is used as it behaves like rubbery material at room temperature but the crystallites present in it prevent any permanent deformation.

(k) *Furniture*: Highly crystalline isotacticpolypropylene due to its high modulus and high gloss is preferred.

(l) *Contact lenses:* Grafted polymethyl methacrylate, which is biocompatible is used for this purpose.

(m) *Electrical insulation:* Materials like Teflon and PVC with low dielectric loss tangents are the most suitable for such applications.

(n) Rellegadla et al.[136a] have reported on the polymers for enhanced oil recovery.

(o) Seymore and Carraher[133] reported about the selection of polymers for some special applications ranging from enhanced oil recovery to components for aerospace vehicles as one can predict properties of polymers to some extent from the knowledge of thair structure.

(p) A. Nussinovitch,[136b] has reported on "Beads and special applications of polymers for agriculture uses" on bead encapsulation of micro organisms for the use in bacterial inoculation technology.

7.12 POLYMERS FOR ENHANCED OIL RECOVERY

Rellegadla et al.[136a] have presented a mini review of Polymers for enhanced oil recovery: fundamentals and selection Criteria. Due to increase in human population the demand for energy has increased over the years. As per the projections, both fossil fuel and renewables will remain as major energy source (678 quadrillion BTU) till 2030 with fossil fuel contributing 78% of total energy consumption. Hence, attempts are continuously made to make fossil fuel production more sustainable and cheaper. From the past 40 years, polymer flooding has been carried out in marginal oil fields and have proved to be successful in many cases. The common expectation from polymer flooding is to obtain 50% ultimate recovery with 15 to 20% incremental recovery over secondary water flooding. Both naturally derived polymers like xanthan gum and synthetic polymers like partially hydrolyzed polyacrylamide (HPAM) have been used for this purpose. Earlier laboratory and field trials revealed that salinity and temperature are the major issues with the synthetic polymers that lead to polymer degradation and adsorption on the rock surface. Microbial degradation and concentration are major issues with naturally derived polymers leading to loss of viscosity and pore throat plugging. Earlier studies also revealed that polymer flooding is successful in the fields where oil viscosity is higher (up to 126 cp) than injection water due to improvement in mobility ratio during polymer flooding. The largest successful polymer flood was reported in China in 1990 where both synthetic and naturally derived polymers were used in nearly 20 projects. The implementation of these projects provides valuable suggestions for further improving the available processes in future.

7.13 ADVANCES IN APPLICATIONS OF POLYMERS IN DIFFERENT AREAS

Research is in progress to identify the suitable polymers for applications in varying area like:

(a) Polymer selection for commonly and uncommonly used natural fibers under uncertainty environment.[140]

(b) Deposit Control Polymer Selection Criteria for High-Temperature Applications.[141]

(c) Polymeric materials for metallic bipolar plates for polymer electrolyte fuel cell.[142]

(d) Polymer selection for and cell design for electric-vehicle super-capacitors.[143]

(e) Role of polymers in water treatment applications and criteria for comparing alternatives.[144]

(a) Thermoplastic

1. *Acrylonitrile-butadiene-Styrene (A-B-S):* Refrigerator lining, lawn and garden equipment, toys, high way safety devices.

2. *Acrylic (poly-methyl-methacrylate):* Lenses, transparent air craft enclosures, drafting equipment, outdoors signs.

3. *Fluorocarbons (PTFE or TFE):* Anticorrosive seals, chemical pipes and valves, bearings, anti adhesive coatings, high temperature electronic parts.

4. *Polyamides (nylons):* Bearings, gears, cams, bushings, handles, and jacketing for wires and cables.

5. *Polycarbonates:* Safety helmets, lenses, light globes, base for photographic film.

6. *Polyethylene:* Flexible bottles, toys, tumblers, battery parts, ice trays, film wrapping materials.

7. *Polypropylene:* Sterilizable bottles, packaging film, TV cabinets, luggage.

8. *Polystyrene:* Wall tiles, battery cases, toys, indoor lighting panels, appliance housings.

9. *Polyester (PET or PETE):* Magnetic recording tapes, clothing, automotive tire cords, beverage containers.

(b) Thermosetting Polymers

1. *Epoxies:* Electrical moldings, sinks, adhesives, protective coatings, used with fiberglass laminates.

2. *Phenolics:* Motor housing, telephones, auto distributors, electrical fixtures.

7.13.1 Cyclic Olefin Polymers: Emerging Materials for Lab-on-a-Chip Applications[146]

Miniaturisation of chemical analysis systems employing microfluidic handling enables rapid and low cost analysis that generate little chemical waste[147]. While designing the microfluidic systems it is important to consider the type of material used to fabricate the device. The material used must be compatible with different chemicals being used, withstand the temperature of test, should be consistent with biochemical and biophysical protocol and methods applied as well as with functional components involved in the miniaturised analysis. The cost of production of the device should be of the level to eneble its mass production.[148]

It is vital to carefully select the polymer material for the fabrication purpose as well as for the chip. Some of the most used polymers for the purpose of fabricating the microfluidic devices are : Poly(dimethyl-siloxane) (PDMS).[149, 150, 151]; Poly(methylmethacrylate) (PMMA)[152, 153]; Polycarbonate (PC)[154, 155]; Polyester[156]; Polystyrene (PS)[157] and SU-8[158, 159, 160]. SU-8 is the material that can be easily strucured using standard lithography technique, with features ranging from a few to hundreds of micrometers[161, 162]. However it need to be noted that SU-8 has high optical losses in *UV* region and high water permeability. PDMS is probably the most popular polymer for soft lithography due to easy fabrication and bonding.

Cyclic olefin polymers (COPs) are increasingly popular as substrate material for microfluidics. This is due to their promising properties, such as high chemical resistance, low water absorption, good optical transparency in the near *UV* range and ease of fabrication. COPs are commercially available from a range of manufacturers under various brand names (Apel, Arton, Topas, Zeonex and Zeonor). Some of these (Apel and Topas) are made from more than one kind of monomer and therefore also known as cyclic olefin copolymers (COCs). In order to structure these materials, a wide array of fabrication methods is available. Laser ablation and micromilling are direct structuring methods suitable for fast prototyping, whilst injection moulding, hot embossing and nanoimprint lithography are replication methods more appropriate for low-cost production. Using these fabrication methods, a multitude of chemical analysis techniques have already been implemented. These include microchip electrophoresis (MCE), chromatography, solid phase extraction (SPE), isoelectric focusing (IEF) and mass spectrometry (MS). Still much additional work is needed to characterise and utilise the full potential of COP materials. This is especially true within optofluidics, where COPs are still rarely used, despite their excellent optical properties.

7.14 POLYMERS IN SPACE APPLICATIONS[163]

Polymers are used in a number of applications in the space systems - space craft, space station - like thermal control coatings, adhesives, tapes, potting compounds, toughening/damping materials, seals, thin film substrates, thermal insulations in addition to composite structures. Space research demands very stringent requirements from these polymers because of the extreme environments experienced by these materials[164]. The polymers used outside the space system are the most affected ones. As we move from space exploration to space commercialization, these polymers have to last for 15-20 years of their designed life. Operational requirement needs not only these materials do their job but also have low outgassing to prevent contamination of surrounding optical systems. Hence, there is a need for understanding the effect of space environments on the polymeric materials in use and improvement/synthesis of new polymeric materials. Table below gives some damaging effects of space environment upon polymeric materials.

Table 7.2 Some Damaging effects of space environment upon polymeric materials[163]

Environment	Damage
Ultraviolet radiations	(a) Creation of lattice defects in crystalline materials (b) Chain scission of organic materials, free radical formation (c) Crosslinking of organic materials
Charged particle radiation	(a) Creation of lattice defects in crystalline materials (Electron/proton) (b) Chain scission of organic materials (c) Secondary radiation damages
Vacuum	(a) Volatilization of low vapor pressure fractions and material (b) Dffusion; and (c) Vaccume Welding

Environment	Damage
Thermal	(a) Mechanical degradation; softening or embrittlement; (b) Chemical degradation; (c) Acceleration/deceleration of above environmental effects.
Micrometeoroid	(a) Mechanical; (b) Fracture/punchure
Atomic oxygen	(a) Oxidation; (b) Surface erosion and (d) Cracking/crazing

7.15 SMART POLYMERS[165]

Smart polymers are the polymers that respond to different stimuli or changes in the environment. Different types of smart polymers are:

(a) Temperature-responsive polymers;

(b) pH-responsive polymers;

(c) Photo-responsive polymers;

(d) Magnetically responsive polymer gels and elastomers;

(e) Enzyme-responsive polymers;

(f) Shape memory polymers;

(g) Smart polymer hydrogels;

(h) Self healing polymer systems;

7.15.1 Applications of Smart Polymers

(i) Smart intrusive polymer substrates for tissue engineering

(ii) Smart polymer nanocarriers for drug delivery,

(iii) The use of smart polymers in medical devices for minimally invasive, diagnosis and other applications,

(iv) Smart polymers for bioseparation and other biotechnological applications,

(v) Smart polymers for textile applications,

(vi) Biopolymers for food packaging applications,

(vii) Smart polymers for optical data storage.

In addition polymer composites find applications in monocoque structures (A structure in which all structural loads are carried by the skin) for aerospace and automobiles as well as some more mundane products like fishing rods and bicycles. The stealth bomber was first all composite air craft but many passenger aircraft like Airbus and the Boeing 787 use an increasing proportion of composites in their fuselages, such as hydrophobic melamine foam[166]. The very different physical properties of composites give designers a much greater freedom in shaping parts , which is why composite products look different to conventional products. On the other hand, some products such as drive shaft , helicopter rotor blades, and propellers look identical to metal precursors owing to the basic functional needs of such components

These are only some of the numerous applications for the polymers in domestic and industrial applications. The final selection after short listing, based on the properties and strength considerations is always based on the economic evaluations.

REFERENCES

1. A. D. Drozdov and A. Dorfmann, Acta Mechanica, **154(1-4)**,189-214 (2002).
2. A. D. Drozdov, S. Agarwal and R. K. Gupta, Int. J. Eng. Sci. **43(3-4)**, 304-320 (2005).
3. A. D. Drozdov, S. Agarwal and R. K. Gupta, J. Appl. Polym. Sci. **94(1)**, 9-23 (2004).
4. A. D. Drozdov, Mechanics of Time-Dependent Materials, **14(4)**, 411-434 (2010).
5. V. Hiremath, M. Singh, D. K. Shukla, Procedia Engineering , **97**, 479-487 (2014).
6. S. Pandini and A. Pegoretti, Polym. Eng. Sci., published online in Wiely Interscience (www.interscience.wiley.com)© 2008 Society of Plastic Engineers.
7. D.Y. Papadogiannis, R. S. Lakes, Y. Papadogiannis, G. Palaghias, M. Helvatjoglu-Antoniades, Dental Materials, **24**, 257-266 (2008).
8. N. Kao,and S. N. Bhattacharya , R. Shanks and I. N. Coopes, J. Rheol. **42**, 493 (1998).
9. P. Driscoll, K. Fujiwara, T. Masuda, Afurukawaand R. W. Lenz, Polymer J.**20**, 351-356 (1988).
10. J. M. L. dos Reis, Materials Reseach, 2012. [scielo.br].
11. M.-R. Caputo, J. Selb, and F. Candau, Polymer, **45**, 231-240 (2004). [deepdyve. com].
12. A. Kumar and S. K. Gupta, Fundamentals of Polymer Science and Engineering, Tata McGraw Hill Pub. Co. Ltd., New Delhi, 1978, pp 442-444.
13. N. G. McCrum, C. P. Buckley and C. B. Bucknall, Principles of Polymer Engineering, Oxford Univ. Press, New York, 1997, p 137.
    (a) Vishu Shah, John Wiley & Sons, Kings Plstic Corporation, December 11, 2012.
14. K. H. Illers and E. Jenckel, Rheol. Acta, **1**, 322 (1958).
15. K. H. Illers and E. Jenckel, J. Polym. Sci., **41**, 528 (1959).
16. K. M. Sinnott, Soc. Plas. Engrs. Trans., **2**, 65 (1962).
17. N. G. McCrum and B. E. Read and G. Williams, Anelastic and Dielectric effects in Polymer Solids, 1ˢᵗ Ed., John Wiley & Sons Inc. New York, 1967.

18. L. E. Nielsen, Mechanical Properties of Polymers, 1st ed. Van Nostrand Reinhold Co. New York, 1962.

19. E. Catsiff and A. V. Tobolsky, J. Collod Sci, **10**, 375 (1955).

20. E. Catsiff and A. V. Tobolsky, J. Polym. Sci, **19**, 111 (1956).

21. L. A. Holmes, S. Kusamizu, K. Osaki and J. D. Ferry, J. Polym. Sci. **9** (A2), 2009 (1971).

22. S. Onogi, T. Masuda and K. Kitagawa, Macromolecules, **3**, 109 (1970).

23. R. H. Colby, L. J. Fetters, W. G. Funk and W. W. Graessley, Macromolecules, **24**, 3873-3882 (1991).

24. M. Barmer, V. Ribitsch, B. Kaffashi, M. Barikani, Z, Sarreshtehdari and J. Pfragner, Colloid and Polym. Sci. **282(5)**, 454-460 (2004).

25. Jing Tao, "Effects of Molecular Weight and Solution Concentration on Electrospinning of PVA". A Master of Science Thesis submtted to the faculty of the Worcester Polytechnic Institute. June 2003.

26. Y. Zhang, X Xu, J. Xu and L. Zhang, Polymer, **48(22)**, 6681-6690 (2007).

27. K. Osaki, T. Inoue and T. Uematsu, J. Polym. Sci. Part B: Polym. Phys., **39 (2)**, 211-217 (Jan. 2001)

28. C. Clasen, M. Verani, J. P. Plog, G. H. McKinley and W.-M. Kulicke, Proc. XIVth Congr. On Rheology, August 22-27, 2004 Seoul, Korea.

29. M. Verani," Effect of Polymer Concentration and Molecular Weight on the dynamics of Visco-Elasto-Capillary Breakup", A thesis submitted to the Department of Aeronautics and Astroautics, at MIT, in Jan. 2004.

30. R. K. Gupta, D. A. Nguyen and T. Sridhar, Physics of Fluids, **12**, 1296 (2000).

31. W. W. Graessley, Polymer, **21(3)**, 258-262 (1980).

32. A. V. Tobolski, J. J. Alkonis and G. J. Akovali, J. Chem. Phys., **42** , 723 (1965).

33. E. Bagley and D. West, J. Appl. Phys., **29**, 1511 (1958).

34. G. C. Berry and T. G. Fox, Adv. Polym. Sci., **5**, 261 (1968).

35. S. Onogi, T. Kobayashi, Y. Kojima and Y. Taniguchi, J. Appl. Polym. Sci., **7**, 847 (1963).

36. E. Maekawa, R. G. Mancke and D. J. Ferry, J. Phys. Chem., **69**, 281 (1965).

37. D. J. Plazek, J. Polym. Sci., **4** (A2), 745 (1966).

38. M. F. Drumm, C. W. H. Dodge and N. E. Nielsen, Ind. Eng. Chem., **48**, 76 (1956).

39. T. Indei, J. D. Schieber, Jun-ichi Takimoto, Rheologica Acta, vol. **51**, issue:11-12, (2012).

40. K. V. Rao, G. S. Ananthapadmanabha and G. N. Dayanada, J. Mat. Eng. And Performance, **25** (12), 5314-5322 (2016).

41. M. Krumova, D. Lo'pez, R. Benavente, C. Mijangos, and J. M. Perena, Polymer, **42** (26) 9265-9272 (2000).

42. D. J. Plazek, J. Polym. Sc. Part B: Polymer Physics, **4** (5), 745-763 (1966).

43. R. Hagen, L. Salme'n and B. Stenerg, J. Polym. Sc. Part B: Polymer Physics, **34** (12), 1997-2006 (1996).

44. E. A. Nielsen, MONSANTO RESEARCH CORP ST LOUIS MO, " Crosslinking-Effects on Physical Properties Of Polymers", May 1968. PDF Url : AD0834683.

45. G. W. Becker, Kollioid Z., **175**, 99 (1961).

46. H. Nakayashu, H. Markovitz and D. J. Plazek, Trans. Soc. Rheol., **5**, 261 (1961).

47. L. Olasz and P. Gudmundson, Mechanics of Time-Dependent Materials, **9(4)** 23-44 (2005).

48. T. Sakai, Y. Hirai and S. Somiya, Time Dependent Constitutive Behaviour and Fracture/Failure Process, **3**, 159-164 (April 2011, online).

    (a) T. Sakai, Y. Hirai and S. Somiya, Mechanics of Advanced Materials and Modern Processes, **4(5)**, 2018.

49. K. Yamada and S. Somiya " Effect of Crystallanity on Creep behaviour of GFRPOM" An article presented online at [sem.org]. (Also Society of Experimental Mechanics, **2** , 980-985 (2008).

50. A. Tanaka, H. Nagano, M. Shinohara and S. Onogi, Polymer J., **20**, 987-1002, (1988).

51. B. Feng, X Fang, H-X. Wang, W Dong and Y-C Li, Polymers, **8**, 356 (2016).

52. Y. T. Sung, C. K. Kum, H. S. Lee, J. S. Kim, H. G. Yoon, and W.N. Kim, Polymer, **46(25)**, 11844-11848 (2005).

53. A. A. Armsgan, " Research Center for Plastic Materials, Univ.of Kocaeli, Kocaeli/ TURKEY. An online article [iccm-central.org].

54. P.T., Curtis, Davies, P., Partridge, I.K., Elsevier App. Sci. Publ., Barking, p.4. 401 (1987)

55. J., Mijovic, Gsell, T.C., SAMPE Q, **21** (2) pp.42 (1990) 56. Olley, R.H., Bassett, D.C., Blundell, D.J., Polymer, **27**, 344, (1986)

57. J.T., Hartness, SAMPE J, **20,** pp.26, (1986)

58. J.E., Manson, Schneider, T.L., Polym.Composites, **11**,114, (1990).

59. S., Saiello, Kenny, J., Nicholais, L., J.Mater.Sci., **25**, 3493, (1990)

60. S. Incardona, et al., Comp.Sci.Tech., 47, 43, (1993).

    (a) R. Khalil, A. G. Chryss, M. Jollands and S. Bhattacherya, J. Materials Sci., **42(24)**, 10219-10227 (2007).

61. N. Dusunceli and O. U. Colak, International J. of Plasticity, **24(7)**, 1224-1242 (2008).

62. M. Toft, " The Effect of Crystalline Morphology on Glass Transition and Enthalpic Relaxation in Poly(Ethjer Ether-Ketone), A Thesis submitted to the Univ. Birgingham for thr degree of Master of Research in Science and Engineering of Materials., Sept. 2011.

    (a) Z. Bartczak and A. Galeski, "Plastycity of Semicrystalline polymers", August 2010. [researchgate.net].

63. K. Osaki and J. L. Schrag, J. Polym. Sci. Polym. Phys. Ed., **11**, 549 (1973)

64. K. Osaki, Y. Mitsuda, J. L. Schrag and J. D. Ferry, Trans. Soc. Rheol., **18**, 395 (1974).

65. T. Fujimoto, H. Narukawa and M. Nagasawa, Macromolecules, **3**, 57 (1970).

66. T. Masuda, Y. Nakagawa, Y. Ohta and S. Onogi, Polym. J., **3**, 92 (1972).

67. A. Mortimer, J. Appl. Polym. Sci., **15**, 1231 (1971).

68. P. M. Wood-Adams, J. M. Dealy, A.W. deGroot and O.D. Redwine, Macromolecules, **33(20)**, 7489-7499 (2000).

    (a) E. Catsiff and A. V. Tobolski, J. Appl. Phys., **25**, 1092 (1954)

    (b) M. Nakatani, K. Iijima, A. Suganuma and H. Kawai, J. Macromol. Sci.- Phys., **B2**, 55 (1968).

69. (The Free Library)"Effect of short chain branching on the viscoelastic behavior during fatigue fracture of medium density ethylene copolymers." The Free Library. 1995 Society of Plastics Engineers, Inc. 12 Sept, 2017.

70. A. Moet, Proc. Benibana International Symposium, p. 255, Yamagata University, Yamagata, Japan (1990).

71. K. Chaoui, J. J. Strebel, A. Chudnovsky, and A. Moet, Proc. 1989 International Gas Research Conference, p. 44. T. L. Cramer, ed., Tokyo, Japan (1989).

72. M. L. Kasakevich, A. Moet, and A. Chudnovsky, J. Macromol. Sci.-Phys., **B28** (3&4), 433 (1989).

73. K. Chaoui, PhD thesis, Case Western Reserve University, Cleveland (1989).

74. J. Krey, K. Friedrich, and A. Moet, Polymer, **29**, 1433 (1988).

75. J. J. Strebel, PhD thesis, Case Western Reserve University, Cleveland (1992).

76. N. Brown and X. Lu, 1989 GRI Annual Report, Gas Research Institute, Chicago (1989).

77. N. Brown and X. Lu, 1990 GRI Annual Report, Gas Research Institute, Chicago, (1990).

78. J. F. Vega, J. Ramos and J. Martinez-Salazar, Rheologica Acta **53**, 1-13 (2014).

79. A. A. Askadskii and O. V. Kovriga, Polymer Science U. S. S. R., **33(9)**, 1821-1831 (1991).

80. C. Former, J. Castro, C. M. Fellows, R. I. Tanner and R. G. Gilbert, J. Polym. Sci. Part A: Polymer Chemistry, **40 (20)**,3335-3349 (2002).

81. J. den Doelder, C. Das and D. J. Read, Rheologica Acta, **50 (5-6)** 469-484 (2011).

82. Von D. Heinze, Macromol. Chem., **101**, Issue 1, 166–187, March 1967.

83. G. Kraus, C. W. Childers and J. T. Gruver, J. Appl. Polym. Sci., **11**, 1581 (1967).

84. P. Ghosh, T. Das, and D. Nandi, Am. J. Polym. Sci., **1(1)**, 1-5 (2011).

85. A. Y. C. Koh, S. Mange, M. Bothe, R. L. Leyrer and R. G. Gilbert, Polymer, **47**, 1159-1165 (2006).

86. M. Riahinezhad, N. McManus, and A. Pendilis, Polym. Reaction Eng. IX, ECI Sympsium Series, (2015).(http://dc.engconfintl.org/polymer_rx_eng_IX/36).

87. S. E. Diaz-Silvestre, C. St. Thomas, C. Rivera-Vallejo, G. Cadenas-Pliego, M. Perez-Alvarez, R. D.de Leon-Gomez, E. J. Jimenez-Regalodo, Polimer Bulletin, 74(10), 4009-4021 (2017).

88. Y. Jahani, M. Farahani, H. Arabi and S. Ahmadjo, J. Macromol. Sci. **52 (7)**, 532-539 (2015).

89. M. Liao, Q. Wang, N. Wang, L. Xu, C. Li and A. Liang, Polymer Sci. Series B, **56(6),** 753-761 (2014).

90. A. Roos and C. Creton, Macromolecules, **38(18), 7807-7818** (2005).

91. A. Garcia-Rejon, L. Rios and L. A. Lopez-Latorre, [researchgate.net], 1987.

92. W. M. Prest and R. S. Porter, J. Polym. Sci., **10 (A2)**, 1639 (1972).

93. H. Y. Kim and I. J. Chung, J. Polym. Sci. Part B; Polym. Physics, **25(10)**, 2039-2957 (1987).

94. D. J. Read, K. Jagannathan S. K. Sukumaran and D. Auhl, J. Rheology, **56(4)**, 823 (2012).
    (a) Graham et al. J. Rheol., **47**, 1171 (2003)

95. D. Auhl, P. Chambon, T C. B. Mcleish and D. J. Read, Physical Review Letters, PRL, **103**, 136001 (Sept. 2009)

96. M. van Gurp and J. Palmen, " Time-Temperature Superposition for Polymeric Blends", DSM Research, Geleen, Netherlands.

97. M. L. Williams, R. F. Landel and J. D. Ferry, J. Amer. Chem. Soc., **77(14)**, 3701-3707 (1955).

98. D.J. Plazek, J. Rheol., **49**, 987 (1996).

99. http://goldbook.iupac.org/CT07581.html

100. L. A. Utracki, Polymer Alloys and Blends, Hanser, Munich, 1989.

101. C. Tsenoglou in: New Trends in Physics and Physical Chemistry of Polymers, L.-H. Lee(ed.), Plenum Press, New York, pp 375-383, 1989.

102. R. H. Colby, Polymer, **30**, 1275 (1989).

103. B. H. Arendt, R. M. Kannan, M. Zewail, J. A. Komfield and S. D. Smith, Rheol. Acta, **33**, 322 (1994)

104. R. Hagen and R. A. Weiss, Polymer, **36**, 4657 (1995).

105. S. K. Kumar, R. H. Kolby, S. H. Anastasiadis and G. Fytas, J. Chem. Physc., **105**, 3777 (1996).

106. H. Watanabe and T. Kotake, Macromol.**16**, 769-774, (1983).

107. C. Wisniewski, G. Marin and Ph. Monge nd, Eur. Polym. J. **21**, 479-484 (1985).

108. Y. S. Lipatov, JAPS, **26**, 499 (1981).

109. N. V. Cassagneau et al. JAPS **58**, 1393, (1995).

110. C. D. Han and J. K. Kim, Polymer, **34**, 2533-2539 (1993).
    (a) Prince Antony, A. K. KBhattacharya and S. K. De, J. Appl. Polym. Sci., **71**, 1257 (1999).
    (b) Swapan K. Ghosh, P. P. De, D. Khastagir and S. K. De, J. Appl. Polym. Sci., **78**, 743 (2000).
    (c) Prince Antony and S. K. De, J. Appl. Polym. Sci., **70**, 483 (1998).
    (d) Swapan K. Ghosh, P. P. De, D. Khastagir and S. K. De, J. Appl. Polym. Sci., **78**, 326 (2000).

111. M. Fujimura, Ivakura, and Kobunshi Ronb., Eng. Ed. **3**, 1864 (1974).

112. C. Feng, Y. Li, J. D. Yang, J. Hu, X. Zhang and X. Huang, Chem. Soc. Reviews, **40 (3)**, 1282-95 (2011).

113. K. Matyjaszewski, " Graft Copolymers, March, 2014.

114. E. M. Pearce, J. Polym. Sci. Part C: Polym. Letters, **25(5)**, 233-234 (1987).

115. al. J. Volker Abetz, {Encyclopedia of Polymer Science and Technology (Wird aktualisiert. ed.)[Hoboken N. J.]: Wiley- Interscience. (ed. 2005).

116. Y. Jahani, F. S. Hosseini and N. Goodarzian, "reserchgate.net", July 2014.

117. Y. Lin, Y. Gu, Z. Chen, S. Zhang, G. Zhang, Y. Wang, H. Xing and T. Tang, Polymer, **122 (28),** 87-95 (July 2017).

118. B. Wong and W.E. Baker, Polymer, **38 (11),** 2781-2789 (1997).

119. B. Lu, K. Lamnawar and A. Maazouz, polymer testing, **61,** 289-299 (2017).

120. M. A. Vargas, A. Sa'nchez, G. Guthausen and O. Manero, Intl J. Pavmt. Eng., **16 (8),** 730-744 (2015).

121. C. Sailer, M. Weber, H. Steininger and U. A. Handge, Rheologica Acta, **48,** 579 (2009).

    (a) V. H. Pino-Ramos, A. Ramos-Ballesteros, F. Lo'pez-Sausedo, J. E. Lo'pez-Barriguete, G. H. C. Barca and E. Bucio,Top Curr Chem (Z), **374,** 63 (2016). [This article is a part of topical collection"Application of Radiation Chemistry". Ed. Margherita Venturi, Mila D' Angelantonio.

122. R. F. Landel, Trans. Soc. Rheol., **2,** 53 (1958).

123. S. Onogi, T. Matsumoto and Y. Warashina, Trans. Soc. Rheol., **17,** 175 (1973).

124. A. R. Pyne, in Rheology of Elastomers, edited by P. Mason and N. Workey, 1st ed., Pargamon Press Ltd., London, 1958

125. H. S. Katz and J. V. Milewski (eds) " *Handbook of fillers and reinforcments",* von Nostrand Reinhold, New York (1978).

126. Aroon V. Shenoy " *Rheology of filled polymer systems"* Kluwer Academic Publishers, Dordrecht/Boston / London.

127. F. W. Maine, B. E. Riseborough and J. E. Theberge, Polymer Structures and Properties, SPE, RETEC, Toronto (1976).

    (a) Hans-Georg Ellias, "Plastics, General survey" in Ulmann's Encyclopedia of of Industrial Chemistry, 2005, Wiley-VH, Weinheim.

    (b) B. C. A. Silvera, R. G. Larson, and A. N. Kottov, Science, **350,** 6257, (2015) :1242477. [en.wickipedia.org].

    (c) EPA Particle Size Terminalogy; Module 3; Characteristics of particles-Particl size catagories. Epa.gov. [http://www.epa.gov/apti/bces/module3/category/ category.htm]

    (d) D. A. MacNaught, D. Alan , and R. A. WillKinson, eds., Compedium of Chemical Terminology: IUPAC Recommendation (2nd e. Blackwell Scince, (1997 ed.).

    (e) *Alemán, J.; Chadwick, A. V.; He, J.; Hess, M.; Horie, K.; Jones, R. G.; Kratochvíl, P.; Meisel, I.; Mita, I.; Moad, G.; Penczek, S.; Stepto, R. F. T. (2007). "Definitions of terms relating to the structure and processing of sols, gels, networks, and inorganic-organic hybrid materials (IUPAC Recommendations 2007)". Pure and Applied Chemistry. 79 (10), 1801.*

    (f) *C. Buzea, I. I. Pacheco, and K. Robbie, Biointerphases,* ***2(4),*** *MR17-MR71 (2007).*

    (g) *ASTM E 2456-06, Standard Terminology Relating to Nanotechnology.*

    (h) *B. D. Fahlman, Materials Chemistr. Springer. Pp282-283 (2007). ISBN 1-4020-6119-6.*

       1. *S. Liang, C. L. John, S. Xu, and J. X. Zhao, Chapter in Book " Advanced Fluorescence Reporters in Chemistry and Biology II", PP229-251, January, 2010.*

2. *S. S. Ray, and M. Okamotop, Progress in Polymer Science, 28, 1-63 (2000)*

(i) Ph. Buffat and J. P. Borel, Physical review **A.13(6)**, 2287-2298 (1976).

(j) J. Wu, P. Yu, S. A. Susha, A. K. Sablon ans A. Kimberlay, H. Chen, Z. Zhou, H. Li, H. Ji, and X. iu; Nano Energy, **13**, 827-835 (April,2015).

(k) A. R. Taylor, E. P. Phelan, E. Patric, P. T. Onttanicar, P. Todd, R. Adrian, and R. Prasher, Nanoscale Research Letters, **6(1)**, 225 (2011).

(l) A. R. Taylor, T. Otanicar and G. Rosengarten, Light: Science and Applications, **1(10)**, e34, (2012).

(m) A. R. taylor, P. T. Otanicar, Y. Herukerrupu, F. Bremond, G. Rosengarten, R. E. Hawkes, X. Jiang and S. Coulombe, Applied Optics, **52(7)**, 1413-22 (2013).

(n) P. Yu, Y. Yao, J. Wu, X. Niu, A. L. Rogach, and Z. Wang, Scientific Reports, **7(1)**, (August, 2017).

(o) Y. L. Hewakurruppu, A. L. Dombrovsky, C. Chen, V. Timchenko, X. Jiang, S. Baek, and R. A. Taylor, Applied Optics, **52(24)**, 6041-6050 (2013).

(p) P. S. Gubin, Magnetic Nanoparticles, (2009); Wiley-VCH ISBN3-527-40790-1.

(q) "The Textile Nanotechnology Laboratory", nanotextiles.human.crnnell.edu. (Dec.2016).

(r) R. Taylor, S. Coulumbe, T. Otanicar, P. Phelan, A. Gunawan, Lv Wei, G. Rosengarten, R. Prasher, H. Tyagi, J. Applied Physics, **113**, 011301(2013).

(s) A. Luchini, D. Geho, B. Bishop, D. Tran, C. Xia, R. Dufour; et al. Nao Letters, **8(1)**, 350-351 (2008).

(t) understanding nano.com

(u) W. H. De Jong and P JA Borm, Int. J.Nanomediciens, **3(20)**, 133-149 ((Jun-2008)

(v) Fangming Du, R. C. Scogna, W. Zou, S. Brand, J. E. Fischer and K. I. Winey. Macromolcules, **37**, 9048-9055 (2004).

(w) R. Krishnamoorti and E. P. Giannelis, Macromolecules, **30**, 4097-4102 (1997).

(x) Q. Zang and L. A. Archer, Lonmuir **18**, 10435-10442 (2002).

(y) K. Lozano, J. Bonilla-Riosand E. V. Barrera, J. Appl. Polym. Sci., **80**, 1162-1172 (2000).

(z) P. Potschke, T. D. Fornes, and D. R. Paul, Polimer, **43**, 3247-3255 (2002).

(A) E. P. Giannelis, R. Krishnamoorti and E. Manias, Advances in Polymer Science, **138**, 107- 147 (1999).

(B) N. S. Enikolopyan, M. L. Fridman, I. U. Stalnova and V. L. Popov, Adv. Polym. Sci. **96**, 1 (1990).

(C) W.S. Chow and Z. A. Mohd Ishak, xPress Polymer Letters, **1(2)**, 77-83 (2007).

(D) Chong Min Koo, Mi Jung Kim, Min Ho Choi, Sang Ouk Kim and In Jae Chung, J. Appli. Polym. Sci., **88**, 1526-1535 (2003).

(E) R. A. Vaia, B. B. Sauer, O.K. Tse, and E. P. Giannelis, J. Polym. Sci. **35**, 59 (1997)

(F) E. Hackett , E. Manias, and E. P. Giannelis, Chem.mater. **12**, 2161 (2000).

(G) J. Bujdak, E. Hackettand E. P. giannelis, Chem. Mater.**12**, 2167 (2000).

(H) R. A. Vaia, K. D. Jandt, E. J. Kramer, and E. P. Giannelis, Macromolecules, **28**, 8080, (1995).

(I) R. Krishnamoorti, R. A. vaia, E. P. Giannelis, Chem. Mater., **8**, 1728 (1996)

(J) R. Krishnamoorti, R. A. Vaia, E. P. Giannelis, macromolecules, **30**, 4097 (1997).

(K) R. A. Vaia, E. P. Giannelis, Macromolecules, **30**, 8000, (1997).

(L) J. Ren, A. S. Silva, and R. Krishnomoorti, macromolecules, **39**, 3739 (2000).

(M) B. Hoffmann, C. Dietrich, R. Thomas, C. fredrich, and R. Mulhaupt, Macro Rapid Commun. **21**, 57 (2000).

(N) Y. T. Lim, O. O. Park, Macro Rapid Commnu., **21**, 231, (2000).

(O) Y. T. Lim, O. O. Park, Rheol. Acta, **40**, 220 (2001).

(P) S. Hambir, N. Bulakh, P. Kodgire, R. Kolgaonkar P. J. Jog, J. Polym. Sci. Polymer Physics **9**, 446 (2001).

(R) S. T. Lim, H. J. Choi, and M. S. Jhon, J. Ind. Eng. Chem., **9(1)**, 51-57 (2003).

(S) Y. H. Hyun, S. T. Lim, S. J. Choi, and M. S. Jhon,Macromolecules, **34**, 8084 (2001).

(T) H. J. Choi, S. G. Kim, Y. H. Hyun, and M. S. Jhon, Macromolecule Rapid Commun., **22**, 20 (2001)

(U) J. K. Mishra, K-J. Hwang and Chang-Sik Ha, Polymer, **46**, 1995-2002 (2005).

(V) L. Incarnato, P. Scarfato, L. Scatteia' D. Acierno, Polymer, **45(10)**, 3487-3496, 2004.

(V) 1. R. K. Gupta and S. N. Bhattacharya, Koea-Australia Rheology J., **22**, 197-203 (2010)

(V) 2. M. M. Ready, R. K. Guptan, S. N. Bhattacharya and R. Parthasarthi, Koea-Australia Rheology J., **19(3)**, 133-139 (2007).

(V) 3. A. Dorigato, A. Pegoretti and A. Penati, eXRESS Polymer Letters, **4(2)**, 115-119 (2010).

(V) 4. S. S. Roy, and M. Okamoto, Macromol. Mater. Eng., **288(12)**, 937-944 (2003).

(V) 5. X. He, J. Yang, L. Zhu, B. Wang, G. Sun, P. Lv, I. Y Phang and T. Liu, J. Appl. Polym. Sci., **102**, 542-549 (2006).

(V) 6. M. E. Mackay, T. Dao, A. Tuteja, D. L. Ho, B. V. Horn, H-C. Kim and C. J. Hawker, Nature Materials, **2**, 762-766 (2003).

(V) 7. M. Okamoto, Polymer/layered silicate nonocomposites, Capter 3, October, pp. 57-78, 2009.

(V) 8. J. Li, C. Zhau, G. Wang, and D. Zhou, J. Appl. Polym. Sci., **89**, 318-323 (2003).

(V) 9. P. Nawani, P. Desai, M. Lundwall, M. Y. Gelfer, B. S. Hsiao, M. Rafailovich, A. Frenkel, A. H. Tsou, J. W. Gimanand S. Khalid, Polymer, **48**, 827-840 (2007).

(V) 10. J. Vermant, S. Ceccia, M. K. Dolgoskij, P.L. Maffettone and C. W. Macosko, J. Rheol. **51(3)**, 429-450 (May-June 2007).

(V) 11. D. Wu, C. Zhou, W. Yu and X. Fan., J. Polym. Sci., Part B, **43** 2807-2818 (2005).

(V) 12. H. Patel, R. S. Somani, H. C. Bajaj and R. V. Jasra, Bullt. Mater. Sci., **29(2)**, 133-145 (2006).

(V) 13. K. Okada, T. Mitsunaga and Y. Nagase, Korea-Australia Rheology J. **15(1)**, 43-50 (2003).

(W) B-T Zhang, X. Zheng, H-F Li, and J-H. Lin, Analica Chemica Acta, **784**, 1-17 (19, June, 2013).

(X) G. Schmidt, and M.M. Malwitz, Current Opinion in colloid and interface science, **8**, 103-108 (2003).

(Y) Y. Y. Huang, S. V. Ahir and E. M. Terentjev, Physical Review **B73**,125422 (2006).

(Y) 1. C. F. Schmid and D. J. Klingenberg, Phys. Rev. Lett., **84**, 290 (2000).

(Z) A. K. Kota, B. H. Ciprino, M. K. Duesterberg, A. lL. Gershon, D. Powell, S. R. Raghavan, and H. A. Bruck, Macromolecule, 40, 7400-7406 (2007).

(Z) 1. M. S. P. Shaffer and J. K. W. Sandler, Chapter 1, " Carbon Nanotube/ Nanofiber Polymer Composites".

(Z) 2. Y. Y. Huang and E. M. Terentjev, International J. of Material Forming, **1(2)**, 63-74 (2008).

(Z) 3. T. Chatterji and R. Krishnamoorti, "Rheology of polymer carbon nanotubes composites" October, 2013.

(Z) 4. D. Wu, J. Wang, M. Zhang and W. Zhou, Ind. Chem. Res. , **51 (19)**, 6705-6713 (2012).

(Z) 5. Q. Zhang, F. Fang, X. Zhao, Y. Li, M. Zhu and D. Chen, J. Phys. Chem. B, **112(40)**, 12606-12611 (2008).

(Z) 6. A. Rochfort and P. Avouris, Cond-Mat.mes-hall, **28** (April 1999).

(Z) 7. B. Kulshreshtha, V. Choughuleand D. A. Zhambrunnen, ANTEC, 441-445 (2006).

128. R. Rothan, "Particulate fillers for polymers", Research and Markets, February, 2003.

129. W. Hofmann, "Rubber Technology Handbook", Hanser Publishers. Munich, Vienna, New York, Oxford Univ. Press, 1989.

130. S. Brunauer, P. H. Emmet and E. Teller, J. Am. Chem. Soc. **60**, 309 (1938).

131. C. DeArmitt and R. Rothan, Polymers and polymeric Composites: A reference series pp 1- 26 (2nd May 2016).

132. S. Marz, " MENU Machine Design", Oct. 2016.

133. R. B. Seymor and C. E. Carraher Jr., Structure Property Relationship in Polymers, Springer, Boston, MA, 1984. ISBN 978-1-4684-4750-7.

134. J. Preston, G. P. Bierwagen, M. P. Stevence, G. B. Kaufman, F. Rodridguez, A. N. Gent, Major Industrial Polymers, Encyclopedia Britannica, Inc., 2020.

135. Essays, UK, (November 2018). Polymers and its applications. Retrieved from [https://www.ukessays.com/essays/chemistry/polymers-and-its-applications.php?vref=1]

136. Vishu Shah, Handbook of Plastic Testing and Failure Analysis, Third Edison, John Wiley & Sons 14 June 2006.
    (a) S. Rellledadia, G. Prajapat and A. Agarwal, Applied Microbiology and Biotechnology, **101**, 4387-4402 (2017).
    (b) A. Nussinovitch, " Polymer macro-and micro-Gel Beads: Fundamentals and Applications" pp 231-253, (August, 2010).

137. F. M. Al-Oqla and S. M. Sapuan, J. Clean. Prod., **66**, 347-354 (2014).

138. Zahid Amjad, Ph.D., and Robert W. Zuhl, P.E., The Lubrizol Corporation and John F. Zibrida, ZIBEX, Inc., the Analyst, **22(3)**, pp 40-46 (2015)

139. A. Shanian,and O. Savadogo, J. Power Sources, **159(2)**, 1095-1104 (2006)

140. M. Mastrgostino, CV. Arbizzani, R. Paraventi and A. Zanelli, Feb 2000.

141. [researchgate.net]

142. R. W. Zuhl, and Z. Amjad, "The role of polymers in water treatment Applications and Criteria for comparing alternatives", Lubrizol Advanced Materials Inc. (Cleveland., OH), 5th Annual Convention & Exposition, Water Technologies'93, 7 to10 November 1993., Caesars Palac, Las Vegas, Navada.

143. "Material Science", Prof. Satish V Kailas, Asso. Prof. Dept. Mech. Eng., II Sc. Bangalore-560012 India, Chapter 11. Applications and Processing of polymers.

144. P. S. Nunes, P. D. Ohisson, O. Ordeig and J. P. Kutter , "Microfluidics and nanofluidics", vol. 9, August 2010, [researchgate.net]

145. H. Becker and C. Gartener, Anal. Bioanal. Chem., **390 (1)**, 89-111 (2008).

146. H. Becker and L. E. Locascio, Talanta, **56 (2)**, 267-287 (2002).

147. J. S. Kee,D. P. Poenar, P. Neuzil and L.Yobas, **134(2)**, 532-538 (2008)

148. S. W. Park, J. H. Yoon, B. W. Kim,S. J. Sim,H. Chae and S. S. Yang, **10(6)**, 859-868 (2008)

149. S. K. Sia and G. M. Whiteside, Electrophoresis, **24(21)**, 3563-3576 (2003).

150. C. M. Chen, G. L. Chang and C. H. Lin, J. Chro. **1194(2)**, 231-236 (2008a)

151. H. Y. Tan, W. K. Loke, Y. T. Tan and N. T. Nguyen, **8(6)**, 885-891 (2008).

152. Y. X. Wang, Y. Zhou, B. M. Bagley, J. W. Cooper, C. S. Lee and D. L. DeVoe, **26(19)**, 3631-3640 (2005).

153. Y. R. Wang, H. W. Chen, Q. H. He, and S. A. Soper, **29(9)**, 1881-1888 (2008).

154. Y. L. Guo, K. Uchiyama, T. Nakagama, T. Shimosaka and T. Hobo, **26(9)**, 1843-1848 (2005).

155. K. N. Toft, B. Vestergaard, S. S. Nielsen, D. Snakenborg, M. G. Jeppesen, J. K. Jecobsen, L. Arleth and J. P. Kutter, Anan. Chem., **80 (1-0)**, 3648-3654 (2008).

156. P. Abgrall, V. Conedera, H. Camon, A. M. Gue, N. T. Nguen, **28 (24)**, 4539-4551 (2007).

157. T. B. Christensen, D. D. Bang and S A. Wolff, Microelctron Eng., **85 (5)**, 1278-1281 (2008).

158. T. Sikanen, S. Tuomikoski,R. A. Ketola, R. Kostiainen, S. Franssila and T. Kotiaho, Ana. Chem., **79 (23)**, 9135-9144 (2007).

159. Microchem. (2009) at {http://www.microchem.com/products/su_eight.htm. Accessed June 2009.

160. Krishnamurthy V.N. (1995) Polymers in Space Environments. In: Prasad P.N., Mark J.E., Fai T.J. (eds) Polymers and Other Advanced Materials. Springer, Boston, MA

161. Ray. A. Cull, "Elastomers", Pp. 23 (1989).

162. "Smart Polymers and their Applications" Edited by M. R. Aguilar De Armas and J. S. Roma'n; 2014, Woodhead Publishing Ltd. [sciencedirect.com]

163. V. N. Krishnamurthy, Polymers and other advanced materials pp 221-226 ; in : Prasad P.N., Mark J.E., Fai T.J. (eds), Polymers and Other Advanced Materials. Springer, Boston, MA

164. http://www.polytechinc.com/news/08232013-recognized-by-theboyeingcompany

165. **Deceleration:** Figure numbers 7.22(a), 7.22(b), 7.22(c), 7.22(d), 7.22(e) are reproduced with permission from Prof. Terentjev (Ref. 127Y).

❑ ❑ ❑

Rheology in Polymer and Rubber Processing

8

The processing of rubbers and other thermoplastics to get the finished products involves a number of steps like mixing, calendering, extrusion, moulding and vulcanization. During processing the material is subjected to the heat transfer to soften and melt the polymer, and flow, undergoing deformation at different deformation rates, temperature and pressure. The mixing is carried out in two-roll mills, internal mixers like Banbury, and mixer- extruder where as the calenders, extruders and moulding (compression, transfer, injection and blow moulding) operations are used for shaping the components. Vulcanization is carried out for imparting the strength to the matrix. The importance of the rheological background to scientifically analyze and understand the technology and engineering involved in all these operations need not be overemphasized. In this chapter the role of rheology in processing the polymers and rubbers is discussed in short.

8.1 RHEOLOGY OF TWO ROLL MILL AND CALENDER

The flow analysis of two-roll mills and calenders has been discussed by a number of authors[1-4]. The two-roll mill is essentially used for mixing rubbers with fillers and other ingredients uniformly whereas the calenders are used for fractioning, topping and sheeting. The viscosity of the rubber melt on a two-roll mill is very high where as in calenders relatively lower viscosities are encountered. In spite of the difference in the purpose of the two machines the operating principles are identical.

The actual operation involves the interplay of machine parameters like roll speed of rotation, nip gap and friction ratio and the physical variables like temperature, pressure, shear stress, rate of shear, drag force and material viscosity. As soon as the material is put on the bank of the rolls it is subjected to the following phenomena[1]:

    (i) The drag force between the material and the roll surface starts pulling it towards the nip area of the roll;

    (ii) As the material moves towards the nip, the area of cross section for the flow decreases progressively increasing the pressure acting on the system till just before the nip,

(iii) The material thus continuously accelerates up to the nip point beyond which, it moves with a uniform velocity,

(iv) The pressure acting on the material falls to the atmospheric value at the exit of the nip

The drag force and pressure force acting on the material act opposite to each other. The pressure force is proportional to the pressure gradient (dP/dx), whereas the drag force is proportional to the velocity difference $(V - v_x)$, where V is the peripheral velocity of the roll surface and v_x is the average velocity in the X direction. The velocity of the material, v_x progressively increases as the material moves towards the nip i.e. as X increases becoming equal to V at the nip, hence the drag force reduces to zero there. The pressure force on the other hand keeps increasing acquiring a maximum value at just before the nip and then decreases as the material comes out of the nip.

Application of equations of continuity and equations of motion to the flow of incompressible fluids under isothermal and steady state condition through the nip area of the calender or two roll mill, the expressions for the velocity distribution, pressure distribution, maximum pressure, shear stress and rate of shear have been derived[1]. In the present text the derivation is not included but the final equations are given. The interested readers may refer to references from 1 to 4. The schematics of the fluid flow around the nip area is given in Fig. 8.1 The velocity distribution for a Newtonian fluid with symmetric roll rotation is given by

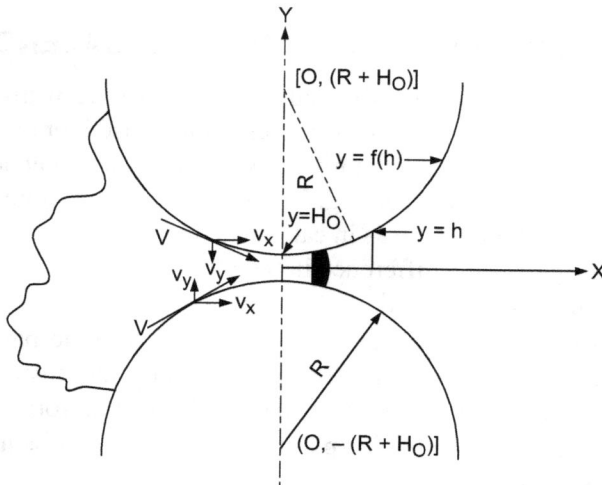

Fig. 8.1 Schematic representation of the flow around nip area

$$v_x = V\left[1 + \frac{3(1 - \delta^2)(\beta^2 - \alpha^2)}{2(1 + \alpha^2)}\right] \qquad \qquad ...(8.1)$$

For a Newtonian fluid with non-symmetric roll rotation the velocity distribution is

$$v_x = V_0 \left[1 + \lambda \delta + \frac{3}{2} \left(\frac{\beta^2 - \alpha^2}{1 + \alpha^2} \right) (1 - \delta^2) \right] \qquad ...(8.2)$$

For non-Newtonian fluid with symmetric roll rotation it is given by

$$v_x = V_0 - \frac{1}{(dp/dx)} \int_\tau^{\tau_w} \frac{\tau}{\eta} d\tau \qquad ...(8.3)$$

The integration can be carried if the viscosity as function of the shear stress is available.

The velocity distribution is shown in Fig. 1.2(d). The pressure gradient for the Newtonian fluid for symmetric roll rotation is

$$\frac{dP}{dx} = 3\eta V \left[\frac{(1 + \alpha^2) - (Q/2H_0 V)}{H_0^2 (1 + \alpha^2)^3} \right] \qquad ...(8.4)$$

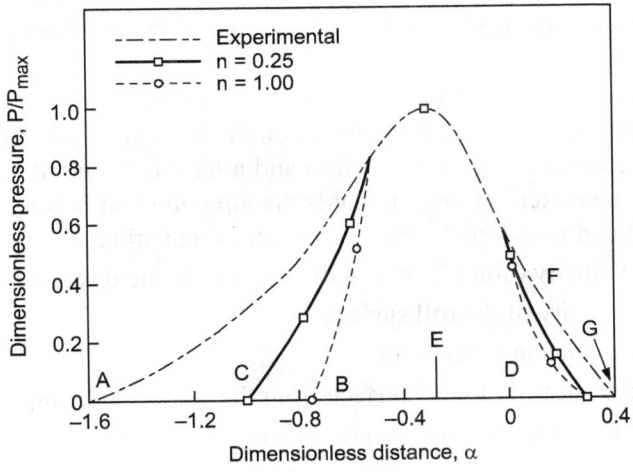

Fig. 8.2   Pressure variation for Newtonian and non-Newtonian fluids along flow direction[3]

The expression for the maximum pressure is

$$P_{max} = \frac{15}{2} \frac{\eta V \beta^3}{H_0} \sqrt{\frac{R}{2H_0}} \qquad ...(8.5)$$

For asymmetric roll rotation the maximum pressure equation is

$$P_{max} = \frac{15}{2} \frac{\eta V_0 \beta^3}{H_0} \sqrt{\frac{R}{2H_0}} \qquad ...(8.6)$$

The variation in the pressure is shown in Fig. 8.2. The maximum pressure occurs some distance away from the nip. The rate of shear and shear stress expressions for the symmetric mill for Newtonian fluids are

$$\dot{\gamma}(\delta) = \frac{3V\delta}{H_0}\frac{(\alpha^2 - \beta^2)}{(1+\alpha^2)^2} \qquad\qquad ...(8.7)$$

and

$$\tau(\delta) = \frac{3V\delta\eta}{H_0}\frac{(\alpha^2 - \beta^2)}{(1+\alpha^2)^2} \qquad\qquad ...(8.8)$$

For non-symmetric roll rotation these expressions are

$$\dot{\gamma}(\delta) = V_0\left[\frac{\lambda(1+\alpha^2) + 3\delta(\alpha^2 - \beta^2)}{H_0(1+\alpha^2)^2}\right] \qquad\qquad ...(8.9)$$

$$\tau(\delta) = V_0\eta\left[\frac{\lambda(1+\alpha^2) + 3\delta(\alpha^2 - \beta^2)}{H_0(1+\alpha^2)^2}\right] \qquad\qquad ...(8.10)$$

For non-Newtonian fluids the apparent viscosity, which is a function of rate of shear must be used. For this purpose one of the non-Newtonian model must be applied

The variation of the velocity in the direction of the flow shows that there is region of melt circulation of the material on the bank and as it approaches the nip area it becomes uniform just before and after the nip point. The pressure increases as the material moves towards the nip point and becomes maximum just before it and then rapidly falls. In the above equations the terms used are

v_x is the velocity of the fluid in × direction or in the direction of the flow

V is the velocity of the roll surface

y is the distance in y direction

h is the distance of the roll surface from the center of the nip

H_1 is the distance h where the pressure is maximum

$$V_0 = \frac{(V_1 + V_2)}{2}$$

V_1 is the peripheral velocity of the fast roll

V_2 is the peripheral velocity of the slow roll

$$\lambda = \frac{(V_1 - V_2)}{(V_1 + V_2)}$$

$$\delta = \frac{y}{h}$$

$$a^2 = \left(\frac{x^2}{2H_0R}\right)$$

R is the radius of the rolls

$$(1 + \beta^2) = \left(\frac{Q}{2 H_0 V} \right)$$

Q is the volumetric flow rate of the fluid

η is the viscosity of the fluid

$2 H_0$ is the nip gap

The rubber, which is mixed in a two-roll mill is very high viscosity material and also has high elastic characteristics where as on a calender the material is relatively of lower viscosity and elasticity. Depending on the temperature the rubber may not deform and may not go to the nip area at low temperature. At moderate temperature, where the viscous and elastic effects are well balanced the rubber forms a tight elastic band on the roll. At still higher temperature the rubber might tear or bag and further increase in temperature gives fluid like band.

8.2 RHEOLOGY OF INTERNAL MIXERS

The rubber industry uses a number of mixers, which may be either batch type like Banbury mixer (Fig. 1.7) and Francis Shaw Intermix or continuous types like Werner and Pfleiderer EVK Contimix, Transfermix (Fig. 1.10) and Pin-Barrel extruders. For efficient mixing of rubber, fillers and other ingredients, it is essential that the rubber mix be simultaneously subjected to cross flows in varying directions inside the mixing chamber. The rotors of the batch mixers and the screws of the mixing extruders are specially designed to achieve this objective. Freakley[5] has reported different types of rotors like Banbury rotors[5] (Fig. 1.8). Figures 8.3 shows different types of rotors[5a] and Fig. 8.4 show the functions of the mixing rotors in an internal mixer[5a1]. The complex design of the rotors makes it difficult to calculate the rate of shear and shear stress acting on the material. The highest rate of shear and shear stress act at the smallest clearance between the rotor tip and the chamber wall and in the other sections of the mixer the material will experience varying rate of shear and shear stress.

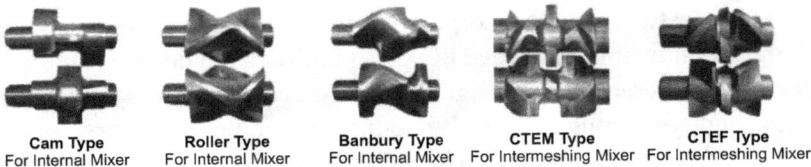

| Cam Type | Roller Type | Banbury Type | CTEM Type | CTEF Type |
| For Internal Mixer | For Internal Mixer | For Internal Mixer | For Intermeshing Mixer | For Intermeshing Mixer |

Fig. 8.3   Different types of rotors [5a, Chareon Tut Co. Ltd. Thailand]

During the mixing operation the elastomer is first fed to the mixer, allowed to get heated and masticated resulting in the reduction of viscosity of the rubber. The filler addition then increases the viscosity due to incorporation of the filler resulting in the increased power consumption. Due to incorporation process the filler enters the bulk of the rubber melt forming the rubber-filler blobs making system highly viscous. The blobs are acted upon by the much higher shear stress

Rotor type		Tangential rotor	Intermeshing rotor
Rotor shape		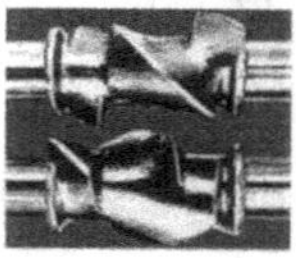	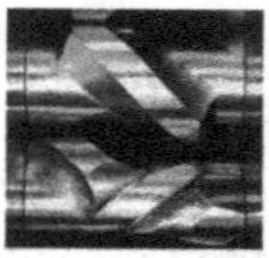
Mixing mechanism		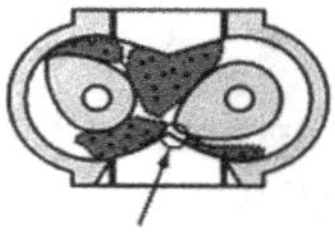 (1) Between rotor and chamber wall	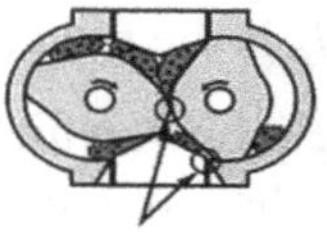 (1) Between rotor and chamber wall (2) Between rotor and rotor
Function	Shearing	◎ Excellent	◎ Excellent
	Distribution	○ Good	◎ Excellent
	Intake	◎ Excellent	○ Good
	Cooling	△ Acceptable	◎ Excellent

Fig. 8.4 Functions of mixing rotors in an internal mixer[5a1]

due to increase in the viscosity giving a sharp rise in the power consumption. The system is first subjected to the dispersive mixing (breakdown of the filler aggregates) and then to the distributive mixing. It causes considerable viscous heat generation giving rise in the system temperature. As a result of temperature rise and reduction of the filler particle size, the viscosity decreases continuously resulting in the reduced power consumption. Due to the complex design of the mixing elements and equally complex rheological behaviour of the mix it is not possible to theoretically analyze the flow behaviour of the system.

8.3 RHEOLOGY OF EXTRUSION

Extruders are by far the most important equipment used in numerous applications in the processing of plastics and rubbers. These may be of ram type or screw type (single screw or twin-screw) incorporating the wide variation in screw design to suit the requirements. Rauwendall[6] has published the details of the type of extruders and different screw designs. The feed to the extruders may be plastic granules, highly viscous rubber compounds or molten polymers, monomer solutions or the liquid monomers (for polymerization). In all the applications the polymer is subjected to the varying degree of shearing action depending on the viscosity of the material under different temperature and rate of deformation. The energy required to make the material flow through the extruder depends on the rate of flow, viscosity of the material, the pressure developed and the die resistance. The information about the rheological behaviour of the material under different rate of shear, shear stress, temperature and pressure is a single most important consideration for the design and optimization of performance of extruders. The theoretical development for the analysis of the polymer melt flow through an extruder is based entirely on the flow of plastic melts. Fig. 8.5 shows a plastic extruder having granular plastic as feed. It has four zones, (i) conveying or the feeding zone, (ii) melting zone, (iii) metering zone and (iv) die zone. In the feeding zone the granules are compacted and sheared as they move ahead. The shearing and the friction between the material and, the barrel and screw surfaces along with the external heating, heat the material to form a deformable solid plug. In the melting zone the temperature is further increased to the melting point of the polymer and it becomes a mixture of high viscosity liquid and the deformable solid particles that keep on reducing in size as the material moves ahead. Over the length of this section the plastic is completely melted before it enters the metering zone.

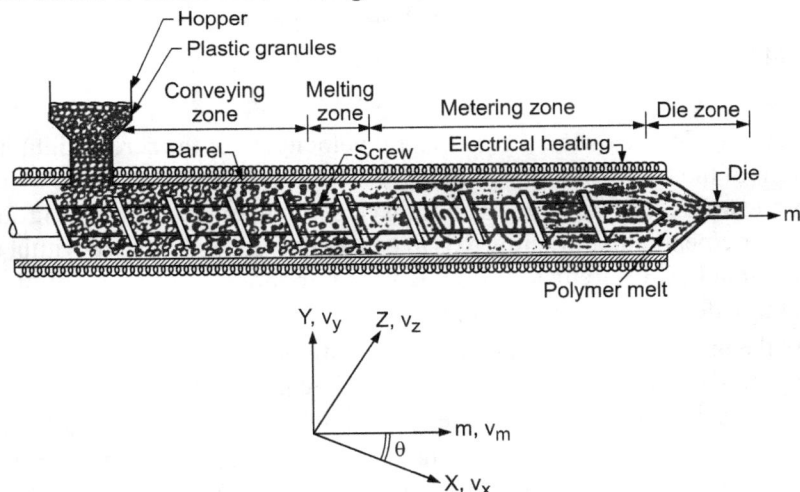

Fig.8.5 Schematic of single screw extruder

The lengths of the feeding zone and the melting zone are essentially determined based on the feed rate (depends on the screw speed and its design), molecular weight (MW) and molecular weight distribution (MWD) of the polymer, and the rate of heat input such that the sufficient residence time is provided for the granules to melt. A polymer with relatively low MW and MWD will melt at lower temperature and require smaller residence time and therefore shorter melting zone than the one with high MW and MWD material. In the metering zone the melt flows through the screw channel and may also circulate within each flight. The pressure inside the extruder slowly rises to the level required for the flow through the die.

The theoretical analysis in the conveying zone has been presented in the literature[7] and is based on the balance between the frictional forces acting on the granules due to the barrel and the screw channel surfaces. However it is rather inadequate to predict the characteristics of this zone due to highly simplified assumptions and therefore never used in the design of the extruder.

The flow analysis in the melting zone is highly complex due to the continuously changing state of the material from a highly compacted solid plug to the molten state over a small zone. The nature of flow in this zone is not fully understood and therefore no theoretical analysis is available. The length of the melting zone is designed on the basis of experience with other extruders. The combined length of the conveying and melting zones constitutes about 2/3rd of the total extruder length irrespective of the mode of heating of the material.

In the metering zone the polymer melt flows as a non-Newtonian fluid with well defined pressure gradient, velocity, rate of shear and shear stress distributions. The Navier -Stokes equation has been applied assuming the melt to behave as a Newtonian fluid to develop the relationship between the volumetric flow rate, pressure gradient and the viscosity of the melt. The net volumetric flow rate Q of the melt consists of three components (i) the drag flow Q_D (ii) the pressure flow Q_P and (iii) the leakage flow Q_L thus

$$Q = Q_D + Q_P + Q_L \qquad \qquad ...(8.11)$$

The drag flow is due to the peripheral velocity, v_z of the screw flight in the direction of the flow (direction m in Fig. 8.5) and therefore causes the material to flow towards the die zone. The transverse component of v_z acting in the direction perpendicular to screw axis m causes the melt to circulate within each flight channel. This circulatory flow is very important from the point of view of heat transfer to the melt and mixing.

As the melt moves ahead the pressure continuously increases and becomes maximum at the die entry resulting in a positive pressure gradient and flow in the opposite direction giving the pressure flow, Q_P.

The leakage flow is the consequence of the progressive build up in the pressure gradient and the gap between the barrel surface and the flight. The continuous rotation of the screw wears the flight surface increasing the gap and hence increasing the leakage flow. The expressions for the volumetric flow rate Q, velocity distribution, shear rate, the shear stress, drag flow, Q_D, and the pressure flow, Q_P, for a Newtonian fluid have been derived and given below,

The volumetric flow rate Q is

$$Q = \frac{WH\pi D \cos \theta N}{2} - \frac{WH^3}{12\eta}\left(\frac{dp}{dz}\right)$$

$$= Q_D - Q_P \qquad \qquad ...(8.12)$$

The velocity distribution in Z direction is

$$v_z = \frac{V_z}{H}y - \frac{y(H-y)}{2\eta}\frac{dP}{dz} \qquad \qquad ...(8.13)$$

The shear rate $\qquad \dot{\gamma}(y) = \frac{V_z}{H} - \left[\frac{H-2y}{2\eta}\right]\left(\frac{dP}{dz}\right) \qquad \qquad ...(8.14)$

The shear stress $\tau(y) = \eta\left[\dfrac{V_z}{H} - \left[\dfrac{H-2y}{2\eta}\right]\left(\dfrac{dP}{dz}\right)\right]$ ...(8.15)

In the above equations, W is the channel width, H is the channel height, D is the barrel diameter, θ is the helix angle, N is the screw rotation rate, η is the melt viscosity, v_z is the velocity in Z direction, and V_z is the Z velocity at $y = H$.

The ratio of the pressure flow rate to the drag flow rate, $(Q_P/Q_D = \phi)$ gives a measure of the throughput of the extruder i.e. if $\phi = 1$ then the net flow rate will be zero and as the screw rotates the melt must keep circulating inside the extruder and Q_P, ϕ and leakage flow must balance with Q_D. The expression for ϕ is

$$\phi = \frac{H^2}{6\eta V_z}\left(\frac{dP}{dz}\right)$$ ...(8.16)

The above equation suggests that the pressure flow will increase with increasing the channel depth and pressure gradient whereas the drag flow will be higher for the high viscosity melt and high screw rpm. The velocity distributions in X, Z and m directions as a function of ϕ are given in Fig. 8.6[8]. The coordinate axes and the velocity components are indicated in Fig. 8.5. It can be seen that the X-component of the velocity is independent of ϕ and both Z and m components vary with it. For $\phi = 1$, $v_m = 0$ indicating that the material does not move forward. As screw keeps rotating the melt must only move with the screw and circulate.

Equation 8.12 is applicable to the Newtonian fluids but the polymers are non-Newtonian fluids having viscoelastic characteristics. Generally power law model is applicable to these fluids to describe the melt viscosity as a function of rate of shear. Wagner et. al. have reported that multiplying Q_D in equation 8.12 by the correction factor $(4 + n)/5$ and Q_P by $3/(1 + 2n)$ where n is the power law index give good approximation of the screw throughput. [J.R. Wagner Jr., E.M. Mount III, H.F. Giles Jr., Extrusion (Second Edition)—The definitive processing guide and handbook, Plastic design library, pp 47-70, 2014].

A plot of volumetric flow rate versus pressure in the extruder head is known as the screw characteristic diagram (Fig.8.7)[5]. The volumetric out put decreases with increase in pressure at the screw head due to increase in Q_P. The performance of a screw extruder is always influenced by die design and its characteristics. The die characteristics are defined by a plot of Q versus ΔP as shown in Fig. 8.7. It can be seen that Q increases with (ΔP). Thus the combined performance of the extruder and die will be decided by the intersection of the two characteristic curves. As the viscosity of the material is a strong function of the temperature, the operation temperature influences both the die and the extruder characteristics. Hence for correct estimation of the extruder and die performance, both characteristic curves must be at the same temperature. From

Fig. 8.7 one can get the range of extruder-die operating points from the triple intersections of screw and die characteristics with the material temperature.

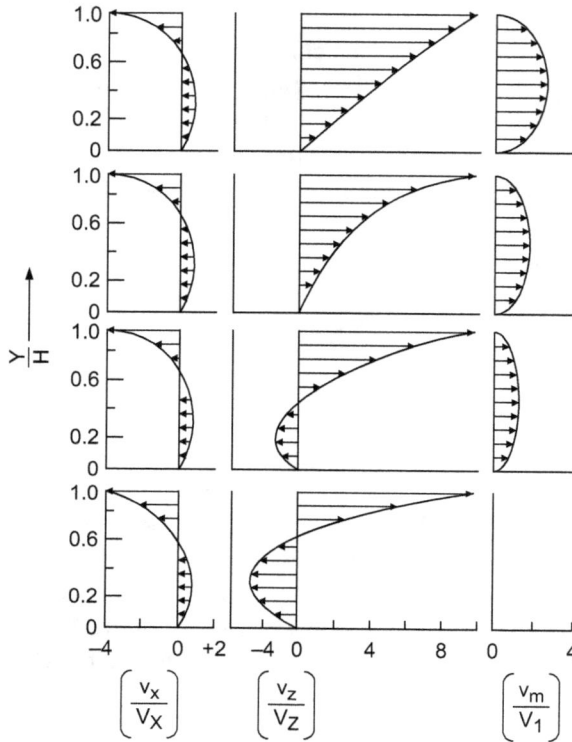

Fig. 8.6   The velocity distribution in a single screw extruder[8]

8.4 RUBBER EXTRUDERS

In rubber processing industries the screw extruders are used for a number of applications like shaping, mixing and forming operations. The rubber industry employs two main categories of extruders i.e. (i) hot feed extruders where the feed to the extruder is pre-warmed in another machine like two-roll mill or an internal mixer and (ii) cold feed extruders, which use the feed at room temperature either in the form of strip from a calender or in the form of granules. The L/D ratio of hot feed extruder is much smaller than that of the cold feed extruder due to absence of the heating and plasticizing zone in the earlier machine.

The viscosity of the rubber compound is very high in comparison to the plastic melt hence the energy requirement for processing the rubber compound in an extruder is normally much higher. The mixer extruders have very complex screw and barrel design giving still higher energy consumption. Some well known mixer extruders used in rubber processing are (i) Werner and Pfleiderer EVK (Contimix) continuous mixer (Fig. 1.9), (ii) Transfermix (Shearmix) (Fig. 1.10), (iii) Pin barrel extruder etc. Due to the complex design of the mixer elements the flow direction of the melt continuously changes giving unusual flow patterns and thus no theoretical flow analysis is possible. However the analysis developed for the plastic extruders in the metering zone can be applied to the hot feed extruders used for shaping and forming operations provided the non-Newtonian rheological behaviour of the material is known.

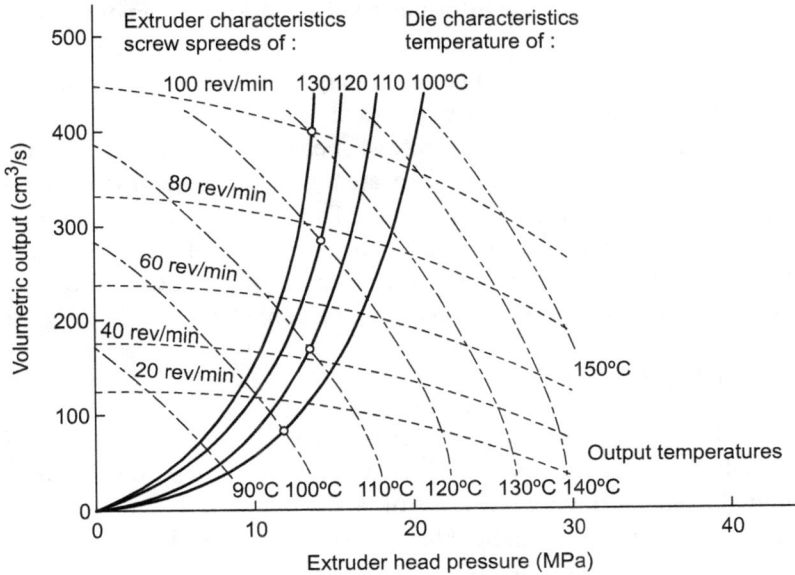

Fig. 8.7 Operating diagram for a 90 mm cold feed rubber extruder[5]

8.5 VENTED EXTRUDERS

In polymer and rubber processing it is often necessary to subject the melt to a devolatilization stage to remove the solvents or monomers or the entrapped air to produce a better quality product. A vent section is provided in the extruder after the melting zone, where the screw design is changed to provide a deeper flighted section enabling the melt to form a thin film in the region of evacuation[9,10]. The flight design in this section has two sections (i) a conical section from the root diameter to the flight diameter to give a smooth change of direction of the melt flow to the vent section and (ii) a cylindrical section of a certain length and diameter for forming a thin cylindrical layer of the melt between the barrel and the cylinder to facilitate the easy removal of the vapour. The length and the diameter of the cylindrical section depend on the extent of the solvent present and the viscosity of the melt. For a high viscosity melt the diameter and the length should be higher and vice versa. However due considerations should be given to the bubbling or foaming in the vent section due to the higher concentration of the volatiles. The pressure in the vent section falls to zero and the residual solvent, monomer or entrapped air bubbles are removed through suction. The twin-screw extruders have been reported[10] to be more efficient in devolatilization of melts than a single screw extruder.

For designing the vent section the flow analysis through an annular section may be applied. The viscosity of the melt will continuously increase as the solvent evaporates and the material will be subjected to a continuously varying stress and viscosity. As the length of the vent section should be kept relatively smaller the viscosity and the stress gradient may become very steep resulting in a high-pressure drop. Thus the information about the variation of the viscosity of the melt with the solvent or monomer concentration is essential. As most of the melts are pseudoplastic in nature it may be advisable to design the vent section as trapezoidal instead of cylindrical. Such a design will subject the material to progressively increasing rate of shear and thus decreasing viscosity as it moves ahead and therefore compensate for the increase in viscosity due to the evaporation of the solvent or monomer. The increase in temperature due to shearing of the melt will decrease the viscosity of the melt. Hence for a proper design and operation of the vent zone the effects of temperature, shear rate and solvent concentration on the viscosity must be considered.

8.6 COEXTRUSION

The coextrusion technology is used for the production of a number of items like multi-layer sheets, tubes and pipes, refrigerator liners, conjugate fibers, extremely thin sheets having hundreds of layers of micro thickness of different polymers, coating of wires and cables, packaging materials etc. It is a highly sophisticated and microprocessor controlled operation where the flow rates and temperature of each melt need to be controlled precisely. A number of theoretical and experimental results related to the coextrusion of polymers have been reported in the literature[2,11-17]. The thickness of the layers in the multi-layered composites may be of the order of 0.1 μm as in the case of iridescent film produced from more than 100 layers of PMMA and PS to get the total composite thickness of 20 μm whereas much larger thickness is encountered with other applications like multi-layer containers. During the extrusion of material of very small thickness the rheological response of the polymers plays an important role. The polymers in general used for the coextrusion are LDPE, LLDPE, HDPE, PS, PMMA etc. These systems have widely varying shear viscosities, which are important during flow through the extrusion dies and also very different elongational viscosities, which play a crucial role during multilayer film blowing. During the flow through the extrusion die the addition of low viscosity component reduces the pressure gradient for the flow and thus the cost of production. It has been reported[2] that the low viscosity component act as a lubricant as it is in contact with the die surface due to its high fluidity.

The shear sensitivity of different polymer melts and extrusion temperature are closely linked with each other and both these parameters play an important role in forming the core and the cover polymers during the co-extrusion process. Han[12] has reported the experimental viscosity of PS and HDPE at 200°C and at

240°C at wide range of shear stress. At 200°C the PS melt shows higher shear sensitivity than HDPE melt and gives a cross over point in viscosity versus shear stress plot. At shear stress higher than 12×10^4 Pa, PS melt viscosity is lower than HDPE whereas at shear stress lower than this its viscosity is higher than that of HDPE melt. At 240°C however the viscosity of PS is lower than that of HDPE melt over a wide range of shear stress. Thus when PS and HDPE are coextruded at 200°C and below 12×10^4 Pa stress the PS will form the core and HDPE will encapsulate it where as at shear stress higher than 12×10^4 Pa HDPE will form the core. The coextrusion at 240°C the PS will encapsulate HDPE over the entire shear stress range due to its lower viscosity.

8.7   EFFECT OF RATIO ON THE EXTRUDER PERFORMANCE-EXTRUSION: HOW MUCH L/D DO YOU REALLY NEED?

(A post by Jim Frankland, President, Frankland Plastics Consulting, LLC), 28/11/2011. This post is reproduced below:

Just like selecting the extruder size and drive combination, the L/D should be carefully evaluated. For most applications, you will need a longer two-stage screw to match the output of a single-state screw.

In the early 1960s, extruders typically had a length/diameter ratio of 20:1, and a machine with a 24:1 L/D was considered long. Since then, extruders have gotten longer, with the 30:1 to 36:1 L/D becoming the industry "standard." Some extruders even exceed 40:1 L/D for special purposes like double venting, compounding, or high-speed processing.

What benefits does the additional length provide? Mostly increased output and improved homogenization. Since the feed section stays approximately the same length, regardless of the L/D, the rest of the screw is devoted to melting and pumping. The deeper the screw channels, or the higher the specific output (lb/rpm), the more length you need to complete melting and develop the pressure necessary to push the polymer out the die. As designers reached these limits, extruders were built longer to handle the economic requirements to pump out more and better product. However, there are actual limits on increasing output as L/D is increased. Usually these limits are due to the inability of the feed section to deliver more polymer. On smaller-diameter screws, that limit often is determined by screw strength. On small screws you can only go so deep in screw channels before the screw is over torqued and fails. On larger extruders, the efficiency of feeding decreases as the channels get deeper until there is no further increase in output. Two-stage screws benefit more with increasing L/D because about 4-6 D is consumed in the vent area, which contributes very little to melting or pressurization. For most applications, you'll need a 30:1 two-stage screw to match the output of a 24:1 single-stage screw.

Are there any disadvantages associated with longer extruders? Some polymers melt much easier and faster than others. Also, some processes typically have low head pressures, while others have much higher discharge pressures. Inherent viscosity differs a lot between polymers, and some shear-thin significantly while others do not (i.e., are more "Newtonian"). As a result screw performance is optimized at a variety of L/Ds rather than at any one standard L/D. A screw that is too long for the overall processing situation can actually limit output. The limitation generally shows up as excessively high melt temperature that can cause polymer breakdown, colour shift, loss of additive effectiveness, and plate out, to name a few issues. For a polymer that melts easily, the melting length should ideally be shorter, as excessively long transitions can actually reduce melting rate. The same is true of pressure development, as widely used melt pumps greatly reduce the need for long metering sections to handle the discharge pressure. As a result, the tendency to buy longer and longer L/D extruders can actually penalize overall performance. Custom extrusion houses may simply have to live with this reality because they never know what they will be running next year, and a longer extruder has more inherent flexibility than a shorter one. But if you have dedicated process, there can be self-imposed limits with an extruder that is too long.

Just like selecting the extruder size and drive combination, the L/D should be carefully evaluated. Everyone wants the most usable output from their extruder, but if the material comes out too hot or too degraded then the singular focus on rate actually reducing the extruder's capability. Data such as diffusivity, power-law coefficients, melting points, head pressure, viscosity, and crystallinity should be part of the evaluation process.

Relative Output Levels

L/D Ratio	Single-stage Screw Output	Two-stage Screw Output
20:1	0.8	0.5
24:1	1.0	0.8
30:1	1.2	1.0
36:1	1.38	1.15

8.8 RHEOLOGY OF MOULDING AND FORMING OPERATIONS

The processing equipment dedicated to the moulding and forming operations are (i) Injection moulding, (ii) transfer moulding, (iii) compression moulding and (iv) blow moulding. The application of rheology to these equipment is discussed next.

8.8.1 Injection Moulding

The injection moulding is a forming operation where a certain mass of polymer at a higher temperature than its melting point (for thermoplastics) or above its glass transition temperature (for elastomer and amorphous polymers) is forced under pressure in to a mould to produce an item of the shape of the mould cavity. The process involves heating the polymer to the required temperature either in a ram extruder or a screw extruder (Fig. 1.12), pressurizing, filling of mould, cooling and ejection of the moulded component. The moulded product must be free of any entrapped air, sink mark and any structural deformation or warpage and must be strong enough to withstand the stresses in the application. The different sections of an injection mould through which the melt flows are, (i) nozzle, (ii) sprue, (iii) runner, (iv) gate and (v) mould cavity as shown in the Fig. 8.8. Each of these parts of the mould has well defined functions. A sprue is a small channel with diverging conical shape, which facilitates the streamline transition between the nozzle and the runner. It suppresses the flow disturbances, which would have been caused at the runner entry due to the sudden change in the cross section area of the flow from nozzle to the runner (in absence of the sprue) resulting in an uneven stress distribution in the moulded product. The runners are the channels, which feed the cavity such that the melt reaches all the sections of the mould cavity uniformly. The flow of the melt should not suffer considerable pressure drop and at the same time the scrap produced should be minimum. In the modern practice the runners are heated to the melt temperature thus avoiding any cooling of melt on the surface of the runner to facilitate lower ΔP, reduced scrap and short cycle time. The gate is a small tapered entry to the mould cavity, where, as the melt enters, its velocity is increased so also the rate of shear acting on it. The rate of shear encountered may be of the order of 10^5 s^{-1} and the viscosity of the melt will go down considerably due to its pseudoplastic nature thus giving considerably low-pressure drop. The gate also facilitates the removal of the material solidified in the runner and in minimization of the out flow of the melt when the pressure is released.

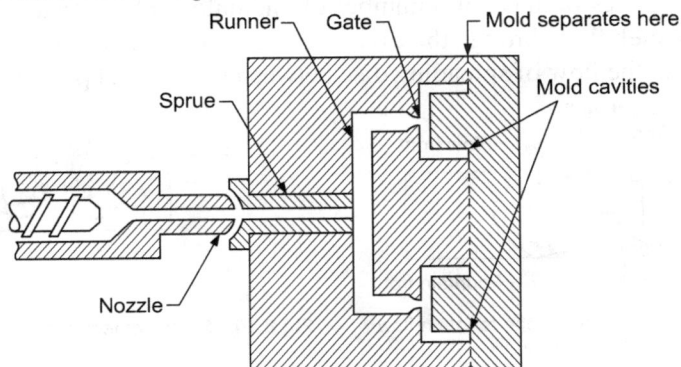

Fig. 8.8 Schematic of a simple injection mould showing its different components[10]

The melt flow inside the mould cavity is very critical from the point of view of the product quality. The smaller diameter of the gate though helpful in the reduction of the ΔP may cause the melt velocity to be high resulting in the melt fracture and jetting of the melt inside the cavity resulting in the uneven stress distribution and unfavorable molecular orientation. For producing a stress free, smooth and good quality moulding it is important that the mould fills at such a rate that sufficient time is available for the melt to relax the stresses. It should also start filling the cavity from the gate onwards and fill it completely without any residual stresses. As the melt starts flowing inside the cavity the following anomalous phenomena are known to occur

(a) A frozen layer starts forming at the cavity walls[18]

(b) A viscosity and the shear stress gradient develops having high viscosity near the walls and low viscosity at the centre. The maximum shear stress occurs at some distance away from the wall[19].

(c) Fountain effect- the melt initially flows down the centre of the cavity to the melt front and then flows out towards the walls (Fig. 8.9)[10, 20].

(d) Jetting- Exiting the melt from the gate in the form of a jet due to the high velocity of the melt without wetting the walls near the gate (Fig. 8.10)[10, 21, 22].

(e) Weld line or Knit line- When two streams of the melt meet either due to the presence of an insert or a slot or in case of use of a number of gates for filling, a large weld line or the knit line forms at the junction of the two fronts. The structural strength of the product is reduced along the weld lines.

The melt flow through sprue, runner and the gate can be analyzed following the standard approach of flow of the power law fluids through a circular cross-section. For the non-circular cross sections the hydraulic mean radius or the substitute radius can be used in place of the diameter. However the melt flow in the mould cavity cannot be analyzed using the analysis of the flow through the parallel plates because of a number of anomalous phenomena mentioned above. The melt flow through the cavity has been analyzed using the approach applicable to the flow of spreading disc between the parallel plates[2].

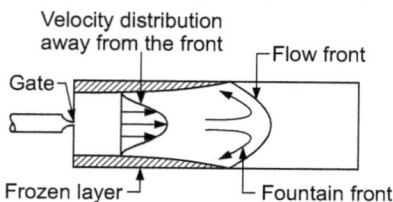

Fig.8.9   The sketch of the fountain flow phenomenon[10]

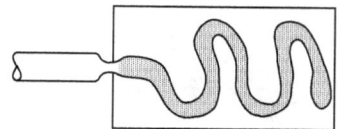

Fig.8.10   Sketch of jetting phenomenon[10]

The volumetric flow rate as a function of ΔP for a power law fluid flowing through a circular tube of uniform diameter under steady, isothermal and laminar flow conditions can be expressed as

$$Q = \frac{n \pi R^3}{3n+1} \left(\frac{R\Delta P}{2kL} \right)^{1/n}$$

...(8.17)

where R is the channel radius, n is the non-Newtonian index, k is the consistency index, L is the length of the channel, and ΔP is the pressure drop.

The volumetric flow rate through the system remains constant but the pressure drop varies in each section of the mould. The total pressure drop through the mould, ΔP_m, is the sum of the individual pressure drops through sprue, ΔP_s, runner, ΔP_r, gate, ΔP_g, and the cavity, ΔP_c i.e.

$$\Delta P_m = \Delta P_s + \Delta P_r + \Delta P_g + \Delta P_c \qquad \text{...(8.18)}$$

For a given melt and the volumetric flow rate using equation 8.17 and rearranging it, the expression for the pressure drop can be written as

$$\Delta P = \left(\frac{2Lk}{R}\right)\left[\frac{Q}{\pi R^3}\left(\frac{3n+1}{n}\right)\right]^n$$

$$= C\left(\frac{L}{R^{3n+1}}\right) \qquad \text{...(8.19)}$$

where

$$C = 2k\left(\frac{Q}{\pi}\frac{3n+1}{n}\right)^n \qquad \text{...(8.20)}$$

The total pressure drop through the mould can be written from equations 8.18 and 8.19 as

$$\Delta P_m = C\left[\left(\frac{L_s}{R_s^{3n+1}}\right)+\left(\frac{L_r}{R_r^{3n+1}}\right)+\left(\frac{L_g}{R_g^{3n+1}}\right)\right] + \Delta P_c \qquad \text{...(8.21)}$$

The expression for ΔP_c is[2]

$$\Delta P_c = \left(\frac{3Q}{\pi}\right)^n\left(\frac{1}{1+n}\right)\left(\frac{2k}{h^{(1+2n)}}\right)\left(R_f^{(1-n)} - r_0^{(1-n)}\right) \qquad \text{...(8.22)}$$

where h is the distance between two plates, R_f is the flow length i.e. the instantaneous radius of the disc and r_o is the radius of the gate.

Hence the total pressure drop through the mould can be estimated using equations 8.21 and 8.22.

For non-circular cross sections the expression for the substitute radius given in Equation 3.90 can be used to calculate the pressure drop.

The flow problem for the transfer moulding is identical to the injection moulding system.

8.8.2 Reaction Injection Moulding (RIM)

Reaction injection moulding is a process where two or more low viscosity reactive liquid streams (liquid monomers or their solutions) are mixed in a mixing head (Fig. 1.5) prior to filling the mould. The monomers react to produce polymer, which solidifies on cooling to produce the shape of the cavity. This process has the following advantages and disadvantages as compared to the injection moulding process due to low viscosity (generally of the order of 1 poise) of the liquid

(a) Accurate metering is achieved,

(b) Highly homogeneous mixing is achieved in the mixing head,

(c) The mould clamping pressure is much smaller,

(d) The pressure drop through the mould is considerably reduced,

(e) The process takes longer time,

(f) It is not suitable for all polymerizing systems, particularly where the rate of polymerization is slow,

(g) Specific volume of the polymer is reduced as compared to the monomers and hence it is necessary to provide the packing flow

(h) As the polymerization reaction proceeds, the viscosity of the system inside the mould cavity increases and therefore high pressure is required for the packing flow. The design of the flow system of the mould must have a provision for the pressurization to the required level.

The flow analysis of the liquids through the RIM system is carried out following the standard approach for the flow through different cross sections discussed under Chapter-3. The flow inside the cavity of the mould does not produce the anomalous phenomena observed in the injection moulding process, although the fountain flow behaviour is a distinct possibility due to the polymerization reaction initiated in the mixer head.

8.8.3 Compression Moulding

In compression moulding the components are formed by introducing in the mould, a preheated and preformed shape approximately having the shape of the cavity and compressing and heating it. The melt flows under pressure to fill the mould cavity and then the cross-linking or setting takes place to produce the component shape. The polymer used may be a thermoplastic, partially polymerized thermosetting polymer, a rubber compound containing curatives or a thermoplastic elastomer. After the preformed shape is placed in the mould and the moulds are closed, the material receives heat from the platens and the melt flows in to the cavity to fill it completely before the curing reaction starts.

The cycle is terminated after the predetermined time for the cross-linking. Fig. 8.11 shows a compression moulding system of a cup shaped cavity[3]. Initially when the plunger or the platen closes, the melt flows radially outwards till the flow radius equals Ri and then it enters the annular space and flows upward to the pre-designed height as shown in the figure.

The flow problem can be treated in two parts viz. (i) an isothermal radial flow of an incompressible fluid flowing between two parallel plates where one of the plates moves towards the other at a constant rate and (ii) the flow through the annulus. The pressure drop, velocity distribution and the force required to move the upper plate downwards can be derived following the standard approach for the power law fluids under the above mentioned flow situations.

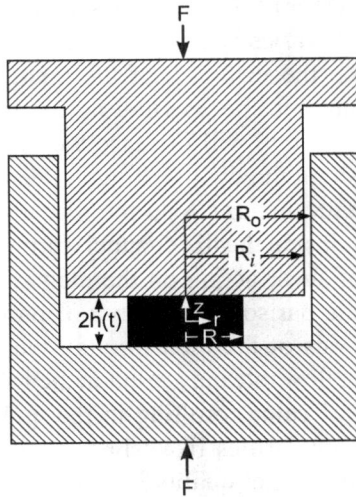

Fig. 8.11   The schematic of a compression moulding system[3]

8.8.4 Blow Moulding

Blow moulding is a process through which the moulded items like containers are produced by inflation of a tube shaped preform known as 'parison' using the pressurized air. The injection blow moulding and extrusion blow moulding are the two important methods to produce such items. In the injection blow moulding a tube like preform with a threaded end is injection moulded and subsequently reheated and inflated inside a mould. In the extrusion blow moulding process a tube or parison of the melt is extruded from a die in the opened two halves of the mould and inflated after closing the halves. As the two halves of the mould close, the parison is pinched off at one end. The inflation allows the parison to grow inside the mould and take the desired shape. It is important that the melt flow is accurately metered and an exact amount of the melt enters the mould. The injection blow moulding is normally suited for making small containers with a high quality surface finish and uniform wall thickness, where as the extrusion blow moulding is used for large containers.

The polymer melt is subjected to the following flow situations on the way to the mould cavity and subsequent inflation.

(i) *Accumulation and pressurization of the melt at desired temperature in the die head:* The pressure build up in the die head depends on the screw speed and the die flow rate. The minimum pressure at this stage must be equal to the pressure drop required for the flow through the die channel. It is therefore important that the pressure drop through the die should be calculated following a suitable procedure using an appropriate rheological model.

(ii) *Flow through the die channel at a predetermined flow rate:* The volumetric flow rate pressure drop relationship depends on the rheological behaviour of the melt and the die design. The rate of shear acting on the melt varies with the volumetric flow rate and the shear stress depends on the viscosity of the melt, which in turn depends on the temperature and the pressure in the die head. The increase in the volumetric flow rate will increase the rate of shear acting on the melt flowing through the die resulting in decrease in its viscosity due to the pseudoplastic nature. Also the viscous flow will increase the melt temperature inside the die channel, further reducing its viscosity. Hence the rate of shear acting on the melt is a critical parameter for the stability of the parison. A number of publications have reported on the flow analysis of the melt through the die channels[23-26]. The higher volumetric flow rate may result in the increase in the shear stress beyond the critical value resulting in the melt fracture phenomenon giving surface irregularities like 'shark skin' on the parison surface and unacceptable product quality. Narrow MWD polymers and high melt viscosity are more likely to give such problems[10]. During the flow through the die the melt is subjected to the elongational flow, the molecular orientation and the molecular disentanglement often referred as the 'shear modification' or the 'shear refining'[27]. The elongational deformation and the molecular orientation result in the storage of the elastic energy, which in turn generate the normal stresses giving rise to parison swell. The parison may also develop some weld lines due the presence of the 'spider legs' used for supporting the mandrel.

(iii) *The gravitational force acting on the melt:* It may result in the parison sagging or the 'draw down' due to the inability of the parison to support the weight of the melt due to the reduced melt strength caused by the different consequences of the higher flow rate. It results in a considerable variation in the thickness and the diameter of the parison along its length. Also the weight in the lower section of the parison may cause high compressive hoop stresses resulting in 'pleating' or 'draping' or 'curtaining' of the upper portion of the parison. To avoid pleating the melt should have reasonably high viscosity and high stiffness.

(iv) *Pressurization and inflation of parison:* Once the parison is inside the mould, it is inflated with the help of pressurized air resulting in the increase in the diameter and reduction in the wall thickness. The process of inflation is controlled by the extensional rheology of the melt, which govern the inflation rate as well as the inflation pressure. Due to the non-uniform thickness of the parison, the inflation may cause the bulging at the centre[28]. Also the non-uniformity in the parison thickness may give extremely thin section at the centre resulting in the blow out if the ratio of the mould diameter to the parison diameter is too high. The strain thickening or the strain hardening behaviour of the melts like LDPE helps in smooth inflation and prevents the blow out whereas the melts with strain softening nature present problems in inflation and have higher probability of the blow out. Thus for producing bigger containers with small wall thickness the strain hardening polymers must be recommended.

In summary it can be said that the process of blow moulding requires an in depth knowledge of (i) both shear and extensional rheological behaviours of the polymers melts, (ii) die swell behaviour, (iii) dependence of viscosity and die swell on the rate of shear, molecular structure of polymers and temperature and (iv) strain softening and strain hardening behaviour of the melts.

8.8.5 Rheology of Film Blowing

The film blowing process is used to produce thin polymer films using pressurized air to form a bubble by inflating the extruded polymer melt, cooling, solidifying and winding off the film. Figure 8.12 gives a schematic diagram of the film blowing process. It shows an extruder, screen pack, die, air inlet, mandrel,

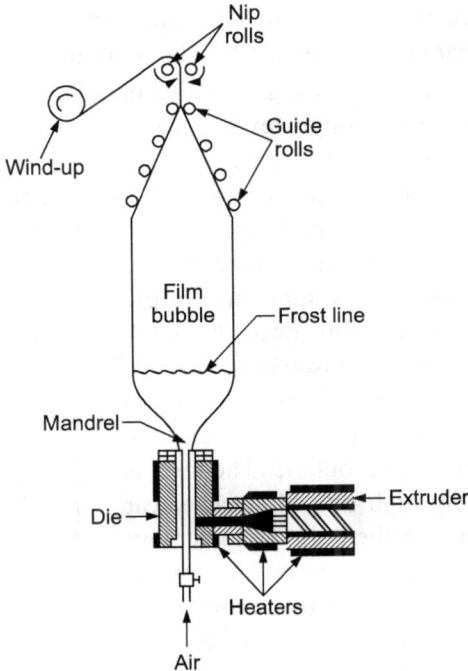

Fig. 8.12   Schematic diagram of the film blowing process[10]

film bubble, guide rolls, nip rolls and wind-up roll. The polymer melt first passes through the screen pack to remove any gel or other particles and then through the annular die. At the exit of the die the pressurized air is injected and the melt moves upwards in the "machine direction" as a cylindrical inflated film under the influence of the force exerted by the nip rolls. The introduction of the pressurized air prevents the collapse of the melt tube and inflates it in the radial direction to form a cylindrical bubble. The bubble is cooled with the help of an air ring, which directs the air along the outer surface of the bubble. During the flow as the melt passes through different stages in the equipment it gets subjected to different forces and experiences varying type of stresses. These are

(i) Flow in extruder and die: The melt flow through the extruder has been discussed under the extruder section. The most commonly used thermoplastics for the film blowing are polyolefins like the polyethylenes such as LDPE, HDPE and LLDPE and polypropylene. The melt flow of the LDPE through the extruder has been completely studied and the extruder design standardized. However the HDPE melt gives the problems due to (i) insufficient solid conveying (ii) decrease in the melt flow rate with increase in the screw speed and (iii) resultant poor film properties. However high molecular weight HDPE (HMW-HDPE) has been used to produce blown films by Formosa Plastics Corporation USA. [https://www.fpcusa.com/products/PE/pdf/PE-FilmProcessingGuide.pdf] LLDPE shows higher viscosity as compared to LDPE under operating shear rates in the extruder hence results in (i) higher torque for the screw, (ii) accelerated screw and barrel wear and (iii) higher melt temperature.

The die used in the film blowing is a spiral mandrel die shown in Fig. 8.13 where the melt flows out through several feeding ports to a series of helical grooved channels made with decreasing cross section along the flow direction, giving progressively increasing land width. Such a design produces a substantial mixing of the melt giving uniform temperature profile. The basic flow pattern is highly complicated involving multidirectional velocity components and as such no simple mathematical flow model can describe it. A number of researchers have published approximate flow analysis[29, 30]. The melt flows out through the annular space between the mandrel and the outer tube of the die to from bubble. The melt is subjected to relatively high rate of shear, shear stress and the pressure gradient. In the lower part of the mandrel the molecular orientation of the melt is prevented due to prevailing cross flows whereas in the upper part the melt molecules get oriented in the flow direction.

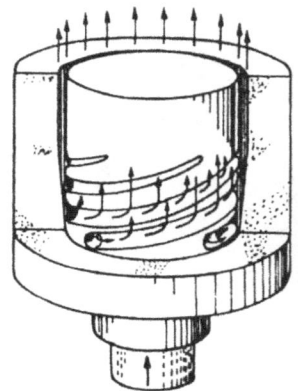

Fig.8.13 The Spiral Mandrel[10]

(ii) The melt flow in the bubble: The melt leaving the die lip undergoes the stress relaxation and gets subjected to the normal stresses due to the release of the elastic energy and thus swells. The forces acting on the bubble (Fig. 8.14) are

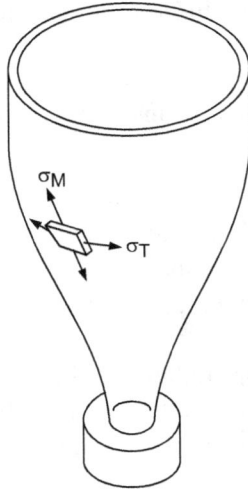

Fig. 8.14 Forces acting on the Bubble[10]

(a) The gravity force acting downwards opposite to the machine direction,

(b) The force due to the rotation of the nip roll acting in the machine direction. It results in the stretching flow causing elongational strain and the orientation of the polymer molecules in the machine direction.

(c) The pressure force acting on the melt surface in the transverse direction due to the air pressure inside the bubble. It causes the peripheral strain resulting in the molecular orientation in the transverse direction.

The polymer melt undergoes the biaxial stretching but it is practically impossible to obtain the equibiaxial stretching of the film, which would have given a uniform molecular orientation in machine as well as in the transverse directions resulting in the film with no anisotropy in the stress distribution. The rate of the peripheral orientation is high, as the time available for the radial expansion of the bubble is relatively smaller as compared to the stretching in the machine direction. The machine direction stretching is expressed by the ratio of linear velocity through the nip roll to the melt velocity at the die lip known as " Draw down ratio (DDR)" where as the radial stretching is expressed by the ratio of the final bubble diameter to that of the die lip known as " Blow up ratio (BUR)". The anisotropy in the molecular orientation in two directions produces a poor quality film with low impact strength that can be torn easily in the machine direction.

(d) The aerodynamic force arising from the interaction between the cooling air and the bubble surface can make the bubble oscillate and deform resulting in the variation of the thickness along the machine direction or even rupture the bubble under the extreme conditions.

Under the influence of the different forces the bubble has to be balanced to produce the film with uniform thickness and diameter and having very good surface finish. The nature of the forces and their interaction is rather complex, which is known to produce the flow instability in the bubble[10, 31, 32]. The flow instability results in

(a) Periodic thickness fluctuations under high DDR and low BUR,
(b) Periodic axisymmetric diameter fluctuations at high values of DDR x BUR and at intermediate inflations,
(c) An abrupt up and down movement of neck of the melt at fixed time intervals,
(d) Wobbling of the entire bubble at high BUR due to the imbalance of the aerodynamic force resulting in the helical bulge in the final film.
(e) High frequency flutter or buffeting due to the turbulence in the cooling air.

Thus for a better quality film it is very important that the forces acting on the melt bubble are well balanced.

8.8.6 Rheology of Sheet Extrusion

Sheet extrusion is a process for producing thin sheet of polymer using a long slot die with a narrow gap. This process is used for slot casting (chill roll casting), embossing and extrusion coating operations. The melt extrudate known as a "Web" is subjected to elongational force due to the rotation of the draw down roll, normal stresses due to the viscoelastic nature, thermal stresses due to the change in temperature of the surface and the surface tension force.

The action of the normal stresses and the surface tension force causes the thickness of the sheet in the edges to increase giving the "Edge bead" formation. It has been reported[33] that the nature of the stress field at the edge is close to the tensile extension whereas at the centre it is more like that of the planner extension. The tensile stress field at the edge may result in the "Neck-in" phenomenon giving reduction in the width of the sheet. The higher draw down velocity of the rolls may result in the draw resonance phenomenon resulting in the periodic variation of the thickness of sheet and even its rupture at high velocity, which has been reported to be due to the hydrodynamic instability phenomena[34].

8.8.7 Rheology of Melt Flow Through Dies

The dies are the essential pieces of equipment in all polymer processing steps involving the extruders. These include film and sheet dies, pipe and tube dies, blow film dies, wire coating dies, profile extrusion dies, coextrusion dies etc. The function of a die is to distribute the melt in the flow channel so that it exits with a uniform velocity. For a given melt the type of die geometry is designed on the basis of the required extrudate shape. The dies normally have a short length in comparison to their aperture size and have a widely varying cross section design from a simple circular type to a complex one.

There are a number of parameters, which influence the die performance. These are (i) channel geometry, (ii) rheological and flow behaviour of melt, (iii) melt flow rate, (iv) pressure and (v) melt temperature. The cross section at the die entry is invariably much higher than at the exit therefore the dies are generally tapered either partially or fully. Hence in addition to the shear deformation, the polymer melt undergoes the high extensional deformation during the flow through the dies. This results in the development of (i) the normal stress differences giving rise to the die swell and (ii) the flow instabilities at high flow rates causing the melt fracture phenomenon, which deforms the extrudate surface. The volumetric flow rate of the melt is limited due to the melt fracture, pressure drop and the temperature rise considerations. The melt flow rate should be decided on the basis of the compatibility between die and extruder characteristics but should not cause high temperature increase in the melt. The temperature rise normally results in uneven temperature distribution within the die, decrease in the melt viscosity resulting in the flow instabilities and some times the polymer degradation particularly for thermally sensitive melts.

The melt flow in the dies may be (i) one dimensional as in the case of circular, annular or thin section dies or (ii) two or three dimensional for different complex profile dies. In general one dimensional flow dies have two sections, first a "Lead-in" section, where the cross section reduces progressively from extruder head to the die head and second the "Land" section, which imparts the required shape to the extrudate. The lead-in section streamlines the convergent melt flow without creating the stagnation points and circulatory flow. The total pressure drop in the die is the sum of the pressure drops in these two sections. A completely tapered die is some times useful when the two sections give excessive pressure drop. However such a die is likely to give increase surface distortions due to the fact that the melt gets no time for the stress relaxation before it comes out of the die.

8.9 PRESSURE DROP THROUGH THE DIES

The lead-in section and the tapered dies: The design of the lead-in section of the tapered dies may be either coni-cylindrical or wedge shaped. The melt undergoes both the extensional and shear deformations while flowing through these sections. Hence the total pressure drop through the section will be the sum of the pressure drops due to both these deformations as suggested by Cogswell[35]. Using the power law model for shear flow and extensional viscosity for convergent flow Cogswell derived the following relation for the extensional pressure drop, $\Delta P_{E,C}$ for the flow through a circular cross section (Fig. 8.15), as

$$\Delta P_{E,C} = \frac{\lambda \dot{\gamma}_1 \tan \theta}{3} \left[1 - \left(\frac{r_1}{r_0} \right)^3 \right] \qquad \qquad ...(8.23)$$

where

$$\dot{\gamma}_1 = \frac{4Q}{\pi r_1^3} \qquad \qquad ...(8.24)$$

and the extensional viscosity, λ, is estimated based on the flow through a zero-length (knife edge) die and is given by

$$\lambda = \frac{9(n+1)^2}{32 \eta_a} \left(\frac{P_0}{\dot{\gamma}} \right)^2 \qquad \qquad ...(8.25)$$

In the Eq.8.25 η_a is the apparent viscosity at shear rate $\dot{\gamma}$ in the knife-edge die, n is the non-Newtonian index and P_0 is the pressure drop for the knife-edge die.

For shear flow the pressure drop through a conicylindrical die, $\Delta P_{S,C}$, is given by

$$\Delta P_{S,C} = \frac{2\tau_1}{3n \tan \theta} \left[1 - \left(\frac{r_1}{r_0} \right)^{3n} \right] \qquad \qquad ...(8.26)$$

where θ is the cone angle and τ_1 is the shear stress at radius r_1 (Fig. 8.15). The total pressure drop can be calculated by adding the two pressure terms.

The pressure drops for the wedge shaped dies or the lead-in sections (Fig. 8.15) due to the extensional flow, $\Delta P_{E,W}$ and shear flow $\Delta P_{S,W}$ are given by Eq. 8.27 and Eq. 8.29 respectively.

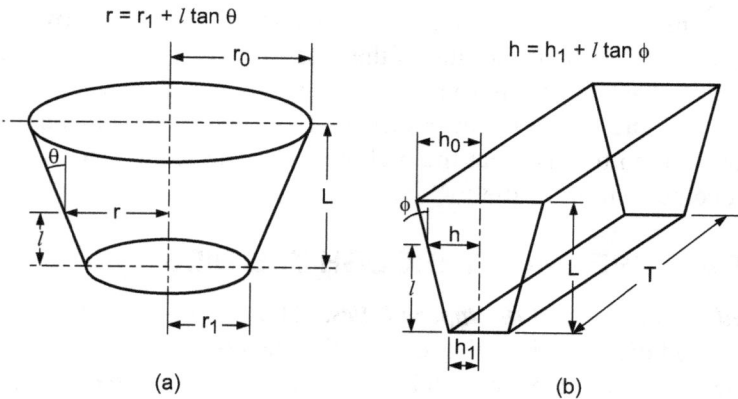

Fig. 8.15 Tapered die lead-in sections, (a) Conicylindrical and (b) Wedge shaped[5]

$$\Delta P_{E,W} = \frac{4\dot{\varepsilon}\lambda}{3}\left[1-\left(\frac{h_1}{h_0}\right)^2\right] \qquad ...(8.27)$$

where
$$\dot{\varepsilon} = \frac{\dot{\gamma}\tan\phi}{3} = \frac{Q\tan\phi}{2Th^2} \qquad ...(8.28)$$

$$\Delta P_{S,W} = \frac{\tau}{2n\tan\phi}\left[1-\left(\frac{h_1}{h_0}\right)^{2n}\right] \qquad ...(8.29)$$

where $\dot{\gamma}$ and τ refer to the rate of shear and shear stress at sections (Fig. 8.15(b)) where half width of the wedge die is h. The total pressure drop is obtained by adding the two pressure terms as in the case of conicylindrical die.

The land sections of these dies will be the circular or rectangular flow channels. The rectangular channel flow can be approximated to the flow through the parallel plates if the end effects are negligible. The pressure drop-volumetric flow rate relationships for these have been discussed in Chapter 3.

8.10 THE DIE SWELL

When a viscoelastic fluid flows out of die, the extrudate diameter is usually greater than the die diameter. This is called die swell, extrudate swell or the Barus effect. The extrudate from the extruder die rapidly increases in the cross section area as soon as it emerges from the die. It is defined as

$$\alpha = \left(\frac{\text{Extrudate cross section area}}{\text{Die cross section area}}\right)^{1/2}$$

$$= \frac{\text{Extrudate diameter}, D_e}{\text{Die diameter}, D_c} \text{ (For a simple circular die)} ...(8.30)$$

It is generally agreed that the die swell is a consequence of elastic stress relaxation behaviour however there is no single theory that can explain the die swell behaviour of all the fluids like Newtonian fluids, polymer solutions and polymer melts. For Newtonian fluids it has been reported[36,37] that in the laminar flow region the die swell, α, varies from 0.87 for Reynolds number, $N_{Re} > 100$ to 1.12 for $N_{Re} < 2$. At the low N_{Re} it should be noted that the inertia forces are almost negligible in comparison to the viscous forces however at higher N_{Re} both these forces are important.

Die swell is primarily driven by release of normal stresses at the die exit. As the polymer is forced into the narrow die channel it is extended in the direction of flow and compressed in the normal direction[37a]. This type of flow results in anisotropies (different properties) in microstructure of polymer chains in the direction of flow (x-direction) and normal to flow direction (y and z directions). It simply means that the polymer chains are unevenly compressed in three directions and the release of compressive forces at the end of the die causes contraction of the polymer chains in flow direction and the expansion in normal directions, resulting in an extrudate diameter larger than the die diameter or the die swell.

There is a large amount of experimental data available on the die swell behaviour of viscoelastic polymers. It is essentially due to the fact that there is no single theory to predict the die swell of a polymer from the knowledge of other polymers. The extensive volume of data has resulted in various interpretations from the macroscopic viewpoint of rheology, such as normal stress effect, an elastic energy effect, an entropy enlargement effect, an orientation effect and memory effect. However it has been pointed out that all these interpretations are not unique but are interrelated.[37b, 37c]

Thus it can be said that the die swell is a consequence of complex interplay between extensional stress, shear stress, compressive stress and stress relaxation process. The polymer molecules orient, disentangle, uncoil, elongate and reorient, re-entangle and recoil under different stresses resulting in the stresses being relaxed giving die swell.

8.10.1 Delayed Die Swell

Brandt and Duwig[37d] have reported about delayed die swell particularly at high Reynolds number flow through the capillary. As the melt flows out of die the extrudate increases in diameter as a result of various forces acting on it. The start of swelling has been reported to depend on the flow rate of melt. At low rates of extrusion the swell appears to start at the pipe exit. At high rates of extrusion the tendency of jet to swell at the exit is suppressed thus swelling gets delayed and occurs at some distance L away from the die exit. It was observed that the delayed die swell is a critical phenomenon, in the sense that it occurs beyond a critical speed of the fluid similar to the wave speed of small perturbations in a viscoelastic fluid.

The parameters, which are used to describe the delayed die swell phenomenon are dimensionless numbers known as (a) Viscoelastic Mach number,

$$M = \frac{\text{Average velocity } V_{avg}}{\text{Speed of the shear waves in to the fluid, } c.}$$

$$= \frac{V_{avg}}{c} = \sqrt{\text{Reynolds no.} \times \text{Weissenberg no.}}$$

$$= \sqrt{Re \times We} \qquad \qquad ...(8.30(a))$$

The speed of shear waves into the fluid at rest,

$$c = \sqrt{\frac{\eta}{\lambda \rho}} \qquad \qquad ...(8.30(b))$$

and (b) Elasticity number, E is defined as

$$E = \frac{\text{Relaxation time, } \lambda}{\text{Viscous time scale}(\rho d^2 / \eta)}$$

$$= \frac{\lambda \eta}{\rho d^2} = We/Re \qquad ...(8.30(c))$$

The important points to understand in the delayed die swell phenomenon are,

(a) Delayed die swell may occur in viscoelastic solution if M is large enough.

(b) It is a critical phenomenon and M seems to be the critical parameter.

(c) Numerical simulations show that inertia induces a change from subcritical to supercritical conditions.

(d) The influence of E has been shown in some experimental conditions; an increasing E corresponds to a decreasing delay distance, L.

(e) As an apparent contradiction the influence of E has not been observed clearly in plots of L/d (delay distance/extrusion diameter) vs mach number.

8.10.2 Die swell and non-Newtonian Index, n

Using a simple combination of mass and momentum balance equations, the lower limit of die swell α can be predicted by theory, which results in

$$\alpha = \frac{\sqrt{3}}{2} = 0.87 \qquad ...(8.31)$$

For power law fluids this analysis yields

$$\alpha = \left(\frac{2n+1}{3n+1}\right)^{1/2} \qquad ...(8.32)$$

where n is the non-Newtonian index.

It is important to understand the role of the viscous forces at low N_{Re} value at and near the die exit. The velocity distribution inside the capillary is a fully developed laminar flow profile, which gets disturbed at the die exit due to the viscous effects. Tanner[38] has solved the dynamic and the continuity equations for Newtonian fluids in the neighborhood of the capillary exit and predicted the $\alpha = 1.13$ for low N_{Re} giving a good agreement with the experimental results.

By using the momentum balance method Metzner et al.[39] have analyzed the problem of polymer solutions, which normally show the values higher than 1.13. They assumed that (i) the velocity profile at the exit of the capillary is that for the fully developed power law flow, (ii) the die swell primarily is a consequence of the first normal stress difference, $(\tau_{11} - \tau_{22})$ and the effect of secondary normal stress difference, $(\tau_{22} - \tau_{33})$ is negligible and (iii) the pressure within the fluid at the capillary exit is equal to the ambient pressure. A relationship between die swell and recoverable shear at the wall for the laminar flow was derived from the above analysis. The recoverable shear, $S_R \equiv \dfrac{\tau_{11} - \tau_{22}}{2\tau_{12}}$

is related to α as follows

$$S_R = \frac{1}{f}\left[\left(\frac{1+n}{n}\right)\left(\frac{1+3n}{1+2n}\right) - \frac{1}{n\alpha^2}\left(1+n-\frac{d\log\alpha}{d\log\dot{\gamma}_{w,a}}\right)\right] \quad ...(8.33)$$

where f is the friction factor and $\dot{\gamma}_{w,a}$ is the apparent rate of shear at the wall for the Newtonian fluids. For the polymer solutions this equation predicted the first normal stress difference, which compared well with the experimental results.

As compared to the polymer solutions the polymer melts give much higher die swell, which may vary between 2 to 3. The normal stress values calculated using Eq. 8.33 for polystyrene melts[40] have been reported to deviate by a large margin (almost by a factor of 10^7) with respect to the experimental values. Hence this equation cannot be applied to the polymer melts. The reasons may be the invalidity of the assumptions particularly those related to conditions near the die exit. Following the Lodge's theory of free recovery[41], Tanner[38] further proposed a simple relationship for the die swell of the polymer melts flowing through a long capillary,

$$\alpha = 0.1 + \left(1 + \frac{1}{2}S_R^2\right)^{1/6} \quad ...(8.34)$$

The factor of 0.1 has been added empirically to compensate the viscous effects as reported by Middleman and Gavis[36].

The experimental results on a number of polymers melts have shown that the Eq. 8.34 is quite inadequate for the prediction of the die swell probably due to the fact that the die swell is a complex function of a number of parameters, which are not taken care of during the derivation of this equation. The different parameters, which affect the die swell are (i) length to diameter ratio of the die, (ii) shear stress and shear rate, (iii) temperature of the melt and extrudate surroundings, (iv) solid fillers loading and (v) molecular weight and molecular weight distribution of the melt. Effect of these parameters has been discussed qualitatively by a number of investigators on the die swell behaviour of thermoplastic melts like polyethylenes[42-52], polypropylene[53,54] and polystyrene[40,55], gum and filled rubbers[56-59], rubber plastic blends[60-66] and gum and filled thermoplastic elastomers. The results and discussions of these investigations are summarized in the next sections.

8.10.3 Effect of Length to Diameter Ratio of the Die[42, 47, 48, 54, 60, 67]

Increase in length to diameter ratio, L/D, of the capillary reduces the die swell and for high L/D value the equilibrium die swell is reached. At the die entry high extensional deformation causes the polymer molecules to be oriented in the direction of the flow at a much faster rate, which does not allow elastic stresses to be relaxed. As a consequence high elastic energy is stored in the system. For short dies this energy manifests in the form of high magnitude of first normal stress difference giving rise to high die swell value. If the die swell measurements are done fairly close to die exit one gets what is known as the running die swell, which is normally much higher than the equilibrium value obtained after sufficient time is allowed to relax the stresses.

Figure 8.16 gives the variation of the percent running die swell ($\%\alpha = [(D_e - D_c)/D_c] \times 100$) with L/D ratio for gum and short carbon fiber filled thermoplastic elastomeric blends of natural rubber and high density polyethylene at 160°C and 200°C and 50.6 s^{-1} apparent rate of shear. The trend of variation of α with L/D is similar to the trend observed for a number of thermoplastics and rubbers i.e. a sharp initial decrease in α, which levels off at high L/D values. For long capillaries the residence time of the melt increases and the elastic stresses are progressively reduced tending to disorient the molecules. However the shear flow tends to orient the molecules and store elastic energy. Hence for the long capillaries the melts acquire an equilibrium-relaxed state, giving rise to the equilibrium die swell, which becomes independent of die length.

Wong[67a] has reported the effect of temperature, load and L/D ratio on flow rate and the die swell of polyethylenes. It was found that while the melt flow rate increased with temperature and load it decreased with increasing L/D ratio. The extrudate swell measured at the die exit has a concave downwards characteristics with temperature. Sagging of the extrudate was responsible for maximum swelling ratio, which appeared to shift as L/D ratio varied.

Musameh and Jodeh[67b] reported that the die swell ratio of EPDM composition decreased with decreasing L/D ratio, where as for HDPE it increased. This reverse trend has been explained on the basis, that the relaxation time of EPDM is more sensitive to the shear than that of HDPE.

Liang[67c] has experimented on extrusion conditions on die swell behaviour of polypropylene/diatomite composite melts. It has been reported that the die swell of the composite increased nonlinearly with increase of shear stress and shear rate while it decreased linearly with temperature. At constant load and temperature it increased nonlinearly with diameter, whereas it reduced nonlinearly with increase in L/D ratio.

Liang and Ness[67d] have worked on the die-swell behaviour during capillary extrusion of LDPE/PP blends. It has been reported that the die-swell ratio of the blends decreased approximately linearly with increase in temperature and the L/D ratio.

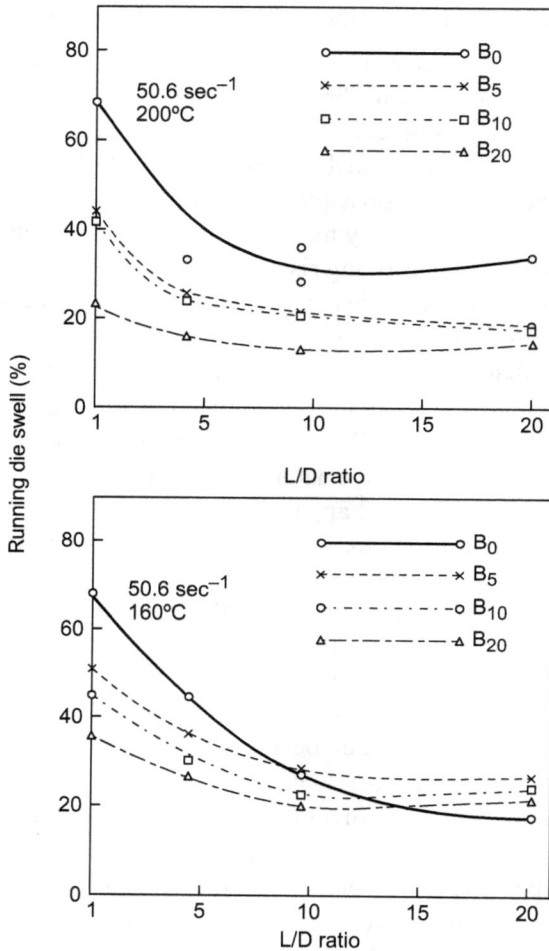

Fig. 8.16   Percent running die swell for NR-HDPE based TPE for
gum and short carbon fiber filled systems[59]
B_0 ... B_{20} refer to the fiber loading (Parts per hundred of rubber, phr)

Behzadfar et al.[67e] have reported that the die swell for HDPEs increases
with decreasing L/D ratio and changes nonlinearly with entrance angle exhibiting
different trends for concave (swell increases from 60° to 180°) and convex
(swell decreases from 180° to 360°) angles.

8.10.4 Effect of Shear Rate and Shear Stress[42,47,48,52-54,56-59,65-67]

Almost all the investigators have reported the variation of the die swell either
with shear stress or shear rate. Die swell increases with increases in both these
parameters for a number of thermoplastics, gum and filled rubbers, rubber-
rubber and rubber-plastic blends. A number of thermoplastic melts show a
gradual increase in the die swell with rate of shear till the onset of flow instability
at high rate of shear leading to extrudate surface irregularities. The rubber

melts however show a maximum in the running die swell at around 1000 to 3000 s^{-1} rate of shear as shown in Fig. 8.17 for both gum and filled natural rubber. Initially at low rate of shear the die swell only increases marginally but at rate of shear value around 1000 s^{-1} it starts increasing very fast goes through a maximum and then decreases. The exact reasons for such a behaviour are not clearly understood.

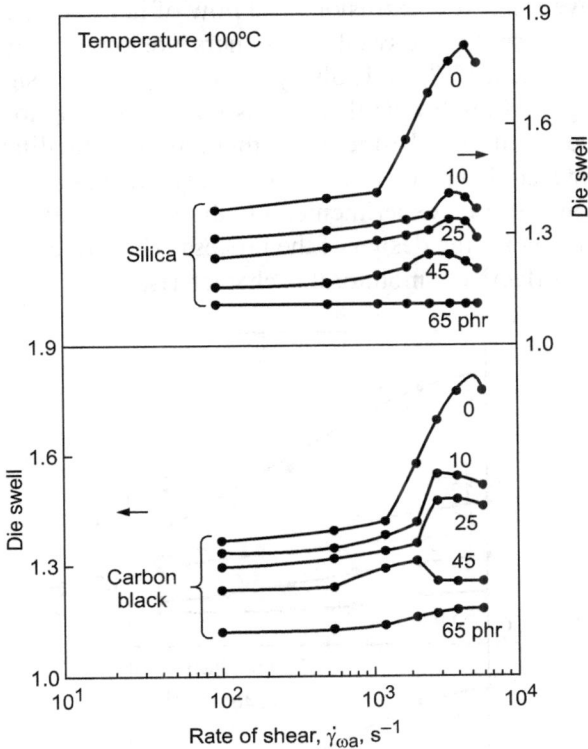

Fig. 8.17   Variation of running die swell on the rate of shear and filler loading for gum and filled NR [56]

8.10.5 Effect of temperature[45,53,56-58,60,61,64,66,67,68,68a]

In general the die swell increases with decrease in temperature of the melt essentially due to the fact that reduction in temperature increases the melt elasticity (Fig. 8.18). Here it is important to note that the temperature of surrounding when the extrudate emerges out of the die has considerable influence on the die swell. Highest die swell is obtained under the isothermal conditions i.e. when the die exit region temperature is maintained equal to the melt temperature either by electrical heating or the extrudate is allowed to float in the hot oil having same temperature and density as that of the melt. In the first case the gravity may affect the die swell whereas in the case of use of oil, the oil should be so selected that it must not swell the polymer. For this the solubility parameters of oil and polymer must be quite different. If the surrounding temperature is much lower than the melt temperature then the extrudate may be frozen giving the lower die swell[45,53], which increases on annealing the extrudate. Reduction in die wall temperature also has been reported to increase the die swell[68].

Henderson and Rudin[68] have reported on the effects of die temperature on extrudate swell in screw extrusion. The flow of hot polymer melt through cold die zone increases the die swell of polymers like low-density polyethylene (*LDPE*), polypropylene (*PP*), and polystyrene (*PS*). Two opposing effects were observed during the flow (a) the thermoplastic melts having low temperature give higher die swell and (b) the same thermal state facilitates molecular disentanglement resulting in reduced melt elasticity and thus the lower die swell. Such a behaviour must be experimented in laboratory to carry out laboratory scale measurements for the design of the process. The authors have described the procedure to effectively monitor this characteristic.

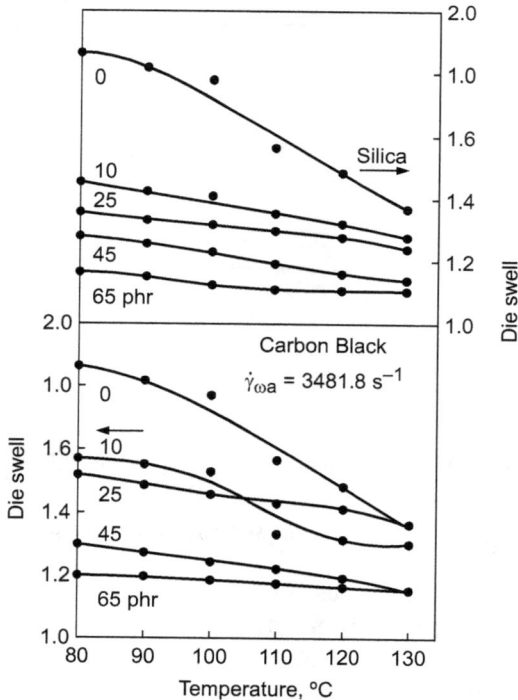

Fig. 8.18   Effect of temperature on the die swell for gum and filled NR[56]

8.11 EFFECT OF THE SOLID FILLERS[56-58,61,65,67,69]

Addition of fillers to the polymers reduces the die swell of the filled compound in comparison to the unfilled system and as the filler loading increases the die swell progressively decreases. Fig. 8.17 shows the effect of the filler loading of carbon black and silica on the running die swell of the natural rubber compounds. At the rate of shear of around 1000 s^{-1} the die swell rapidly increase and goes through the maximum and then decreases. This feature is more or less common for both filled and unfilled systems except for high filler loading when the die swell is independent of the rate of shear. This behaviour is attributed to a number of factors such as polymer-filler interaction, reduced rubber content per unit volume of the compound as the filler dose increases thereby reducing the capacity of the matrix to store the elastic energy and increase in the viscosity of the matrix as suggested by White[69].

8.12 EFFECT OF MOLECULAR WEIGHT AND MOLECULAR WEIGHT DISTRIBUTION [49-52,55,59]

The molecular weight (MW) and the molecular weight distribution (MWD) or the polydispersity index of the melt influence the die swell considerably. It is generally observed that increase in the MW increases the die swell due to increase in the elasticity of the melt. Increase in the MWD also increases the die swell for the similar reasons. At low rate of shear values $< 115 \text{ s}^{-1}$ the die swell attains a maximum value for $\bar{M}_w/\bar{M}_n > 10$ possibly due to relative decrease in viscosity as well as elasticity of the melt due to broadening of the MWD. For polymers of different MWD prepared by blending the samples of same polymer but different MW, it has been reported[51] that the die swell of such blends is highly sensitive to the MW of high MW fraction. Mastication of the natural rubber[59] reduces the die swell considerably due to reduction in the MW and broadening of the MWD.

8.13 RELATIONSHIP BETWEEN THE MAXIMUM RECOVERABLE DEFORMATION AND DIE SWELL[56-58,70,71]

The die swell is a result of the elastic recovery due to the relaxation of the stored elastic stresses. The elastic energy stored per unit volume, W, of the polymer melts flowing through a capillary is proportional to the maximum recoverable deformation, γ_m, and maximum deformational stress, τ_m and can be expressed as[70]

$$W = C\gamma_m\tau_m \qquad \qquad ...(8.35)$$

where C is a constant.

The elastic energy associated with the die swell essentially arises due to the extensional deformation of polymer matrix. Under this deformation the principal extension ratios in X, Y and Z directions are λ_1, λ_2 and λ_3 respectively. The work of deformation or the work energy function is given as[71]

$$W = \frac{G}{2}\left(\lambda_1^2 + \lambda_2^2 + \lambda_3^2 - 3\right) \qquad \qquad ...(8.36)$$

where G is the modulus of elasticity. When the extrudate emerges out of the die it behaves like an elastic body. Due to relaxation the diameter of the extrudate increases by a factor of α, and the length decrease by a factor of α^2 (Fig. 8.19). The principal extension ratios can be expressed in terms of α as

$$\lambda_1 = \alpha^{-2} \text{ and } \lambda_2 = \lambda_3 = \alpha \qquad \qquad ...(8.37)$$

Assuming that the elastic properties of the polymer can be described by the Hooke's law under the low strain region then

$$\tau_m = G\gamma_m \qquad \qquad ...(8.38)$$

Using Eqs. 8.35 to 8.38, the maximum deformation γ_m can be expressed as

$$\gamma_m = \sqrt{\left[\frac{1}{2C}(\alpha^{-4} + 2\alpha^2 - 3)\right]} \qquad \qquad ...(8.39)$$

The constant C can be estimated from the following considerations

The general expression for the elastic energy stored per unit volume during deformation of the fluid while flowing through the capillary is[70]

$$W = \int_0^{r_0} 2\pi r V(r)\left[\int_0^{\gamma_e} \tau\, d\gamma_e\right] dr \qquad \qquad ...(8.40)$$

where r_0 is the capillary radius.

Applying the Hooke's law for the elastic energy

$$\int_0^{\gamma_e} \tau\, d\gamma_e = \frac{\tau^2}{2G} \qquad \qquad ...(8.41)$$

Assuming that the power law is applicable for the polymer the velocity distribution of the melt flow is given as

$$V(r) = \frac{3n' + 1}{n' + 1}\left(1 - \left(\frac{r}{r_0}\right)^{\frac{n'+1}{n'}}\right)V_{avg} \qquad \qquad ...(8.42)$$

where V_{avg} is the average velocity of the fluid and n' is the flow behaviour index obtained from pseudo-shear plots (as discussed in Chapter 3). Substituting Eqs. 8.41 and 8.42 in Eq. 8.40 and solving one gets on comparison with Eq. 8.35

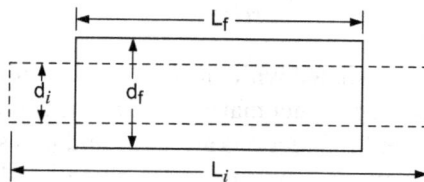

Fig. 8.19 Effect of die swell on extrudate dimensions.[57]

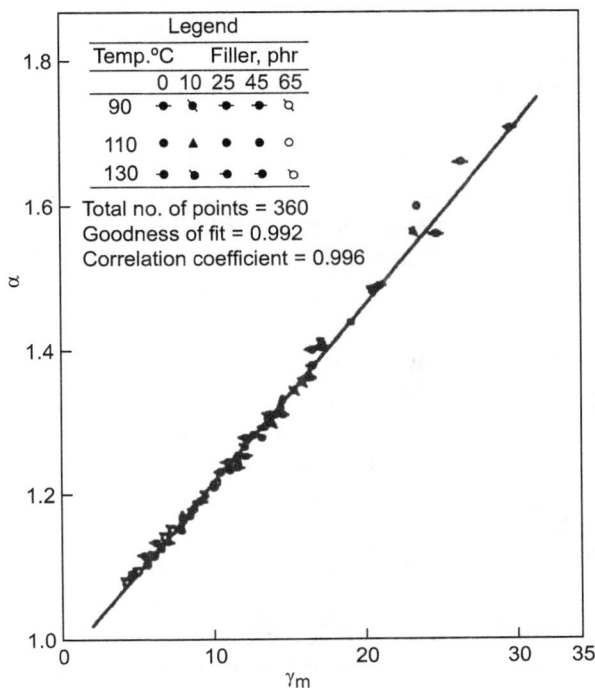

Fig. 8.20 Die swell as a function of maximum deformable deformation[56]
(Unfilled symbols are for carbon black and filled ones for silica filled systems)

$$C = \frac{3n' + 1}{4(n' + 1)} \qquad \qquad ...(8.43)$$

The final expression for maximum recoverable strain is derived as

$$\gamma_m = \sqrt{\left[\frac{2(n' + 1)}{3n' + 1} \left(\alpha^{-4} + 2\alpha^2 - 3 \right) \right]} \qquad ...(8.44)$$

Thus the maximum recoverable deformation can be calculated from the knowledge of the die swell and the non-Newtonian index of the system. A plot of γ_m versus α for number of filled and gum rubbers like natural rubber[56], neoprene and acrylic rubbers[57] and epoxidized and acrylic rubbers[58] has been plotted, which gave a straight line (Fig. 8.20) with a very simple relationship, which is

$$\alpha = 0.26\gamma_m + 0.96 \qquad ...(8.45)$$

8.14 MELT FRACTURE

In plastics industry there are essentially two ways of processing a plastic for production of plastic components. These are, (i) by injection moulding, and (ii) by extrusion. Injection moulding refers to the process in which polymer melt is injected in to a mould of the component to be produced. Extrusion is essentially a process in which the plastic components are made by pumping a polymer melt through a die and then cooling and solidifying. The occurrence of die swell and melt fracture are the common problems faced by the polymer industries in general.

The occurrence of the die swell has been accepted to be a consequence of elastic recovery or the elastic stress relaxation. Another phenomenon believed to be associated with the elastic response of the polymer melts is melt fracture. As the melt is extruded through the dies with progressively increasing flow rate the extrudate surface undergoes a number of transformations. These include, (i) "Matte or sharkskin", a fine scale surface roughness, (ii) "Ripple", a fairly regular wavy appearance on the surface, (iii) "Bamboo or palm tree", extrudate containing alternatively relatively smooth longer section followed by a smaller section having appearance of bamboo knot, hence the name, (iv) "Screw or helix", having the appearance of helical screw or a rope and (v) "Gross melt fracture", a totally random surface distortion having absolutely no regularity in appearance (Fig. 8.21). Melt fracture therefore represents a real problem in polymer processing affecting the product quality in almost all extrusion and film blowing industries.

Tordela[72], Petrie and Denn[73], and White[74] have published the reviews of melt fracture phenomenon. The parameters responsible for the melt fracture are (i) Die characteristics like entrance geometry, L/D ratio and material of construction, which may cause the wall slip and stick and slip, (ii) temperature, (iii) flow rate, rate of shear and shear stress and (iv) material parameters like chemical structure, molecular weight, molecular weight distribution, branching, entanglement and the relaxation time. However it is believed that the melt fracture is the consequence of flow instability caused during the fluid flowing from a larger cross section to the smaller cross section area.

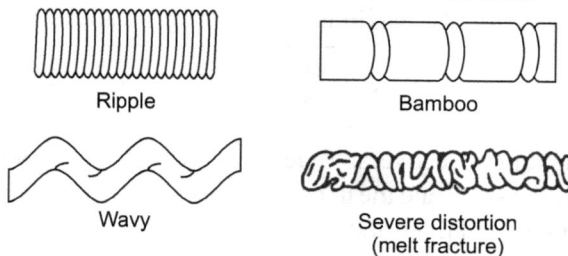

Fig. 8.21   Different extrudate irregularities[37]

Ho and Denn[74a] studied the linear stability of flow of polymers through a pipe (Poiseuille flow). They found that the flow was linearly stable. This has been the reason for most researchers to discard the possibility that the undulations could result from a flow instability in the die, i.e., in pipe or slit through which the polymer flows. However it has been reported by Saarloos[74b] that even if the flow is linearly stable, it could well be nonlinearly unstable. This means that while the flow is stable to any small perturbation, it is unstable to perturbations of small but finite size, a phenomenon similar to that takes place during transition from laminar flow to turbulant flow in Newtonian fluids.

It has been more or less accepted that the melt fracture originates at the die entry and the die exit helps only to aggravate it. The streamline patterns obtained by streak photography of the melt flow of LDPE and PS reported by Baired and Collias[17] and observations of Clegg[75] for polyethylenes confirms it. At the die entry the melt flows from a bigger cross section of extruder to a smaller cross section of the die undergoing high extensional deformation. The flow streamlines change the flow direction by a varying degree, which is the highest for those near the wall and least near the center. The flow lines develop a natural conical entry profile surrounded by large vortices all around the cross section for LDPE (Fig. 8.22). For LLDPE and the PS these vortices are small and the streamlines curve almost by 90° over a short distance[17]. As the flow rate or the rate of shear is increased the circulating fluid is periodically drawn into the main flow causing the flow instabilities resulting in the melt fracture.

For the short dies, high surface irregularities like severe ripples, bamboo or gross melt fracture appear. Whereas for the long dies at relatively small flow rates the disturbance may die down in the die and only small-scale sharkskin or waviness might result. Nevertheless all materials do not show all forms of melt fracture.

Lenk[2] has reported that the formations of deformed shapes on the extrudate surface are the result of "Snapback" action of melt at the die exit. Snapback is a cyclic oscillatory elastic retraction due to strain recovery. Higher the flow rate higher will be the strain giving faster recovery and severe deformations. Bergem[76] reported that the screw thread like deformation resulted at the die exit. Cogswell[77] suggested that the abrupt change in the boundary conditions at the die exit causes a high degree of stretching of the melt resulting in the fracture of the extrudate. Ramaswamy[78] and Kalika[79] have suggested the wall slip to be the main cause for the screw thread appearance.

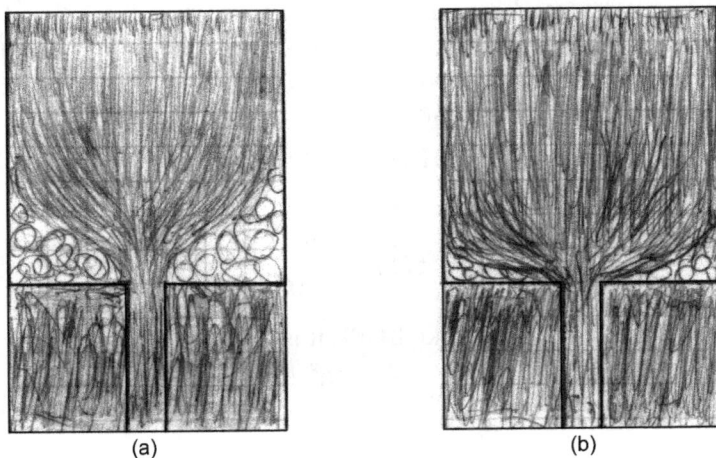

Fig.8.22   Die entry flow patterns (a) LDPE and (b) PS[17]
[(a) Temp = 150°C and $\dot{\gamma}$ = 80 s^{-1} and (b) Temp. = 190°C and $\dot{\gamma}$ = 20 s^{-1}]

The instability caused at the entry of the die tends to oscillate the pressure as well as flow rate particularly for polyethylenes resulting in the extrudate having rough helical and rope-like appearance followed by smooth zones[76,78,79-84]. Vinogradov[80, 81] reported this phenomenon and called it as a " Spurt effect", which has been attributed to the stick and slip at the capillary wall. Lupton and Regester[82] have suggested that the limited compressibility of the polymer melts under the processing conditions is the main reason for such fluctuations in the pressure and flow rates. Bergem[76] and Piau et al.[84] have shown that the helical pattern is related to the swirling motion at the die entry. Ramamurthy's[78] studies with LLDPE and HDPE have shown that the oscillatory flow occurs at wall shear stress of 0.43 M Pa.

Increase in the flow rate beyond the region of the oscillatory flow causes the gross melt fracture or intensely random surface distortion in the extrudate. Kalika and Denn[79] termed it as "Waviness". Ballenger et al[83] and, Spencer and Dillon[85] reported the similar observations for polystyrene and others[83, 86] for polypropylene at high flow rates of the melts. Low-density polyethylene does not show the waviness before the onset of gross melt fracture as observed by Ramamurthy[78] and Yamane and White[89].

Studies on the effects of the molecular structure of the polymers on the melt fracture have been reported by a number of workers[72-74,83,85,87]. The branched polymers behave differently than the linear once in the extent and nature of the melt fracture. Low-density polyethylenes show progressively reduced melt fracture with increase in the die length where as HDPE and PS, show the opposite effect. However the PP is an exception to this generalization. Vlachopoulos and Alam[87] have investigated a large number of melts of both linear and branched polymers by varying the MW and MWD. The analysis of their data gave a good straight-line correlation when the ratio of critical stress, τ_{cr}, to absolute temperature, T_{abs}, was plotted against inverse of the MW (Fig. 8.23). The critical stress refers to the stress at the onset of the melt fracture. These correlations are given below.

For linear polymers like PP, HDPE and PS of narrow and broad MWD the equation is

$$\frac{\tau_{cr}}{T_{abs}} = 1717 + \frac{2.7 \times 10^8}{\bar{M}_w} \qquad ...(8.46)$$

And for branched polymer like LDPE it is

$$\frac{\tau_{cr}}{T_{abs}} = 1317 + \frac{10^8}{\bar{M}_w} \qquad ...(8.47)$$

Cogswell and Lamb[88] proposed a relationship for the estimation of the critical stress from the data of the capillary rheometer for the power law fluids using a knife-edge die. It is given as

$$\tau_{cr} = \frac{3(3n+1)P_0}{8} \qquad ...(8.48)$$

where P_0 refers to the pressure, which gives unacceptable extrudate quality.

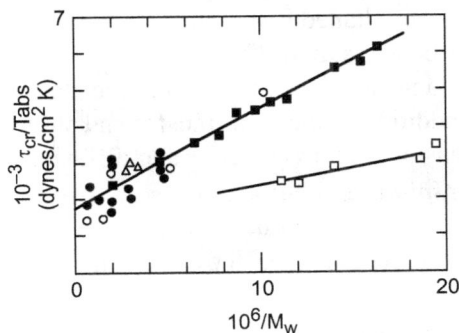

Fig. 8.23 Correlations of critical stress with MW[87]

In the figure 8.23 different symbols refer to different polymer with different MW as:

Δ	Polypropylene	MW = 287,000–660,000	M_w/M_n = –
■	HDPE	MW = 60,000–150,000	M_w/M_n = 2.4–1.9
•	Broad distribution PS	MW = 200,000–1,350,000	M_w/M_n = 2.4–19
○	Narrow distribution PS	MW = 97,000–180,000	M_w/M_n = 1.1–1.2
□	LDPE	MW = 50,000–90,000	M_w/M_n = 3–6

From a careful study of the literature about the melt fracture and the explanations given by different workers it can be said that the mechanism of the occurrence of this phenomenon is still not clearly understood. It is not possible to give definite quantitative correlations between different parameters and the extent of the melt fracture. The experimental data for the onset of the melt fracture has to be obtained to decide the upper limit of flow rate for individual polymer melts for good quality extrudate.

8.15 FIBER SPINNING

The fiber spinning is the process to produce a fiber, which is the fundamental unit of textiles. It is a process, which involves different methods where die swell and extensional flow play important roles. It is mainly achieved by using any of the three approaches

(i) *Melt spinning:* It involves the cooling to solidify and simultaneous stretching of the melt extrudate flowing out of the spinneret. This process is applied to the polymers like polyamides, polyesters, polystyrene, polyolefins and the inorganic glasses, which show good thermal stability and high fluidity at the processing temperature. The heat transfer and the extensional rheological behaviours are the main considerations for the analysis of the process.

(ii) **Solution dry-spinning:** It involves the extrusion of the polymer solution in a volatile solvent through the spinneret, which is a plate containing a large number of small holes, evaporation of the solvent using a stream of hot air and simultaneous stretching to form the fiber. The solvent is recovered by condensation. The fibers from the polymers like cellulose acetate, acrylonitrile, vinyl chloride and acetate are produced using this process. In addition to the heat transfer and stretching, it involves the considerations of the mass transfer and diffusion.

(iii) **Solution wet spinning:** The thermally sensitive polymers and also those, which cannot be dissolved in a volatile solvent are treated using this process. The process involves the extrusion of the polymer solution (in a relatively nonvolatile solvent) into a liquid bath of coagulating agents that drive the solvent out of the filament. In this process also the mass transfer and diffusion are the additional considerations for the process analysis.

Other processes, which are also sometimes used for the fiber spinning, are phase separation spinning, emulsion spinning, gel spinning and reaction spinning.

The melt spinning process consists of (a) pumping the melt through a screen pack by means of an extruder, (b) filtering of melt through screens and sand, (c) passing it through a spinneret, to divide the flow into many small streams, (d) cooling of the these streams by cold air or water vapor flowing perpendicularly to the filament axis, (e) solidification of the melt streams and simultaneous cold drawing to form the fibers and (f) winding at the take up rolls as shown in Fig. 8.24. The holes in the spinneret may be circular, trilobal, square or cross-shaped depending on the requirement of the filament type.

During the flow of the filament following phenomena occur as it moves away from the spinneret

(i) Melt extrusion through the die having low L/D ratio where the fluid velocity distribution changes rapidly.

(ii) Swelling of the extrudate immediately after die exit. Very fast change in the length of the filament due to the application of take up force.

(iii) Rapid change of the filament temperature due to flow of the cold air thus very quick change in the rheological nature of the melt. Melt crystallization under high stress and rapid cooling.

These phenomena occur so rapidly that there is no time to attend either the equilibrium temperature or the cross section and therefore only the transient conditions exist. The mathematical analysis and modeling of this problem is extremely difficult and only highly simplified case of Newtonian, isothermal and steady state case has been modeled[17] and solved analytically for the filament radius and velocity distributions. Kase and Matsuo[90, 91] and Kase[92] have modeled the dynamics of the non-isothermal Newtonian process. Their analysis resulted in a set of transient equations, which must be solved numerically. Denn et al.[93] and Fisher and Denn[94] evaluated the effects of melt viscoelasticity on the melt spinning process. The analyses mentioned above are applicable for low and moderate speed (from 1000 m/min to 3000 m/min) spinning process, where very little degree of crystallinity can be generated in the spun fiber. High degree of orientation and crystallinity are obtained in a single stage high speed spinning process above 5000 m/min. Lin et al.[95] reported the effect of degree and rate of crystallinity on the spinning process and the physical characteristics of the system.

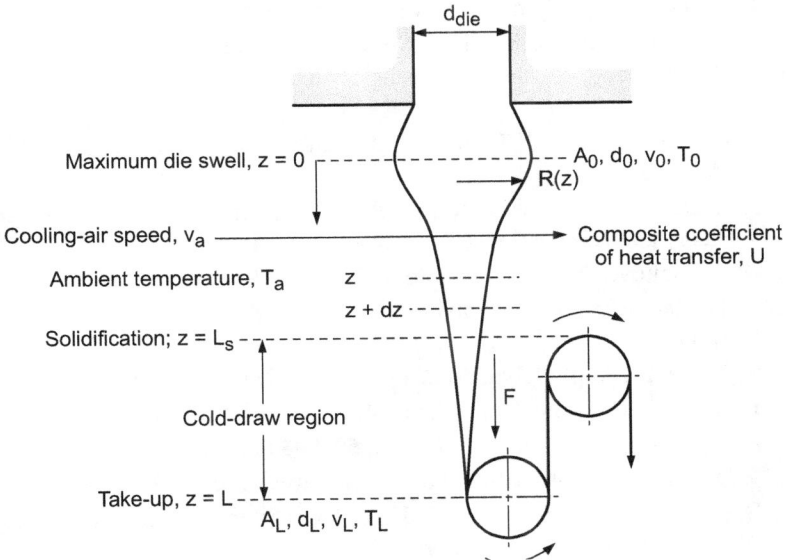

Fig. 8.24 Schematics of melt spinning process[17]
(A, d, v and T are area, diameter, velocity and temperature respectively)

Various forces, which act on the filament and influence the spinning process, are (i) air drag (ii) inertia, (iii) rheological, (iv) gravitational and (v) surface tension forces. Ziabicki and Kawai[96] reported that the influence of the air drag is dominating over the other forces. The effects of rheological forces remain more or less constant from low to high take-up speeds. Both air drag and inertia force increase substantially as the take-up speed increases.

Thus it can be said that both viscous and the elastic rheological responses of the melt play an important roll in understanding and solving the polymer processing problems resulting in better quality control and optimum production rate.

REFERENCES

1. B. R. Gupta, Rubber Processing on a two-roll mill, Allied Publishers Ltd., New Delhi, 1988.
2. R. S. Lenk, Polymer Rheology, Applied Science Publishers Ltd., London, 1978.
3. Z. Tadmor and C. G. Gogos, Principles of Polymer Processing, John Wiley and Sons, New York, 1979.
4. A. Kumar and S. K. Gupta, Fundamentals of polymer Science and Engineering, Tata-P. K. Freakley, Rubber Processing and Production Organization, Plenum Press, New York, 1985McGraw Hill Pub. Co. Ltd., New Delhi, 1978.
5. P. K. Freakley, Rubber Processing and Production Organization, Premium Press , New York, 1985.
    (a) Chareon Tut Co. Ltd. Thailand [chareontut.com]
    (a1) Mitsubishi Heavy Industries Technical Review Vol. 53(3), (September 2016).
6. C. Rauwendaal, Polymer Extrusion, Hanser, New York, 1986.
7. W. H. Darnell and E. A. J. Mol, S P E J., **21**, 20 (1956).
8. J. M. McKelvey, Polymer Processing 1st ed. John Wiley and Sons Inc, New York, 1962.
9. J. A. Biesenberger, ed., Devolatilization of Polymers, Hanser Publications, New York, 1983.
10. J. M. Dealy and K. F. Wissbrun, Melt Rheology and its Role in Plastic Processing – Theory and Application, Chapman and Hall, London, 1995.
11. M. H. Naitove, Plastics Technology, **23(2)**, (1997).
12. C. D. Han, Rheology of Polymer Processing, Academic Press, New York and London, 1976.
13. C. D. Han, J. Appl. Polym. Sci., **17**, 1289 (1973).
14. J. Meissner, Pure and Appl. Chem., **42**, 553 (1975).
15. H. H. Winter, Pure and Appl. Chem., **55**, 943 (1983).
16. W. J. Schrenk and T. Alfrey, Polym. Eng. Rev., **2(4)**, 363 (1983).
17. D. G. Baired and D. I. Collias, Polymer Processing- Principles and Design, John Weiley and Sons Inc., New York, 1998.
18. H. Janeschitz - Kriegel, Rheol. Acta, **16**, 327 (1977).
19. H. Van Vijngaarden, J. F. Dijksman and P. Wesselling, J. Newt.Fl.Mech., **11**, 175 (1982).
20. W. Rose, Nature, **191**, 242 (1961).
21. H. Yokoi, T. Hayashi, N. Morikita and K. Toda, SPE Tech. Papers, **34**, 329 (1988).
22. K. Oda, J. L.White and E. S. Klark, Polym. Eng. Sci., **16**, 585 (1976).
23. H. T. Kim, J. P. Darby and G. F. Wilson, Polym. Eng. Sci., **13**, 372 (1973).
24. R. A. Worth and J. Parnaby, Trans. Instn. Chem. Engrs., **52**, 368 (1974).
25. R. J. Pritchalt, J. Parnaby and R. A. Worth, Plastics and Polymers, P 55 (April 1975).
26. H. H. Winter and H. J. Fritz, Polym. Eng. Sci., **26**, 543 (1986).
27. A. Rudin and H. P. Schreiber, Polym. Eng. Sci., **23**, 422 (1983).

28. M. E. Ryan and A. Dutta, Polym. Eng. Sci., **22**, 569 (1982).
29. C. Rauwendaal, Polym. Eng. Sci., **27**, 186 (1987).
30. D. M. Kalyon, J. S. Yu and C.-C. Du, Polym. Proc. Eng., **5**, 179 (1987).
31. C. D. Han and R. Shetty, I E C Fund., **16**, 49 (1977).
32. J. L. White and H. Yamane, Pure Appl. Chem., **59**, 193(1987).
33. T. Debroth and L. Erwin, Polym. Eng. Sci., **26**, 62 (1986).
34. N. R. Anturkar and A. Co, J. Non-Newt. Fl. Mech., **28**, 287 (1988).
35. F. N. Cogswell, Polym. Eng. Sci., **12(1)**, 64 (1972).
36. S. Middleman and J. Gavis, Phys. Fluids, **4**, 355 (1961).
37. S. Middleman, Fundamentals of Polymer Processing, McGraw Hill Book Co., New York, 1979.
    (a) T. Kealy, Rheology Solutions Pty. Ltd., "Tim's Top Tips- Explanations and Evaluation of Die Swell". [info@rheologysolutions.com]
    (b) J. Z. Liang, Plast. Rubber Compos. Process Appl. 1991157579
    (c) J. Z. J. Liang, Process Mater. Techol., 199552207215.
    (d) L. Brandt and C. Duwig, "Delayed die swell in Fluid Mechanics for graduate students-Non-Newtonian fluid Mechanics" [mech.kth.se]
38. R. I. Tanner, J. Polym. Sci., **A-2(8)**, 2067 (1970).
39. A. B. Metzner, W. T. Houghton,R. A. Sailer and J. L. White, Trans. Soc. Rheol., **5**, 133 (1961).
40. W. W. Graessley, S. D. Glasscock and R. L. Crawley, Trans. Soc. Rheol.,**14**, 519 (1970).
41. A. S. Lodge, Elastic Liquids, Academic Press, New York, 1964, P. 131.
42. C. R. G. Allen and M. Rides, NPL UK report No. CMMT (A) 9, !996.
43. N. Orbey and J. M. Dealy, Polym. Eng. Sci., **24**, 511 (1984).
44. A. Gracia-Rejon and J. M. Dealy, Polym. Eng. Sci., **22**, 158 (1982).
45. J. L. White and J. F. Roman, J. Appl. Polym. Sci., **20**, 1005 (1976).
46. L. A. Utracki, Z. Bakerdjian and M. R. Kamal, J. Appl. Polym. Sci., **19**, 481 (1975).
47. C. D. Han, M. Charles and W. Phillippoff, Trans. Soc. Rheol., **14**, 393 (1970).
48. E. B. Bagley, S. H. Storey and D. C. West, J. Appl. Polym. Sci., **7**, 1668 (1963).
49. M. G. Rogers, J. Appl. Polym. Sci., **14**, 1679 (1970).
50. R. A. Mendelso and F. L. Finger, J. Appl. Polym. Sci., **19**, 1061 (1975).
51. R. J. Koopmans, J. Polym. Sci., **A-26**, 1157 (1988).
52. S. K. Goyal, I. B. Kazatchkov, N. Bohnet and S. G. Hatzikiriakos, Annu. Tech. Conf.- Soc. Plst. Eng., 55th (Vol.1), 1076,(1997).
53. J. Brandao, E. Spieth and C. Lekakou, Polym. Eng. Sci., **36(1)**, 49 (1996).
54. Y. Mori and K. Funatsu, J. Appl. Polym. Sci. (Appl. Polym Symp.), **20**, 209 (1973).
55. J. Vlachopoulos, M. Morie and S. Lidorikis, Trans. Soc. Rheol., **16**, 669 (1972).
56. N. R. Kumar, A. K. Bhowmick and B. R. Gupta, Kautsch. Gummi, Kunstst., **45(7)**, 531 (1992).

57. Nalini R. Kumar, B. R. Gupta and Anil Bhowmick, Polym. Eng. Sci., **33(15)**, 1002 (1993).

58. N. R. Kumar, B. R. Gupta and A. K. Bhowmick, J. Polym. Eng., **13(2)**, 105 (1994).

59. B. R. Gupta, Ind. J. Nat. Rubber Res., **2**(1), 38 (1989).

60. D. Roy, A. K. Bhattacharya and B. R. Gupta, J. Elast. Plast., **25**, 46 (1993).

61. S. Mukhopadhyay and B. R. Gupta, Kautsch. Gummi, Kunstst., **52(2)**, 80 (1999).

62. Sujata Mukhopadhyay and B. R. Gupta, J. Elast. Plast., **30**, 340 (1998).

63. S. Thomas, B. Kuriakose, B. R. Gupta and S. K. De, Plast. Rubber Proc. Appl., **6**(1), 85 (1986).

64. Tapan K. Bhoumik, Prajna P. De, Anil K. Bhowmick and B. R. Gupta, Ind. J. Technol., **28**, 33 (1990).

65. T. K. Bhoumik, P. P. De, A. K. Bhowmick and B. R. Gupta, Kautsch. Gummi, Kunstst., **42(2)**, 115 (1989).

66. R. N. Jana, A. K. Bhattacharya, G. B. Nando and B. R. Gupta, Kautsch. Gummi, Kunstst., **55**(12), 660 (2002).

67. D. Roy and B. R. Gupta, J. Appl. Polym. Sci., **49**, 1475 (1993).

   (a) A.C.-Y. Wong, J. Mate. Proc. Tech., **79(1-3)**, 163-169, (1998)

   (b) S. Musameh and S. Jodeh, An-Najah Univ. J. Res. (N. Sc.), **16(1)**, 2002

   (c) J. Z. Liang, Polym. Testing, **27(8)**, 936-940, (2008).

   (d) J. Z. Liang and J. N. Ness, Polym. Testing **17(3)**, 179-189, (1998)

   (e) E. Behzadfar, M. Ansari, V. K. Konaganti, S. G. Hatzikiriakos, J. Non-Newt. Fluid Mech., **225**, 86-93, 2015.

68. A. M. Henderson A. Rudin, J. Appl. Polym. Sci., **31**, 353 (1986).

   (a) B. Yang and L. J. Lee, Polym. Eng. Sci., **27(14)**, 1088-1094 (1987).

69. J. L. White in Proc. Int. Conf. Polym. Process., Nam P. Suh and Nak-Ho Sund (Eds.), Massachusets, Inst. Tech., Cambridge, 1977, P.26.

70. G. V. Vinogradov and A. Ya. Malkin, Rheology of Polymers, Mir Publishers, Moscow, 1979, P106 and 304.

71. L. R. G. Treloar, The Physics of Rubber Elasticity, Clarendon Press, Oxford, 1975, P. 60.

72. J. P. Tordella, Rheology, Vol. 5, F. R. Eirich ed., Academic Press, New York (1969).

73. C. J. S. Petrie and M. M. Denn, A. I. Ch. E. J., **22**, 209 (1976)

74. J. L. White, Appl. Polym. Symp., **20**, 155 (1973).

   (a) T. C. Ho, and M. M. Denn, J. Non-Newtonian Fluid Mech., 3, 179-195 (1977).

   (b) Wim von Saarloos, Institut-Loentz, June 30, 2003. [http://wwwhome.lorentz.leidenuniv.nl/~saarloos/Patternf/meltfracture.html]

75. P. Clegg, in Rheoelogy of Elastomers, p. 174, Welwyn Garden City Conf., Pergaman Press, Oxford (1958).

76. N. Bergem, Proc.VIIIth Int. Conf. Rheol., Goethenberg, 1976, p.50.

77. F. N. Cogswell, Polymer Melt Rheology, John Wiley and Sons, New York, 1981, P.101.

78. A. V. Ramamurthy, J. Rheol., **30**, 337 (1986).

79. D. S. Kalika and M. M. Denn, J. Rheol., **31**, 815 (1987).

80. G. V. Vinogradov, N. I. Insarova, B. B. Boiko and E. K. Borisenkova, Polym. Eng. Sci., **12**, 323 (1972).

81. G. V. Vinogradov, Polymer, **18**, 1275 (1977).

82. J. M. Lupton and R. W. Regester, Polym. Eng. Sci., **5**, 235 (1965).

83. T. F. Bellenger, I.-Jen Chen, J. W. Crowder, G. E. Hagler, D. C. Bogue and J. L. White, Trans. Soc. Rheol., **15**, 195 (1971).

84. J. M. Piau, N. El. Kissi and B. Tremblay, J. Non-Newt. Fl. Mech., **34**(2), 145, (1990).

85. R. S. Spencer and R. E. Dillon, J. Colloid Sci., **4**, 241 (1949).

86. M. Fujiyama and H. Awaya, J. Appl. Polym. Sci., **16**, 275 (1972).

87. J. Vlachopoulos, M. Alam, Polym. Eng. Sci., **12**, 184 (1972).

88. F. N. Cogswell and P. Lamb, Trans. Plast. Inst., **35**, 809 (1967).

89. H. Yamane and J. L. White, J. Rheol. Japan, **15**, 131 (1987).

90. S. Kase and T. Matsuo, J. Polym. Sci., **A**, **3**, 2541 (1965).

91. S. Kase and T. Matsuo, J. Appl. Polym. Sci., **11**, 251 (1967).

92. S. Kase, in High Speed Fiber Spinning, A. Ziabicki and H. Kawai, Eds., Wiley, New York,1985.

93. M. M. Denn, C. J. S. Petrie and P. Avenas, AIChE J., **21**(4), 791 (1975).

94. R. J. Fisher and M. M. Denn, AIChE J., **22(4)**, 236 (1976).

95. C.-Y. Lin, P. A. Tucker and J. A. Cuculo, J. Appl. Polym. Sci., **46**, 531 (1992).

96. A. Ziabicki and H. Kawai, Eds., High Speed Fiber Spinning, Wiley, New York,1985.

❑   ❑   ❑

Annexure-I(A)

Table 1 Modified power law constants (Eq. 2.31) for various polymer systems

Sl. No.	Systems/Source	Temp. (°C)	Modified power law constants[2]				% δ	Applicable $\dot{\gamma}$ range, s^{-1}
			a (Pas^m)	b	c	d		
(1)	(2)	(3)	(4)	(5)	(6)	(7)	(8)	(9)
1.	Acrylonitrile-Butadiene Styrene (ABS) Terluran 877T (BASF)	170	94000.0	0.055	0.315	0.092	+10, −2	1–400
2.	Acrylonitrile-Styrene Acrylester (ASA) Lu-ranS-757R (BASF)	170 210 250	114691.0 13694.0 2557.3	0.056 0.217 0.286	0.33 0.59 0.819	0.123 0.15 0.15	+3, −5 +2, −2 +1, −2	1–200 8–10000 100–10000
3.	ASA-LuranS 776S (BASF)	170 210 250	90202.0 17414.0 4669.3	0.57 0.19 0.255	0.34 0.61 0.73	0.149 0.174 0.154	+1, −1 +8, −5 +4, −3	1–200 4–10000 10–10000
4.	Polyamide (PA)-Ultra-mid-A5 (BASF)	270	1339.4	0.658	2.509	0.406	+5, −10	20–1000
5.	PA-Ultramid-B3 (BASF)	250	36.0	0.57	1.70	0.146	+9, −5	20–2000
6.	PA-Ultramid-B35 (BASF)	250	6124.0	0.346	1.168	0.18	+7, −9	1–100
7.	PA-Ultramid-B4 (BASF)	250	632.7	0.547	1.484	0.20	+10, −5	1–1500
8.	PA-Vestamid L-1600 (CWH)	200 220 240 260	57.8 6.70 6.89 2.76	0.592 0.768 0.59 0.61	1.648 2.054 1.682 1.649	0.155 0.168 0.119 0.10	+8, −10 +1, −3 +2, −4 +9, −1	20–4000 30–6000 60–4000 100–8000
9.	Polyethylene (PE) Lupo-len-1800H (BASF)	150 190 230	8778.0 3385.4 915.6	0.116 0.207 0.303	0.613 0.67 0.878	0.123 0.13 0.143	+9, −8 +9, −9 +5, −4	1–1000 5–3000 10–10000
10.	PE-Lupolen 1800S (BASF)	150 190 230	1510.0 533.8 179.3	0.065 0.117 0.168	0.691 0.748 0.819	0.033 0.044 0.050	+2, −2 +6, −4 +6, −3	2–2000 10–6000 20–10000
11.	PE-Lupolen 1812D (BASF)	150 190 230	26903.0 15214.4 9641.0	0.045 0.062 0.073	0.411 0.42 0.458	0.046 0.045 0.051	+5, −7 +6, −7 +7, −9	1–1000 1–3000 1–10000
12.	PE-Lupolen 2410H (BASF)	150 190 230	11790.0 6310.7 2402.3	0.073 0.085 0.160	0.482 0.517 0.616	0.064 0.063 0.079	+5, −9 +5, −8 +5, −5	1–1500 3–3000 20–10000

Sl. No.	Systems/Source	Temp. (°C)	Modified power law constants[2]				% δ	Applicable $\dot{\gamma}$ range, s^{-1}
			a (Pas^m)	b	c	d		
(1)	(2)	(3)	(4)	(5)	(6)	(7)	(8)	(9)
13.	PE-Lupolen 3020D (BASF)	150	28246.0	0.093	0.434	0.070	+5, −7	1–1000
		190	15502.0	0.112	0.517	0.088	+4, −7	1–2000
		230	10655.0	0.106	0.533	0.082	+8, −9	10–3000
14.	LDPE-Lotrene FA-0240 (CdF)	190	2779.4	0.283	0.719	0.127	+6, −8	10–10000
15.	LDPE-Lotrene FB3010 (CdF)	190	9182.0	0.20	0.566	0.109	+5, −5	10–10000
16	LDPE-Lotrene FD0274 (CdF)	190	1677.4	0.290	0.76	0.118	+3, −6	10–10000
17	LDPE-Lotrene FD5005 (CdF)	190	7517.6	0.16	0.494	0.07	+3, −9	10–10000
18	LDPE-Lotrene FD0374 (CdF)	190	2400.0	0.20	0.68	0.083	+4, −1	10–10000
19	LDPE-Lotrene FE7040 (CdF)	190	6503.0	0.182	0.568	0.087	+6, −2	10–10000
20	LDPE-Lotrene MA0701 (CdF)	190	950.0	0.303	0.887	0.128	+3, −5	10–10000
21	LDPE-Lotrene MA2004 (CdF)	190	410.0	0.31	0.844	0.093	+8, −10	10–10000
22	LDPE-Lotrene MA7007 (CdF)	190	78.2	0.365	1.03	0.088	+4, −8	60–8000
23	HDPE-Vestolen A3512 (CWH)	190	12708.0	0.120	0.638	0.057	+4, −5	10–1000
		230	7151.0	0.155	0.684	0.059	+4, −3	10–1000
24	LDPE, Alkathene-11 (ICI)	170	6315.0	0.167	0.691	0.136	+10, −10	1–900
		210	2265.0	0.20	0.763	0.117	+10, −10	1–1000
25	LDPE, Alkathene-21 (ICI)	170	845.6	0.211	0.823	0.087	+9, −8	2–10000
		210	185.0	0.32	1.01	0.102	+8, −8	10–10000
26	LDPE, Alkathene-23 (ICI)	170	8.1	0.64	1.70	0.161	+4, −6	30–10000
		210	0.53	0.792	2.34	0.177	+9, −10	100–10000
27	LDPE, Alkathene-33 (ICI)	170	20333.0	0.032	0.463	0.077	+5, −10	1–100
		210	10093.0	0.09	0.48	0.068	+7, −7	1–1000
28	LDPE, Alkathene-46 (ICI)	170	5854.7	0.165	0.752	0.137	+7, −8	1–300
		210	1904.0	0.208	0.87	0.13	+10, −10	1–1000
29	Polyether Sulfon Victrex-300P (ICI)	320	6974.3	0.27	1.078	0.125	+5, −8	1–300
		370	992.3	0.217	1.03	0.068	+4, −6	1–300
30	Polymethylmethacry-late PMMA-Diakon LG (ICI)	210	1600.0	0.385	1.105	0.165	+5, −2	10–1000
		240	314.2	0.335	1.18	0.12	+8, −2	5–1000
31	PMMA-Diakon MH (ICI)	210	3684.2	0.259	0.779	0.246	+2, −4	1–100
		240	7000.0	0.240	0.852	0.157	+10, −9	1–1000
32	PMMA-Plexiglas 5N (Roehm)	200	6503.0	0.325	0.937	0.212	+5, −5	1–6000
		220	944.0	0.445	1.27	0.204	+7, −1	10–10000
33	PMMA-Plexiglas 6H (Roehm)	200	37419.5	0.158	0.589	0.186	+9, −10	1–5000
		220	11159.0	0.247	0.795	0.196	+9, −5	1–10000
		240	2724.4	0.317	0.942	0.169	+5, −5	5–10000
34	PMMA-Plexiglas 6N (Roehm)	200	14285.0	0.243	0.835	0.201	+10, −10	1–10000
		220	2368.5	0.372	1.116	0.197	+2, −4	2–10000
		240	544.6	0.450	1.336	0.197	+3, −7	5–10000

Sl. No.	Systems/Source	Temp. (°C)	Modified power law constants[2]				% δ	Applicable γ̇ range, s⁻¹
			a (Pas^m)	b	c	d		
(1)	(2)	(3)	(4)	(5)	(6)	(7)	(8)	(9)
35	PMMA-Plexiglas 7H (Roehm)	200	66171.6	0.143	0.482	0.202	+6, −10	1–1000
		220	26623.7	0.176	0.549	0.158	+7, −7	10–10000
		240	6003.0	0.288	0.914	0.201	+5, −4	1–10000
36	PMMA-Plexiglas 7N (Roehm)	200	29732.6	0.175	0.67	0.164	+9, −6	1–3000
		220	5431.7	0.31	0.97	0.188	+10, −6	1–10000
		240	875.7	0.448	1.33	0.21	+7, −2	10–1000
36	PMMA-Plexiglas 8H (Roehm)	200	91583.0	0.12	0.445	0.177	+9, −10	1–2000
		220	27584.0	0.22	0.651	0.208	+6, −4	1–6000
		240	7903.0	0.308	0.896	0.227	+6, −4	2–10000
38	PMMA-Plexiglas 8N (Roehm)	200	66171.0	0.158	0.549	0.175	+9, −7	1–1000
		220	14765.0	0.273	0.827	0.206	+5, −2	2–5000
		240	3789.5	0.35	1.03	0.202	+8, −2	10–10000
39	PMMA-Plexiglas ZK43 (Roehm)	220	25591.0	0.22	0.56	0.182	+4, −8	2–1000
40	Polypropylene-PP Novo-len-1120H (BASF)	170	11849.0	0.137	0.543	0.141	+3, −7	1–1000
		210	6974.4	0.175	0.607	0.148	+8, −10	3–2000
		250	4024.0	0.212	0.741	0.165	+11, −10	1–3000
41	Polypropylene-PP Novo-len-1120L (BASF)	170	5431.7	0.197	0.708	0.157	+7, −7	1–2000
		210	2480.0	0.25	0.861	0.177	+10, −10	1–3000
		250	862.5	0.332	1.062	0.185	+7, −10	2–8000
42	Polypropylene-PP Novo-len-1320L (BASF)	170	6124.0	0.215	0.726	0.164	+2, −7	6–1500
		210	2165.0	0.302	0.878	0.17	+4, −2	10–3000
		250	943.9	0.34	0.933	0.155	+2, −3	20–8000
43	Polypropylene-PP Propathene GSE-16 (ICI)	210	7920.0	0.158	0.644	0.15	+10, −10	1–1000
		240	6248.0	0.125	0.631	0.125	+10, −10	1–1000
44	Polypropylene-PP Propathene GWE-23 (ICI)	210	4675.0	0.191	0.638	0.142	+9, −9	1–9000
		240	3197.0	0.185	0.604	0.112	+2, −5	1–9000
45	Polystyrene-PS Polystyrol 143E (BASF)	170	18398.0	0.163	0.549	0.213	+8, −8	1–500
		210	1043.0	0.367	1.073	0.203	+8, −4	10–10000
		250	167.0	0.48	1.377	0.204	+8, −4	60–10000
46	Polystyrene-PS Polystyrol 165H (BASF)	170	35419.0	0.11	0.411	0.232	+2, −9	1–500
		210	25848.0	0.108	0.436	0.197	+3, −10	1–500
		250	1153.0	0.343	1.062	0.215	+8, −8	30–10000
47	PVC Welvic-R7/622 (ICI)	170	16647.0	0.282	0.823	0.164	+5, −2	10–1000
		190	8350.0	0.312	0.844	0.156	+7, −4	10–1000
48	SAN Luran-378P (BASF)	170	64216.0	0.061	0.449	0.134	+4, −8	1–100
		210	4582.0	0.294	0.932	0.184	+10, −2	2–3000
		250	200.0	0.545	1.65	0.218	+9, −6	20–100000
49	Styrene-Butadiene Copolymer SB-Polystyrol 456M (BASF	210	7708.0	0.203	0.657	0.174	+6, −10	2–10000
		250	1480.0	0.338	0.99	0.203	+1, +10	6–10000
50	Styrene-Butadiene Copolymer SB-Polystyrol 475K (BASF)	170	27723.0	0.155	0.42	0.134	+5, −9	1–2000

Annexure-I(B)

Table 2 Two range power law constants for various polymer systems

Sl. No.	Systems / Source	Temp. (oC)	Two Ranges Power Law constants		%, δ	Applicable, $\dot{\gamma}$ range, s^{-1}
			n	K, Pas^n		
(1)	(2)	(3)	(4)	(5)	(6)	(7)
1.	Acrylonitrile-Butadiene Styrene (ABS) Terluran 877T (BASF)	170	0.216	1×10^5	+9, −5	1–400
			0.3265	9.1×10^4	+3, −2	5–500
2.	Acrylonitrile-Styrene Acrylester (ASA) LuranS-757R (BASF)	170	0.2012	1.415×10^5	+7, −4	3–200
			0.4494	$1.15 \text{ x x } 10^5$	+14, −8	0.1–3
		210	0.2196	5.8×10^4	+12, −3	100–10000
			0.4179	2.3×10^4	+11, −7	6–100
		250	0.2869	1.7×10^4	+4, −4	50–300
			0.4229	9.0×10^3	+5, −5	300–10000
3.	ASA-LuranS 776S (BASF)	170	0.1858	1.08×10^5	+6, −6	5–500
			0.4356	8.6×10^4	+5, −5	0.3–5
		210	0.1856	6.0×10^4	+8, −8	60–10000
			0.40	2.5×10^4	+8, −8	2–60
		250	0.2524	2.5×10^4	+7, −7	100–10000
			0.4375	1.1×10^4	+5, −5	10–100
4.	Polyamide (PA)-Ultramid-A5 (BASF)	270	0.2324	7.0×10^4	+12, −10	90–1250
			0.531	1.8×10^4	+9, −3	20–90
5.	PA-Ultramid-B3 (BASF)	250	0.658	1.24×10^3	+7, −2	140–2000
			0.973	2.98×10^2	+7, −3	2–140
6.	PA-Ultramid-B35 (BASF)	250	0.5773	2.1×10^4	+2, −6	10–100
			0.873	1.1×10^4	+15, −8	0.1–10
7.	PA-Ultramid-B4 (BASF)	250	0.4615	1.48×10^4	+9, −3	100–1500
			0.819	3.1×10^3	+18, −7	1–100
8.	PA-Vestamid L-1600 (CWH)	200	0.5106	5×10^3	+8, −3	350–5000
			0.8652	620.0	+8, −5	20–350
		220	0.575	1950.0	+8, −10	360–6000
			0.9876	170.0	+7, −7	30–360
		240	0.74	400.0	+4, −4	400–4000
			1.0	85.0	+5, −3	50–400
		260	0.82	175.0	+5, −3	400–6000
			1.0	60.0	+3, −3	100–400
9.	Polyethylene (PE) Lupo-len-1800H (BASF)	150	0.3348	1.45×10^4	+10, −4	10–1600
			0.5286	9.1×10^3	+3, −6	1–10
		190	0.315	1.05×10^4	+6, −5	30–3000
			0.5315	5.0×10^3	+5, −2	6–30
		230	0.3252	7.05×10^3	+8, −3	100–10000
			0.5506	2.48×10^3	+5, −5	10–100

Sl. No. (1)	Systems / Source (2)	Temp. (oC) (3)	Two Ranges Power Law constants		%, δ (6)	Applicable, $\dot{\gamma}$ range, s^{-1} (7)
			n (4)	K, Pas^n (5)		
10.	PE-Lupolen 1800S (BASF)	150	0.565	2.05×10^3	+4, −4	20–2000
			0.65	1.5×10^3	+6, −8	3–100
		190	0.5688	1×10^3	+10, −6	20–6000
			0.677	6.7×10^2	+6, −6	10–100
		230	0.571	5.5×10^2	+7, −3	100–10000
			0.7208	280.0	+8, −2	20–200
11.	PE-Lupolen 1812D(BASF)	150	0.3407	3.0×10^4	+3, −3	4–1000
			0.4378	2.65×10^4	+11, −6	0.2–10
		190	0.3316	1.98×10^4	+3, −5	20–3000
			0.4135	1.55×10^4	+5, −5	0.6–25
		230	0.3355	1.35×10^4	+7, −2	30–10000
			0.4396	9.8×10^3	+3, −3	2–50
12.	PE-Lupolen 2410H (BASF)	150	0.3495	1.62×10^4	+5, −5	10–1500
			0.44	1.23×10^4	+9, −9	0.7–50
		190	0.359	0.95×10^4	+3, −5	30–3000
			0.4732	6.9×10^3	+9, −1	3–30
		230	0.349	7.0×10^3	+4, −4	80–10000
			0.48	3.95×10^3	+10, −4	6–100
13.	PE-Lupolen 3020D (BASF)	150	0.2883	4.3×10^4	+2, −5	10–1000
			0.4427	3.1×10^4	+5, −5	0.2–10
		190	0.295	3.02×10^4	+5, −5	30–1500
			0.4421	1.9×10^4	+8, −5	0.6–30
		230	0.3093	2.1×10^4	+10, −5	50–3000
			0.456	1.15×10^4	+9, −3	1–50
14.	LDPE-Lotrene FA-0240 (CdF)	190	0.294	1.85×10^4	+3, −1	300–10000
			0.4324	8.6×10^3	+3, −2	10–300
15.	LDPE-Lotrene FB3010 (CdF)	190	0.2356	4.4×10^4	+3, −3	500–10000
			0.3593	2.1×10^4	+3, −3	20–500
16.	LDPE-Lotrene FD0274 (CdF)	190	0.317	1.47×10^4	+5, −5	400–10000
			0.465	6.05×10^3	+5, −5	10–400
17.	LDPE-Lotrene FD5005 (CdF)	190	0.2984	2.28×10^4	+3, −3	150–10000
			0.3969	1.36×10^4	+3, −3	10–150
18.	LDPE-Lotrene FD0374 (CdF)	190	0.378	9.3×10^3	+10, −2	100–10000
			0.5284	4.86×10^3	+7, −7	10–100
19.	LDPE-Lotrene FE7040 (CdF)	190	0.2967	2.48×10^4	+4, −4	500–10000
			0.3985	1.34×10^4	+4, −4	10–500
20.	LDPE-Lotrene MA0701 (CdF)	190	0.3123	1.19×10^4	+5, −5	1000–10000
			0.489	3.65×10^3	+7, −5	10–1000
21.	LDPE-Lotrene MA2004 (CdF)	190	0.37	6.2×10^3	+9, −6	1000–10000
			0.5736	1.52×10^3	+8, −5	10–1000
22.	LDPE-Lotrene MA7007 (CdF)	190	0.539	1.18×10^3	+10, −5	500–8000
			0.709	4.05×10^2	+6, −4	60–500
23.	HDPE-Vestolen A3512 (CWH)	190	0.476	2.2×10^4	+10, −6	10–1000
		230	0.5263	1.31×10^4	+10, −4	10–1000

Sl. No.	Systems / Source	Temp. (oC)	Two Ranges Power Law constants		%, δ	Applicable, $\dot{\gamma}$ range, s^{-1}
			n	K, Pas^n		
(1)	*(2)*	*(3)*	*(4)*	*(5)*	*(6)*	*(7)*
24.	LDPE, Alkathene-11 (ICI)	170	0.3582	1.31×10^4	+5, –3	15–1000
			0.5698	7.5×10^3	+7, –5	1–15
		210	0.4167	6.0×10^3	+7, –5	35–1000
			0.6441	2.61×10^3	+8, –8	1–35
25.	LDPE, Alkathene-21 (ICI)	170	0.4621	3.25×10^3	+10, –4	100–10000
			0.6719	1.27×10^3	+11, –7	3–100
		210	0.5107	1.43×10^3	+17, –5	100–1000
			0.833	3.4×10^3	+14, –9	8–100
26.	LDPE, Alkathene-23 (ICI)	170	0.4803	1.25×10^3	+10, –5	1000–10000
			0.7525	1.95×10^2	+13, –6	30–1000
		210	0.5716	3.2×10^2	+10, –5	1000–10000
			0.8974	35.3	+5, –5	100–1000
27.	LDPE, Alkathene-33 (ICI)	170	0.3760	2.21×10^4	+10, –4	3–100
			0.4852	$\times 10^4$	+9, –5	0.2–5
		210	0.329	1.5×10^4	+11, –1	20–1000
			0.4386	1.15×10^4	+7, –6	1–20
28.	LDPE, Alkathene-46 (ICI)	170	0.45	9.71×10^3	+14, –8	5–400
			0.776	5.85×10^3	+13, –7	0.2–5
		210	0.4914	4.2×10^3	+14, –7	10–1000
			0.801	2.05×10^3	+8, –6	0.5–10
29.	Polyether Sulfon Victrex-300P (ICI)	320	0.6646	1.83×10^4	+13, –6	10–300
			0.9644	9.1×10^3	+10, –6	0.1–10
		370	0.6081	5.85×10^3	+14, –8	80–3000
			0.935	1.39×10^3	+7, –5	1–80
30.	Polymethylmethacry-late PM-MA-Diakon LG (ICI)	210	0.4032	1.45×10^4	+13, –10	40–3000
			0.8315	3.15×10^3	+15, –12	0.6–40
		240	0.6108	1.77×10^3	+7, –6	40–1500
			0.9548	5.25×10^2	+6, –7	3–40
31.	PMMA-Diakon MH (ICI)	210	0.4034	6.3×10^4	+9, –8	3–100
			0.8049	4.1×10^4	+7, –8	0.1–3
		240	0.4158	1.78×10^4	+12, –8	10–1000
			0.8185	7.5×10^3	+13, –10	0.1–10
32.	PMMA-Plexiglas 5N (Roehm)	200	0.2863	3.55×10^4	+10, –10	4–7000
			0.6464	1×10^4	+14, –10	1–40
		220	0.3063	2×10^4	+10, –8	200–10000
			0.6443	3.55×10^3	+12, –7	6–40
33.	PMMA-Plexiglas 6H (Roehm)	200	0.2235	7.3×10^4	+10, –12	10–6000
			0.5329	3.55×10^4	+7, –11	0.4–10
		220	0.2305	4.9×10^4	+10, –8	40–10000
			0.5737	1.415×10^4	+12, –7	1–40
		240	0.2784	2.5×10^4	+9, –8	100–10000
			0.5587	6.25×10^3	+8, –8	4–100
34.	PMMA-Plexiglas 6N (Roehm)	200	0.242	6.0×10^4	+12, –6	50–10000
			0.5646	1.81×10^4	+14, –11	1–50
		220	0.2905	2.96×10^4	+11, –6	100–10000
			0.6603	5.7×10^4	+16, –8	3–100
		240	0.3287	1.35×10^4	+13, –8	150–10000
			0.682	2.35×10^3	+17, –9	6–150

Sl. No.	Systems / Source	Temp. (oC)	Two Ranges Power Law constants		%, δ	Applicable, $\dot{\gamma}$ range, s^{-1}
			n	K, Pas^n		
(1)	(2)	(3)	(4)	(5)	(6)	(7)
35.	PMMA-Plexiglas 7H (Roehm)	200	0.2044	1.14×10^5	+12, −8	10–2000
			0.401	7.35×10^4	+8, −5	0.5–10
		220	0.2039	7.8×10^4	+9, −7	100–10000
			0.3459	4.05×10^4	+10, −6	8–100
		240	0.249	4.0×10^4	+12, −8	80–10000
			0.5569	1.03×10^4	+10, −10	2–80
36.	PMMA-Plexiglas 7N (Roehm)	200	0.2743	6.5×10^4	+10, −11	10–3000
			0.6154	2.95×10^4	+9, −6	0.6–10
		220	0.2618	4.4×10^4	+9, −9	100–10000
			0.5826	1.05×10^4	+12, −12	3–100
		240	0.3355	1.6×10^4	+14, −7	100–10000
			0.6972	3.1×10^3	+10, −6	6–100
37.	PMMA-Plexiglas 8H (Roehm)	200	0.1891	1.5×10^5	+5, −5	10–2000
			0.3991	9.4×10^4	+8, −6	0.3–10
		220	0.1878	1.03×10^5	+10, −10	40–6000
			0.461	3.7×10^4	+10, −13	1–40
		240	0.20	6.0×10^4	+10, −8	100–10000
			0.504	1.5×10^4	+11, −11	3–100
38.	PMMA-Plexiglas 8N (Roehm)	200	0.2283	1.25×10^5	+9, −9	10–1000
			0.4933	6.8×10^4	+8, −6	0.6–10
		220	0.2233	7.8×10^4	+9, −8	50–5000
			0.5323	2.235×10^4	+10, −10	2–50
		240	0.2481	4.25×10^4	+6, −6	200–10000
			0.4698	1.35×10^4	+3, −3	15–200
39.	PMMA-Plexiglas ZK43 (Roehm)	220	0.1468	1.16×10^5	+9, −7	100–3000
			0.3508	4.2×10^4	+9, −6	3–200
40.	Polypropylene-PP Novo-len-1120H (BASF)	170	0.2803	2.05×10^4	+8, −8	10–1000
			0.502	1.25×10^4	+9, −8	1–10
		210	0.253	1.90×10^4	+6, −4	50–2000
			0.433	9.3×10^3	+10, −8	3–50
		250	0.2616	1.37×10^4	+10, −10	80–4000
			0.4832	5.4×10^3	+10, −10	2–100
41.	Polypropylene-PP Novo-len-1120L (BASF)	170	0.3402	1.135×10^4	+10, −10	10–1500
			0.6496	5.8×10^3	+11, −9	0.6–10
		210	0.3211	9.2×10^3	+9, −7	40–3000
			0.6122	3.2×10^3	+10, −10	1–40
		250	0.3083	8.1×10^3	+10, −6	100–8000
			0.6558	1.64×10^3	+11, −12	3–100
42.	Polypropylene-PP Novo-len-1320L (BASF)	170	0.24	2.8×10^4	+3, −3	100–2000
			0.392	1.25×10^4	+5, −3	6–200
		210	0.276	1.75×10^4	+9, −5	200–3000
			0.466	6.2×10^3	+7, −5	10–200
		250	0.297	1.1×10^4	+5, −5	300–8000
			0.481	4.03×10^3	+5, −5	20–300
43.	Polypropylene-PP Propathene GSE-16 (ICI)	210	0.3255	1.45×10^4	+10, −7	10–1000
			0.596	8.05×10^3	+12, −10	0.1–10
		240	0.3377	1.1×10^4	+10, −5	10–1000
			0.6208	5.9×10^3	+13, −8	0.1–10

Sl. No.	Systems / Source	Temp. (oC)	Two Ranges Power Law constants		%, δ	Applicable, $\dot{\gamma}$ range, s^{-1}
			n	K, Pas^n		
(1)	(2)	(3)	(4)	(5)	(6)	(7)
44.	Polypropylene-PP Propathene GSE-23 (ICI)	210	0.2526	1.52×10^4	+12, −8	50–10000
			0.5377	5.25×10^3	+14, −8	1–50
		240	0.2907	1.02×10^4	+15, −7	10–10000
			0.5763	3.45×10^3	+17, −13	1–50
45.	Polystyrene-PS Polystyrol 143E (BASF)	170	0.2388	3.1×10^4	+7, −6	10–400
			0.4383	2×10^4	+15, −8	0.6–10
		210	0.2448	1.4×10^4	+8, −8	100–10000
			0.5453	3.65×10^3	+12, −8	10–100
		250	0.3187	5.2×10^3	+10, −7	300–10000
			0.5649	1.28×10^3	+7, −6	60–300
46.	Polystyrene-PS Polystyrol 165H (BASF)	170	0.180	4.8×10^4	+8, −7	5–500
			0.3969	3.35×10^4	+7, −7	0.3–7
		210	0.2703	3.6×10^4	+10, −8	5–500
			0.3783	2.6×10^4	+6, −7	0.5–10
		250	0.2477	1.175×10^4	+11, −9	100–10000
			0.41	5.0×10^3	+10, −6	20–500
47.	PVC Welvic-R7/622 (ICI)	170	0.336	7.3×10^4	+7, −5	50–1250
			0.527	3.35×10^4	+6, −3	7–55
		190	0.334	5.1×10^4	+3, −3	100–1100
			0.551	1.85×10^4	+5, −5	6–100
48.	SAN Luran-378P (BASF)	170	0.3221	6.9×10^4	+10, −7	1–100
			0.596	6.9×10^4	+8, −8	0.1–1
		210	0.296	2.5×10^4	+7, −9	50–3000
			0.5938	8.15×10^3	+8, −9	2–50
		250	0.3363	1.1×10^4	+13, −7	200–10000
			0.6497	2.1×10^3	+9, −10	20–200
49.	Styrene-Butadiene Copolymer SB-Polystyrol 456M (BASF)	210	0.219	2.61×10^4	+10, −8	40–10000
			0.4588	1.05×10^4	+12, −9	1–50
		250	0.2576	1.36×10^4	+10, −7	100–10000
			0.515	4.3×10^3	+12, −7	6–100
50.	Styrene-Butadiene Copolymer SB-Polystyrol 475K (BASF)	170	0.1935	5.1×10^4	+5, −5	40–2000
			0.3516	3×10^4	+9, −5	0.5–40

Annexure-II
Rheology Program

```java
import java.awt.*;
import java.awt.event.*;
import java.applet.*;
import java.io.*;
    public class Rheology extends Applet implements ActionListener
    {
        DataOutputStream f0,f1;
        Panel Data,value,last[],dummy[],comand[],result,bgraph;
        int n=10,i;
        double K,N;
        double graph[][];
        TextField text[];
        Button buton[],gr,save,exit;
        Label labgdot,labtau;
        TextField text1[];
        double temp1[],temp2[];
        String source,source1;
        String cons[] = { "K = ","n = ","phi0 = ","phi1 = ","Alfa = ","Tauhaf
        = ","A = ","B = ","C = ","Eta0 = ","EtaInf = ","Alfa = ","TauM = "};
        String label[]={"System Used","Equipment","Capillary
        entry","Temperature","Piston speed range","Piston dia","Capillary
        dia","Capillary length","Number of Data"};
        String bon[]={"Next","Close","Refresh"};

// LEASTSQUARES METHOD CLASS DEFINITION *******************

class Leastsq
    {
        double X[],Y[];
        double M,C;
```

```
        int num;
        Leastsq(double x[],double y[],int number)
        {
            X=x; Y=y; num = number;
        }
        void calculate(Leastsq o)
        {
            double sigY=0,sigX=0,sigXY=0,sigX2=0,D=0;
            for(i=0;i<num;i++)
            {
                sigX+= o.X[i];
                sigY+= o.Y[i];
                sigXY+= o.X[i] * o.Y[i];
                sigX2+= o.X[i] * o.X[i];
            }
            D = o.num * sigX2 - sigX * sigX;
            o.C = (sigX2 * sigY - sigX * sigXY)/D;
            o.M = (o.num * sigXY - sigX * sigY)/D;
        }
    }


// RMS ERROR METHOD CLASS DEFINITION **********************
class Error
    {
        double Exp[],Cal[];
        double Cerror;
        int num;
        Error(double exp[], double cal[],int number)
        {
            Exp=exp; Cal=cal; num=number;
        }
        void finderror(Error o)
        {
            double deviation= 0;
            for(i=0;i<num;i++)
            deviation += Math.pow(((Exp[i] - Cal[i])/Exp[i]),2.0);
```

```
            o.Cerror = Math.sqrt(deviation/num);

       }
   }
double constants[],error[],data[];
Leastsq powlaw,careu,ellis,meter;
Error powerr,carerr,eliserr,meterr;
public void init() {


// USER INTERFACE COMPONENTS ***************************
Data = new Panel();
text = new TextField[9];
buton= new Button[6];
last = new Panel[3];
dummy = new Panel[3];
comand = new Panel[3];
Data.setLayout(new GridLayout(9,2,6,6));
for(i=0;i<9;i++)
    {
        text[i] = new TextField(10);
        Data.add(new Label(label[i],Label.CENTER));
        text[i].addActionListener(this);
        Data.add(text[i]);
    }
    int k=0;
    for(int j=0;j<2;j++)
    {
        last[j] = new Panel();
        dummy[j] = new Panel();
        comand[j] = new Panel();
            dummy[j].add(new Label(" "));
            dummy[j].add(new Label(" "));
            dummy[j].setLayout(new GridLayout(2,1,0,4));
            for(i=0;i<3;i++)
            {
                buton[i+k] = new Button(bon[i]);
                comand[j].add(buton[i+k]);
```

```
            buton[i+k].addActionListener(this);
         } k=k+3;
         comand[j].setLayout(new GridLayout(1,3,6,0));
         last[j].setLayout(new BorderLayout(0,10));
         last[j].add("South",comand[j]);
      }
      last[0].add("Center",Data);
      add(last[0]);
   }
   public void actionPerformed(ActionEvent e)
   {
      if(e.getSource()== buton[0]){
      n = Integer.parseInt(text[8].getText());
         super.remove(last[0]);
         repaint();
         value = new Panel();
         value.add( labgdot = new Label("Shear Rate",Label.CENTER));
         value.add( labtau = new Label("Shear Stress",Label.CENTER));
         text1 = new TextField[n * 2];
         for(i = 0; i < 2 * n ; i++) {
         text1[i] = new TextField(10);
         value.add(text1[i]);
   }
   value.setLayout(new GridLayout(n+1,2,0,6) );
   last[1].add("Center",value);
   add(last[1]);
   }
   if(e.getSource()==buton[3])
   {
      data = new double[n*2];


//TRANSFERING DATA FROM THE USER INTERFACE TO ARRAY***
for(i=0;i<n;i++)
   {
      data[i] = Double.parseDouble(text1[i*2].getText());
      data[n+i] = Double.parseDouble(text1[i*2+1].getText());
```

```java
        }
        super.remove(last[1]);
        repaint();
        constants = new double[13];
        error = new double[4];
        temp1 = new double[n];
        temp2 = new double[n];
        graph = new double[n][6];

// POWERLAW MODEL CONSTANTS CALCULATION **************
for(i=0;i<n;i++)
    {
        temp1[i] = Math.log(data[i]);
        temp2[i] = Math.log(data[n+i]);
    }
    powlaw = new Leastsq(temp1,temp2,n);
    powlaw.calculate(powlaw);
    K = Math.exp(powlaw.C);
    N= powlaw.M;
    constants[0] = Math.ceil(K);
    constants[1] = Math.ceil(N * 1000)/1000;
    for(i=0;i<n;i++)
    {
        temp1[i]= data[n+i];
        temp2[i]= K * Math.pow(data[i],N);
    }
    powerr = new Error(temp1,temp2,n);
    powerr.finderror(powerr);
    error[0]=Math.ceil(powerr.Cerror *100)/100;

// CARREAU MODEL CONSTANTS CALCULATION **************
    double A,B,C;
    double Lerror=0;
    double Af=0,Bf=0,Cf=0;
    for(int theta = 10; theta<82;theta+=3)
    {
        B = theta * 3.14/180;
        for(i=0;i<n;i++)
```

```
    {
        temp1[i] = Math.log(1+ Math.tan(B) * data[i]);
        temp2[i] = Math.log(data[n+i]/data[i]);
    }
    careu = new Leastsq(temp1,temp2,n);
    careu.calculate(careu);
    A = Math.exp(careu.C);
    C = -careu.M;
    for(i=0;i<n;i++)
    {
        temp1[i]= data[n+i];
        temp2[i]= data[i] * A/ Math.pow((1+B * data[i]),C);
    }
    carerr = new Error(temp1,temp2,n);
    carerr.finderror(carerr);
    if(theta==10)
    {
        Lerror=carerr.Cerror;
        Af=A; Bf=B; Cf=C;
    }
    else if((carerr.Cerror) < Lerror)
    {
        Lerror=carerr.Cerror;
        Bf=B; Af=A; Cf=C;
    }
}
for(i=0;i<n;i++)
{
    graph[i][0]= Math.log(data[i]);
    graph[i][1]= Math.log(data[n+i]/data[i]);
    graph[i][2]= Math.log(K * Math.pow(data[i],N-1));
    graph[i][3]=Math.log(Af/Math.pow((1+Bf * data[i]),Cf));
}
constants[6]=Math.ceil(Af);
constants[7]=Math.ceil(Bf * 100)/100;
constants[8]=Math.ceil(Cf * 100)/100;
error[1]=Math.ceil(Lerror * 100)/100;
```

```
//ELLIS MODEL CONSTANTS CALCULATION ***********************
double phi0,phi1,alfa;
double PHI0=0,PHI1=0,ALFA=0,EtaZ=0;
final double STEP = 100;
final double LIMIT = data[n]/data[0]+STEP;
EtaZ = LIMIT;
phi0 = 1/EtaZ;
while(EtaZ < 1.5 * LIMIT)
    {
        for(i=0;i<n;i++)
        {
            temp2[i] = Math.log(data[i]/data[n+i] - phi0);
            temp1[i] = Math.log(data[n+i]);
        }
        ellis = new Leastsq(temp1,temp2,n);
        ellis.calculate(ellis);
        phi1 = Math.exp(ellis.C);
        alfa = 1 + ellis.M;
        for(i=0;i<n;i++)
        {
            temp1[i]= data[n+i];
            temp2[i]= data[i] /(phi0 + phi1 * Math.pow(data[n+i],alfa-1));
        }
        eliserr = new Error(temp1,temp2,n);
        eliserr.finderror(eliserr);
        if(phi0==1/LIMIT)
        {
            Lerror=eliserr.Cerror;
            PHI0=phi0; PHI1=phi1; ALFA=alfa;
        }
        else if((eliserr.Cerror) < Lerror)
        {
            Lerror=eliserr.Cerror;
            PHI0=phi0; PHI1=phi1; ALFA=alfa;
        }
        EtaZ+=STEP;
        phi0 = 1/EtaZ;
```

```
    }
    for(i=0;i<n;i++)
    graph[i][4] = Math.log(1/(PHI0 + PHI1 * Math.pow(data[n+i],ALFA-1)));
    constants[2]=PHI0;
    constants[3]=PHI1;
    constants[4]=Math.ceil(ALFA * 100)/100;
    constants[5]=Math.ceil(Math.pow((PHI0/PHI1),(1/(ALFA-1))));
    error[2]=Math.ceil(Lerror * 100)/100;
// METER MODEL CONSTANTS *******************************
double eta0,etain,alfam,taum,eta;
double ETA0=0,ETAIN=0,ALFAM=0,TAUM=0;
int flag=1;
eta0 = LIMIT;
etain = Math.exp(flag);
while(etain < data[2*n-1]/data[n-1])
    {
        while(eta0 < 1.5 * LIMIT)
        {
            for(i=0;i<n;i++)
            {
                eta = data[n+i]/data[i];
                temp1[i] = Math.log(data[n+i]);
                temp2[i] = Math.log((eta0 - eta)/(eta-etain));
            }
            meter = new Leastsq(temp1,temp2,n);
            meter.calculate(meter);
            alfam = 1 + meter.M;
            taum = Math.exp((meter.C)/(1-alfam));
            for(i=0;i<n;i++)
            {
                temp1[i]= data[n+i];
                temp2[i]= data[i] * ( etain + (eta0 - etain)/(1+Math.
                pow((data[n+i]/taum),(alfam-1))));
            }
            meterr = new Error(temp1,temp2,n);
            meterr.finderror(meterr);
            if(eta0==LIMIT)
```

```
            {
                Lerror=meterr.Cerror;
                ETA0=eta0; ETAIN=etain; ALFAM=alfam; TAUM=taum;
            }
            else if((meterr.Cerror) < Lerror)
            {
                Lerror=meterr.Cerror;
                ETA0=eta0; ETAIN=etain; ALFAM=alfam; TAUM=taum;
            }
            eta0 = eta0+100;
        }
        flag+=1;
        etain = Math.exp(flag);
    }
    for(i=0;i<n;i++)
    graph[i][5] = Math.log( ETAIN + (ETA0 - ETAIN)/(1+Math.pow((data[n+i]/
    TAUM),(ALFAM-1))));
    constants[9]=Math.ceil(ETA0);
    constants[10]=Math.ceil(ETAIN * 100)/100;
    constants[11]=Math.ceil(ALFAM * 100)/100;
    constants[12]=Math.ceil(TAUM);
    error[3]=Math.ceil(Lerror * 100)/100;

// USER INTERFACE COMPONENTS ***************************

String model[] = {"Power Law","Carreau","Ellis","Meter"};
String heading[] = {"Models","Rmserr(%)","Constants"};
Panel result1 = new Panel();
Panel result11 = new Panel();
result11.add(new Label(" "));
    for(i=0;i<8;i++)
    {
        if(i<2)
        result11.add(new Label(label[i] +" "+ text[i].getText(),Label.LEFT));
        else
        {
            result1.add(new Label(label[i],Label.LEFT));
            result1.add(new Label(text[i].getText(),Label.LEFT));
        }
    }
```

```
     result11.setLayout(new GridLayout(3,1,0,4));
     result1.setLayout(new GridLayout(3,4,2,4));
     Panel result2 = new Panel();
     result2.add(new Label(labgdot.getText(),Label.CENTER));
     result2.add(new Label(labtau.getText(),Label.CENTER));
     for(i=0;i<text1.length;i++)
     result2.add(new Label(text1[i].getText(),Label.CENTER));
     result2.setLayout(new GridLayout(text1.length+1,2,4,2));
     Panel result3 = new Panel();

// SENDING THE DATA FROM ARRAYS TO THE USER INTERFACE ****
for(i=0;i<3;i++)
     {
          result3.add(new Label(heading[i],Label.CENTER));
     }
     int k=0,b=2;
     /* for(i=0;i<3;i++)
     result3.add(new Label(" ")); */
     for(i=0;i<4;i++)
     {
          result3.add(new Label(model[i],Label.CENTER));
          result3.add(new Label(""+error[i],Label.CENTER));
          result3.add(new Label(cons[k]+constants[k],Label.CENTER));
          for(int l=1;l<b;l++)
          {
               result3.add(new Label(" "));
               result3.add(new Label(" "));
               result3.add(new Label(cons[l+k]+constants[l+k],Label.CENTER));
          }
          k=k+b;
          if(i==0) b=3;
          else b=4;
     }
     result3.setLayout(new GridLayout(19,3));
     Panel result12 = new Panel();
     result12.setLayout(new BorderLayout());
     /* result12.add("North",result11);*/
     result12.add("North",result1);
     result = new Panel();
```

```
result.setLayout(new BorderLayout(5,15));
result.add("North",result12);
result.add("West",result2);
result.add("Center",result3);
last[2] = new Panel();
last[2].setLayout(new BorderLayout());
last[2].add("Center",result);
bgraph = new Panel();
bgraph.setLayout(new GridLayout(2,1,0,6));
double best;
String bestm;
best=error[0];
bestm=model[0];
for(i=1;i<4;i++)
if(best>error[i])
{
    best=error[i];
    bestm =model[i];
}
bgraph.add(new Label("THE BEST MODEL IS " +bestm+"
MODEL",Label.CENTER));
bgraph.add(gr = new Button("Graph Data"));
gr.addActionListener(this);
last[2].add("North",bgraph);
last[2].add("West",new Label(" "));
last[2].add("East",new Label(" "));
add(last[2]);
}
if(e.getSource()==gr)
{
    Panel pgraph = new Panel();
    pgraph.setLayout(new GridLayout(n+1,6,3,3));
    for(i=0;i<n;i++)
    for(int j=0;j<6;j++)
    pgraph.add(new Label(" "+Math.ceil(graph[i][j]*100)/100));
    super.remove(last[2]);
    repaint();
    pgraph.add(save= new Button("SAVE"));
    pgraph.add(exit= new Button("EXIT"));
```

```
        save.addActionListener(this);
        exit.addActionListener(this);
        add(pgraph);
    }
    if(e.getSource()==save)
    {
        Save();
     }
    if(e.getSource()==exit)
    {
        System.exit(0);
    }
    validate();
    repaint();
    }
    public void Save()
    {
        try
    {
        f0 = new DataOutputStream(new FileOutputStream("data.txt"));
        f1 = new DataOutputStream(new FileOutputStream("data.xls"));
    }
    catch(IOException e)
    {
        System.err.println("File not opened properly\n" + e.toString() );
        System.exit(1);
    }
    for(i=0;i<n;i++)
    source = source+" "+data[i];
    source=source+"\n";
    for(i=0;i<n;i++)
    source=source+" "+data[n+i];
    source=source+"\n";
    for(i=0;i<13;i++)
    source = source+" "+constants[i];
    source=source+"\n";
    for(i=0;i<4;i++)
    source=source+" "+error[i];
```

```java
for(i=0;i<n;i++)
{
    source1 = source1+"\n";
    for(int j=0;j<6;j++)
    source1 = source1+" "+Math.ceil(graph[i][j]*100)/100;
}
try
{
    f0.writeBytes(source);
    f1.writeBytes(source1);
}
catch(IOException io)
{
    System.err.println("writing error\n" + io.toString() );
    System.exit(1);
}
try
{
    f0.close();
    f1.close();
}
catch(IOException io)
{
    System.err.println("File not closed properly\n" + io.toString() );
    System.exit(0);
}
}
public static void main(String args[]) throws Exception
{
    Frame app = new Frame("Rheology software");
    app.setSize(600,800);
    Rheology Rh = new Rheology();
    Rh.init();
    Rh.start();
    app.add(Rh,BorderLayout.CENTER);
    app.setVisible(true);
}
}
```

Annexure-III

COMPARISON BETWEEN THE MODELS USING THE RHEOLOGY SOFTWARE

1. System Used — LDPE , BASF
   Temperature — 150°C
   Equipment — High pressure capillary viscometer

Shear Rate (1/sec)	Shear Stress (Pa)	Model	Rms(%) error	Model's constants
2.24	18,600.0	Power Law	0.05	$K = 14,932.0$ Pa
4.77	25,600.0			$n = 0.36$
13.00	37,900.0	Ellis	0.08	$\Phi_0 = 7.93$ E–5 $Pa^{-1}.S^{-1}$
27.20	50,200.0			$\Phi_1 = 7.9$ E–14 $Pa^{-\alpha}.S^{-1}$
56.50	65,900.0			$\alpha = 3.07$
146.00	93,900.0			$\tau_{1/2} = 22,743$ Pa
296.00	121,000.0	Carreau	0.08	$A = 11,842.0$ Pa.S
594.00	154,000.0			$B = 0.44$ S
1,450.0	202,000.0			$C = 0.69$
2,850.0	247,000.0	Meter	0.08	$\eta_0 = 12,604$ Pa.S
5,770.0	320,000.0			$\eta_\infty = 2.72$ Pa.S
				$\alpha = 3.09$
				$\tau_m = 22,816.0$ Pa

The Best Model is Power Law Model

2. System Used — LDPE , BASF
   Temperature — 170°C
   Equipment — High pressure capillary viscometer

Shear Rate (Pa)	Shear Stress (1/sec)	Model	Rms (%) error	Model's constants
2.41 5.05	14,700.0 19,900.0	Power Law	0.03	$K = 10,702.0$ Pa $n = 0.387$
12.90 26.60 54.20 138.00	28,200.0 38,000.0 50,700.0 74,000.0	Ellis	0.11	$\Phi_0 = 1.08$ E–4 $Pa^{-1}.S^{-1}$ $\Phi_1 = 1.49$ E–12 $Pa^{-\alpha}.S^{-1}$ $\alpha = 2.86$ $\tau_{1/2} = 17,036.0$ Pa
278.00 563.00 1,370.0	96,600.0 129,000.0 173,000.0	Carreau	0.09	$A = 8,471.0$ Pa.S $B = 0.44$ S $C = 0.66$
2,690.0 5,400.0	219,000.0 287,000.0	Meter	0.11	$\eta_0 = 9,200.0$ Pa.S $\eta_\infty = 2.72$ Pa.S $\alpha = 2.88$ $\tau_m = 17,100.0$ Pa

The Best Model is Power Law Model

3. System Used — LDPE , BASF
   Temperature — 190°C
   Equipment — High pressure capillary viscometer

Shear Rate (1/sec)	Shear Stress (Pa)	Model	RMS (%) error	Model's constants
4.44 12.00	14,200.0 22,300.0	Power Law	0.04	$K = 8,433.0$ Pa $n = 0.394$
25.30 52.50 137.00 269.00	30,600.0 41,200.0 61,300.0 77,100.0	Ellis	0.07	$\Phi_0 = 2.04$ E–4 $Pa^{-1}.S^{-1}$ $\Phi_1 = 4.32$ E–12 $Pa^{-\alpha}.S^{-1}$ $\alpha = 2.81$ $\tau_{1/2} = 17,562.0$ Pa
556.00 1,370.0 2,720.0	106,000.0 145,000.0 187,000.0	Carreau	0.08	$A = 5,327.0$ Pa.S $B = 0.34$ S $C = 0.64$
54,20.0	242,000.0	Meter	0.06	$\eta_0 = 4,899.0$ Pa.S $\eta_\infty = 2.72$ Pa.S $\alpha = 2.83$ $\tau_m = 17,653.0$ Pa

The Best Model is Power Law Model

4. System Used — LDPE , BASF
   Temperature — 210°C
   Equipment   — High pressure capillary viscometer

Shear Rate (1/sec)	Shear Stress (Pa)	Model	RMS (%) error	Model's constants
2.00	6,640.0	Power Law	0.09	$K = 5,733.0$ Pa
4.12	9,970.0			$n = 0.431$
11.20	16,700.0	Ellis	0.06	$\Phi_0 = 2.26$ E–4 Pa^{-1}.S^{-1}
24.30	24,700.0			$\Phi_1 = 4.64$ E–12 Pa$^{-\alpha}$.S^{-1}
51.00	34,000.0			$\alpha = 2.65$
133.00	50,800.0			$\tau_{1/2} = 11,918.0$ Pa
271.00	68,000.0	Carreau	0.05	$A = 4,507.0$ Pa.S
548.00	90,000.0			$B = 0.39$ S
1,360.0	127,000.0			$C = 0.62$
2,710.0	164,000.0	Meter	0.06	$\eta_0 = 4,320.0$ Pa.S
5,230.0	201,000.0			$\eta_\infty = 2.72$ Pa.S
				$\alpha = 2.68$
				$\tau_m = 12,337.0$ Pa

The Best Model is Carreau Model

5. System Used — LDPE, BASF
   Temperature — 230°C
   Equipment   — High pressure capillary viscometer

Shear Rate (1/sec)	Shear Stress (Pa)	Model	RMS (%) error	Model's constants
2.00	4,930.0	Power Law	0.08	$K = 4,067.0$ Pa
4.10	7,410.0			$n = 0.462$
11.10	12,700.0	Ellis	0.06	$\Phi_0 = 2.80$ E–4 Pa^{-1}.S^{-1}
23.40	18,500.0			$\Phi_1 = 7.92$ E–12 Pa$^{-\alpha}$.S^{-1}
49.20	26,600.0			$\alpha = 2.43$
128.00	41,100.0			$\tau_{1/2} = 7,966.0$ Pa

Shear Rate (1/sec)	Shear Stress (Pa)	Model	RMS (%) error	Model's constants
259.00	55,900.0	Carreau	0.05	$A = 3,239.0$ Pa.S
523.00	76,300.0			$B = 0.39$ S
1,320.0	114,000.0			$C = 0.59$
2,560.0	145,000.0	Meter	0.06	$\eta_0 = 3,465.0$ Pa.S
4,960.0	183,000.0			$\eta_\infty = 2.72$ Pa.S
				$\alpha = 2.46$
				$\tau_m = 8,335.0$ Pa

The Best Model is Carreau Model

6. System Used — PP , BASF
   Temperature — 190^0C
   Equipment — High pressure capillary viscometer

Shear Rate (1/sec)	Shear Stress (Pa)	Model	RMS (%) error	Model's constants
3.67	22,400.0	Power Law	0.13	$K = 18,761.0$ Pa
7.73	32,700.0			$n = 0.302$
22.80	53,600.0	Ellis	0.16	$\Phi_0 = 1.56$ E–4 Pa^{-1}.S^{-1}
51.40	69,800.0			$\Phi_1 = 1.07$ E–12 Pa$^{-\alpha}$.S^{-1}
114.00	87,100.0			$\alpha = 4.25$
306.00	111,000.0			$\tau_{1/2} = 47,693.0$ Pa
682.00	148,000.0	Carreau	0.05	$A = 9,545.0$ Pa.S
1,380.0	173,000.0			$B = 0.23$ S
3,340.0	202,000.0			$C = 0.77$
6,390.0	221,000.0	Meter	0.17	$\eta_0 = 6,404.0$Pa.S
				$\eta_\infty = 2.72$ Pa.S
				$\alpha = 4.27$
				$\tau_m = 47,653.0$ Pa

The Best Model is Carreau Model

7. System Used — PP , BASF
   Temperature — 210°C
   Equipment — High pressure capillary viscometer

Shear Rate (1/sec)	Shear Stress (Pa)	Model	RMS (%) error	Model's constants
3.58	16,700.0	Power Law	0.17	$K = 14,951.0$ Pa
7.32	24,700.0			$n = 0.311$
21.20	42,900.0	Ellis	0.21	$\Phi_0 = 2.09$ E–4 Pa^{-1}.S^{-1}
48.50	59,300.0			$\Phi_1 = 4.42$ E–12 Pa$^{-\alpha}$.S^{-1}
109.0	76,000.0			$\alpha = 4.17$
294.00	94,100.0			$\tau_{1/2} = 42,875$ Pa
658.00	125,000.0	Carreau	0.05	$A = 6670.0$ Pa.S
1,400.0	156,000.0			$B = 0.18$ S
3,510.0	190,000.0			$C = 0.76$
6,740.0	208,000.0	Meter	0.22	$\eta_0 = 4,765$ Pa.S
12,700.0	224,000.0			$\eta_\infty = 2.72$ Pa.S
				$\alpha = 4.21$
				$\tau_m = 42,787.0$ Pa

The Best Model is Carreau Model

8. System Used — PP, BASF
   Temperature — 230°C
   Equipment — High pressure capillary viscometer

Shear Rate (1/sec)	Shear Stress (Pa)	Model	RMS (%) error	Model's constants
7.35	20,400.0	Power Law	0.11	$K = 14,345.0$ Pa
20.90	35,600.0			$n = 0.294$
47.80	49,900.0	Ellis	0.07	$\Phi_0 = 3.25$ E–4 Pa^{-1}.S^{-1}
107.0	63,700.0			$\Phi_1 = 1.03$ E–12 Pa$^{-\alpha}$.S^{-1}
294.0	82,800.0			$\alpha = 4.15$
623.0	101,000.0			$\tau_{1/2} = 40,085$ Pa
1260.0	117,000.0	Carreau	0.03	$A = 5,431.0$ Pa.S
3,280.0	153,000.0			$B = 0.18$ S
6,730.0	186,000.0			$C = 0.75$

Shear Rate (1/sec)	Shear Stress (Pa)	Model	RMS (%) error	Model's constants
13,000.0	207,000.0	Meter	0.08	$\eta_0 = 3,076$ Pa.S
				$\eta_\infty = 2.72$ Pa.S
				$\alpha = 4.22$
				$\tau_m = 40,106.0$ Pa

The Best Model is Carreau Model

9. System Used — PP , BASF
   Temperature — 250°C
   Equipment   — High pressure capillary viscometer

Shear Rate (1/sec)	Shear Stress (Pa)	Model	RMS (%) error	Model's constants
19.10	27,300.0	Power Law	0.09	$K = 13,810.0$ Pa
43.60	42,100.0			$n = 0.288$
97.40	55,500.0	Ellis	0.06	$\Phi_0 = 5.78$ E–4 Pa^{-1}.S^{-1}
276.0	73,700.0			$\Phi_1 = 7.0$ E–12 Pa$^{-\alpha}$.S^{-1}
612.0	94,600.0			$\alpha = 4.21$
1,250.0	109,000.0			$\tau_{1/2} = 44,595$ Pa
3,270.0	141,000.0	Carreau	0.05	$A = 4,636.0$ Pa.S
6,690.0	170,000.0			$B = 0.18$ S
13,100.0	192,000.0			$C = 0.74$
		Meter	0.08	$\eta_0 = 1,730$ Pa.S
				$\eta_\infty = 2.72$ Pa.S
				$\alpha = 4.3$
				$\tau_m = 44,623.0$ Pa

The Best Model is Carreau Model

10. System Used — PP, BASF
    Temperature — 270°C
    Equipment   — High pressure capillary viscometer

Shear Rate (1/sec)	Shear Stress (Pa)	Model	RMS (%) error	Model's constants
42.60	32,800.0	Power Law	0.04	$K = 11,004.0$ Pa
93.80	44,300.0			$n = 0.306$
256.0	59,900.0	Ellis	0.09	$\Phi_0 = 7.87$ E–4 Pa^{-1}.S^{-1}
560.0	79,200.0			$\Phi_1 = 2.84$ E–16 Pa$^{-\alpha}$.S^{-1}
1,170.0	99,700.0			$\alpha = 3.73$
2,980.0	129,000.0			$\tau_{1/2} = 37,007.0$ Pa
6,030.0	158,000.0	Carreau	0.03	$A = 3,585.0$ Pa.S
11,600.0	179,000.0			$B = 0.18$ S
				$C = 0.71$
		Meter	0.01	$\eta_0 = 1,270.0$ Pa.S
				$\eta_\infty = 2.72$ Pa.S
				$\alpha = 3.82$
				$\tau_m = 37,266.0$ Pa

The Best Model is Carreau Model

11. System Used — Polystyrene, BASF
    Temperature — 210°C
    Equipment — High pressure capillary viscometer

Shear Rate (1/sec)	Shear Stress (Pa)	Model	RMS (%) error	Model's constants
7.96	34,800.0	Power Law	0.09	$K = 26,439.0$ Pa
23.30	52,600.0			$n = 0.219$
54.30	67,000.0	Ellis	0.14	$\Phi_0 = 2.09$ E–4 Pa^{-1}.S^{-1}
126.00	82,400.0			$\Phi_1 = 4.65$ E–24 Pa$^{-\alpha}$.S^{-1}
369.00	105,000.0			$\alpha = 5.57$
790.00	122,000.0			$\tau_{1/2} = 55,497.0$ Pa
1620.0	138,000.0	Carreau	0.05	$A = 8,849.0$ Pa.S
4,120.0	161,000.0			$B = 0.18$ S
8,170.0	179,000.0			$C = 0.83$
16,100.0	196,000.0	Meter	0.22	$\eta_0 = 4,772.0$ Pa.S
				$\eta_\infty = 2.72$ Pa.S
				$\alpha = 5.67$
				$\tau_m = 55,582.0$ Pa

The Best Model is Carreau Model

12. System Used — Polystyrene, BASF
    Temperature — 230°C
    Equipment — High pressure capillary viscometer

Shear Rate (1/sec)	Shear Stress (Pa)	Model	RMS (%) error	Model's constants
20.10	35,200.0	Power Law	0.08	$K = 20{,}103.0$ Pa
45.00	48,100.0			$n = 0.232$
105.00	62,700.0	Ellis	0.10	$\Phi_0 = 5.12$ E–4 $Pa^{-1}.S^{-1}$
319.00	83,800.0			$\Phi_1 = 1.05$ E–24 $Pa^{-\alpha}.S^{-1}$
704.00	98,700.0			$\alpha = 5.38$
1,510.0	116,000.0			$\tau_{1/2} = 54{,}161.0$ Pa
3,930.0	138,000.0	Carreau	0.04	$A = 6{,}160.0$ Pa.S
7,900.0	155,000.0			$B = 0.18$ S
15,600.0	172,000.0			$C = 0.80$
		Meter	0.13	$\eta_0 = 1{,}952.0$ Pa.S
				$\eta_\infty = 2.72$ Pa.S
				$\alpha = 5.50$
				$\tau_m = 54{,}157.0$ Pa

The Best Model is Carreau Model

13. System Used — Polystyrene, BASF
    Temperature — 250°C
    Equipment — High pressure capillary viscometer

Shear Rate (1/sec)	Shear Stress (Pa)	Model	RMS (%) error	Model's constants
18.40	21,200.0	Power Law	0.12	$K = 11{,}545.0$ Pa
40.50	32,900.0			$n = 0.283$
91.30	45,500.0	Ellis	0.09	$\Phi_0 = 7.98$ E–4 $Pa^{-1}.S^{-1}$
276.00	64,700.0			$\Phi_1 = 4.28$ E–20 $Pa^{-\alpha}.S^{-1}$
617.00	79,100.0			$\alpha = 4.53$
1,320.0	92,600.0			$\tau_{1/2} = 41{,}335.0$ Pa
3,570.0	118,000.0	Carreau	0.08	$A = 3{,}871.0$ Pa.S
7,260.0	136,000.0			$B = 0.18$ S
14,300.0	152,000.0			$C = 0.75$

Shear Rate (1/sec)	Shear Stress (Pa)	Model	RMS (%) error	Model's constants
		Meter	0.13	$\eta_0 = 1,253.0$ Pa.S $\eta_\infty = 2.72$ Pa.S $\alpha = 4.64$ $\tau_m = 41,185.0$ Pa

The Best Model is Carreau Model

14. System Used   — Standard Malaysian Rubber SMR-CV-70 (Unmasticated)
    Temperature   — 120°C
    Equipment   — Davenport Extrusion Rheometer
    Piston Data   — Dia = 22.5 mm, Speed range = 2 to 200 mm/mt
    Capillary Data — Dia = 2.5 mm, Length = 12.5 mm, Entry = Flat

Shear Rate (1/sec)	Shear Stress (Pa)	Model	RMS (%) error	Model's constants
1.73	207,000.0	Power Law	0.02	$K = 183,593.0$ Pa
3.46	241,000.0			$n = 0.216$
5.18	262,000.0	Ellis	0.15	$\Phi_0 = 5.56$ E–6 $Pa^{-1}.S^{-1}$
6.91	283,000.0			$\Phi_1 = 3.23$ E–29 $Pa^{-\alpha}.S^{-1}$
8.64	293,000.0			$\alpha = 5.36$
17.30	338,000.0			$\tau_{1/2} = 216,874.0$ Pa
34.60	386,000.0	Carreau	0.12	$A = 158,777.0$ Pa.S
51.80	421,000.0			$B = 0.44$ S
69.10	448,000.0			$C = 0.91$
86.40	476,000.0	Meter	0.15	$\eta_0 = 179,554.0$ Pa.S
130.00	517,000.0			$\eta_\infty = 2.72$ Pa.S
173.00	569,000.0			$\alpha = 5.36$
216.00	603,000.0			$\tau_m = 216,880.0$ Pa

The Best Model is Power Law Model

15. System Used   — Standard Malaysian Rubber SMR-CV-70 (Masticated for – 5 mts)
    Temperature   — 120°C
    Equipment   — Davenport Extrusion Rheometer

Piston Data   — Dia = 22.5 mm , Speed range = 2 to 200 mm/mt
Capillary Data — Dia = 2.5 mm , Length = 12.5 mm , Entry = Flat

Shear Rate (1/sec)	Shear Stress (Pa)	Model	RMS (%) error	Model's constants
1.73	138,000.0	Power Law	0.04	$K = 131,976.0$ Pa
4.32	193,000.0			$n = 0.244$
8.64	241,000.0	Ellis	0.09	$\Phi_0 = 8.34$ E–6 Pa^{-1}.S^{-1}
17.30	269,000.0			$\Phi_1 = 8.80$ E–26 Pa$^{-\alpha}$.S^{-1}
25.90	296,000.0			$\alpha = 4.83$
34.60	317,000.0			$\tau_{1/2} = 166,689.0$ Pa
43.20	338,000.0	Carreau	0.09	$A = 108,134.0$ Pa.S
51.80	345,000.0			$B = 0.39$ S
60.50	355,000.0			$C = 0.89$
69.10	362,000.0	Meter	0.09	$\eta_0 = 119,769.0$ Pa.S
77.80	372,000.0			$\eta_\infty = 2.72$ Pa.S
86.40	396,000.0			$\alpha = 4.83$
130.00	421,000.0			$\tau_m = 166,694.0$ Pa
173.00	465,000.0			
216.00	476,000.0			

The Best Model is Power Law Model

16. System Used   — Standard Malaysian Rubber SMR-CV-70 (Masticated for 10 mts)

Temperature   — 120°C
Equipment   — Davenport Extrusion Rheometer
Piston Data   — Dia = 22.5 mm , Speed range = 2 to 200 mm/mt
Capillary Data — Dia = 2.5 mm , Length = 12.5 mm , Entry = Flat

Shear Rate (1/sec)	Shear Stress (Pa)	Model	RMS (%) error	Model's constants
6.48	172,000.0	Power Law	0.02	$K = 117,829.0$ Pa
8.64	190,000.0			$n = 0.222$

Shear Rate (1/sec)	Shear Stress (Pa)	Model	RMS (%) error	Model's constants
17.30	228,000.0	Ellis	0.08	$\Phi_0 = 2.50$ E–5 Pa^{-1}.S^{-1}
25.90	241,000.0			$\Phi_1 = 8.90$ E–30 Pa$^{-\alpha}$.S^{-1}
34.60	259,000.0			$\alpha = 5.63$
43.20	272,000.0			$\tau_{1/2} = 193,770.0$ Pa
51.80	283,000.0	Carreau	0.05	$A = 61,736.0$ Pa.S
60.50	293,000.0			$B = 0.28$ S
69.10	303,000.0			$C = 0.87$
77.80	310,000.0	Meter	0.08	$\eta_0 = 39,944.0$ Pa.S
86.40	324,000.0			$\eta_\infty = 2.72$ Pa.S
130.00	345,000.0			$\alpha = 5.63$
173.00	365,000.0			$\tau_m = 193,772.0$ Pa
216.00	386,000.0			
259.00	393,000.0			

The Best Model is Power Law Model

17. System Used — Standard Malaysian Rubber SMR-L (Unmasticated)
    Temperature — 120°C
    Equipment — Davenport Extrusion Rheometer
    Piston Data — Dia = 22.5 mm , Speed range = 2 to 200 mm/mt
    Capillary Data — Dia = 2.5 mm , Length = 12.5 mm , Entry = Flat

Shear Rate (1/sec)	Shear Stress (Pa)	Model	RMS (%) error	Model's constants
8.64	345,000.0	Power Law	0.02	$K = 186,237.0$ Pa
17.30	400,000.0			$n = 0.27$
25.9	434,000.0	Ellis	0.13	$\Phi_0 = 1.66$ E–5 Pa^{-1}.S^{-1}
34.60	469,000.0			$\Phi_1 = 3.51$ E–26 Pa$^{-\alpha}$.S^{-1}
43.20	503,000.0			$\alpha = 4.72$
51.80	552,000.0			$\tau_{1/2} = 361,637.0$ Pa
60.50	565,000.0	Carreau	0.05	$A = 98,910.0$ Pa.S
69.10	586,000.0			$B = 0.28$ S
77.80	607,000.0			$C = 0.81$

Shear Rate (1/sec)	Shear Stress (Pa)	Model	RMS (%) error	Model's constants
86.40	621,000.0	Meter	0.13	$\eta_0 = 60,031.0$ Pa.S
130.00	689,000.0			$\eta_\infty = 2.72$ Pa.S
173.00	758,000.0			$\alpha = 4.73$
				$\tau_m = 361,633.0$ Pa

The Best Model is Power Law Model

18. System Used — Standard Malaysian Rubber SMR-L (Masticated for 5 min.)

   Temperature — 120°C

   Equipment — Davenport Extrusion Rheometer

   Piston Data — Dia = 22.5 mm , Speed range = 2 to 200 mm/mt

   Capillary Data — Dia = 2.5 mm , Length = 12.5 mm , Entry = Flat

Shear Rate (1/sec)	Shear Stress (Pa)	Model	RMS (%) error	Model's constants
8.64	248,000.0	Power Law	0.02	$K = 166,091.0$ Pa
17.30	283,000.0			$n = 0.19$
25.9	310,000.0	Ellis	0.10	$\Phi_0 = 2.31$ E–5 Pa^{-1}.S^{-1}
34.60	331,000.0			$\Phi_1 = 7.88$ E–37 Pa$^{-\alpha}$.S^{-1}
43.20	338,000.0			$\alpha = 6.81$
51.80	345,000.0			$\tau_{1/2} = 265,909.0$ Pa
60.50	359,000.0	Carreau	0.04	$A = 72,476.0$ Pa.S
69.10	372,000.0			$B = 0.23$ S
77.80	379,000.0			$C = 0.92$
86.40	386,000.0	Meter	0.10	$\eta_0 = 43,204.0$ Pa.S
95.00	390,000.0			$\eta_\infty = 2.72$ Pa.S
104.00	396,000.0			$\alpha = 6.81$
130.00	414,000.0			$\tau_m = 265,908.0$ Pa
173.00	448,000.0			

The Best Model is Power Law Model

19. System Used   —   Standard Malaysian Rubber SMR-L (Masticated for 10 mts.)
    Temperature   —   120°C
    Equipment   —   Davenport Extrusion Rheometer
    Piston Data   —   Dia = 22.5 mm, Speed range = 2 to 200 mm/mt
    Capillary Data —   Dia = 2.5 mm, Length = 12.5 mm, Entry = Flat

Shear Rate (1/sec)	Shear Stress (Pa)	Model	RMS (%) error	Model's constants
17.30	241,000.0	Power Law	0.02	$K = 148,932.0$ Pa
25.90	259,000.0			$n = 0.168$
34.60	269,000.0	Ellis	0.11	$\Phi_0 = 4.75$ E–5 $Pa^{-1}.S^{-1}$
43.20	276,000.0			$\Phi_1 = 2.27$ E–42 $Pa^{-\alpha}.S^{-1}$
51.80	286,000.0			$\alpha = 7.91$
60.50	293,000.0			$\tau_{1/2} = 252,496.0$ Pa
69.10	303,000.0	Carreau	0.03	$A = 56,483.0$ Pa.S
77.80	310,000.0			$B = 0.23$ S
86.40	317,000.0			$C = 0.90$
130.00	345,000.0	Meter	0.11	$\eta_0 = 21,031.0$ Pa.S
173.00	352,000.0			$\eta_\infty = 2.72$ Pa.S
216.00	362,000.0			$\alpha = 7.92$
				$\tau_m = 252,490.0$ Pa

The Best Model is Power Law Model

20. System Used   —   Styrene-Acrylonitrile, BASF
    Temperature   —   190°C
    Equipment   —   High pressure capillary viscometer

Shear Rate (1/sec)	Shear Stress (Pa)	Model	RMS (%) error	Model's constants
1.33	32,600.0	Power Law	0.15	$K = 37,174.0$ Pa
2.76	48,600.0			$n = 0.323$
7.88	79,800.0	Ellis	0.17	$\Phi_0 = 3.96$ E–5 $Pa^{-1}.S^{-1}$
18.2	110,000.0			$\Phi_1 = 4.65$ E–21 $Pa^{-\alpha}.S^{-1}$
42.3	147,000.0			$\alpha = 4.26$
117.0	187,000.0			$\tau_{1/2} = 78,519.0$ Pa

Shear Rate (1/sec)	Shear Stress (Pa)	Model	RMS (%) error	Model's constants
252.0	231,000.0	Carreau	0.06	$A = 30,416.0$ Pa.S
510.0	266,000.0			$B = 0.34$ S
1,220.0	300,000.0			$C = 0.80$
		Meter	0.18	$\eta_0 = 25,212.0$ Pa.S
				$\eta_\infty = 2.72$ Pa.S
				$\alpha = 4.26$
				$\tau_m = 78,495.0$ Pa

The Best Model is Carreau Model

21. System Used   —   Styrene-Acrylonitrile, BASF
    Temperature   —   210°C
    Equipment   —   High pressure capillary viscometer

Shear Rate (1/sec)	Shear Stress (Pa)	Model	RMS (%) error	Model's constants
2.53	20,500.0	Power Law	0.24	$K = 23,678.0$ Pa
6.76	41,900.0			$n = 0.314$
14.50	62,300.0	Ellis	0.26	$\Phi_0 = 1.21$ E–4 $Pa^{-1}.S^{-1}$
33.30	89,100.0			$\Phi_1 = 6.12$ E–21 $Pa^{-\alpha}.S^{-1}$
103.0	128,000.0			$\alpha = 4.37$
239.0	158,000.0			$\tau_{1/2} = 69,304.0$ Pa
534.0	189,000.0	Carreau	0.06	$A = 11,701.0$ Pa.S
1,420.0	225,000.0			$B = 0.18$ S
2,930.0	258,000.0			$C = 0.79$
5,690.0	275,000.0	Meter	0.26	$\eta_0 = 8,203.0$ Pa.S
				$\eta_\infty = 2.72$ Pa.S
				$\alpha = 4.39$
				$\tau_m = 69,191.0$ Pa

The Best Model is Carreau Model

22. System Used   —   Styrene-Acrylonitrile, BASF
    Temperature   —   230°C
    Equipment   —   High pressure capillary viscometer

Shear Rate (1/sec)	Shear Stress (Pa)	Model	RMS (%) error	Model's constants
6.40	22,200.0	Power Law	0.19	$K = 16,362.0$ Pa
13.40	36,800.0			$n = 0.333$
28.90	55,700.0	Ellis	0.19	$\Phi_0 = 2.8$ E–4 $Pa^{-1}.S^{-1}$
87.90	91,100.0			$\Phi_1 = 2.46$ E–19 $Pa^{-\alpha}.S^{-1}$
204.0	118,000.0			$\alpha = 4.14$
471.0	146,000.0			$\tau_{1/2} = 63,055.0$ Pa
1280.0	175,000.0	Carreau	0.10	$A = 6,978.0$ Pa.S
2,720.0	207,000.0			$B = 0.18$ S
5,510.0	234,000.0			$C = 0.73$
		Meter	0.19	$\eta_0 = 3,569.0$ Pa.S
				$\eta_\infty = 2.72$ Pa.S
				$\alpha = 4.16$
				$\tau_m = 62,896.0$ Pa

The Best Model is Carreau Model

23. System Used　　— Styrene-Acrylonitrile, BASF
　　Temperature　　— 250°C
　　Equipment　　— High pressure capillary viscometer

Shear Rate (1/sec)	Shear Stress (Pa)	Model	RMS (%) error	Model's constants
12.30	15,000.0	Power Law	0.21	$K = 7,451.0$ Pa
26.20	28,500.0			$n = 0.415$
72.00	53,700.0	Ellis	0.20	$\Phi_0 = 7.57$ E–4 $Pa^{-1}.S^{-1}$
161.0	78,000.0			$\Phi_1 = 1.24$ E–14 $Pa^{-\alpha}.S^{-1}$
356.0	99,500.0			$\alpha = 3.29$
1,090.0	145,000.0			$\tau_{1/2} = 53,214.0$ Pa
2,330.0	171,000.0	Carreau	0.16	$A = 3,259.0$ Pa.S
4,900.0	202,000.0			$B = 0.18$ S
				$C = 0.63$
		Meter	0.21	$\eta_0 = 1,320.0$ Pa.S
				$\eta_\infty = 2.72$ Pa.S
				$\alpha = 3.31$
				$\tau_m = 52,924$ Pa

The Best Model is Carreau Model

24. System Used   — N R – HDPE (50:50) Blend
    Temperature   — 160°C
    Equipment   — Monsanto Processability Tester (MPT)
    Capillary Data — L/D = 20.0 , Length = _____ , Entry = Conical

Shear Rate (1/sec)	Shear Stress (Pa)	Model	RMS (%) error	Model's constants
50.60	137,000.0	Power Law	0.02	$K = 63,009.0$ Pa
101.20	164,000.0			$n = 0.203$
202.40	187,000.0	Ellis	0.01	$\Phi_0 = 2.49$ E–4 $Pa^{-1}.S^{-1}$
401.20	209,000.0			$\Phi_1 = 1.09$ E–36 $Pa^{-\alpha}.S^{-1}$
				$\alpha = 7.24$
				$\tau_{1/2} = 154,116.0$ Pa
		Carreau	0.02	$A = 18,545.0$ Pa.S
				$B = 0.18$ S
				$C = 0.84$
		Meter	0.01	$\eta_0 = 4,008.0$ Pa.S
				$\eta_\infty = 2.72$ Pa.S
				$\alpha = 7.25$
				$\tau_m = 154,075.0$ Pa

The Best Model is Ellis Model

25. System Used   — N R – HDPE (50:50) Blend
    Temperature   — 180°C
    Equipment   — Monsanto Processability Tester (MPT)
    Capillary Data — L/D = 20.0 _____ Length = _____ ,
                  Entry = Conical

Shear Rate (1/sec)	Shear Stress (Pa)	Model	RMS (%) error	Model's constants
50.60	108,000.0	Power Law	0.02	$K = 38,597.0$ Pa
101.20	132,000.0			$n = 0.266$
202.40	161,000.0	Ellis	0.04	$\Phi_0 = 2.99$ E–4 $Pa^{-1}.S^{-1}$
401.20	186,000.0			$\Phi_1 = 2.88$ E–26 $Pa^{-\alpha}.S^{-1}$
				$\alpha = 5.33$
				$\tau_{1/2} = 121,926.0$ Pa

Shear Rate (1/sec)	Shear Stress (Pa)	Model	RMS (%) error	Model's constants
		Carreau	0.02	A = 12,499.0 Pa.S
				B = 0.18 S
				C = 0.77
		Meter	0.04	η_0 = 3,335.0 Pa.S
				η_∞ = 2.72 Pa.S
				α = 5.34
				τ_m = 121,874.0 Pa

The Best Models are Carreau and Power law Models

26. System Used   — Hytrel-PVC (60:40) Blend
    Temperature   — 170°C
    Equipment   — Instron Capillary Viscometer (Model-1195)

Shear Rate (1/sec)	Shear Stress (Pa)	Model	RMS (%) error	Model's constants
3.06	540,000.0	Power Law	0.07	K = 275,181.0 Pa
6.12	840,000.0			n = 0.641
12.23	1,350,000.0	Ellis	0.07	Φ_0 = 3.77 E–6 $Pa^{-1}.S^{-1}$
30.59	2,520,000.0			Φ_1 = 6.25 E–11 $Pa^{-\alpha}.S^{-1}$
61.17	4,080,000.0			α = 1.8
122.33	6,500,000.0			$\tau_{1/2}$ = 1,033,406.0 Pa
305.84	1.14 E 7	Carreau	0.06	A = 214,023.0 Pa.S
611.69	1.675 E 7			B = 0.28 S
1223.3	2.3 E 7			C = 0.41
		Meter	0.07	η_0 = 264,958.0 Pa.S
				η_∞ = 2.72 Pa.S
				α = 1.8
				τ_m = 1,033,383.0 Pa

The Best Model is Carreau Model

27. System Used   — Hytrel-PVC (60:40) Blend
    Temperature   — 190°C
    Equipment   — Instron Capillary Viscometer (Model-1195)

Shear Rate (1/sec)	Shear Stress (Pa)	Model	RMS (%) error	Model's constants
3.06	250,000.0	Power Law	0.15	$K = 143{,}032.0$ Pa
6.12	430,000.0			$n = 0.676$
12.23	940,000.0	Ellis	0.16	$\Phi_0 = 9.75$ E–6 Pa^{-1}.S^{-1}
30.59	1,720,000.0			$\Phi_1 = 1.13$ E–10 Pa$^{-\alpha}$.S^{-1}
61.17	2,250,000.0			$\alpha = 1.8$
122.33	3,550,000.0			$\tau_{1/2} = 1{,}630{,}650.0$ Pa
305.84	0.77 E 7	Carreau	0.11	$A = 102{,}292.0$ Pa.S
611.69	1.225 E 7			$B = 0.18$ S
1223.3	1.75 E 7			$C = 0.38$
3,058.4	2.58 E 7	Meter	0.16	$\eta_0 = 102{,}540.0$ Pa.S
				$\eta_\infty = 2.72$ Pa.S
				$\alpha = 1.8$
				$\tau_m = 1{,}630{,}489.0$ Pa

The Best Model is Carreau Model

28. System Used — mEPDM (60) – mHDPE (40)
    Temperature — 190°C
    Equipment — Monsanto Processability Tester (MPT)

Shear Rate (1/sec)	Shear Stress (Pa)	Model	RMS (%) error	Model's constants
61.45	159,160.0	Power Law	0.01	$K = 36{,}424.0$ Pa
122,90	207,425.0			$n = 0.361$
245.80	267,181.0	Ellis	0.04	$\Phi_0 = 2.5$ E–4 Pa^{-1}.S^{-1}
491.60	336,131.0			$\Phi_1 = 1.06$ E–19 Pa$^{-\alpha}$.S^{-1}
				$\alpha = 3.91$
				$\tau_{1/2} = 192{,}260.0$ Pa
		Carreau	0.01	$A = 13{,}401.0$ Pa.S
				$B = 0.18$ S
				$C = 0.67$

Shear Rate (1/sec)	Shear Stress (Pa)	Model	RMS (%) error	Model's constants
		Meter	0.04	$\eta_0 = 3,991.0$ Pa.S $\eta_\infty = 2.72$ Pa.S $\alpha = 3.92$ $\tau_m = 192,158.0$ Pa

The Best Models are Carreau and power law Models

29. System Used — MEPDM (60) – mHDPE (40) + ZnO (10 phr) + St. Acid (1 phr)
    Temperature — 190°C
    Equipment — Monsanto Processability Tester (MPT)

Shear Rate (1/sec)	Shear Stress (Pa)	Model	RMS (%) error	Model's constants
61.45	534,363.0	Power Law	0.03	$K = 275,191.0$ Pa $n = 0.168$
122.90	638,937.0			
245.80	698,119.0	Ellis	0.04	$\Phi_0 = 9.43$ E–5 Pa^{-1}.S^{-1}
491.60	764,196.0			$\Phi_1 = 1.64$ E–58 Pa$^{-\alpha}$.S^{-1} $\alpha = 10.27$ $\tau_{1/2} = 631,094.0$ Pa
		Carreau	0.02	$A = 74,993.0$ Pa.S $B = 0.18$ S $C = 0.87$
		Meter	0.04	$\eta_0 = 10,596.0$ Pa.S $\eta_\infty = 2.72$ Pa.S $\alpha = 10.28$ $\tau_m = 631,043.0$ Pa

The Best Model is Carreau Model

30. System Used — SBR + CB (25 phr)
    Temperature — 80°C
    Equipment — Goetfert- 1000 Capillary Viscometer
    Piston Data — Dia = 12 mm, Speed range = 2 to 200 mm/mt
    Capillary Data — Dia = 2.5 mm, Length = 25 mm, Entry = Flat

Shear Rate (1/sec)	Shear Stress (Pa)	Model	RMS (%) error	Model's constants
0.74	130,000.0	Power Law	0.12	$K = 166,236.0$ Pa
1.47	170,000.0			$n = 0.248$
3.69	227,500.0	Ellis	0.28	$\Phi_0 = 5.28$ E–6 $Pa^{-1}.S^{-1}$
7.37	285,500.0			$\Phi_1 = 1.05$ E–26 $Pa^{-\alpha}.S^{-1}$
17.18	377.500.0			$\alpha = 4.89$
36.86	502,500.0			$\tau_{1/2} = 215,687.0$ Pa
73.73	547,500.0	Carreau	0.14	$A = 189,908.0$ Pa.S
147.50	595,000.0			$B = 0.49$ S
368.60	670,000.0			$C = 0.87$
737.30	775,000.0	Meter	0.28	$\eta_0 = 189,376.0$ Pa.S
1475.0	890,000.0			$\eta_\infty = 2.72$ Pa.S
				$\alpha = 4.89$
				$\tau_m = 215,690.0$ Pa

The Best Model is Power Law Model

31. System Used   —   SBR + CB (25 phr)
     Temperature   —   100°C
     Equipment   —   Goetfert- 1000 Capillary Viscometer
     Piston Data   —   Dia = 12 mm , Speed range = 2 to 200 mm/mt
     Capillary Data —   Dia = 2.5 mm , Length = 25 mm , Entry = Flat

Shear Rate (1/sec)	Shear Stress (Pa)	Model	RMS (%) error	Model's constants
0.74	82,500.0	Power Law	0.14	$K = 110,094.0$ Pa
1.47	112,500.0			$n = 0.292$
3.69	162,500.0	Ellis	0.32	$\Phi_0 = 8.67$ E–6 $Pa^{-1}.S^{-1}$
7.37	210,000.0			$\Phi_1 = 2.39$ E–23 $Pa^{-\alpha}.S^{-1}$
17.18	280,000.0			$\alpha = 4.36$
36.86	360,000.0			$\tau_{1/2} = 168,215.0$ Pa
73.73	467,500.0	Carreau	0.13	$A = 119,660.0$ Pa.S
147.50	542,500.0			$B = 0.49$ S
368.60	612,500.0			$C = 0.82$

Shear Rate (1/sec)	Shear Stress (Pa)	Model	RMS (%) error	Model's constants
737.30 1475.0	670,000.0 737,500.0	Meter	0.32	$\eta_0 = 115,287.0$ Pa.S $\eta_\infty = 2.72$ Pa.S $\alpha = 4.37$ $\tau_m = 168,207.0$ Pa

The Best Model is Carreau Model

32. System Used   — SBR + CB (25 phr)
    Temperature   — 120°C
    Equipment   — Goetfert- 1000 Capillary Viscometer
    Piston Data   — Dia = 12 mm , Speed range = 2 to 200 mm/mt
    Capillary Data — Dia = 2.5 mm , Length = 25 mm , Entry = Flat

Shear Rate (1/sec)	Shear Stress (Pa)	Model	RMS (%) error	Model's constants
0.74 1.47	62,500.0 90,000.0	Power Law	0.14	$K = 84,834.0$ Pa $n = 0.319$
3.69 7.37 17.18 36.86	130,000.0 172,500.0 235,000.0 300,000.0	Ellis	0.26	$\Phi_0 = 1.12$ E–5 Pa^{-1}.S^{-1} $\Phi_1 = 6.99$ E–21 Pa$^{-\alpha}$.S^{-1} $\alpha = 3.97$ $\tau_{1/2} = 134,832.0$ Pa
73.73 147.50 368.60	377,500.0 465,500.0 580,000.0	Carreau	0.12	$A = 91,904.0$ Pa.S $B = 0.49$ S $C = 0.79$
737.30 1475.0	635,000.0 682,500.0	Meter	0.26	$\eta_0 = 88,660.0$ Pa.S $\eta_\infty = 2.72$ Pa.S $\alpha = 3.97$ $\tau_m = 134,825.0$ Pa

The Best Model is Carreau Model

Annexure-IV
Answers to the Problems

CHAPTER – 2

8. (a) Pseudoplastic fluid

   n-values: 170°C – 0.65; 180°C – 0.7; 190°C – 0.77; 200°C – 0.82

   (b) Students should try to find the physical reason for variation of n with temperature.

9. (i) Power law model: $n = 0.7$; $k = 32.86$ kN / m^2 s^n

   (ii) Ellis model: $\eta_0 = 40.14$ kNs/m^2 ; $\alpha = 1.62$, $\tau_{1/2} = 327.75$ kN/m^2

   (iii) Meter model: $\eta_0 = 40.14$ kNs/m^2; $\eta_\infty = 2.21$ kNs/m^2; $\alpha = 1.72$

10. $\tau_y = 9 \times 10^4$ N/m^2; $\beta = 1000$ Ns/m^2

11. $n = 0.247$; $k = 304.1$ kNs^n/m^2

12. n-values. No single straight line is obtained for ln τ vs ln γ^{dot}. Twi lines have been ploted (i) for $\gamma^{dot} < 55$ s^{-1}; and another for $\gamma^{dot} > 55$ s^{-1}.

    The n values are – for EPDM (i) 0.47 and (ii) 0.22; for blend (i) 0.55 and (ii) 0.21;

    For BIIR (i) 0.58 and (ii) 0.167.

15. Power law:

	$L/D = 12.5$		37.1		75.1	
	LDPE	LLDPE	LDPE	LLDPE	LDPE	LLDPE
n	0.406	0.375	0.395	0.3	0.356	0.253
k, (kPs^{-1})	14.88	38.47	11.47	49.4	14.88	63.43

Ellis model –

η_0(kPas)	101.5	297.3	Students must work out for other L/D ratio capillaries
$\tau_{1/2}$(kPa)	16.86	29.4	
α	2.25	2.85	

20. For LDPE (L/D = 12.5) B = 37.25 s^-1 ; A = 378 kPa.

CHAPTER – 3

15. Q for

   (i) 1.69×10^{-3} m^3/s ;

   (ii) 0.8×10^{-3} m^3/s;

(iii) 1.89×10^{-3} m^3/s.

$\Delta P/L = 5.33$ M Pa/m.

16. $u_m = 8.43 \times 10^{-3}$ m/s; $r_y = 0.5$ cm.

23. 1.35 cc/s per unit width.

CHAPTER – 4

6. $T_g = 359$ K

8. $T_g = 235$ K

9. $T_g = 225$ K and $T_a = 275$ K

CHAPTER – 5

1. Pseudoplastic; The power law equation is $\tau = 1043.0\,(\gamma^{dot})^{0.96}$.

7. Pseudoplastic; $n = 0.3$; $k = 14764.8$ N s^n/m^2

CHAPTER – 6

1. $G_1 = 7466.7$ MN/m^2; $\eta_1 = 1.123 \times 10610$ MNs/m^2

$G_2 = 18665$ MN/m^2; $\eta_1 = 56.0 \times 10^8$ MNS/m^2

2. Strain $\gamma(100) = 8.65 \times 10^{-3}$ and strain $\gamma(230) = 9.9 \times 10^{-3}$

3. Strain $\gamma(3000) = 0.0178$; $G = 312.5$ MN/m^2; $\eta = 424.7$ GNs/m^2

5. $G = 1.52$ GN/m^2

6. For M-model6.93 s; Z-model 9.16 s ; V-K model 6.93 s.

7. Strain $\gamma(60) = 0.39$ %; $\gamma(120) = 0.79$%; $\gamma(160) = 0.45$%

13. 1.88 watts

❑   ❑   ❑

For Product Safety Concerns and Information please contact our EU
representative GPSR@taylorandfrancis.com
Taylor & Francis Verlag GmbH, Kaufingerstraße 24, 80331 München, Germany